Springer Proceedings in Earth and Environmental Sciences

The series Springer Proceedings in Earth and Environmental Sciences publishes proceedings from scholarly meetings and workshops on all topics related to Environmental and Earth Sciences and related sciences. This series constitutes a comprehensive up-to-date source of reference on a field or subfield of relevance in Earth and Environmental Sciences. In addition to an overall evaluation of the interest, scientific quality, and timeliness of each proposal at the hands of the publisher, individual contributions are all refereed to the high quality standards of leading journals in the field. Thus, this series provides the research community with well-edited, authoritative reports on developments in the most exciting areas of environmental sciences, earth sciences and related fields.

More information about this series at http://www.springer.com/series/16067

Nihal Anwar Siddiqui ·
Kanchan Deoli Bahukhandi · S. M. Tauseef ·
Nirmala Koranga
Editors

Advances in Environment Engineering and Management

Proceedings of the 1st National Conference on Sustainable Management of Environment and Natural Resource Through Innovation in Science and Technology

Springer

Editors
Nihal Anwar Siddiqui
University of Petroleum and Energy Studies
Dehradun, India

Kanchan Deoli Bahukhandi
University of Petroleum and Energy Studies
Dehradun, India

S. M. Tauseef
University of Petroleum and Energy Studies
Dehradun, India

Nirmala Koranga
DBS (PG) College
Dehradun, India

ISSN 2524-342X ISSN 2524-3438 (electronic)
Springer Proceedings in Earth and Environmental Sciences
ISBN 978-3-030-79064-6 ISBN 978-3-030-79065-3 (eBook)
https://doi.org/10.1007/978-3-030-79065-3

This Springer imprint is published by the registered company Springer Nature Switzerland AG
The registered company address is: Gewerbestrasse 11, 6330 Cham, Switzerland

Preface

The concept of development is a multidimensional one, encompassing the economic, social, cultural and political aspects of human society. Here, development is seen to include not only growth, but also welfare improvement for all groups of the population. New development strategies must therefore be shaped by the following components: the satisfaction of basic human requirements; the eradication of poverty; self-reliant and participatory development; and environmental consciousness. The environmental resources of water, soil, plant and animal life constitute the natural capital on which development itself must be based. At the same time, it is through the development process that humans interact with and affect the natural environment. Thus, environment cannot be seen in isolation from other factors which affect the development process. It must be seen and dealt with in relation to population, its growth, its distribution and its employment, natural resources, their availability, exploitation and use. The development and deployment of technologies are major and interacting elements within the large framework of development.

Society needs to adapt in order to provide the wealth that an increasing part of the world population is getting used to. We are on a track to ecological and resource collapse if actions are not taken soon. The environmental degradation is happening at a very high rate due to over exploitation of natural resource and environmental pollution. In order to save our environment, there is an urgent need to implement sustainable development in all dimension of environment. Technology will have to play a key role in the process of changing industrial society. But innovation has to be embedded in social and organizational innovation. We need sociotechnical change. Environmentally conscious design has been practiced in engineering design for more than a decade. Its merits are sometimes blamed as futile, as the world has not witnessed a significant contribution to the solution of the larger (global) problems.

Through this publication, the reader can update himself/herself with the advances in environmental science and management. This book provides a wide range of research article in the area of science and technology, sustainability, natural resource management, ecology and its environmental fields, geosciences and geology, atmospheric sciences, sustainability, climate change and extreme weather, global warming and environmental change, effect of climate change on ecosystem, environment and pollution.

This volume presents select papers on advances in environmental engineering and management which were presented at the Ist National Conference on Sustainable Management of Environment and Natural Resource through Innovation in science and technology in the field of Natural resource management, sustainability and environment, allied sciences and engineering (SMTST 2020), August 7 and 8, 2020, organized in collaboration with DBS (PG) College and funded by Uttarakhand Council of Science and Technology (DST) at UPES. The conference was attended by leading academic scientists, leading engineers, policy makers, budding scholars and graduate students.

Kanchan Deoli Bahukhandi
Convener, SMTST 2020
Dehradun, India

Acknowledgements

This volume has benefitted from contributions from participants at the Ist National Conference on Sustainable Management of Environment and Natural Resource through Innovation in Science and Technology, SMTST 2020, August 07–08, 2020, organized in collaboration with DBS (PGO College, H. N. B. Garhwal University and supported by funded by Uttarakhand Council of Science and Technology (DST) at UPES.

We thank Dr. S. J. Chopra (Hon'ble Chancellor, UPES) and Prof. Dr. Sunil Rai (Vice Chancellor, UPES), for their support and encouragement. We are grateful to the Chief Guests for SMTST 2020, Dr. Vandna Shiva, eminent Indian scholar, environmental activist, food sovereignty advocate and anti-globalization author presided the conference as the chief guest.

Conference witnessed five keynote address by eminent researchers in the country—Dr. Sushree Swarupa Tripathy, Principal Scientist, National Physical Laboratory, New Delhi; Dr. Suresh Jain, Associate Professor, Department of Civil Engineering, IIT Tirupati; Dr. Virender Sethi, Professor, IIT Bombay; Dr. Vir Singh, Professor, G. B. Pant University; and Dr. Nitin Endlay, Senior Scientist, CPPRI, Saharanpur—on various aspects of sustainable development.

The organizers of SMTST 2020 wish to thank all the reviewers for their valuable time and comments on the quality of the papers.

We acknowledge the support of our sponsors Uttarakhand Council of Science and Technology (UCOST).

We also thank the chairs and members of various committees as under:

Steering Committee

Chief Patron

Dr. S. J. Chopra, Chancellor, UPES, Dehradun (India)

Patron

Dr. Sunil Rai, Vice Chancellor, UPES, Dehradun (India)

General Chairs

Dr. Kamal Bansal, Dean, School of Engineering (SoE), UPES, Dehradun (India)
Dr. Jitendra Pandey, Head, R&D, UPES, Dehradun

Program Chair

Dr. Nihal Anwar Siddiqi, Programme Director Civil and HSE Engineering, UPES

Organizing Chairs

Dr. S. M. Tauseef, Professor and Head, HSE and Civil Engineering, UPES
Dr. Kanchan Deoli Bahukhandi, Assistant Professor, HSE and Civil Engineering, UPES

Publicity Chairs

Dr. Kanchan Deoli Bahukhandi, Assistant Professor (SG), HSE and Civil Engineering, UPES
Dr. Shefali Arora, Assistant Professor, Chemistry Department, UPES
Dr. Mamta Latwal, Assistant Professor, Chemistry Department, UPES

Public Relation Chairs

Dr. Jitendra Kumar Pandey, Head (R&D), UPES
Prof. P. Mondal, Assistant Professor (SG), HSE, UPES
Prof. Akshi Singh, Assistant Professor (SS), HSE, UPES

Session Management Chairs

Mr. Abhishek Nandan, Assistant Professor (SS), HSE, UPES
Prof. P. Mondal, Assistant Professor (SG), HSE, UPES
Dr. Shefali Arora, Assistant Professor (SG), HSE, UPES
Dr. Shilpi Aggrawal, Assistant Professor (SG), Chemistry Department, UPES
Dr. Masehswar Pathak, Assistant Professor (SG), Maths, UPES
Mr. Aksi Kuwar Singh, Assistant Professor (SS), HSE and Civil Engineering Department, UPES
Mr. Arun P. A., Assistant Professor (SS), HSE and Civil Engineering Department, UPES
Dr. Paritosh Singh, Assistant Professor, DBS (PG) College, Dehradun

Organizing Committee

Conveners

Dr. Kanchan Deoli Bahukhandi, Assistant Professor (SG), HSE and Civil Engineering Department
Dr. Nirmala Koranga, Assistant Professor (SG), Department of Botany, DBS (PG) College

Co–conveners

Dr. Madhuben Sharma, Assistant Professor, HSE and Civil Engineering Department, UPES
Dr. Anupma Tripathi, Assistant Professor, DBS (PG) College

Organizing Secretaries

Dr. Shilpi Agarwal, Assistant Professor, UPES
Dr. Shefali Arora, Chemistry Department, UPES
Dr. Paritosh Singh, Assistant Professor, DBS (PG)

Finance Committee

Mr. Venkat Krishankant, Assistant Professor (SS), HSE and Civil Engineering Department, UPES
Mr. Survinder, Assistant Professor (SS), HSE and Civil Engineering Department, UPES
Ms. Tabbasum Abbasi, Assistant Professor, HSE and Civil Engineering Department, UPES

Advisory committee

Dr. Pankaj Kumar Sharma, UPES
Dr. D. K. Gupta, UPES
Dr. Manish Prateek, UPES
Dr. Sanjeev Kumar, UPES
Dr. Shailey Singhal, UPES
Dr. Piyush Kuchhal, UPES
Dr. Vijay Raj, UPES
Dr. Sunil Kumar Khare, UPES
Dr. Ajay Kumar, UPES
Dr. Bhawna Yadav Lamba, UPES
Dr. Shyam Pandey, UPES
Dr. Ajay Mittal, UPES
Dr. Sunil Kumar Khare, UPES
Dr. P. Vijay, UPES
Dr. Santosh Dubey, UPES
Dr. Ashutosh Mishra, UCOST
Dr. Anita Rawat, FRI

Technical Committee

Mr. Abhishek Nandan, UPES
Dr. Mukesh Kumar Dubey, UPES
Dr. Madhu Sharma, UPES
Dr. Bikarama Prasad Yadav, UPES
Dr. Harsingra Patel, UPES
Dr. Mukesh Singhal, UPES
Dr. Gagan Anand, UPES
Dr. Rajeeve Gupta, UPES
Dr. Pankaj Mishra, UPES

Dr. Maheshwar Pathak, UPES
Dr. Sanjeev Kumar Dubey, UPES
Dr. Ravi Tomar, UPES
Dr. Hitesh Kumar Sharma, UPES
Mr. Prasenjit Mondal, UPES
Mr. Anand Kumar Singh, UPES
Dr. S. K. Joshi, UPES
Dr. Prashant Shambhudayal, UPES
Dr. Shalini Vohra, UPES
Dr. Nirlipta Priyadarshini Nayak, UPES
Dr. Sudhir Kumar Chaturvedi, UPES

Organizing Committee

Mr. Abhishek Nandan, UPES
Mr. Mopidevi Vijai Kishore, UPES
Mr. Akshi Kunwar Singh, UPES
Mr. Arun P. A., UPES
Mr. Surender, UPES
Mr. Ankur Chaudhary, UPES
Venkat Krishnakanth V., UPES
Susanta Kumar Sethy, UPES
Mr. A. N. Shankar, UPES
Mr. Durga Prasad Pandey, UPES
Mr. Rahul Silori, UPES
Ms. Tabbasum Abbasi, UPES
Dr. Nishu Goyal Bansal, UPES
Dr. Annapurna Boruah, UPES
Ms. P. Rose, UPES
Ms. Leena Kapoor, UPES
Ms. Shweta Sachdeva, UPES
Dr. Sapna Jain, UPES
Dr. Pratibha Pathak, UPES
Mamta Latwal, UPES
Dr. Rishi Dewan
Dr. Manjeet Singh Goyat, UPES

Contents

Water Quality Assessment of Streams of Doon Valley Dehradun, Uttarakhand

Deepali Rana, S. K. Gupta, and Rahul Rana

Abstract Water quality of any stream depends on a variety of physico-chemical parameters. The present study was conducted to assess the water quality of various streams in Eastern and Western Doon. It was carried out on two river systems comprising of five main rivers—Baldi, Song, Suswa, Tons and Asan. Sampling was done regularly at 20 sampling stations established along these rivers. 15 physical and chemical parameters like depth, width, water velocity, air temperature, water temperature, dissolved oxygen, carbon dioxide, pH, turbidity, alkalinity, hardness, BOD, nitrate, phosphate and total dissolved solids were analyzed. The study reveals that the magnitude of different parameters is related to the climatic conditions, seasons and river discharge. The interplay of these parameters determines the water quality of the streams. The fluctuation pattern all through 3 different seasons reflected an increment pattern from summer to rainy in the parameters like depth, width, water velocity, CO_2, turbidity and TDS. The declining trend in the values of the aforesaid parameters was noticed beyond rainy months. From rainy to winters, an increment in values was noticed in DO, pH, hardness, alkalinity, BOD, nitrate and phosphate. The parameters which showed increment in values from winters to summers include AT, WT, CO_2. Width, depth and WV have been the chief physical factors with wide range of variations. BOD, Hardness, NO_3^-, TDS, DO and CO_2 values seemed more important from the quality of water chemistry point of view. Seasonal variation in physical and chemical parameters have also been observed.

Keywords Water quality · Doon Valley · Physico-chemical parameters · Eastern and Western Doon

D. Rana (✉)
Department of Zoology, Dolphin (P.G.) Institute of Biomedical and Natural Sciences, Dehradun, Uttarakhand, India

S. K. Gupta · R. Rana
D. B. S. (PG) College, Dehradun, Uttarakhand 248001, India

N. A. Siddiqui et al. (eds.), *Advances in Environment Engineering and Management*, Springer Proceedings in Earth and Environmental Sciences,
https://doi.org/10.1007/978-3-030-79065-3_1

1 Introduction

The water quality of any aquatic ecosystem, whether lentic or lotic has a direct bearing upon the welfare of fish and other aquatic life (plankton/plants/micro- or macroinvertebrates etc.), which are vital links in the food chain of fishes. Therefore, these and related aspects have been the subject of study for various workers abroad and in different regions of Indian subcontinent.

After the pioneering works of F.A. Forel [*vide*1] on the ecology of lakes, the time gone by has witnessed a striking growth of limnological accomplishments outside India, particularly in America. In this regard a mention may be made of important contributions on limnology, including a number of researches done in theis field, viz., [1–9]. Besides, the researches, which deal with various aspects of riverine ecology, are of [10–26] etc.

In India, much progress has been made on freshwater ecology after 1950 when various ecological studies were carried out on different rivers in various parts of country. Special mention may be made here of the works done on North Indian rivers viz., [27–30] etc. Few more surveys on different sections of the rivers were conducted in order to assess the extent of damage caused to fish and other animal communities due to pollution conditions. The water quality analysis and factors controlling water chemistry of Doon valley has been carried out by [31–41].

It is an established fact that to keep the aquatic habitat congenial for existence of fish and other biota, physical and chemical parameters of water exercise their influence individually or synergistically. Therefore, a comparative judgement of water quality parameters at various stations in Eastern and Western Doon was undertaken in the present endeavour with a view to work out the annual fluctuation regime in water quality parameters and to estimate their impact on the fish fauna dwelling therein.

2 Material and Methods

Doon Valley is part of district Dehradun (Latitude—29° 58′ and 30° 32′ N and Longitude—77° 35′ and 78° 20′ E) comprises of 2 main river system, namely, the Ganga river system and the Yamuna river system. The present enterprise was accomplished on these two river systems which comprises of five main rivers—Baldi, Song, Suswa in the East and Tons and Asan in the West. Sampling was regularly done for a period of 2 years at the 20 sampling stations established along the rivers mentioned above. The parameters taken into consideration were Depth, Width, Water Velocity, Air Temperature, Water Temperature, Dissolved Oxygen, Carbon dioxide, pH, Turbidity, Alkalinity, Hardness, BOD, Nitrate, Phosphate and Total Dissolved Solids.

The estimation of physical and chemical parameters (Air Temperature, Water Temperature, Dissolved Oxygen, Carbon dioxide and pH), were analyzed in the field by taking the help of field water and soil analysis kit. Also, parameters which could not be analyzed in the field (like Hardness (Calcium + Magnesium), Alkalinity (as bicarbonates), Turbidity, Biological Oxygen Demand, Nitrates as total nitrogen, Phosphates as total phosphorous and Total Dissolved Solids) were analyzed in the laboratory by following standard protocol [42].

Concurrently, the water samples were also submitted to the Central laboratory of Central Soil and Water Conservation Research and Training Institute (CSWCRTI), Dehradun and Central Pollution Control Board (CPCB), Dehradun for data verification analyzed in the field/laboratory before reaching to any final result (Fig. 1).

3 Results and Discussion

The study of water quality parameters of the streams of Doon Valley reveals that the magnitude of different parameters is certainly related to the climatic conditions, season and river discharge. The interplay of these parameters determines the water quality of the rivers.

The range values (minimum and maximum) along with the values of mean and standard deviation of all the parameters during the study period have been incorporated in Table 1.

The pattern of fluctuation all through 3 different seasons in streams of Eastern and Western Doon reflected an increment pattern from summer to rainy (Figs. 2, 3 and 4) in the parameters like depth, width, water velocity, CO_2, turbidity and TDS.

The declining trend in the values of the aforesaid parameters was noticed beyond rainy months. From rainy to winters, the increment in values was noticed in DO, pH, hardness, alkalinity, BOD, nitrate and phosphate (Figs. 2, 3 and 4).

The parameters which showed increment in values from winters to summers include AT, WT, CO_2 (Figs. 2, 3 and 4).

The present findings, definitely, needed to be evaluated in the light of earlier observations on physico—chemical parameters of various streams of Doon valley. It is worth emphasizing that a variety of objectives were involved with these earlier works e.g., [43–47] have evaluated water quality of river Song. Sharma and Mishra [48] studied seasonal variations in water quality of river Ganga around Hardwar including Song and Suswa. Ecological study of river Suswa was taken up [49] whereas Mothronwala swamp part of Suswa river was also studied [50].

The physico—chemical parameters have also been worked out for Tons river in the West with reference to macroinvertebrate density and diversity [51]. The water quality parameters with their seasonal trend for Asan river have been given [52] in the West.

A comparison of the seasonal trend recorded in the present observation with the earlier observations have revealed similar increasing and decreasing trend based on the fundamentals of climate, geography and geology of tropical regions in general and Western Himalayan region [53–57] in particular, barring differences in the values recorded.

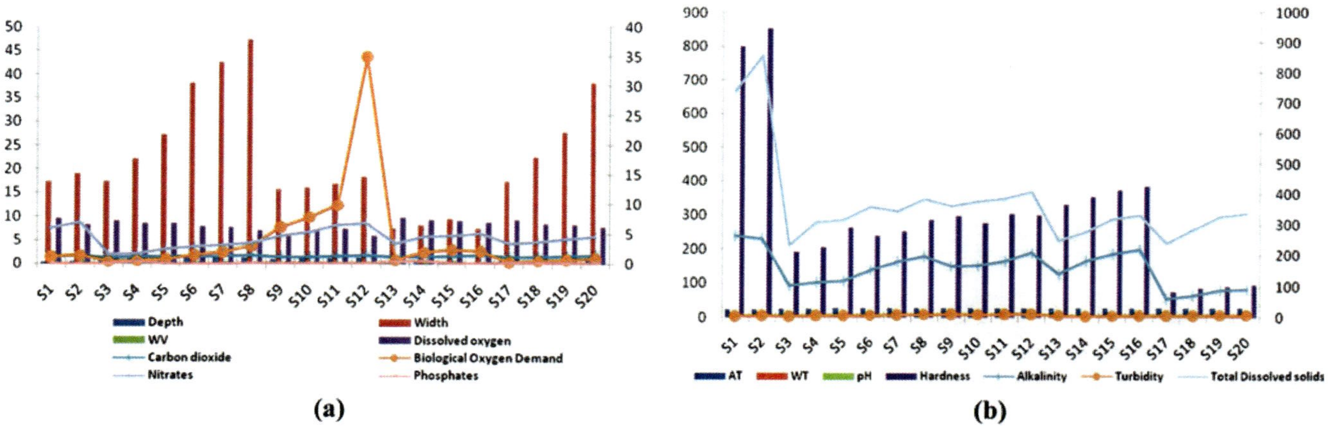

Fig. 2 Variations in water quality parameters (mean values for 2 years) at different stations during Summers. **a** Depth, Width, WV, DO, CO_2, BOD, nitrate, phosphate. **b** AT, WT, pH, hardness, alkalinity, turbidity, TDS

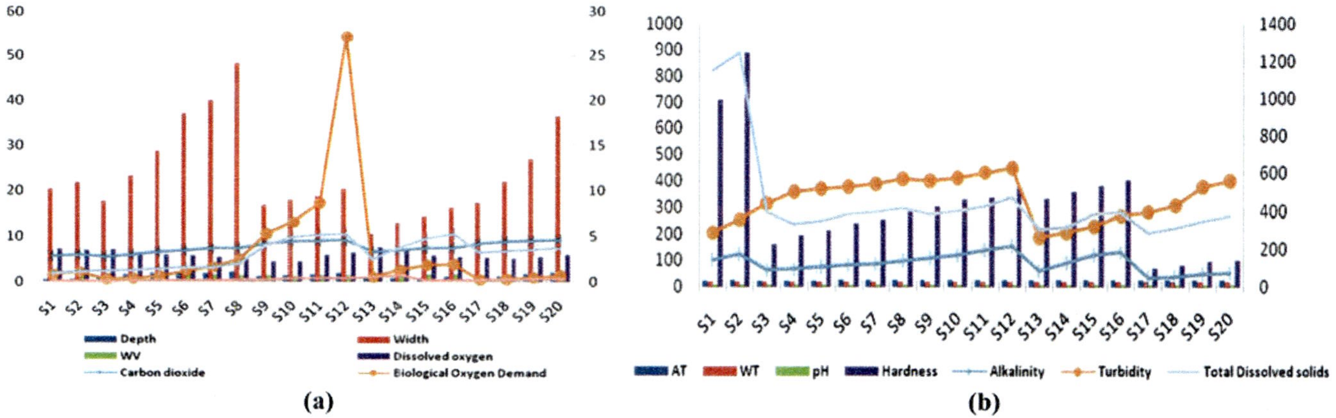

Fig. 3 Variations in water quality parameters (mean values for 2 years) at different stations during rainy Season. **a** Depth, Width, WV, DO, CO_2, BOD, nitrate, phosphate. **b** AT, WT, pH, hardness, alkalinity, turbidity, TDS

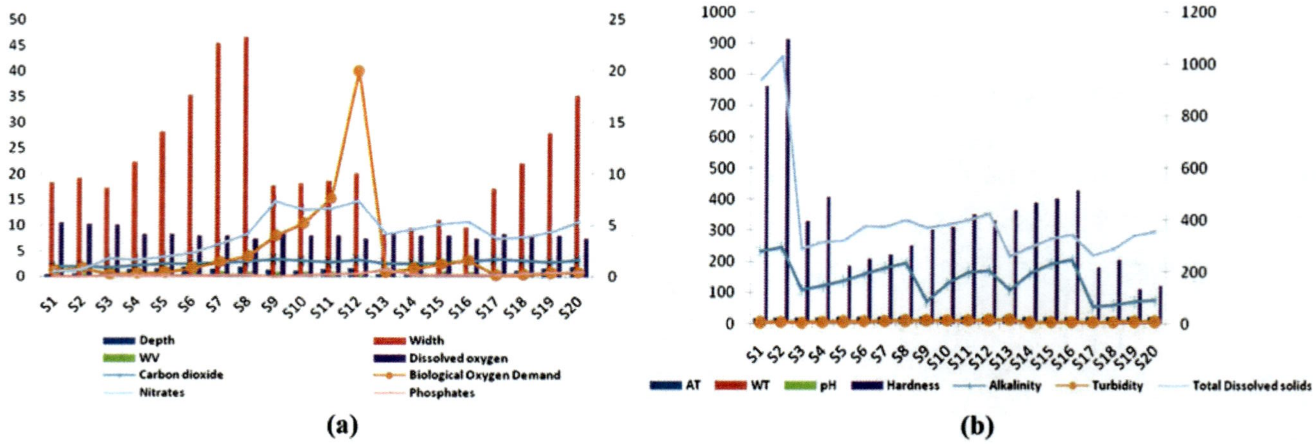

Fig. 4 Variations in water quality parameters (mean values for 2 years) at different stations during winters. **a** Depth, Width, WV, DO, CO_2, BOD, nitrate, phosphate. **b** AT, WT, pH, hardness, alkalinity, turbidity, TDS

[49, 50], respectively. Hardness values for the streams of West, similarly were not available for comparison from the published works mentioned in the foregoing.

Also, Nitrate (total nitrogen) and Phosphate (total phosphrous) values also appeared higher for the streams of Eastern Doon (values of river Suswa) (Figs. 2, 3 and 4).

4 Conclusion

The present study highlighted the fluctuation pattern of various water quality parameters in different seasons and interplay amongst them. Also, the results reflected that there exists a variation in the overall water quality of Eastern and Western Doon Valley streams.

Acknowledgements Authors are indebted to Uttarakhand State Council for Science and Technology, Dehradun (U-COST) for carrying out a Major Research Project. Thanks are due to Zoological Survey of India (Northern Regional Station), Dehradun, Forest Research Institute (FRI), Dehradun and Central Soil and Water Conservation Research and Training Institute (CSWCRTI), Dehradun for providing library facilities. Thanks to DBS (PG) College, Dehradun for providing necessary resources. Special thanks are due to Laboratory staff at CSWCRTI, Dehradun for carrying out sample analyses.

References

1. Welch, P.S.: Limnology. Mc Graw Hill, New York. (1952) 538 pp.
2. Carpenter, K. E.: Life in Inland waters. New York (1928) 267 pp.
3. Hutchinson, G. E.: A Treatise on Limnology I. Geography, Physics and Chemistry, John Wiley and Sons, New York (1957).
4. Hutchinson, G.E.: A treatise on Limnology, John Wiley and Sons, New York (1993).
5. Odum, E.P.: *Fundamentals of Ecology*. Philadelphia: W.B. Saunders. (1959) 546 pp.
6. Reid, G. K.: Ecology of Inland waters and estuaries. Reinhold, New York (1961).
7. Macan, T. T.: Freshwater Ecology. Longmans, London. (1963) 338 pp.
8. Ruttner, F.: The Fundamentals of Limnology. University of Toronto Press, Toronto (1963) 295 pp.
9. Hynes, H. B. N.: The Ecology of running waters, Liverpool University Press. (1970) 555 pp.
10. Kofoid, C. A.: Plankton studies IV. The plankton onf Illinois river, 1894–1899, with introductory notes upon the hydrography of the Illinois river and its basin. Part I. Quantitative investigations and general results. *Bull. Ill. Nat. Hist. Surv.*, 6(2). (1903) 95–624.
11. Kofoid, C. A.: The Plankton of the Illinois river, 1894–1899. With introductory notes upon the hydrography of Illinois river and its basin. Part II. Constituent organism and their seasonal distribution. *Bull. Ill. Nat. Hist. Surv.*, 8 (1&2). (1908)1–361.
12. Shelford, V. E.: Ecological succession. I. Stream fishes and method of Physiography analysis. *Biol. Bull.*, **21.** (1911) 9–35.
13. 13. Allen, W. E.: A quantitative and statistical study of the plankton of San Joaquin River and its tributaries in and near Stockhom, California University California. Publ. Zool., 22(1). (1920) 1–292.

14. Powers, E. B.: Fresh water studies. I. The relative Temperature, Oxygen Content, Alkali Reserve, the Carbon Dioxide Tension and pH of waters of certain Mountain streams at different altitudes in the Smoky Mountain National Park. Ecology, 10 (1929) 97–111.
15. Butcher, R. W., Pantelow, F. T. K. and Woody, J. W. A.: Variations in composition of river water. *Int. Rev. Hydrobiol.*, 24 (1930) 47–80.
16. Buchner, R. W.: Studies on the ecology of rivers. II. The microflora of rivers with special reference to the algae on the river bed. *Ann. Bot.*, 46 (1932) 813–861.
17. Buchner, R. W.: Studies on the ecology of rivers. I. On the distribution of macrophytic vegetation in the rivers of Britain. *J. Ecol.*, 21 (1933) 58–91.
18. Ricker, W. E.: An ecological classification of certain Ontario streams. *Univ. Toronto Studies, Biol. Ser.*, 37 (1934) 114 pp.
19. Berner, L. M.: Limnology of lower Missouri River. *Ecology*, 32(1) (1951) 1–12.
20. Holden, M. J. and Green, J.: The hydrology and plankton of the river Sokoto. *J. Anim. Ecol.*, 29(1) (1960) 65-84.
21. Vannote, R. L., G. W. Minshall, K. W. Cummins, J. R. Sedell, and C. E. Cushing.: The river continuum concept. *Canadian Journal of Fisheries and Aquatic Sciences* 37(1980) 130–137.
22. Welcomme, R. L.: *River fisheries.* FAO Fish Tech Pap. 262 (1985) 1–318.
23. Bellamy, K., Beebe, J. T., Saunderson, H. C. and Imhof, J.: River morphology, sediments and fish habitats. Erosion and Sediment transport Monitiring Programmes in River Basisn (Proccedings of the Oslo Symposium) Publ. No. 210 (1992) 309–315.
24. Rawi, S. M. and Shihab, A. S.: Application of Factor analysis as atool for water quality management of Tigris River within Mosul city. *Raf. Jour. Sci.,* 16 (1) (2005) 56 – 64.
25. Phyllis K. Weber-Scannell and Lawrence K. Duffy.: Effects of Total Dissolved solids on Aquatic Organism: A Review of Literature and Recommendation for Salmonid Species; American Journal of Environmental Sciences, 3(1) (2007) 1–6.
26. Prathumratana, L. Sthiannopkao, S. and Kim, K. W.: The relationship of climate and hydrological parameters to surface water quality in the lower Mekong River. *Environmental International*, 34 (2008) 860–866.
27. Chakraborty, R.D., Ray, P. and Singh, S. B.: A quantitative study of the plankton and the physico-chemical conditions of the river Jumna at Allahabad in 1954-1955. *Indian J. fish.*, 6(1) (1959) 186–203.
28. Lakshminarayana, J. S. S.: Studies on the phytoplankton of river Ganges, Varanasi, Indian Part I. Physico-chemical characteristics of river Ganges. Part II. The seasonal growth and sucession of the plankton algae in the river Ganges. *Hydrobiologia*, 25 (1965) 119–137 and 138–165.
29. Pahwa, D.V. and Malhotra, S. N.: Observations on fluctuations in the abundance of plankton in relation to certain hydrobiological conditions of river Ganges. *Proc. Nat. Acad. Sci. India*, 36(B)II (1966) 157–189.
30. Ray, P., Singh, S. B. and Sehgal, K.L.: A study of some aspects of ecology of the river Ganga and Jamuna at Allahabad (U.P.) in 1958–1959. *Proc. Nat. Acad. Sci. India*, 36(B)III (1966) 235–272.
31. Bahukhandi, K.D., Kumar, S., Siddiqui, N.A., Singh, R., Arora, S., Mondal, P.: Assessment of Major Ion Chemistry and Spatial Variation of Water Quality of Ganga River of Uttarakhand. *Journal of Chemical and Pharmaceutical Research*, 10 (7) (2014)145–150.
32. Bahukhandi, K. D., Siddiqui, N.A.: Impact assessment of Urbanization on water quality. *Environmental Pollution Control Journal*, 12 (1) (2008).
33. Bahukhandi, K. D., Bartarya, S. K.: Hydrochemistry of Surface and groundwater of Doon valley. In: Mayilswami, C. Thangarajan, M. Kulkarni, P.S. Singh, V.P. Proceeding Volume "Fifth International Groundwater Conference (IGWCC), 3, 74, (2012) 1109–1119.
34. Bahukhandi, K.D., Bartarya, S.K.: Major Ion Chemistry of Asan River Catchment of Dehradun district (Impact Assessment of Anthropogenic Sources). *Octa Journal of Environmental Research.* 2(2) (2014) 168–177.
35. Bahukhandi, K. D., Mondal, P., Singh, S.: Water Quality Assessment of River Ganga Health Hazard Identification and Control Singh. *International Journal of Scientific and Research Publications*, 5 (5) (2015) 1–8.

36. Bahukhandi, K.D., Aaron, A.: Impact of Improper Disposal of Municipal Solid Waste on ground water quality in and around the solid waste dumping sites of Visakhapatnam, Andhra Pradesh. *Indian Journal of Global Ecology and Environment*, 5(3) 133–143.
37. Bahukhandi, K. D., Bartarya, S. K., Siddiqui, N.A. Assessment of surface and ground water Quality of Haridwar district of Uttarakhand. *International Journal of Chem Tech Research*,10(10) (2017) 95–118.
38. Bahukhandi, K.D, Siddiqui, N.A., Arora S.: Assessment of Heavy Metal and Physiochemical Parameter in Surface and Groundwater Quality of Dehradun District of Uttarakhand, Springer Transactions in Civil and Environmental Engineering, Advances in Water Pollution Monitoring and Control, Proceedings from HSFEA 2018, (2020). 978-981-32-9955.
39. Bartarya, S.K., Bahukhandi, K.D.: Impact Assessment of Urbanization and Industrialization on surface and groundwater quality of Dehradun district of Uttarakhand. *Global Journal of Engineering, Design and Technology*, 1(1) (2012) 11:22.
40. Thakur, D., Bartarya, S.K. Nainwal, H.C.. Tracing ionic sources and geochemical evolution of groundwater in the Intermountain Una basin in outer NW Himalaya, Himachal Pradesh, India. *Environmental Earth Sciences*, **77**(20) (2018a) 720.
41. Thakur, D., Bartarya, S.K., Nainwal, H.C. Groundwater quality assessment of Soan Basin in Outer Himalaya, Himachal Pradesh, India. *Himalayan Geology*, **39**(2) (2018b) 197–211.
42. American Public Health Association (APHA): Standard Methods for the Examination of Water and Wastewater, *21st ed. Washington, DC: American Public Health Association* (2005).
43. Mathur, H. N. and Kumar, Om.: Environmental impact of distillery effluents on the forest recreation spot at Lacchiwala. *Indian Forester* (1986) 1021–1025.
44. Bisht, S., Grover, S. P. and Bhatt, A. M.: Hydrobiology of the river Song in Eastern Doon. *Uttar Pradesh Journal of Zoology*, 9 (1) (1989)121–123.
45. Dudeja, D., Bartarya, S.K., Biyani, A.K. 2011. Hydrochemical and water quality assessment of ground water in Doon Valley of Outer Himalaya, Uttarakhand India. *Environmental Monitoring Assessment*, **181**(1–4), 183–204.
46. Kumar, Om., Bisht, S. and Singh, N.: Studies on water quality and fish of Song river in Eastern Doon valley forests. *Indian Forester* (1990) 35–42.
47. Gupta, Hari Om and Sharma, Brij Mohan: Quality of water at Laltappar- An industrial area of Doon valley. *Indian Journal of Forestry*, 16(14) (1983) 360–365.
48. Sharma, A. P. and Mishra, A. Present status and prospect of mahaseer fishery in Garhwal region of Central Himalaya. In: Fish and fisheries at higher altitudes: Asia. (Ed. T. Petr.) FAO Fisheries Technical Paper. No. 385. Rome, FAO (2001) 304 p.
49. Kaur, S. and Joshi, B. D.: Seasonal variation in some physico – chemical parameters of river Ganga in and around Hardwar. *Him. J. Env. Zool.*, 17 (1) (2003) 45–55.
50. Bhutiani, R. and Khanna, D. R.: Ecological study of river Suswa: modelling DO and BOD. *Environ. Monit. Assess.*, 125 (2007)183–195.
51. Gupta, N., Sharma, R. C. and Tripathi, A. K.: Study of bio-physico-chemical parameters of Mothronwala swamp, Dehradun (Uttarakhand), *J. Environ. Biol.* 29(3) (2008) 381–386.
52. Sharma, R. C., Arambam, R. and Sharma, R.: Surveying macro – inveretebrate diversity in the Tons river, Doon valley, India. *Environmentalist*, 29 (2009) 241–254.
53. Uniyal, D. P. and Kumar, A.: Fish diversity in the selected streams of Chakrata and Shiwalik hills (District Dehradun, Uttarakhand), India *Rec. Zool. Surv., India*, Occ. Paper No. 253 (2006) 1–120.
54. Gosain, O. P.: Migration routes: River Bhilangana (physico-chemical profile). In: *Mahseer, The Game Fish.* Ed. Prakash Nautiyal. Publisher: RACHNA (1994) 147–168.
55. Gosain, O. P. and Gosain, M.: Applied Fisheries. In: *Garhwal Himalaya (Nature, Culture and Soc.) Status and Scope.* Transmedia, Media House, Srinagar (Garhwal) (2001) 201–215.
56. Joshi, B.D., Dishi and Joshi, K.: Seasonal variation in some physical characteristics of three hill streams, between Byasi and Rishikesh in Uttaranchal Himalaya. *Him. J. Env. Zool.*, 15(2) (2001) 167–172.
57. Negi, R. K., Negi, T. and Joshi, P. C.: Study on physico – chemical parameters of Hinval freshwater stream and Ganga river water at Shivpuri in the Garhwal region. *J. Env. Bio – Sci.*, 22 (2) (2008) 203–212.

58. Sharma, R. C., Bahuguna, M. and Chauhan, P.: Periphytonic diversity in Bhagirathi L preimpoundment study of Tehri dam Reservoir. *Journal of Environ. Science & Eng.*, 50 (4) (2008) 255–262.
59. Sharma, A. P. and Mishra, A. Present status and prospects of mahaseer fishery in Garhwal region of Central Himalaya. In: Fish and fisheries at higher altitudes: Asia. (Ed. T. Petr.) AO Fisheries Technical Paper. No. 385. Rome, FAO (2001) 304 p.
60. Pande, R. K. and Mishra, A.: Water quality study of freshwater of Dehradun (Sahastradhara stream and Mussoorie Lake). *Aquacult*, 1(1) (2000) 57–62.

Synthesis and Medicinal Importance of Benzoxazine, Benzoxazinone and Their Derivatives: A Short Review

Manisha Singhal and Vinay Prabha Sharma

Abstract The research in the pharmacology and medicine is in various ways related with synthesis of medicinally important new compounds. The structural basic skeleton of many drugs and compounds with biological activity contains heterocyclic ring with oxygen, nitrogen and sulphur. Benzoxazin, benzoxazinone and many other compounds having these skeleton are heterocyclic and are of great medicinal importance. The synthesis of benzoxazine, benzoxazinone and their derivatives has been able to draw attention of researchers due to their various medicinal uses and physiological activities. Chemically benzoxazine is a heterocyclic compound. It has an oxazine ring fused with a benzene ring. Oxazine ring is a six membered heterocyclic ring with oxygen and nitrogen atom. Benzoxazine and benzoxazinone are reported to exhibit many useful pharmacological properties including anti-inflammatory, analgesic, antifungal, neuroprotective, and antibacterial activities etc. This review is an attempt to throw some light on synthesis and biological activities of benzoxazines, benzoxazinones and their derivatives. This review motivates researchers to synthesize newer benzoxazine derivatives that can be found to be better drug than already available with lesser toxicity.

Keyword Benzoxazine · Anti-inflammatory · Analgesic · Antifungal · Antibacterial

M. Singhal (✉)
Department of Chemistry, RGPG College, Meerut, India

V. P. Sharma
Department of Chemistry, Meerut College, Meerut, India

N. A. Siddiqui et al. (eds.), *Advances in Environment Engineering and Management*, Springer Proceedings in Earth and Environmental Sciences,
https://doi.org/10.1007/978-3-030-79065-3_2

1 Introduction

Although there are a wide variety of antibiotics available that are used commercially as medicines, but synthesis of new compounds is always of great use due to increasing drug resistance. Heterocyclic compounds having nitrogen, oxygen or sulphur possess numerous biological activities.

In past few decades a number of benzo[1,4]oxazin -3-one **1a** and its isomer **1b** and many other compounds having these systems have been emerged out as compounds of high medicinal values. Like benzoxazinyl-3-chromones **2** have been found to be antimicrobial [1]. 3,1-benzoxazine are inhibitors of interleukin-1 that exhibits a number of immune and anti-inflammatory action [2]. Other activities of benzoxazine have also been reported [3]. Benzo[3,1-] oxazin-4-ones are used as starting compound for the synthesis of fused heterocycles like quinazolin-4-one **3** [4, 5]. Many benzo[1,4]oxazin-3(4H)-one **1a** derivatives including a compound with cardiotonic activity **4** [6], anticancer activity **5** [7], antiparasitic activity **6** [8], antibacterial activity **7** [9], antianxiety activity **8** [10] have been synthesized and screened for their activities in past few years.

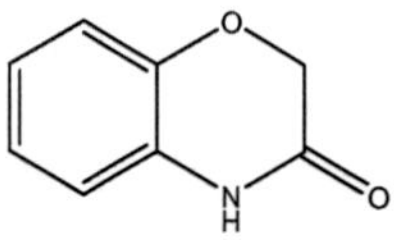

1a

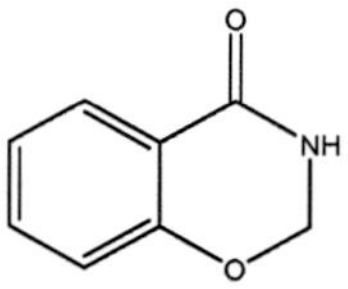

1b

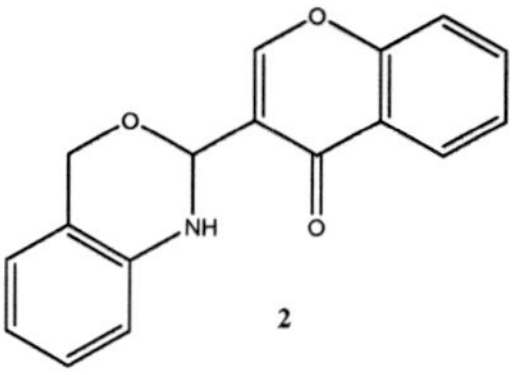

2

Benzoxazine and its biologically active derivatives

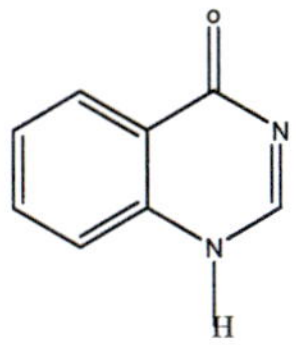

3

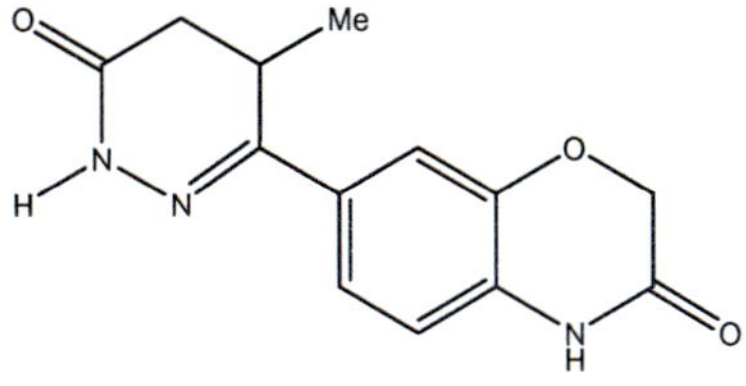

4

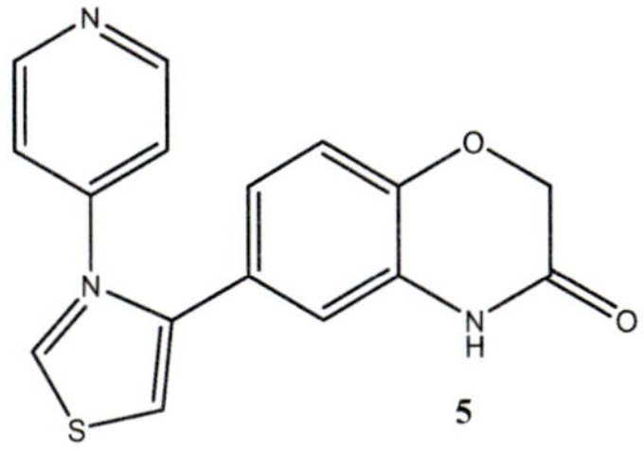

5

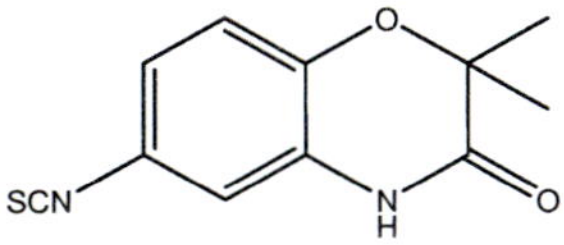

6

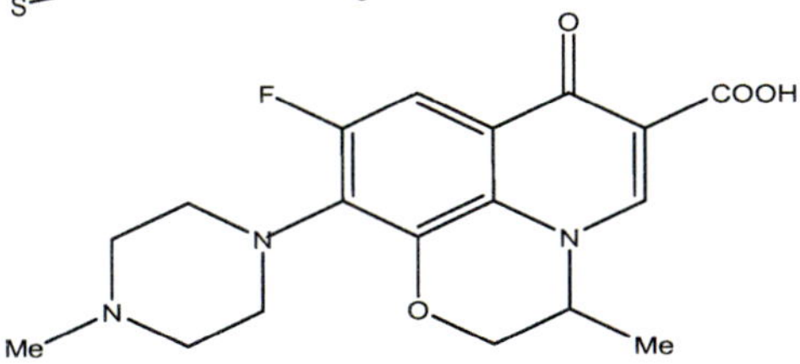

7

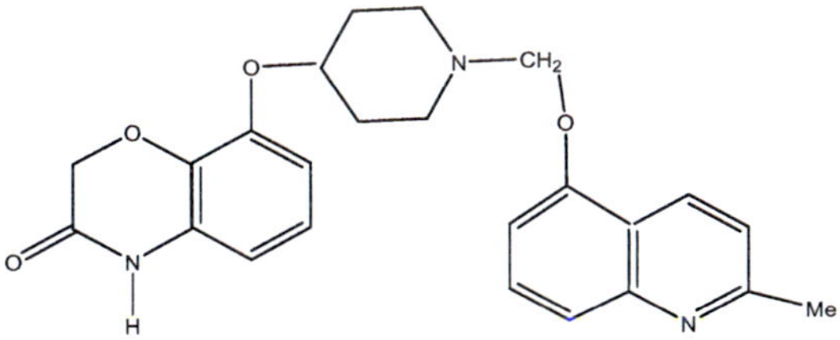

8

Benzoxazine and its biologically active derivatives

Similarly many 1,3-benzoxazines and 1,3-benzoxazinones have been found to play a very important role in various biological activities. A number of benzo[1.3] oxazine derivatives were found to show anti-inflammatory [11], analgesic [11], antifungal [12, 13] and antibacterial activity [14, 15]. A number of benzoxazinone derivatives were synthesized and were found to possess analgesic [16], antimicrobial [17, 18], antibacterial, antifungal [19–22], antiplatelet aggregation activity [23], hypnotic activities [24–26] etc.

In this paper a review of synthesis and medicinal values of various benzoxazine, benzoxazinone and their derivatives have been given.

2 Medicinal Value of Benzoxazines, Benzoxazinones and Their Derivatives

Various benzoxazine and benzoxazinone derivatives were reviewed for their medicinal activities. Some 1,4 benzoxazine imidazole derivatives **9** were synthesized by Fringuelli et al. [27] and were examined by their antifungal activity. These 1,4 benzoxazine derivatives showed immuno modulating activity and invivo activity against a murine experimental model of candidiasis.

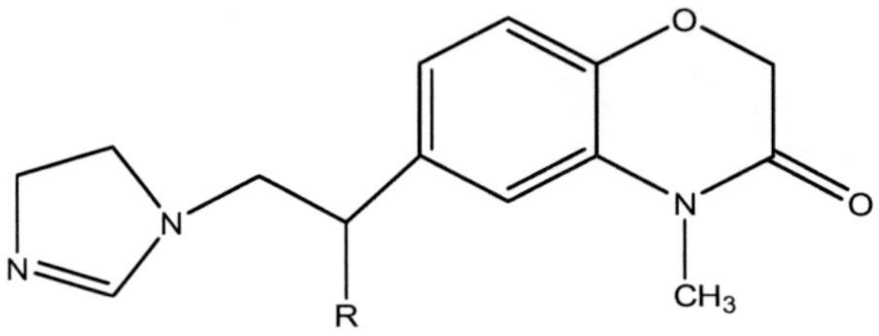

R=-OH,-C_6H_4Cl

9

Ozden et al. [28] have synthesized few 4-hydroxy-2 H- benzo[1,4]oxazine-3 (4H) -ones **10**. Antimicrobial activities of these compounds were tested against *E. coli, S. aureus* and *C. albicans.* Benzoxazine rings with long alkyl chain on 2-position were found to have good antifungal activity.

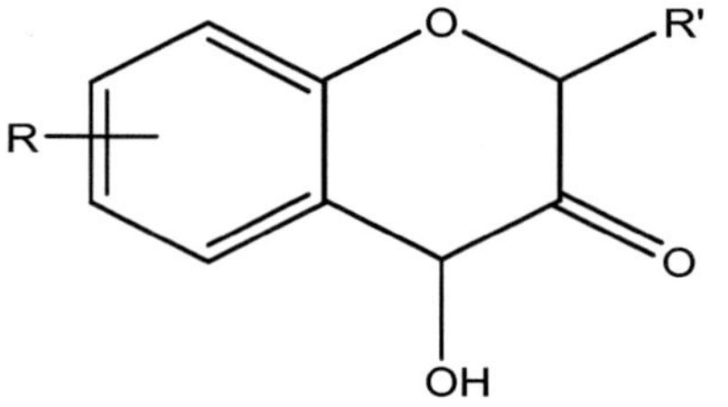

R=-Cl,-F,-OCH_3
R'=-CH_3,-C_2H_5

10

Zamani et al. [29] synthesized a new class of benzo[b][1,4]oxazin 3(4H)-one derivatives **11** by introducing different alkyl or aryl substituents at N-Position. Antifungal as well as cytotoxic activities of these compounds were examined against some standard strains of *Candida dubliniensis* (ATCC 8500) *Candida ablicans* (ATCC 10261) *Candida krusei* (ATCC 6258) *Candida parapsilosis* (ATCC 4344) *Candida tripcalis* (ATCC 750). These were found to show medium to good activity.

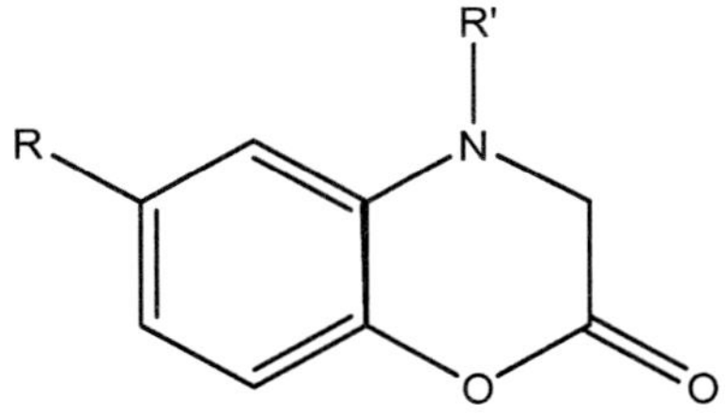

R=H or Cl
R'=alkyl or aryl halide
11

Waisser et al. [30] synthesized a series of 6-chloro-3-phenyl-4-thioxo-2H-Benzo[1,3]oxazine-2 (3H)-ones and a series of 6-chloro-3-phenyl-2H-benzo[1,3]oxazine-2,4 (3H)-diones **12**. These compounds were then tested for their in vitro activity against *M. kansasil, M. tuberculosis* and *M. avium.*

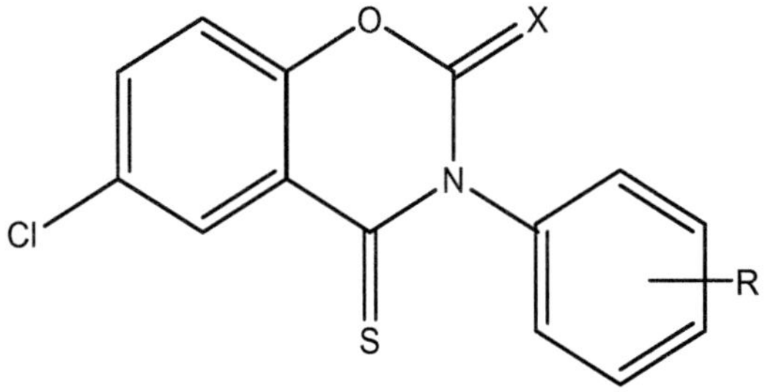

X=O,S
R=H,CH_3,Cl
12

Varshney et al. [31] synthesized 1,3-benzoxazine-4-one derivatives substituted with fatty acids residue **13** and tested these compounds against gram positive (*Staphylococcus aureus* SA 22, *Bacillus subtillis* MTCC 121) gram negative (*Esherichia coli* K 12 *Klebsiella pneumonial*) and fungal strains (*Candida albicans* IOA-109). These compounds were found to show moderate antibacterial and antifungal activity.

Patil et al. [32] synthesized many Benzoxazine derivatives **14** in excellent yields at ambient temperature in PEG in solvent free conditions. They tested these compounds for antifungal and antibacterial activity against Trichothecium, Rhizopus, *E. coli* and *S. auresis.* It was found that presence of chloro and nitro gp decreases the antibacterial activity of benzoxazines but increases antifungal activity of these compounds.

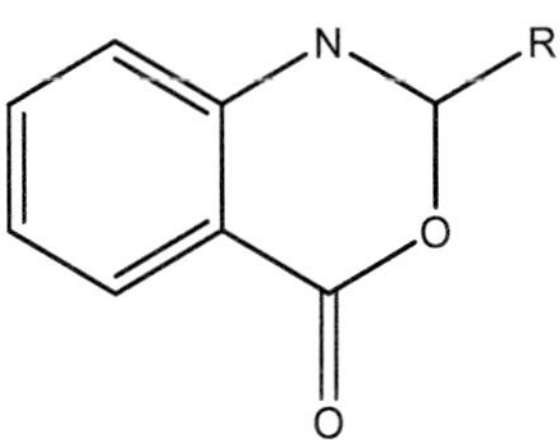

13

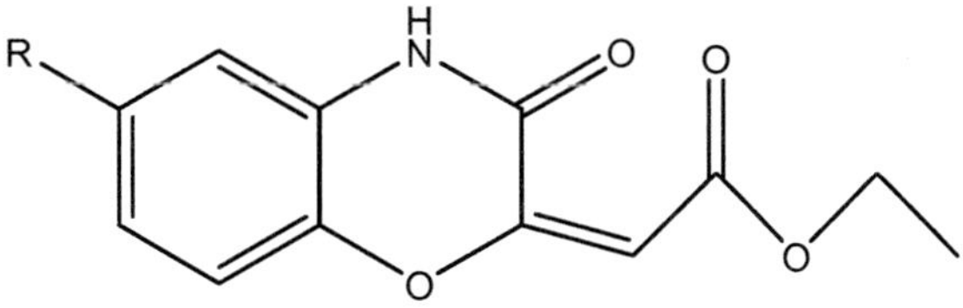

R=-Cl,-NO_2,-CH_3,-H
14

Eissa et al. [33] synthesized 6-iodo-2 methyl-4H-3,1-Benzoxazine-4-one **15** and tested their antimicrobial activity against gram positive bacteria (*Bacillus megathorium*) and gram negative bacteria (*E. coli*) and some fungi. Gilmore et al. [34] synthesized a series of 2-aryl-4H-benzo[1,3]oxazin-4-ones **16** and tested these compounds for their inhibitory action against C1r serine protease.

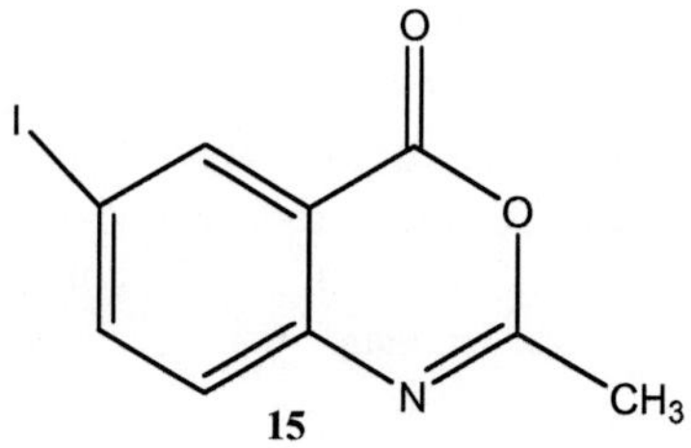

15

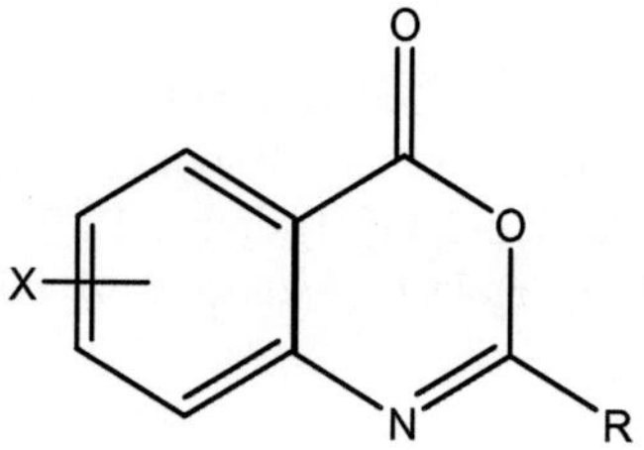

X=H,-CH_3
R=Ph

16

Hasieh et al. [35] synthesized a series of benzoxazinones having substitutions at 2-positions **17** and showed their proper effect to antihuman corona virus.

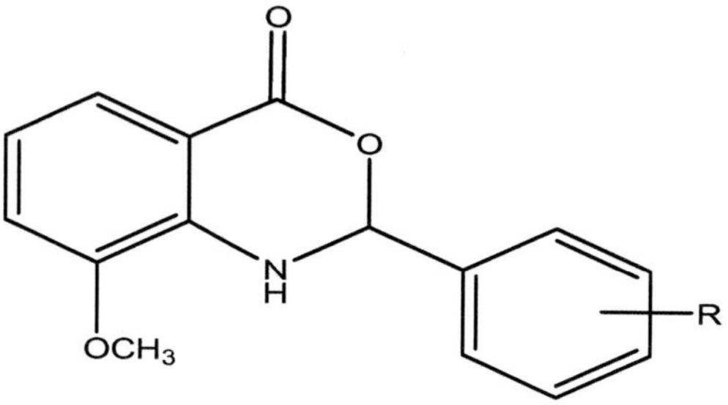

R=Br,Cl,F

17

Dong et al. [36] discovered a new benzoxazine 6-amino-2, 3- dihydro-3-hydroxymethyl-1, 4-benzoxazine. It was found that this compound can improve proliferation of human umblical vein endothelial cells.

Benzoxazine -3-indole alkyl amine **18** and Benzoxazine -3-indole tetrahydro pyridine analogs **19** were synthesized by Zhou et al. [37]. They showed that benzoxazine system is utilised to embrace $5HT_{1A}$ pharmacophore along with the SSRI and $5HT_{1A}$ receptor activities. Most of compounds of this category were examined to function as $5HT_{1A}$ receptor agonists.

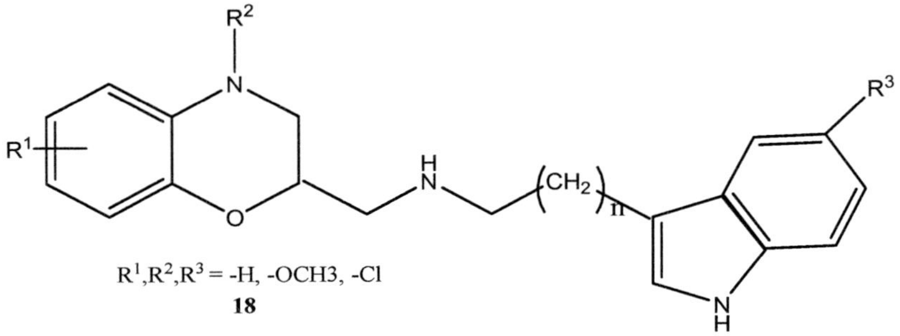

R^1,R^2,R^3 = -H, -OCH3, -Cl

18

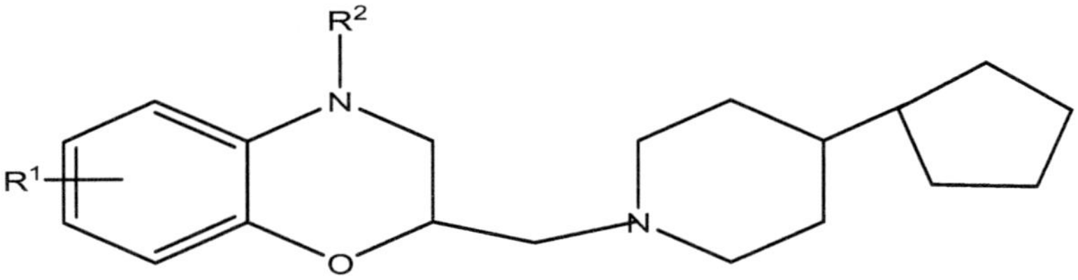

R^1= -H,-OCH_3;R^2=-H,-CH_3,-C_2H_5
R3=-H,-F

19

3 Various Synthetic Routes of Benzoxazine, Benzoxazinone and Their Derivatives

Shridhar et al. [38] reported a method which was very condusive for the synthesis of benzo[1,4]oxazinone **1a**. It involves the reaction of ortho-aminophenol with chloroethanoyl chloride by refluxing in the presence of $NaHCO_3$ and methylisobutylketone (Scheme 1).

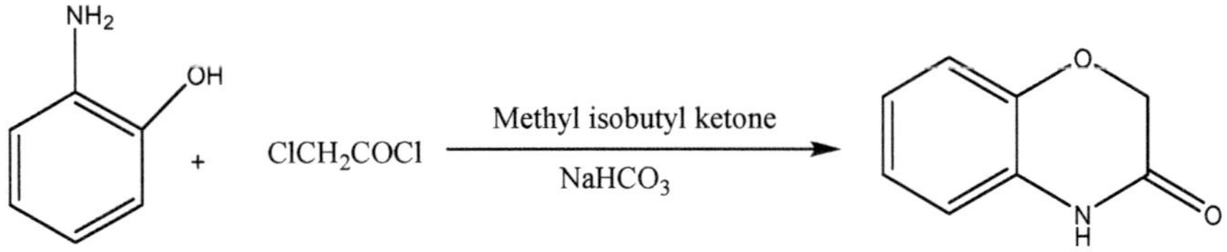

Scheme 1 .

In one method 2-aminophenols were reacted with Alpha-bromo-gama-butyrolactone in DMF to give 2-hydroxyethyl-2,3-dihydro(2H)-benzo[1,4]oxazinone **20** [39]. Reaction was done at room temperature in the presence of NaH or K_2CO_3 followed by reduction [40] (Scheme 2).

Shridhar et al. [41] reacted ortho aminophenols with maleic anhydride to prepare benzo[1,4]oxazinones having- CH_2COOR at 2-position **21** in single step (Scheme 3).

Holly and Cope [42] reported the first synthesis of benzoxazine derivative **22** involving Mannich Condensation that involves reaction of HCHO and an amine having acidic protons. In this method ortho-hydroxy benzylamine was reacted with HCHO or any other aldehydes using H_2O or benzene as solvent (Scheme 4).

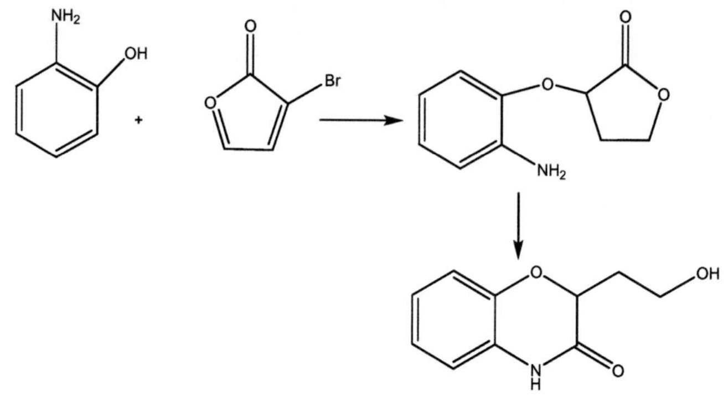

20

Scheme 2 .

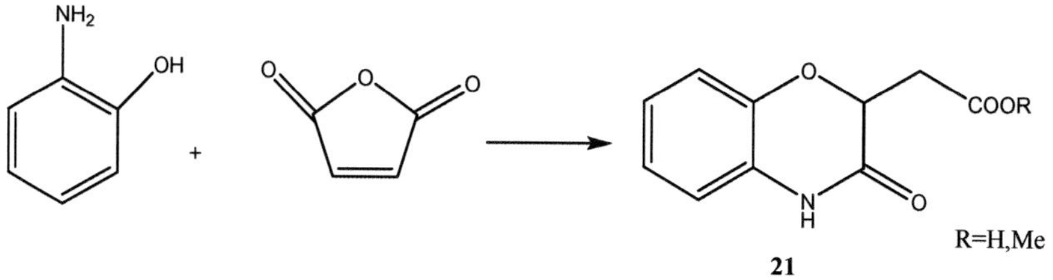

21

Scheme 3 .

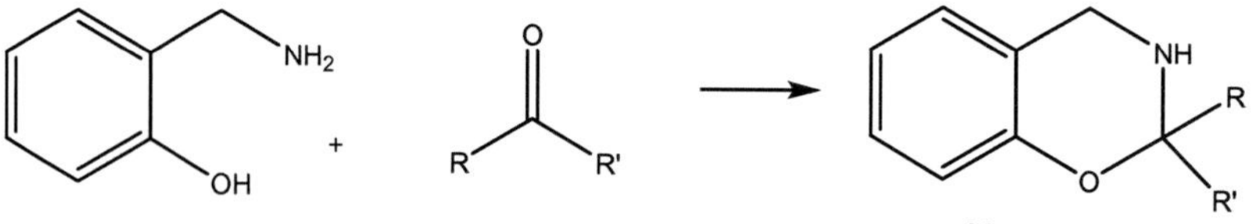

22

Scheme 4 .

Burke [43] in 1949 proposed a simple method for synthesis of benzoxazine derivative **23**. It was a one pot reaction that involved amine, phenol and formaldehyde in 1:1:2 molar ratio (Scheme 5) This method was studied for many years since the same was convenient method as well as there was scope of groups that can be adopted on benzene ring of benzoxazine by using functional group on phenol e.g. nitro [44] aldehyde [44, 45] cyano [46] alkynyl [47] carboxyl [48] etc. were introduced on benzene ring of benzoxazine by using functional phenol.

Burke et al. [49] synthesized hydroquinone based bifunctional benzoxazine **24, 25** from 1,3,5 hexahydrotriazine (Scheme 6).

Bilman and Dorman in 1963 [50] prepared difunctional benzoxazine **26** from bis (ortho-hydroxybenzylamino) ethane and formaldehyde as shown in Scheme 7. Bis (orthohydroxybenzyl amine is prepared by the reduction of Schiff's base obtained from salicyldehyde and primary amine [51, 52]. This method has an advantage that various functional groups can be substituted on oxazine ring.

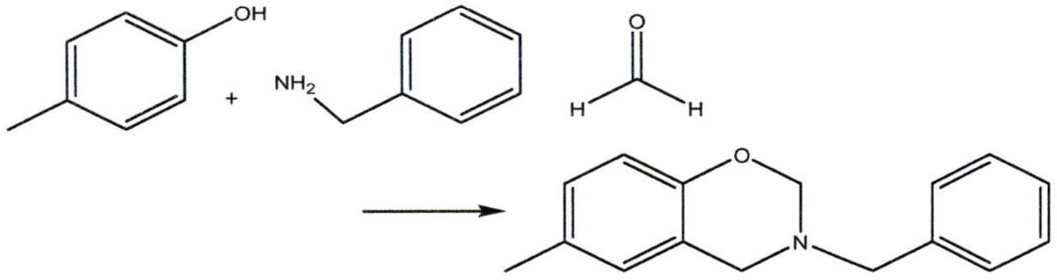

23

Scheme 5 .

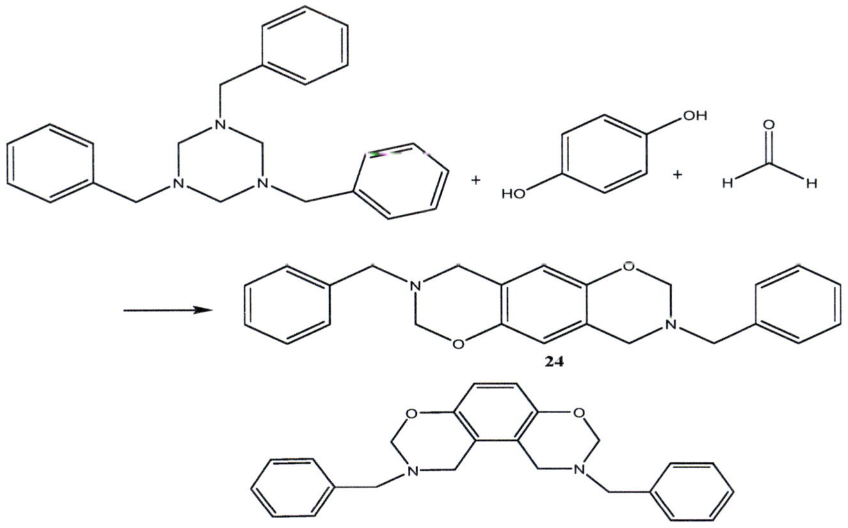

24

25

Scheme 6 .

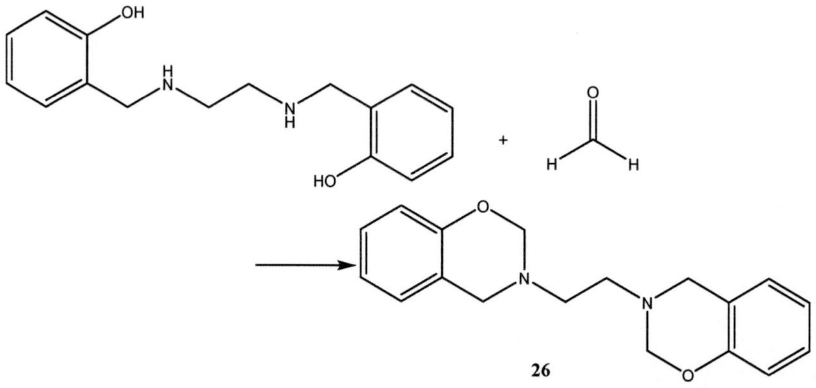

26

Scheme 7 .

Linda Akerbladh et al. [53] prepared 2-aminobenzo[1,3]oxazin-4-ones from orthoidophenols by Domino carbonylation cyclization process. In this method orthoidophenols subjected tocarbonylative coupling with cyanamide $Mo(Co)_6$ in the presence of pd catalyst followed by an intramolecular spontaneous cyclization to give 2-amino benzo[1,3]oxazin-4-ones, **27** (Scheme 8).

Gao et al. [54] synthesized *N*-dichloro acetyl-3,4-dihydro-3-methyl-6-chloro-2H-benzo[1,4]oxazine from **28**. In this method **28** was mixed with toluene and isopropanol and stirred at 60° C under hydrogen atmosphere for 9 h using Pt/C as catalyst. The crude product was purified by column chromatography. It was 3,4-dihydro-3-methyl-6-chloro-2H-benzo[1,4]oxazine **29**. Compound **29** was then mixed with Na_2CO_3 and benzene and dichloroacetyl chloride was slowly added to this mixture. Stirring was done for 3 h at 25° C to give *N*-dichloroacetyl-3,4-dihydro-3-methyl-6-chloro-2H- benzo[1,4]oxazine **30** (Scheme 9).

Alkhathlon [55] synthesized 4-alkoxy-4-methyl-benzo[1,3]oxazinones and 4-alkoxy-4-fluoromethylbenzo[1,3]oxazinones from 2-hydroxyacetophenoneshydrazones **31**. Compound **31** on reaction with triphosgene in

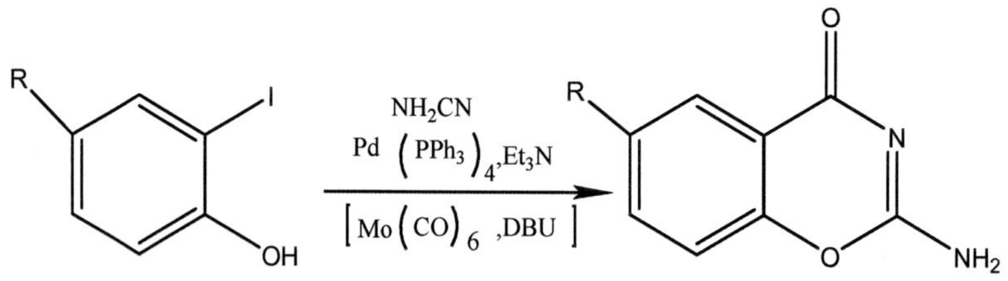

Scheme-8 **27**

Scheme 8 .

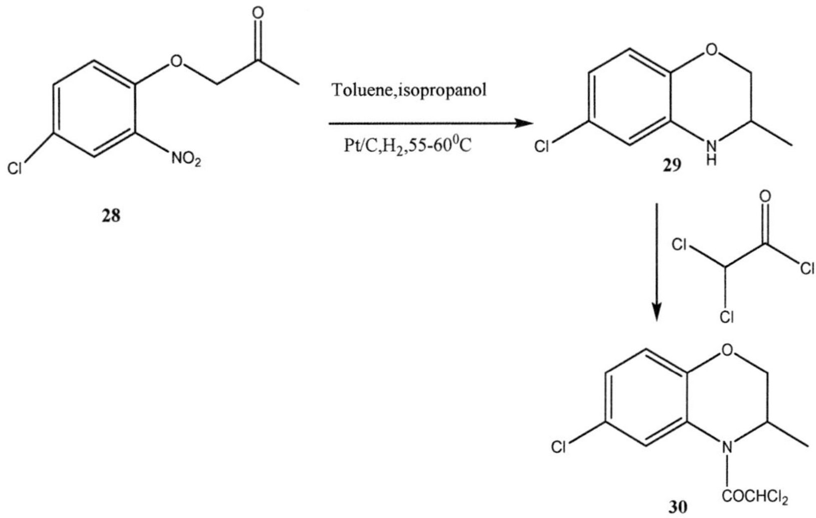

Scheme 9 .

presence of triethylamine and triphosgene undergo cyclization to give 4-methylene-benzo[1,3]oxazinones **32** as product (Scheme 10). Compound **32** was refluxed with ethanol to give 4-alkoxy-4-methyl-benzo[1,3]oxazinones**33** (Scheme 11) and **32** was reacted with 1-chloromethyl-4-fluro-1,4-diazabicyclo[2.2.2] octane bis(tetrafluoroborate) (F-TEDA-BF_4) in CH_3CN in the presence of CH_3OH to give 4-alkoxy-4-fluoromethyl-benzo[1,3]oxazinones **34** (Scheme 12).

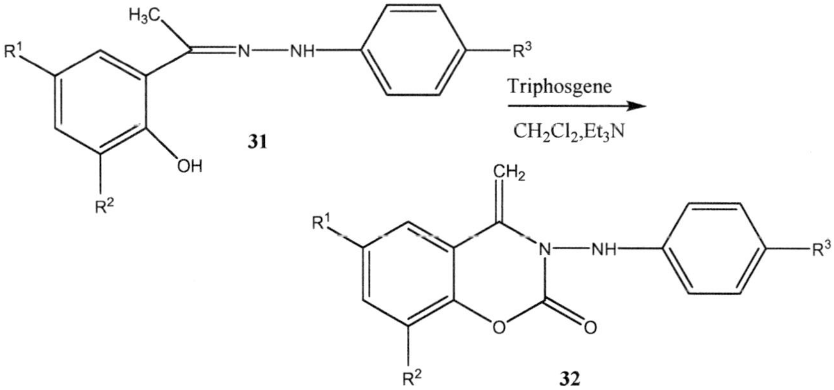

Scheme 10 .

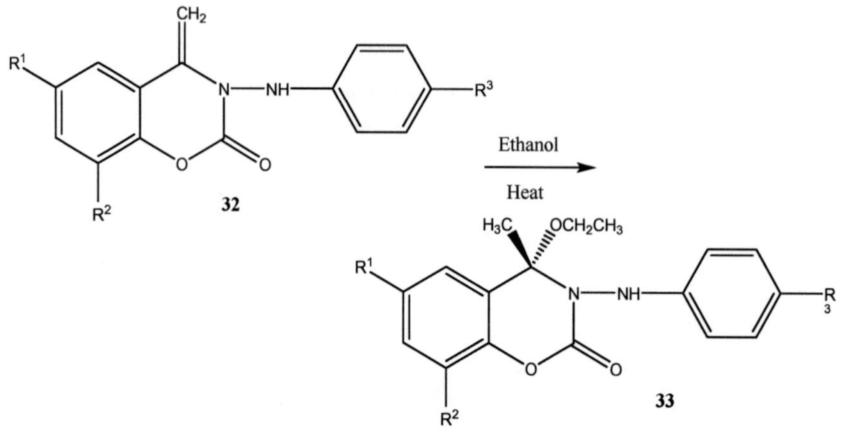

Scheme 11 .

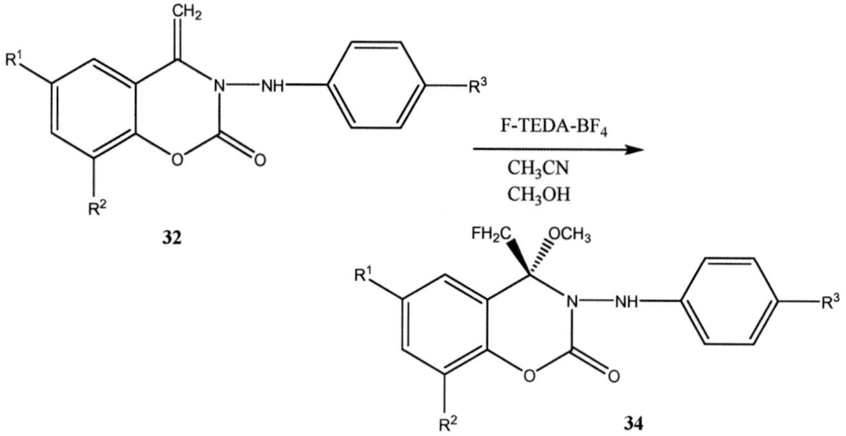

Scheme 12 .

4 Conclusion

A big number of derivatives of benzoxazine and benzoxazinones have been synthesised and studied for their biological values.It can be noticed that various modifications can be done on benzoxazine and benzoxazinone moieties to develop compounds of potential biological importance. This review develops an interest to explore many more new derivatives of this system.

References

1. Sharma V.P., Kumar P., Sharma M., Int. J. Essential Sci. Volume 7(2013) 1–5.
2. Dinarello C.A., Rev. Inf. Dis. 6(1984) 51.
3. Proisl K., Kafka S., Urankar D., Gazvod M., Kimmel R., Kosmrlj J., Org. Biomol. Chem. 12(2014) 9650.
4. Alanzi A.M., Abdel-Aziz A.A.M., Al-Suwaidan I.A., Abdel-Hamide S.G., Shawer T.Z., El-Azab A.S., Eur. J. Med. Chem. 79(2014) 446–454.
5. El-Hashash M.A., El-Badry Y.A., Helv. Chim. Acta, 94(2011) 389–396.
6. Combs D. W., Rampulla M. S., Bell S. C., 6-Benzoxazi-nylpyridazin-3-ones: potent longacting positive inotrope and peripheral vasodilatoragents. J. Med. Chem., 33(1990) 380–386.
7. Bierer D., McCluze, A., Fu, W., Furahi, A., Gaetan, H. L., Burke, M.J., Cheng, B., Barry, H., Jacques, D., Sibley, R., et al. Preparation of 3-pyridyl or 4-iso quinolinylthiazoleas C17, 20 lyase inhibitors. WO0327085. Chem. Abstr. 138(2003) 287663v.
8. Shridhar, D. R., Srinivasa Rao, K., Singh, A. N., Rastogi, K., Jain, M. L., Gandhi, S. S., Krishnan, V. H., Jogibhukta, M., Lovekar, C. D., Tripathi, H. N., et al. Synthesis and anthelmintic activity of some new 6- and 7-isothiocynato- 2H-1,4-benzoxa(thia)zin-3(4H)-ones and benzoxa(thia)zin-3(4H)-thiones. Indian J. Chem. 24B (1985) 1263–1267.
9. Wentland, M.P., Comett, J.B., Chapter15. Quinolone antibacterial agents. Ann. Rep. Med. Chem. 20(1985) 145–154.
10. Otkinson, P. J., Bromidge, S. M., 3,4-Dihydro-2H-benzoxazinones are 5-HT 1A receptor antagonists with potent 5-HT reuptake inhibitory activity. Bioorg. Med. Chem. Lett. 15(2005) 737–741.
11. Berger F.M., Bates H.M., Diamantis V., Kletzkin M., JPlekss O., Sofia R.D., Spencer H.J., Pharmacology, 9, 164 (1973).
12. Iranpoor N., Firouzabadi H. and Shaterian H.R. Syn. Comm., 32(2002) 3653.
13. Billman J.H. and Dorman L.C., J. Med. Chem. 6(1963) 701.
14. Sharma S.C., Swami M.P. and Sharma R.C., Ind. J. Chem Soc., 60(1983) 1002.
15. Tetsuo O., Teruji T., Tadahiko T., Tadashi Y. and Shinzo M., J. Heterocyclic Chem., 28(1991) 1061.
16. Germaine T., Jacqueline L., Pierre B., Jacqueline B. and Jean T., European Journal of Medicinal Chemistry, 10(1975) 37.
17. Thierry B., Charles R.W., Gilles C. and Marie P.A., Bioorganic & Medicinal Chemistry Letters., 6(1996) 2343.
18. Khan R.H. and Rastogi R.C., Indian J. Chem., 32B (1993) 595.
19. Bouillant, M. L., Farre-Bonvin, J. and Ricci, P., Novel applications of the “t-amino effect” in heterocyclic chemistry; synthesis of 5H-pyrrolo- and 1H,6H-pyrido[1,2-a][3,l]benzoxazines Tetrahedron Lett., 24(1983) 51.
20. Kadi, A. A. ,Synthesis and antimicrobial activity of some new quinazolin-4(3H)-onederivatives. Journal of Saudi Chemical Society, 15(2) (2011) 95.
21. Mayama, S., Tani, T., Uneo, T. and Hirabayasjo, K., Isolation and structure elucidation of genuine oat phytoalexin, avenalumin I. Tetrahedron Lett., 22(1981) 2103.
22. Eissa, A. M. F., El-Metwally, A. M., El-Hashash, M. A. and El-Gohary, A, M. F., Synthesis and biological evaluation of some new 2-propyl-4(3H)-quinazolin one derivatives as anti-bacteria. Journal of the Korean Chemical Society, 52(3) (2008) 328.
23. Hsieh, P., Chong, F., Chang, C., Zheng, F., and Lin, K.H., 2-Substituted benzoxazinone analogues as anti-human Coronavirus (anti-HCoV) and ICAM-1 expression inhibition agents. Bioorg. Med. Chem. Lett., 14(2004) 4751.
24. Drummond, G. I. and Severson, D. L., Cyclic nucleotides and cardiac function. Circ. Res., 44(1979) 1945.
25. Bulluci, C., Gualtieri, F. and Chiarini, A., Negative inotropic activity of para-substituted diethylbenzyl phosphonates related to fostedil. Eur. J. Med. Chem., 22(1987) 473.

26. Arfan, M., Khan, R., Imran, M., Khan, H. and Mehmood, J., One-pot synthesis and antimicrobial activities of some 2-artl/alkyl, 3-aminoquinazolin-4(3H)-ones. J. Chem. Soc. Pak., 30(2008) 2.
27. Fringuelli R., Pietrella D., Schiaffella F., Guarraci A, Perito S, Bistoni F, Vecchiarilli A, Bioorg. Med. Chem., 10(2002) 1681–1686.
28. Ozden S., Ozturk A., Goker H., Altanlar N., IL Farma.c, 55(2000) 715–718.
29. Zamani L. et al Docking, Synthesis, Antifungal and Cytotoxic. Activities of Some Novel Substituted 4HBenzoxazin-3-one Polycyclic Aromatic Compounds (2019).
30. Waisser K., Gregor J., L Kubicova L., Klimesova V., Kunes J., Machacek M., Kaustova J. Eur. J. Med. Chem., 35(2000) 733–741.
31. Varshney H. et al. Synthesis and antimicrobial evaluation of fatty chain substituted 2,5-dimethyl pyrrole and 1,3-benzoxazin-4-one derivatives Journal of Saudi Chemical Society, 21(2017) 394–S402.
32. Dipti R. Patil, Sonali M. Salunkhe, Mayur M. Aitawade, Madhukar B. Deshmukh, Govind B. Kolekar and Prashant V. Anbhule *Ecofriendly synthesis of benzoxazines and benzothiazines at ambient temperature without catalyst and their anti-bacterial and anti-fungal activity Der Pharma Chemica, 3(1)(2011) 207–214.
33. Abdel-Monem. M. F. Eissa, Kouser. A. Hebash, Mohamed. Abo Riya and Sherif. I. M. Ramadan Synthesis and Reactivity of 6-Iodo-4H-3,1-Benzoxazin-4-one Towards Nitrogen Nucleophiles and Their Antimicrobial Activities. Chemical and Process Engineering Research www.iiste.org ISSN 2224–7467 (Paper) ISSN 2225–0913 (Online)Vol.15, (2013).
34. Gilmore J.L., Hay S.S., Caprathe W., Lee C., Emmering R., Michael W., Bioorg. Med. Chem. Lett., 6(1996) 679–682.
35. Hsieh P.W., Hwang T.L., Wu C.C., Chang F.R., Wang T.W.,Wu. Y.C., Bioorg. Med. Chem., 15(2005) 2786–2789.
36. Dong Z., Cheng Y., Zhao J., Su. L., Zhao B., Zhang Y., Zhang S., Miao J., J. Cell Physiol., 1(2010) 202–208.
37. Zhou D., Harrison B.L., Shah U., Andree T.H., Hornby G.A., Scerni R., Schechter L.E., Smith D.L., Sullivan KM, Mewshaw RE., Bioorg. Med. Chem. Lett, 16(2006) 1338–1341.
38. Shridhar, D. R., Jogibhukta, M., Krishnan, V. S. H. A general and convenient synthesis of 2H-1,4-benzoxazin-3(4H)-ones. Org. Prep. Proc. Int. 14(1982) 195–197.
39. Frechette, R. F., Beach, M. J., The preparation of 2-hydroxyethyl- 2,3-dihydro-2H-1,4-benzoxazin-3(4H)-one derivatives. Synth. Commun. 28(1998) 3471–3478.
40. Wentland, M.P., Comett, J.B., Chapter15. Quinolone anti bacterial agents. Ann. Rep. Med. Chem. 20(1985) 145–154.
41. Shridhar, D. R., Srinivasa Rao, K., Jain, M. L., A convenient and one pot synthesis of methyl-(3,4-dihydro-3-oxo-2H- 1,4-benzoxazin-2-yl) acetates. Indian J. Chem., 24B(1985) 992–994.
42. Holly F.W., Cope A.C., Condensation products of aldehydes and ketones with o-aminobenzyl alcohol and o-hydroxybenzylamine. J. Am. Chem. Soc. 66(1944) 1875–1879.
43. Burke W.J., "3,4-Dihydro-l,3,2H-benzoxazines. Reaction of p-substituted phenols with N,N-dimethylolamines, J. Am. Chem. Soc. 71 (1949) 609–612.
44. Andreu R., Reina J.A., Ronda J.C., Studies on the thermal polymerization of substituted benzoxazine monomers: electronic effects, J. Polym. Sci. A Polym. Chem. 46(2008) 3353–3366.
45. Ran Q., Tian Q., C. Li, Y. Gu, Investigation of processing, thermal, and mechanical properties of a new composite matrix-benzoxazine containing aldehyde group, Polym. Adv. Technol. 21(2010) 170–176.
46. Qi H., Ren H., Pan G., Zhuang Y., Huang F., Du L., Synthesis and characteristic of polybenzoxazine with phenyl nitrile functional group, Polym. Adv. Technol. 20(2009) 268–272.
47. Agag T.,Takeichi T.,Synthesis and characterization of novel benzox-azine monomers containing ally groups and their high performance thermosets, Macromolecules 36(2003) 6010–6017.
48. Mahfud R., Agag T., Ishida H., Shaikh S., Qutubuddin S., Synthesis and evaluation of novel anionic polymeric surfactants based on poly- benzoxazines, J. Colloid Interface Sci. 407 (2013) 339–347.

49. Burke W.J., Hammer C.R., Weatherbee C., Bis-m-oxazines from hydroquinone, J. Org. Chem. 26(1961) 4403–4407.
50. Billman J.H., Dorman L.C., Imidazolidines. 111. 2-Substituted 1,3- bis-(o-hydroxybenzyl)-imidazolidines, J. Med. Chem. 6(1963) 701–705.
51. Andreu R., Ronda J.C., Synthesis of 3,4-dihydro-2H-1,3-benzoxazines by condensation of 2-hydroxyaldehydes and primary amines: application to the synthesis of hydroxy-substituted and deuterium-labeled compounds, Synth. Commun. 38(2008) 2316–2329.
52. Lin C.H., Lin H.T., Chang S.L., Hwang H.J., Hu Y.M., Taso Y.R., Su W.C., Synthesis of a benzoxazine with precisely two phenolic OH linkages and the properties of its high-performance copolymers, J. Polym. Sci. A Polym. Chem. 51(2013) 2686–2694.
53. Akerbladh L., Shiao Y. Chow, Luke R. Odell, and Mats Larhed. Synthesis of 4H-Benzo[e][1,3]oxazin-4-ones by a Carbonylation–Cyclization Domino Reaction of ortho-Halophenols and Cyanamide" chem. Pub soc Europe 2017, 00, 0–0, Wiley online library.
54. Shuang Gao, Hai-Tao Qu, Fei Ye, and Ying Fu Synthesis and Crystal Structure of N-Dichloroacetyl-3,4-dihydro-3-methyl-6-chloro-2H-1,4-benzoxazine,hindawi publishing corporation, journal of chemistry, 2015.
55. Alkhathlan Hamad Z.* Synthesis of 4-alkoxy-4-methyl- and4-alkoxy-4-fluoromethyl-1,3-benzoxazinones Tetrahedron 59(2003) 8163–8170.

A Review on Onshore Drilling for Oil and Gas Production with Its Safety Barriers and Well Control Methods

Aarushi, Akshi Kunwar Singh, and V. Venkata Krishnakanth

Abstract Throughout the world, Oil and gas industries are a highly risky platform to work in. There is a high risk of injury to personnel engaged in drilling operations and production of crude oil. During drilling and production of hydrocarbons many operations occur electrically, manually or through the power system, so the probability of occurrence of incidents is very high throughout the process. Some incidents occur in the industries because of unsafe conditions but most of the incidents occur due to unsafe act, lack of knowledge, lack of attention and experience. If in any condition all these factors are not the major reason of the incidents so we have to know what are or could be reasons of incidents in oil and gas industries. So, at the time of working at rig site everybody should know the reason of any failure or any unsafe activity could be and after knowing the reason behind of that some serious precautions and preventive actions or equipment's should take and available at the time of working at the rig site. The major accidents at the oil and gas rigs are related to pressure underbalance and kick or blowout while performing drilling operations. So, there are some safety barriers that should be available at the rig site during drilling (Durrant in Safety Management in Drilling Operations, [1]). These safety barriers are kind of well control equipment to prevent the unintentional flow from the wellbore and to secure other equipment, rig personnel and the environment etc.

Keywords Drilling · Kick · Safety barriers · Well control

1 Introduction

In the paper an attempt has been made to review the onshore drilling with its safety barriers and well control methods.The purpose of the oil and gas industries is to extract and produce a good number of hydrocarbons by effective operations. These industries are divided into 3 parts—Upstream, Midstream and Downstream. Drilling, Rigless and Surface facilities are the parts of Upstream operations [2]. Drilling is

Aarushi (✉) · A. K. Singh · V. V. Krishnakanth
University of Petroleum and Energy Studies, Dehradun, India

N. A. Siddiqui et al. (eds.), *Advances in Environment Engineering and Management*, Springer Proceedings in Earth and Environmental Sciences,
https://doi.org/10.1007/978-3-030-79065-3_3

the process to drill a well to produce hydrocarbons. The well is drilled using drill bit which is attached to drill pipe. Well is drilled up to a depth calculated based on location, conditions and physical properties of the reservoir which are determined by the geological surveys. For drilling, seismic survey is done to identify the reservoir conditions.

While drilling, the main aim should be the safety at the drilling rig site. Safety of personnel who are working on the rig, safety of equipment, safety of environment and safety of any property which is present on the site. Some common reasons of the occurrence of accidents are lack of knowledge, lack of competency, lack of experience and unskilled nature, behavioural problems, not following the procedures and standards related to the specific activities etc. These major causes could lead to an accident. Apart from these some operational and technical problems can also become a serious issue like pressure underbalance, improper density and weight of the drilling fluid, equipment failure, reservoir conditions which could lead to the kick and blowout [1]. Kick can be controlled by some operational measures such as by maintaining the density of mud or maintaining the hydrostatic weight greater than development pressure, but if there is any situation of blowout it becomes very difficult to stop it. To mitigate blowout some specific well control equipment should be available and ready to work every time on rig site. If these safety barriers fail due to any reason, then it can have catastrophic results at the site. So the maintenance of these well control equipment should be done periodically to perform safe activities and operations [3].

2 Literature Review

2.1 Drilling

An oil or gas well is drill by use the of drill bit to form a hole to the certain depth to bring hydrocarbons to the surface. The hole is drilled when drill bit rotates under the earth surface with drill pipe or drill string attached to it. After a section is drilled, casing (steel pipe) is placed into the hole. The casing provides strength and structural integrity to the hole to prevent the leakage of drill fluid from the wellbore and isolate dangerous pressure zones from the surface [4]. After putting casing, cement is placed between the casing and wellbore which is called as annulus which provides strength to the wellbore. This process continues for each section until the required depth of wellbore is drilled.

3 Types of Drilling

In onshore there are different types of drilling which are used to drill a well. In onshore, the following two types of drilling are generally used

- Rotary drilling
- Directional and Horizontal drilling.

Rotary drilling: Rotary drilling is often used to drill large holes like petroleum extraction holes, open large pits etc. by using the three-cone drill bits which have many teeth to crush the rock by applying enough weight on the rocks [5]. Electric or hydraulic motors are used for the rotation of drill bit and rotation speeds vary with certain rpm. Rotary drilling is used to drill the well vertically for the extraction of hydrocarbons when the location of the reservoir is easily achievable vertically.

Directional drilling: Directional drilling is used when the target depth of the crude oil or gas reservoir is difficult to achieve by vertical drilling [6]. This type of drilling is used when the reservoir is located far away from the wellbore and the depth of the well is larger than the vertical well. So, to increase the length of the pay zone and for safe and effective production directional drilling must be used to drill such oil or gas well.

4 Types of Onshore Drilling Rigs

- Carrier mounted manual rigs
- A-frame electric rigs
- A frame 3 stand rigs
- Joystick rigs.

5 Types of Systems

There are four types of systems at drilling rigs to perform the drilling operations. These are:

- Power
- Hoisting
- Circulating
- Rotary.

Power system: It is the most important structure on drilling rig because it provides energy and capacity to all the systems on a rig such as hoisting, circulating, rotary.

Energy is usually generated by the diesel engines and then transmitted to the other systems to perform the drilling related activies [1]. So, there are two types of powers:

- Diesel-electric drive power
- Mechanical direct drive power.

Hoisting system: It is used to perform all lifting activities at rig site. While drilling, this activity is performed by lifting and lowering the assesories of the into and out of the wellbore at the time of drilling the well. This structure is a chain of various lifting apparatus such as—Derrick, Drawworks, Block and Tackle and some other like Hooks, Tie-down Anchor.

Circulating system: It is used to circulate the water drilling mud and oil drilling mud into the wellbore and to separate the cuttings and rocks from the total length of the wellbore and surface at the time of drilling process. This structure and process is completed by the different apparatus connected in series such as—Mud pumps, Mud pits, Mud-Mixing apparatus, Contaminant removal apparatus. These systems having connection of series for circulation of fluid to maintain the hydrostatic weight.

Rotary system: It is used for the purpose of drilling having a drill bit attached and combination of various components which are connected to each other in series at the rig floor to provide speed to drill bit. Basically drill bit works to eliminate the rocks by applying excessive weight [6]. Without applying proper weight rotary system would not work and it would not be able to maintain a deeper hole below the earth surface. Process is done by Swivel, Kelly, Rotary drive, Rotary table and Drill pipe.

6 Procedure of Drilling

- Seismic survey
- Exploration well
- Appraisal well
- Development well
- BHA (bottom hole assembly)
- Casing
- Cementing
- Liner.

Seismic survey: These surveys are performed to determine the physical conditions of the earth subsurface to identify the location of the oil and gas reservoir. Sound waves are bounced off earth's subsurface and reflect back to the surface in between the time taken to reflect to the surface determine the conditions of underground rock formations and oil or gas reservoir present [6]. It is also called as geophysical survey.

Exploration well: Exploration well is drilled to identify the presence of oil and gas reservoir in a particular place for the information gathering purpose.

Appraisal well: It is drilled to gather more information about the geological conditions of earth subsurface, rock formations and the size and properties of the reservoirs. Once the conditions of appraisal well is met it can be drilled further successfully [5].

Development well: Development well or Drilling is a next step after the appraisal well which is used to for the production of oil and gas after identifying the geological conditions of the reservoir under the earth's surface.

BHA Components: Drilling is done with the help of some assesories which are fixed with drill pipe for the creation of an oil well. BHA components are:

Drill bit
Mud motor
Heavy weight drill pipe
Drill string
Drill pipe.

Casing: It is a pipe having a sufficient diameter which is brought down into the wellbore and camented in place to protect the drilled sections from pressure and forces which are present beneath the earth [7]. It provides strength and protect against the shallow gas and also prevent contamination of fresh water well zones. There are four types of casing:

- Surface casing
- Conductor casing
- Intermediate casing
- Production casing.

6.1 *Causes of Well Control Incidents*

- Lack of skills and knowledge about drilling and rig operations
- Lack of training in well control procedures
- Not following policies, standards and procedures
- Lack of proper maintenance and periodic inspection
- Lack of proper risk management
- Unintentional flow from the wellbore to the surface called kick.

Causes of kick

- Swabbing
- Lost circulation
- High pore pressure
- Reduction in hydrostatic pressure

- Underbalanced drilling fluid weight
- Failure to fill the hole while tripping
- Failure to maintain the mud properties
- Mechanical failure.

6.2 Distribution of Incidents at Drilling Rig Site

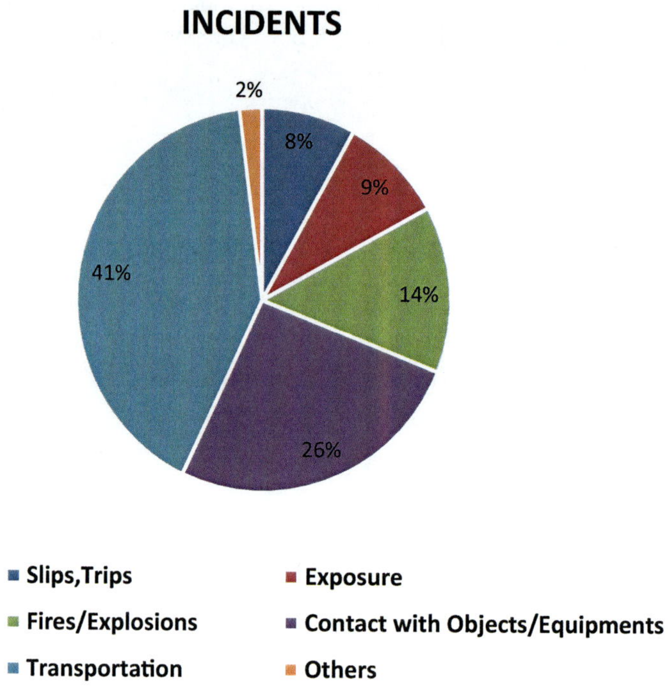

Source https://www.smartdrive.net/industries/oil-gas

7 Methodology

7.1 Safety Barriers

There are some safety barriers to prevent any kind of accident at the rig site, Accidents related to kick inside the wellbore happen when the development pressure is higher than the hydrostatic weight of the drilling mud. There are two types of barrier that should be present at the rig site for well control in case any accident occurs [1].

Two types of barriers

- Primary barrier—Drilling fluid
- Secondary barrier—Well control equipments.

7.2 Mud Types

Water-based mud: It is used at first casing which is conductor casing, to protect the water zones and it is also economically effective.

Synthetic oil-based mud: This type of mud is used on the surface casing and further used throughout the drilling process [8]. This is utilize to expel the drill cuttings, concrete or rocks from the well at the surface.

7.3 Testing of Drilling Fluid

Routine testing of the following

- Density or mud weight
- Viscosity
- Gel strength
- Filtration rate
- Sand content, solid, oil and water content
- Chemical properties.

8 Formula for Calculating the Mud Weight

$$\text{Mud weight} = \text{Hydrostatic pressure}/(\text{TVD} * 0.052)$$

Functions of drilling fluid

- Maintain the wellbore stability
- Seals permeable formations
- Cooling of drill bit
- Support and lubricate the drilling equipment's
- Pressure control
- Corrosion control
- Expel the concrete or rocks from the wellbore
- Clear the surface of the well from the solids
- Transmit the energy to bit and tools
- Reduce the environmental impact

- Maintain hydrostatic pressures
- Support and facilitate the cementing
- Minimize the formation damage inside the wellbore.

8.1 Secondary Barrier

If the primary barrier fails due to any reason then secondary barrier should be available at the rig floor to prevent the unintentional flow from the wellbore to the surface [7]. Well control equipment are—Wellhead, Diverter Line, BOP, Choke Manifold, BOP Control unit etc.

BOP (blowout preventer)

It is a device which is used for well control when the drilling process is going on. It seals the well around any kind of pipe, casing and collar to avoid the undisciplined liquid of reservoir from the wellbore although the primary safety barrier has failed [9].

BOP types

- Ram type preventer
- Annular preventer.

Ram type preventer: It seals the annulus by the contact of two front packing elements with each other and seals around the object in the wellbore to prevent the blowout at the time of kick. Different types of rams are:

Pipe rams: It seals the wellbore/annulus around the fixed size of the drill pipe for which it is designed.

Blind rams: It seals the well if there is no pipe in the wellbore. Front packers come in the contact of each other to seal the annulus.

Shear blind rams: It used to shear the pipe in the well with the help of steel blades and seal the annulus/wellbore simultaneously.

Annular preventer

- There are four basic segments of annular preventer:
  - Body, piston, head, and rubber packing element
- Annular preventer can be closed on drill pipe, casing pipes, drill collar, logging tool and around the tool joint etc. [2].
- It is closed by applying the hydraulic pressure to the piston to move it upward. This forces the packing element towards the wellbore and closes around the object in the wellbore.

- Rubber element is opened by applying the hydraulic pressure to the piston in the opposite direction.

Wellhead

- Wellhead is installed at the surface of the wells which is at the first casing.
- It ensures safe operation and manage the unwanted flow and pressure which is coming from the wellbore at the time unwanted kick.
- Provide enough support to the casing, tubing and to the BOP stack [2].
- It provide means of casing, tubing suspension and for sealing the pressure of the wellbore.

Diverter

- It is installed on a conductor casing with diverter line attached to it.
- It is used only when it is impossible to shut the well because of the chances of lost circulation and formation damage.
- It diverts the flow away from the rig to protect the environment, rig personnel and equipment.

Choke manifold

- Choke manifold is an arrangement of the valves, lines and chokes to control the flow and pressure of mud and kick from the annulus during the well killing process [3].
- It has two remote-controlled and two manually operated chokes and a straight choke bypass line (to trip tank, poor boy degasser and to the pit).

Poor boy degasser

- It is an essential equipment which is used to remove the high percentage of the gas from the mud which is coming out from the chokes during kicks. It provides the means to vent out the gas in a safe manner away from the rig.

BOP Control unit

- It consists of accumulator bottles, a reservoir tank, charging system, pressure regulating valves etc.
- It is used to regulate the pressure of ram preventers, annular preventers, casing and drill pipe pressure while drilling a well [3].
- The purpose of the BOP control unit is to provide nitrogen pressurised hydraulic fluid which will operate the BOP easily to seal the well.

FOSV/IBOP

FOSV/IBOP should always be available on the rig floor for the safety of the drill pipe at the time of any accidental situation.

9 Well Control Procedures

SHUT IN METHODS: If a kick is detected while drilling, then the main objective is to shut the well to prevent unintentional flow from the wellbore [3]. So, there are two methods to shut the well, those methods are:

Soft Shut in	Hard Shut in
• Choke closed all the time • Open HCR valve • Close blowout preventer • Gradually close both types of choke • Record SIDPP, SICP and pit gain	Choke closed all the time Close BOP Open HCR valve Record SIDPP, SICP

9.1 Well Kill Methods

There are mainly three types of techniques:

- Wait and weight
- Driller's
- Concurrent.

Wait and weight method: The wait and weight method involves circulating out the kick with kill weight mud in one circulation. Using this method kick is being removed from the wellbore by increasing the kill mud weight, it requires longest time to circulation if the annulus is full of drill cuttings [10].

Driller's method: In this method, there are two processes. First, remove all the influx from the well and shut in well and then maintain the mud weight up to desired kill mud weight which is then pumped into the well [11]. This is also called Two circulation method.

These two methods are mostly preferable as a well kill methods.

10 Safety at Rig Site

- PPE—safety helmet, safety shoes, safety gloves, safety glasses, earplugs, SCABA, bodysuit.
- First Aid
- Fire exit/egress
- Fire resistive buildings
- Fire alarms

- Fire extinguisher
- Fire hydrants
- Fire hoses
- Firewater tank
- Water pump.

As per OISD 189, requirements of firefighting systems at rig site

Area	Fire extinguisher requirement
At rig floor	2 no. 10 kg DCP
Engine area	1 no. 10 kg DCP
Mud preparation pump area	1 no. 6.8 kg CO_2 or 10 kg DCP
Mud mixing tank area	1 no. 10 kg DCP
Control room	2 no. 10 kg DCP
Diesel storage area	50 L foam mechanical extinguisher
DIC office, Bunk house area	3 no. 10 kg DCP or 3 no. and 6.8 kg CO_2

Source https://www.scribd.com/doc/191283487/OISD-STD-189-2000-Fire-Fighting-Equipment-for-Drilling-Rigs-Work-Over-Rigs-and-Production-Installations

10.1 Standards (Maintenance and Inspection)

Maintenance and inspection, both are important for the safety of equipment, the safety of personnel and property and for the safety of the environment. Periodic inspection of every device or equipment should be done as per the standards. Following are the standards which are to be used:

- OMR (Oil mines and Regulations)
- OEM (Original Equipment Manufacturer) prescriptions
- OISD (Oil Industry Safety Directorate)
- NDT (Non-Destructive test)
- API (American Petroleum Institute).

10.2 Conclusion

Overall, while performing drilling there is a possibility of the failure of well barriers or well integrity during the extraction of hydrocarbons from the well, so two safety barriers policy at every stage of drilling operations should be followed to prevent any unintentional flow from the wellbore. If primary barrier (drilling fluid/mud) fails because of any reason (For e.g., if hydrostatic weight of the mud is not maintained according to the development pressure of the reservoir, it may cause the failure of primary barrier), there should be secondary barrier available at the rig to prevent any hazardous situation. Apart from the barriers of well control, other mandatory safety devices should be available at the rig site every time, because accident/incident can also be related to any equipment damage, property damage, fire, environmental pollution etc. So as per OISD 189 required amount of firefighting equipment (fire extinguisher, fire hydrants, fire hoses, fire water tank, fire pump etc.) should be present at the rig site and proper maintenance and inspection should be done of every equipment on time for the safety of environment, property and personnel.

References

1. Durrant, J. (1991). Safety Management in Drilling Operations. https://doi.org/10.2118/23006-ms. *SPE Asia-Pacific Conference.*
2. Chaney, M. (1990). Planning the Operations for Drilling a Well. https://doi.org/10.2118/19930-ms. *SPE/IADC Drilling Confrence.*
3. Santos, O. (1991). Well Control Operations in Horizontal Wells. https://doi.org/10.2118/21105-pa. *SPE Drilling Engineering.*
4. Hollis, J. W. (1968). Onshore Drilling Operations in Abu Dhabi. https://doi.org/10.2118/2372-ms. *Regional Technical Symposium.*
5. Well Drilling Methods. https://doi.org/10.3133/wsp257. (1911).
6. Simpson, D. A. (2017). Well-Bore Construction. https://doi.org/10.1016/b978-0-12-813022-3.00002-x. *Practical Onshore Gas Field Engineering.*
7. Dodds, R. (1989). Remote and Difficult Onshore Drilling Logistics. https://doi.org/10.2118/18668-ms. *SPE/IADC Drilling Operations.*
8. Ramalho, J. (2006). Well Control aspects of Underbalanced Drilling Operations. https://doi.org/10.2118/106367-ms. *IADC/SPE Asia Pacific Drilling Technology Conference and Exhibition.*
9. Maduro, W. (1989). Enchova Blowout: Record Relief Time. https://doi.org/10.2118/18717-ms. *SPE/IADC Drilling Conference.*
10. IADC Well Classification System for Underbalanced Operations and Management Pressure Drilling. https://doi.org/10.1016/b978-1-933762-24-1.50020-6. (2008). *Managed Pressure Drilling.*
11. 2. Cormack, D. (2017). An Introduction to Well Control Calculations for Drilling Operations. https://doi.org/10.1007/978-3-319-63190-5_4.

Cultivating Effectiveness and Efficiency Using 5S Methodology

Abhishek Nandan, Ankita Sharma, K. Praveen Kumar, Akshi K. Singh, and N. A. Siddiqui

Abstract 5S is a systematic technique used by organizations that comes from five Japanese words: Seiri (Sort), Seiton (Set in order), Seiso (Shine), Seiketsu (standardize), and Shitsuke (sustain) (Patel and Thakkar in J Eng Res Appl 4(3):774–779, [1]). The common problem that arises in an organization is defects in material, quality of the human resource, downtime in production, improper safety for the workers, working condition, etc. This paper deals with the implementation of 5S in a small-scale industry and to analyze the effectiveness in terms of search time, safety, productivity, cost, etc. The implementation of 5S can also result in considerable improvements in efficiency, maintenance, quality, better, more committed employees, and a stepwise reduction in unnecessary items. The 5S also intend to build a stronger work culture within management and workers and keeps them moving towards to have continual improvement.

Keywords 5S · Efficiency · Quality · Safety · Reliability

1 Introduction

The level of competition among the industries has been largely affected by the fast and rapid growing globalization and innovation. The secret of any organization to survive in this competition is depends upon their product and quality. Improvements and innovations breakthroughs have become necessity to stay in growth. Normally this can be achieved by implementing best practices such as lean manufacturing, JUST-IN-TIME (JIT), six sigmas. Lean manufacturing is fact is survival and success of any organization by minimizing the waste. The 5S implementation is the best methodology to achieve lean manufacturing [2].

The detailed review of the implementation in various organizations such as bank, health care, industry, hotels has been discussed. Patel and Thakkar [1] explains the methods and techniques of 5S uses to increase the efficiency of all processes in

A. Nandan (✉) · A. Sharma · K. Praveen Kumar · A. K. Singh · N. A. Siddiqui
University of Petroleum and Energy Studies, Energy Acres, Bidholi Via Premnagar, Dehradun, Uttarakhand 248007, India

N. A. Siddiqui et al. (eds.), *Advances in Environment Engineering and Management*, Springer Proceedings in Earth and Environmental Sciences,
https://doi.org/10.1007/978-3-030-79065-3_4

the company. Thus, this results effective improvement in various organization and detailed explanation of various organization implementation. The definitions of these terms are explained below [3–6].

1. **SORT** is the basic for the standard, which follows the principle: "Just What Is Needed, In Quantities It Is Needed Only and Only When Needed". This refers to removing any unnecessary items from the organization to perform the task.
2. **SET-IN-ORDER** refers to the principle: "A Place for Everything and Everything in Its Place". This is a means by which keeping all our equipment in order and label both our equipment and shelf in order, so that anyone in the organization can figure the equipment at required time.
3. **SHINE** takes in every organization, which includes cleaning, sweeping and removal of dust/dirt/waste. It is to be done in a daily basis to make a healthy and regular live environment.
4. **STANDARDIZE** refers to establish standards of the best practice in the organization and to ensure that the standards are compiled and to undertaking that the organization is clean and tidy at all times. It is to ensure that "Everyone Should Know Their Responsibility". It is also to make document of record and to ensure 5S as a part of organization culture.
5. **SUSTAIN** refers to the process of forming a habit of continuous improvement. Without implementation of other 4S, it is impossible to make sustain. In this the top management should take the responsibility. The management shall help the workers to know about the importance of 5S. This can be done by the help of training and keep updating about the importance of 5S and displaying 5S boards in all the prominent areas.

The five Japanese words and there English meanings are Marshettiwar and Sangode [6] (Fig. 1).

Fig. 1 5S

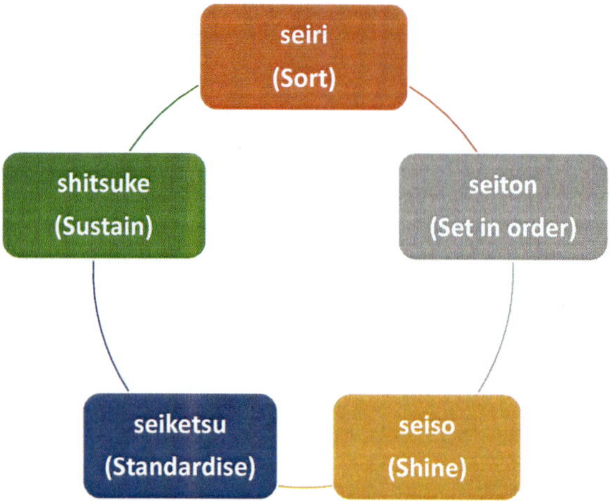

5S is a technique used as path for improve the performance and to organize the whole system, which has been used first time by Japanese. It comes from five Japanese words start with S, which is translated into English words to give the best explanation for them [7]. The methodology of 5S helps in the organization to create and maintain a well-organized, clean, safe, high efficiency and high quality place [8]. Implementing 5S lean manufacturing training has a number of benefits [2, 9] such as:

- The main objective of implementing 5S in an organization is to increase the productivity. It works as an enabler to improve performance by reducing the turnover time which, in turn increases the productivity of the organization, thus increases the overall return on investment [10].
- The organized and segregated workplace especially a manufacturing line results in significant reduction in turnover time leading to less pressure of work due to which number of injuries is reduced, resulting increased confidence level of the artisans [8].
- By implementing 5S aids in on-time deliveries, reducing processing time eventually by the flattened curve of lost or defected goods at the workplace. This enables the improved turnover of the organization.
- The artisans' commitment is the main goal in the implementing 5S lean manufacturing training for the ground level implementation.

2 Methodology: Implementation of 5S

The gain of 5S implementation can be achieved by the joining hands of both management and the entire team. Thus, before the process of implementation, all the employees are briefed about the 5S and should be made aware of effectiveness and benefits of implementation. They should be made to believe and understand that the implementation is going to serve good for both the employees and the organization.

The objectives can be achieved by the combined effort of training and the attitude of the employees in the organization. This also helps in the satisfaction of both the organization and the workers.

The way it is done:

1. The numbers of items, their name and size in the storeroom is determined.
2. The data of time studies for one worker to search SEVEN different items in the storeroom before and after set in order implementation is recorded.
3. The collection of historical data of accidents before 5S implementation from December 2017 and recent data of minor injuries after 5S implementation until march 2018.
4. Data analysis has done to measure the improvement after 5S implementation processes. 5S weekly audit data is obtained based on 5S audit checklist for seven weeks and analyzed with line chart to evaluate the overall 5S implementation,

measure the improvement and improve 5S condition at the storeroom. Measurements have been done three improvement during these 5S implementation activities.
5. The cost data has been analyzed before and after implementation of 5S. The cost is analyzed in terms of search time.
6. The Likert scale questionnaire is made and given to 10 employees and analyzed using SPSS software in order to do reliability test.

3 How It Is Implemented

We had been implemented 5S in both office and industry different organizations. The techniques and methodology used in these organizations vary with respect to time and location.

3.1 Administrative Office

The various steps that has been taken to implement 5S in an administrative office in order to decreases the search time of any file or data and also to provide a peaceful environment for the employees and visitors.

The way it was done is:

1. The various questions regarding the management are asked to the head. Some of them include

   - What is the main motive of the office
   - What is the various equipment and tools that are stored here?
   - Total number of rooms and what are they
   - Number of employees
   - Frequency of cleaning and sweeping
   - How the documents are retained either file or e-base.

2. Once these questions are asked. Then the walkthrough survey of the office is done and the various process are analyzed.
3. The photos and pictures of shelf, rooms, tables, are taken.
4. Once it is done, next is to implement the first s by removing unnecessary items from the office and to make a RED-TAG AREA.
5. Once the first S is implemented then to arrange and label the various available items. The office consists of various files such as feedback file, attendance file, trainer document, courses document etc. These documents are arranged in a proper order depending on frequency use in different color file. The various tools that are available are motherboard, solder, microchips etc.

6. The rooms and the necessary locations are labeled. This includes labeling the place where shoes should be removed, place where files, tools should be placed etc.
7. It is too implemented third S, in this the frequency of cleaning and sweeping done in the office is determined and a person should be kept responsible for inspection and proper checklist and made and given. E.g.

   - Checklist for cleaning
   - Checklist for generator maintenance
   - Checklist for AC maintenance
   - Checklist for restroom cleaning.

8. The final step is to standard the way of approach within the office and taken final image after implementing and analyzed the improvement.

3.2 Industry

Various process is available in an industry in order to complete the product. Since 5S cannot be implemented in each process due to time concern. The major process, which includes major cause to improvement, is determined and then implemented in that process. PARETO DIAGRAM is also known as '80/20' analysis, which defines the process to determine the process, which causes a major problem, and to eliminate it [11, 12].

Example: In this search time, tool missing, safeties are major factors, which results in 80% of problems. Hence, priority must be given to these factors and must be eliminated (Fig. 2).

The various steps included are:

1. Determining the major process which affects the major problem

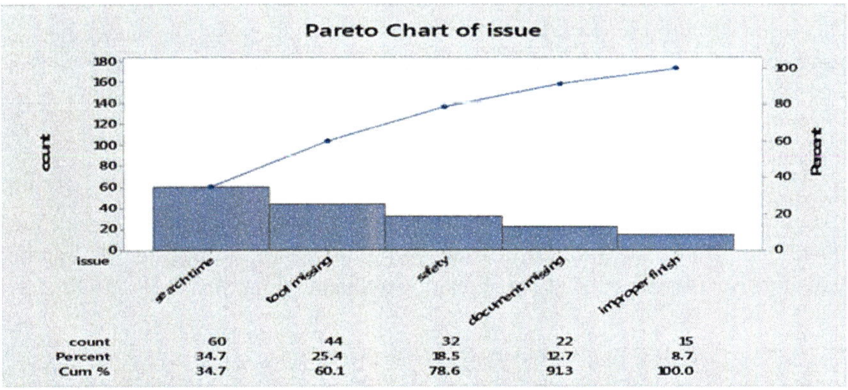

Fig. 2 Pareto chart

2. The manufacturing and assembly unit is selected
3. The various processes are determined
4. The various tools, equipment and necessary safety accessories used in the process are determined.
5. Then first s in implemented by removing unnecessary scrap and making a RED-TAG AREA. These scraps are removed in a proper disposal method.
6. The second s is implemented by labeling and keeping the tools such as grinder, abrasive, driller, surface finisher is kept in a proper-labeled place.
7. And the dangerous place is labeled with cautions.
8. The third s is implemented by having a proper checklist for cleaning process.
9. Then the importance of 5S is explained to workers and help to understand the benefits of achieving it.

4 Result and Discussion

The various results are obtained by making a proper evaluation of data that are obtained before and after implementation. Some of the data are search time, safety, audit score, cost analysis in search time, area, and process and production quantity.

4.1 Evaluation in Terms of Search Time

Waste of time for searching items, waiting and motion are occurred because of the increasing in delay during the parts searching where the workers need to move around to find out the items needed among unnecessary items. The proper labelling system has been done for every rack based on the respective categories in order to reduce the delays. Thus, a worker has done the time study in order to measure the improvement in reducing the delays and waste of time for searching, waiting and motion by 50%. The time study has been conducted to measure searching time for seven different items in the storeroom before and after 5S implementation and the data collected as shown in the Table 1. The timings are obtained from evaluating a single worker (Fig. 3).

The graph represents the data that has been obtained in table. The blue indicated the search time before implementation and red indicates the implementation after 5S (Fig. 4).

Highest time is consumed for searching basic tools such as hammer, screwdriver, spanner etc. It is reduced to a percent of 65 by proper arrangement depending on frequency use. It is done by a help of making a Shadow Board.

Table 1 Material search time

Material/equipment	Before (s)	After (s)	Reduction in time (%)
Valve	190	105	43.58
Flange	200	93	53.5
Filter Bag	113	90	20.35
Filter cartridge	123	95	22.75
Bolts and nuts	285	150	47.3
Hammer/screw-driver/spanner	400	140	65
Electrode	240	110	54.1

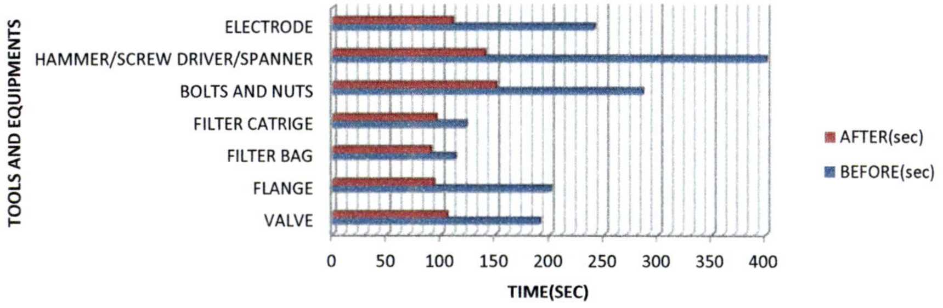

Fig. 3 Material search time

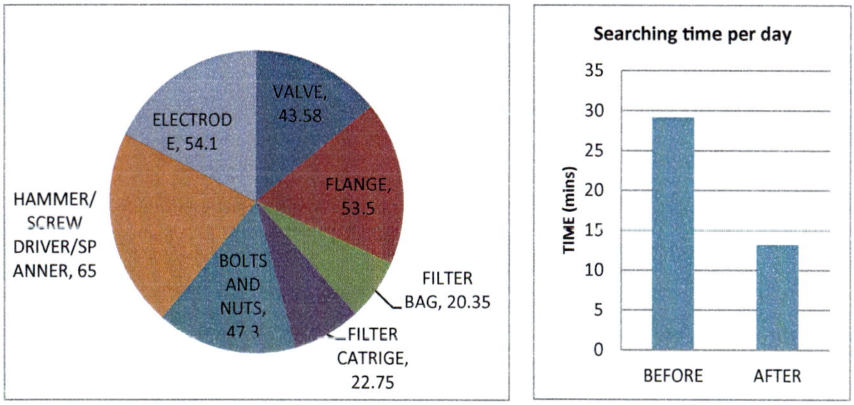

Fig. 4 Reduction time (LHS); search time per day (RHS)

4.2 Evaluation in Terms of Cost (Search Time)

The cost improvement is shown by decreasing the number of defect parts and by increasing the production rate and red-tag improvement. Thus by the help of RED-TAG some of the parts that are to be thrown away are sold to nearby customers. Also before implementation of 5S there has been a number of defect and poor surface finish products which results in decrement of cost. After 5S his has been overcame by proper abrasive material usage.

Cost analysis in terms of search time:

Worker wage per month = Rs 7500/-
The average number of working hours per day = 8
The average labour rate per minute = 7500/ (30 * 8 * 60 *) = Rs.0.52/-
Minutes saved per day = (25.84 − 13) = 12.84 min
Total amount of money saved per day = 12.84 * 0.52 = Rs 6.687/-
The total amount of monthly saving per worker = 6.68 *3 0 = Rs 200/-

4.3 Evaluation in Terms of Relaibality

Ghodrati and Zulkifli [13] the accuracy of hypothesis for any type of research relies on the collected data and analysis of the gathered information. Under this section, real-time collected data will be analysed using SPPS software.

By calculation α in reliability test, the reliability of the collected data is checked. There is unanimity that database with Cronbach alpha greater than 0.7 is reliable. Greater values of Cronbach alpha represent repetitive questions but natural same issue on the 5S system for many of the responses. Based on the 5S system, ten questions were asked to each employee and the responses were analysed for reliability.

Cronbach's Alpha value:

Cronbach's Alpha value	Number of samples (N)
0.733	10

Thus, the value of alpha is greater than 0.7; the implementation of 5S is stable and good.

4.4 Evaluation in Terms of Safety

Safety importance is lacked due to sudden accident, felled and tripped by workers because of their careless. It caused by the items that are placed on the floor without proper arrangement and jammed the work area and due improper evaluation of safety procedure. The storeroom is packed and messy consisted with necessary and unnecessary items. The improvement measurement in terms of safety importance by 50% has achieved. Improvement in term of safety importance after 5S implementation achieved successfully by reducing the number of incidents (Table 2 and Fig. 5).

Table 2 Number of Incidents

	Days	No. of incidents	Total	Reduction rate (%)
Before 5S	December (15–31) 2017	2	5	80
	January (1–20) 2018	1		
	January (21–31) 2108	2		
After 5S	February (1–15) 2018	1	1	
	February (16–28)	0		
	March (1–7)	0		

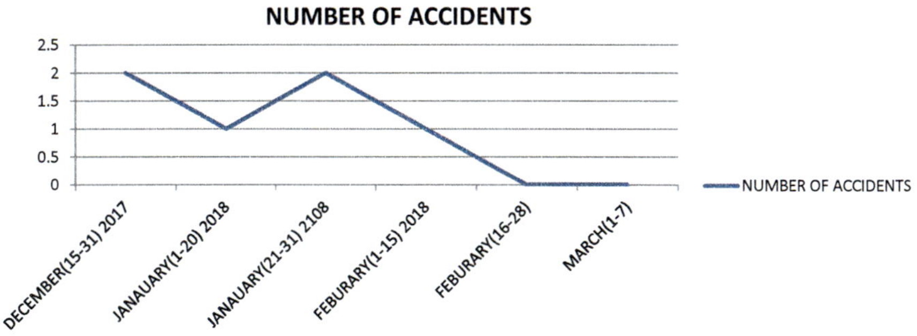

Fig. 5 Number of accidents

The percentage of accidents has been reduced more than 50%, which is 80%. By sorting activities, unnecessary items, which are stored on the rack, are sorted and removed out based on red tag strategy.

4.5 5S Evacuation and Scoring

The audit process at the workshop storeroom is enabled the organization to identify the potential level of improvement [14]. The 5S auditing criterion that is used is relevant to the working environment. The 5S Audit Team does the 5S weekly audit. The 5S audit checklist was structured in which a specific assessment criteria and control point based on five pillars of 5S. Each assessment criteria has its own score, which is: **0 = worst, 1 = bad, 2 = average, 3 = good and 4 = very good** (Table 3 and Fig. 6).

The curve shows the classification in incremental level until the week 7 where the sorting activity is done systematically where the useful items placed on their previous place, unknown items moved to red tag area, and useless items have disposed. So as compared to the initial week and the final week in the graph, it is seen that the workers have accepted the implementation of 1st S i.e., the classification of materials (Fig. 7).

Table 3 5S weekly audit data for 7-weeks

5S activities (weeks)	Sort	Set-in-order	Shine	Standardize	Sustain	Total	Average
1	0	1	0	1	0	2	0.4
2	3	1	2	1	2	9	1.8
3	8	3	3	2	2	17	3.4
4	11	4	4	2	4	25	5
5	12	9	8	3	4	36	7.2
6	12	13	12	6	5	48	9.6
7	15	15	15	9	6	60	12

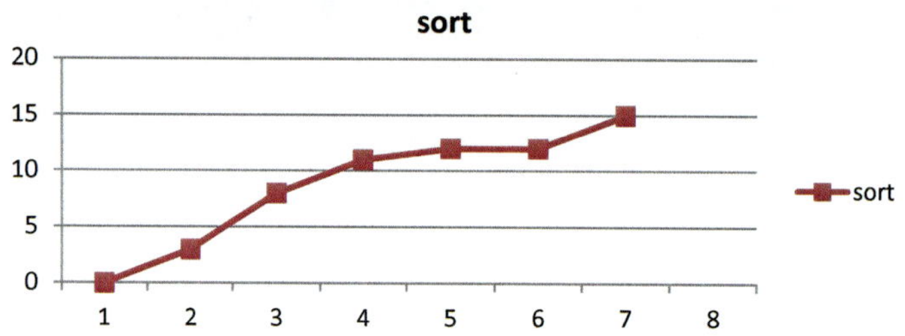

Fig. 6 Sort weekly audit

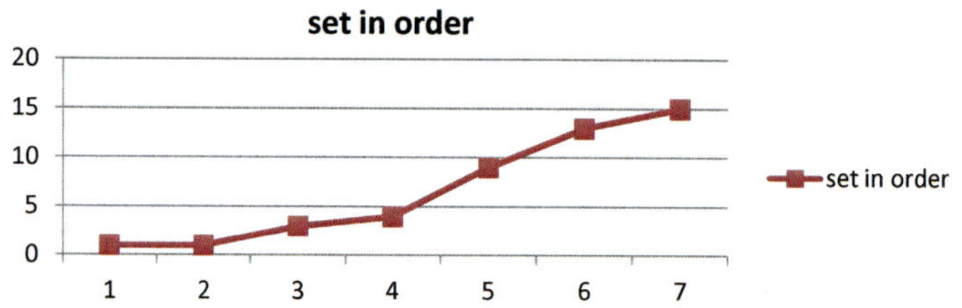

Fig. 7 Set-in-order weekly audit

Set in order has been implemented within the week 3–4. Thus, in the graph it can be seen that there is a linear line between week 3 and 4. Set in order is carried out by items arrangements based on specific rack and categories, systematic labelling system for racks, items and document files, systematic floor tapping and renovated document racks (Fig. 8).

The curve showing the classification score which is indicate a very vertical curve in incremental level till the week 7 since the shine activity is done systematically by setting the cleaning as a habit and daily routine. The shine is improved continuously with the application of proper shine schedule where the shine areas are specifically divided to workers by using a 5S schedule. This has been evaluated by the help of checklist (Fig. 9).

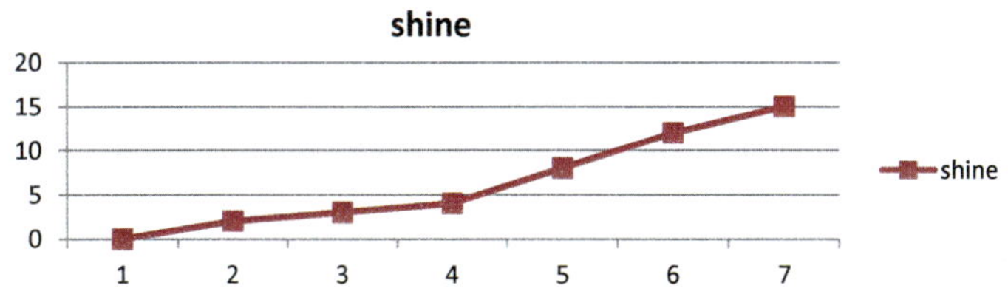

Fig. 8 Shine weekly audit

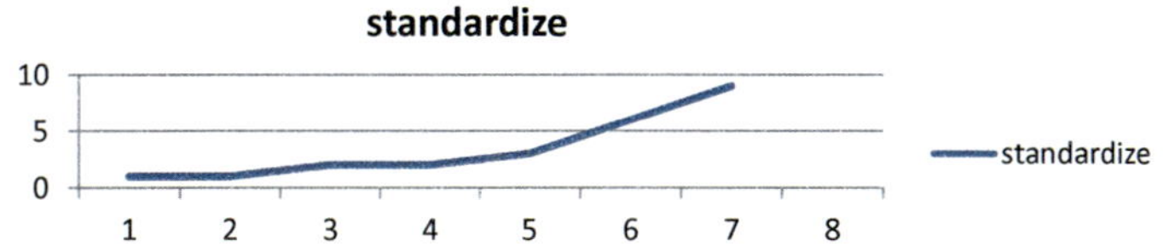

Fig. 9 Standardize weekly audit

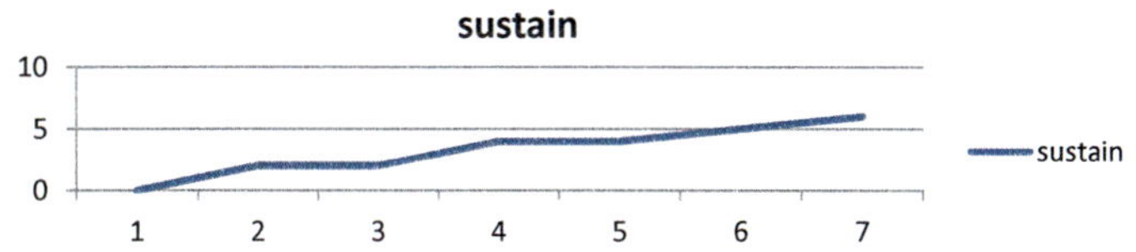

Fig. 10 Sustain weekly audit

Though the scores do not show great differences between the first few weeks, after the proper implementation of other 3S the curve has a standardized increment. This reflects that the first three criterion activities conducted under 5S activities are applied in a well manner. This standardized is evaluated by the help of maintaining checklist for various factors such as safety, cleaning, attendance etc. (Fig. 10).

The graph shows a well understanding of application of rules and procedures, utilization of appropriate equipment and materials and interaction between workers achieved good progress.

5 Conclusion

Thus, implementation of 5S varies from organization to organization depending on the process they carry out and their task to achieve and time. Anyway, it results in improvement within organization by making it a suitable place for work.

Thus, on implementing these 5S methodology, the following advantages are analysed and proved efficiently by the proper evaluation of various terms and the test is said to be stable by the help of sustainability test.

References

1. Patel, V. C., & Thakkar, D. H. (2014). Review on Implementation of 5S in Various Organization. *Journal of Engineering Research and Applications, 4*(3), 774–779. www.ijera.com
2. Omogbai, O., & Salonitis, K. (2017). The Implementation of 5S Lean Tool Using System Dynamics Approach. *Procedia CIRP, 60*, 380–385. https://doi.org/https://doi.org/10.1016/j.procir.2017.01.057
3. Falkowski, P., & Kitowski, P. (2007). The 5S methodology as a tool for improving the organisation of production. *Journal of Achievements in Materials and Manufacturing Engineering, 24*. http://sdpg.pg.gda.pl/pij/files/2013/10/03_2013_18-falkowski.pdf
4. Ho, S. K., & Cicmil, S. (1996). Japanese 5-S practice. *TQM Magazine, 8*(1), 45–53. https://doi.org/10.1108/09544789610107261
5. Lingareddy, H., Sahitya Reddy, G., & Jagadeshwar, K. (2013). 5S As a Tool and Strategy for Improvising the Work Place. *International Journal of Advanced Engineering Technology*, 5–7.
6. Mrunal Marshettiwar & Pallawi B. Sangode. (2018). Implementation of 5S Methodology in the Banking Sector. *IMPACT: International Journal of Research in Humanities, Arts and Literature (IMPACT: IJRHAL), 6*(8), 627–636. http://www.impactjournals.us/archives.php?year=2018_11_2&id=11&jtype=2&page=32
7. Ho, S. K., Cicmil, S., & Fung, C. K. (1995). The Japanese 5-S practice and TQM training. *Training for Quality, 3*(4), 19–24. https://doi.org/10.1108/09684879510098222
8. Sharma, R., & Singh, J. (2015). Impact of implementing japanese 5S practices on total productive maintenance. *International Journal of Current Engineering and Technology, 55*(22), 2277–4106. http://inpressco.com/category/ijcet
9. Agrahari, R. S., Dangle, P. A., & Chandratre, K. V. (2015). Implementation Of 5S Methodology In The Small Scale Industry A Case Study. *International Journal of Scientific & Technology Research, 4*(4), 180–187.
10. Rahman, M. M. T., & Tanvir, M. H. (2017). the Practice of Kaizen Tool in the Apparel Industry of Bangladesh for Process Improvement and Development of the Practice of Kaizen Tool in the Apparel Industry of Bangladesh for Process Improvement and Development of Ergonomics Standard. *International Research of Journal of Engineeing and Techonology (IRJET), 4*(10), 1747–1754. https://doi.org/10.13140/RG.2.2.33341.15844
11. Gawdzińskap, K. (2011). Application of the Pareto chart and Ishikawa diagram for the identification of major defects in metal composite castings. *Archives of Foundry Engineering, 11*(2), 23–28. http://www.afe.polsl.pl/index.php/pl/832/application-of-the-pareto-chart-and-ishikawa-diagram-for-the-identification-of-major-defects-in-metal-composite-castings.pdf
12. Patel, P. J., Shah, S. C., & Makwana, S. (2014). Application of Quality Control Tools in Taper Shank Drills Manufacturing Industry: A Case Study. *Journal of Engineering Research and Applications ISSN, 4*(1), 129–134. www.ijera.com
13. Ghodrati, A., & Zulkifli, N. (2013). The Impact of 5S Implementation on Industrial Organizations' Performance. *International Journal of Business and Management Invention ISSN, 2*(3), 43–49. www.ijbmi.org
14. Gupta, A., Verma, S., & Gupta, S. (2015). an Application of 5S Concept To Organize the Workplace At a Small Scale Manufacturing Company. *International Journal of Engineering Sciences & Research Technology, 4*(1), 713–728.

COVID19: Impact on Environmental Parameters During the Lockdown Period in India

V. Balaji Venkateswaran and Devender K. Saini

Abstract The coronavirus pandemic has entirely changed the current situation and has led people to realize the living scenario is no longer similar to before. The present-day situation is majorly subjugated with studies on developing strategies that can defer the spread of this virus, and to invent the vaccination. One of the prominent solution followed by many countries is to apply regional or country-level lockdown. This has affected the lives of many industries and become a challenge for its economic development. In this paper, we examine the correlation between the air pollutants (such as $PM_{2.5}$, PM_{10}, Ozone, CO, SO_X and NO_X) and weather parameters (such as temperature, humidity and dew point) with the coronavirus disease 2019 (COVID-19) by considering the six major red-zone hotspots identified in India. The effect of these parameters on major hotspots is examined based on Spearman's correlation coefficients for the lockdown period announced by the Government of India. From the results, it is evident that the highest correlation is obtained for different parameters for different red-zone districts. The study results may guide the authorities to develop a decentralized approach for effective implementation of lockdown and take appropriate measures in these red-zone hotspots.

Keywords Coronavirus · Air pollutants · Temperature · Humidity · Dew point · Sustainable development goals

1 Introduction

COVID-19 is a highly infectious and communicable disease that was initially found in December 2019 in Wuhan, China. Later, this virus was spread to many countries across the globe and therefore, the World Health Organization (WHO) has declared

V. Balaji Venkateswaran (✉)
School of Technology, Woxsen University, Hyderabad, Telangana, India

D. K. Saini
Electrical and Electronics Engineering Department, University of Petroleum and Energy Studies, Dehradun 248001, India

N. A. Siddiqui et al. (eds.), *Advances in Environment Engineering and Management*, Springer Proceedings in Earth and Environmental Sciences,
https://doi.org/10.1007/978-3-030-79065-3_5

this as a pandemic. As of May 3rd, 2020, there were about 3,272,202 confirmed cases, 230,104 deaths, worldwide [1]. In a country like India, where the population is more than 1.32 billion, handling this situation is very challenging [2]. Even though the first case reported in India was on 31st January 2020, the significant spread of this virus was noticed on 14th March 2020. As of 2nd May 2020, the total number of tests carried was 11,529,433, and it is identified that the confirmed cases were 26,535 and the number of deaths recorded was 1223 (https://www.mygov.in/covid-19). From these data, it is clear that the ratios of confirmed cases are 0.2% and the death rate is very low. However, the number of cases recorded in major cities like Mumbai, Chennai, Hyderabad, Bengaluru, Kolkata, and New Delhi is high and therefore these are identified under red-zone hotspot region [3] The ratio of confirmed cased in these cities (or district) to the total number of cases in the state is Fig. 1. There are about 3761 confirmed cases in Delhi, where the district wise numbers are under reconciliation.

Initially, many researchers were developing a growth model for prediction of the effect of COVID-19. In [4] a simple growth model is proposed based on cumulative distribution function (CDF) to predict the spread of this virus. Meanwhile, many researchers confirmed that this virus can be transferred from human-to-human through droplets and direct contact of the materials used by the affected ones. Concerning this, many articles were found in recent times, where the correlations between the entire number of cases and weather conditions are identified to model the spread of this virus. In [5], a study is conducted in major cities of Turkey and was found that the temperature is highly correlated with the total number of cases. Conversely, a study conducted in New York City reveals that there is no significant scientific evidence found to conclude that, the warmer conditions will suppress

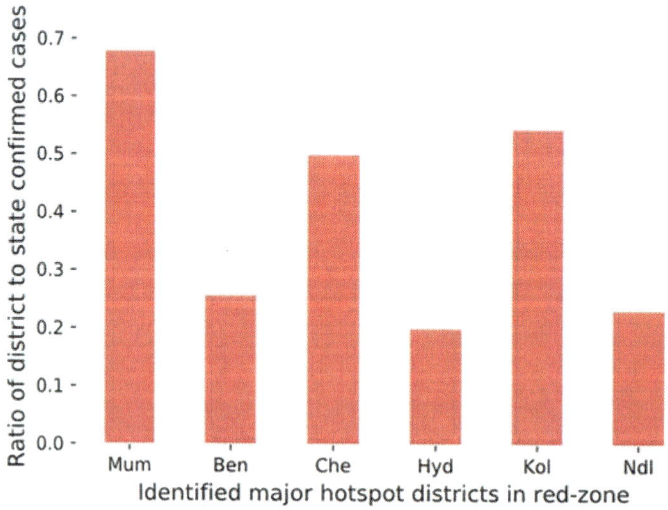

Fig. 1 Ratio of confirmed cases in major red-zone hotspot to the total number of cases in that State of India

the effect of COVID19 [6]. In [7], a study was performed to evaluate the effect of climatic parameters such as average temperature, precipitation, humidity, wind speed, and solar radiation on COVID-19. Here, the Partial correlation coefficient (PCC) and Sobol'-Jansen methods are used to analyze the spreading rate. From the analysis, it was identified that the population density and intra-state movement are the direct cause for the outbreak of infection. In [8], the positive and negative effects on the environment due to COVID-19 were identified for countries like China, France, Germany, Spain, and Italy. A similar kind of research based on suspended particle matter (SPM) was performed in Vembanad Lake, in India with the help of satellite images [9]. It is identified that during the lockdown period, the SPM level has decreased on an average by 15.9% compared to pre-lockdown conditions.

Similar to other countries, in containing the spread of this infection, the Government of India has announced a nationwide lockdown that started on 24th March 2020. However, considering the spread of this virus, the lockdown was extended for the third time for two weeks starting from 4th May 2020. To improve and enhance the financial situation of the country, the government has classified the COVID-19 affected districts into red, orange, and green zones and has provided the guidelines to be followed in these regions [3]. According to these guidelines, most of the activities are not permitted in red-zones. Nonetheless, few relaxations are provided to orange and green zones. Even after taking various steps to contain the infection, the confirmed cases are in the increasing trend, especially in some red-zone districts.

The major objective of this paper is to analyze the correlation between the climatic parameters and the COVID-19 confirmed cases in major red-zone hotspot districts in India. This study has been carried for the entire duration of the lockdown period. Thus, from this study, corrective measures can be derived for the identified red-zone districts to convert them into orange and then green.

2 Methodology

India is a highly populated country and has less per capita land space, especially in major cities. For example, Delhi and Mumbai are the second and the fifth highly populated city in the world [10]. Therefore, it is essential to identify unique ways in which the virus can be contained.

2.1 Dataset

The dataset for $PM_{2.5}$, PM_{10}, Ozone, CO, SO_X, NO_X, temperature, humidity and dew point are taken from the archives of U.S. Embassy and Consulates' Air Quality Monitor in India, National Air Monitoring Programme (NAMP) of Central Pollution Control Board and World Air Quality Index Project (NAMP; "U.S. Embassy and Consulates' Air Quality Monitor in India,"; WAQI Project.). Figure 2 shows the

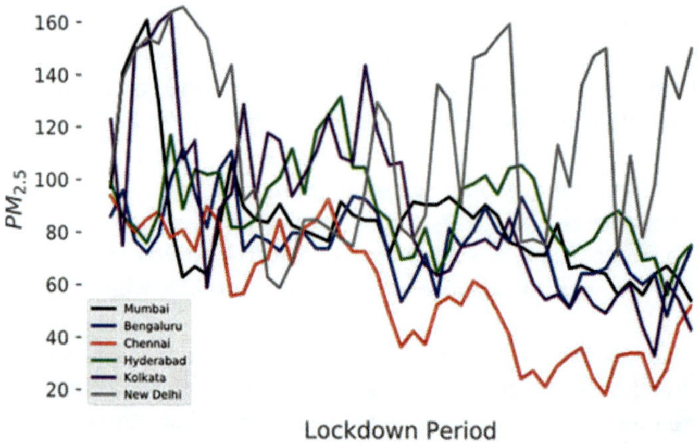

Fig. 2 Variation of $PM_{2.5}$ in identified red-zones of India

dissimilarity of $PM_{2.5}$ during the lockdown period (from 24th March till 4th May 2020) in the identified major red-zone districts of India. It refers to particulate matter in the atmosphere which has a diameter of less than 2.5 μm.

Figure 3 shows the variation of PM_{10} during the lockdown period in the identified major red-zone districts of India. It refers to particulate matter in the atmosphere which has a diameter of less than 10 μm. The variations of air pollutants such as O3, CO, NO_X, and SO_X during the lockdown period are shown from Figs. 4, 5, 6 and 7 respectively.

Figures 8, 9 and 10 shows the temperature, humidity, and dew point variations during the lockdown period in the identified major red-zone districts of India. Table 1 shows the total confirmed cases in major states such as Maharashtra, Karnataka,

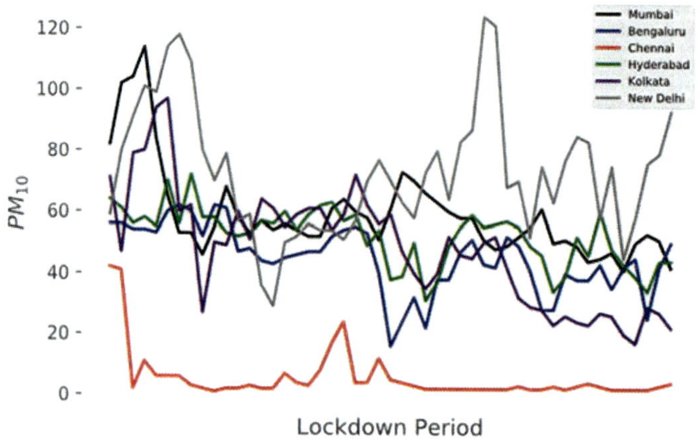

Fig. 3 Variation of PM_{10} in identified red-zones of India

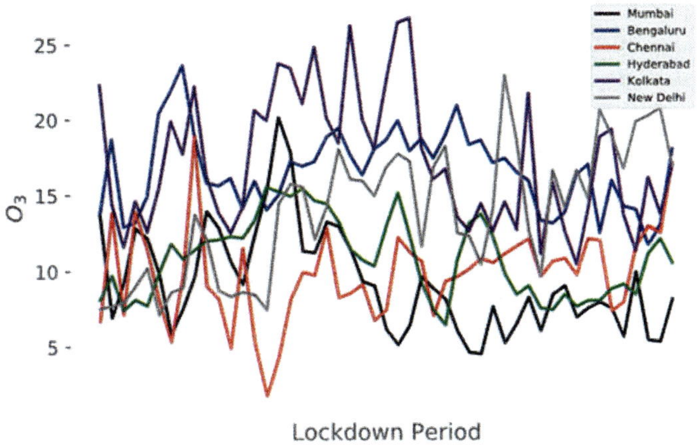

Fig. 4 Variation of O_3 in identified red-zones of India

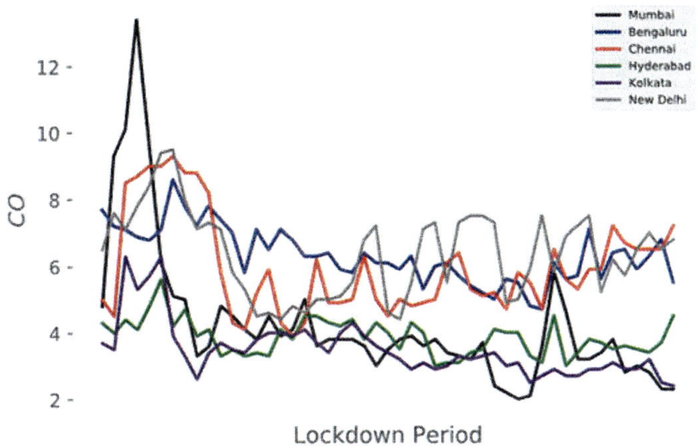

Fig. 5 Variation of CO in identified red-zones of India

Tamil Nadu, Telangana, West Bengal, and Delhi till 4th May, and Fig. 11 shows the trend of the number of confirmed cases between 14th March 2020 to 4th May 2020 in these states. Figure 12 shows the population density (in per km^2 scaled by 100) of identified red-zone districts and corresponding means sea level (MSL) in meters.

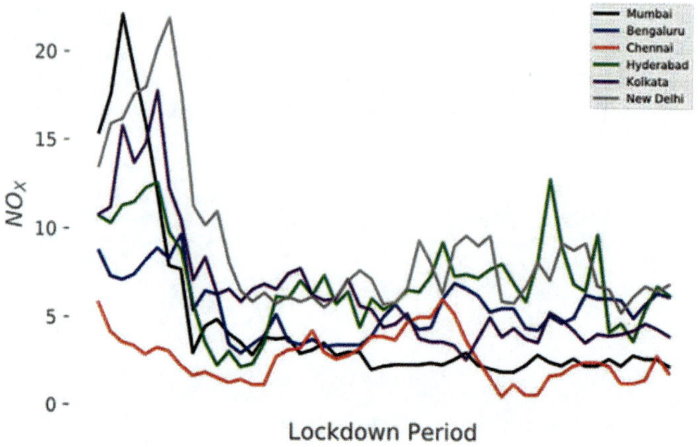

Fig. 6 Variation of NO_X in identified red-zones of India

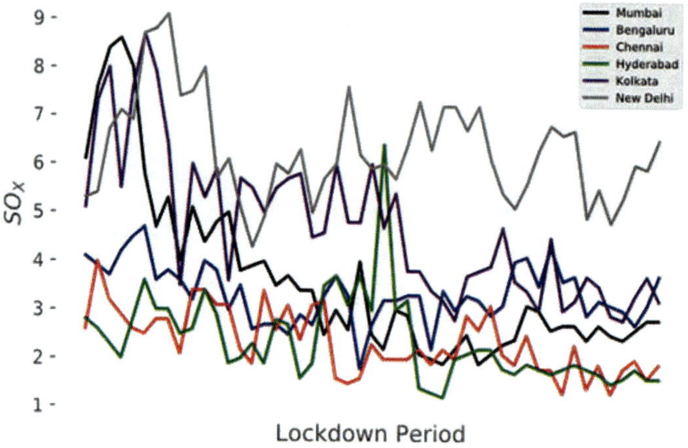

Fig. 7 Variation of SO_X in identified red-zones of India

2.2 *Spearman's Correlation Analysis*

The Spearman's correlation is a non-parametric analysis from which the correlation rank is derived. This rank provides a measure of strength and direction of relationship between the chosen two variables. Here, the correlation coefficients are derived between the two datasets x and y, if the coefficient is greater than zero, it implies that x increases with an increase in y and if the coefficient is less than zero, it implies that x increases with decrease in y. The Spearman's coefficient can be calculated using the Eq. (1).

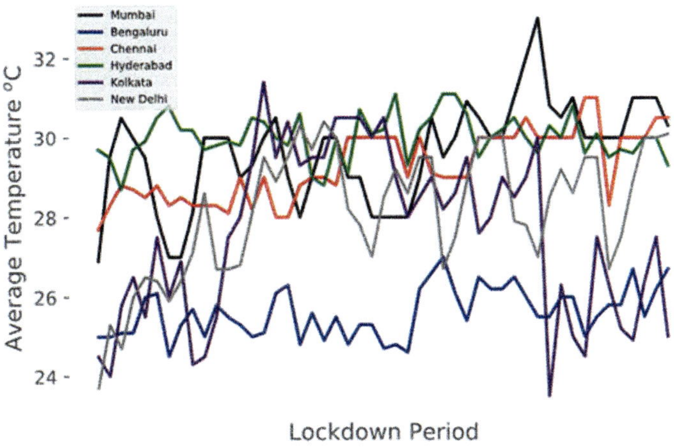

Fig. 8 Variation of average temperature (°C) in identified red-zones of India

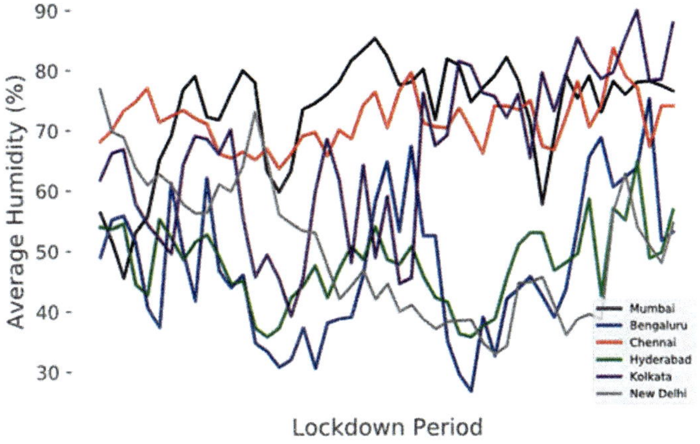

Fig. 9 Variation of average humidity (%) in identified red-zones of India

$$r_s = 1 - 6 \times \frac{\sum d_r{}^2}{n\left(n^2 - 1\right)} \tag{1}$$

where d_r denotes the change in rank between the variables, n is the total quantity of cases. Therefore, in this study, the Spearman rank coefficient test is used to observe the correlation between the air pollutants, weather parameters with the confirmed cased of COVID-19 in six major red-zone hotspot regions of India.

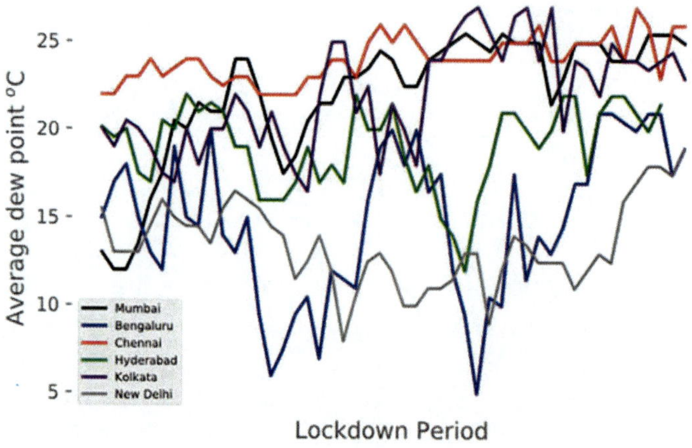

Fig. 10 Variation of average dew point (°C) in identified red-zones of India

Table 1 Total confirmed COVID cases in the Major States of India

State	Total confirmed cases
Maharashtra	12,974
Karnataka	642
Tamil Nadu	3023
Telangana	1082
West Bengal	963
Delhi	4549

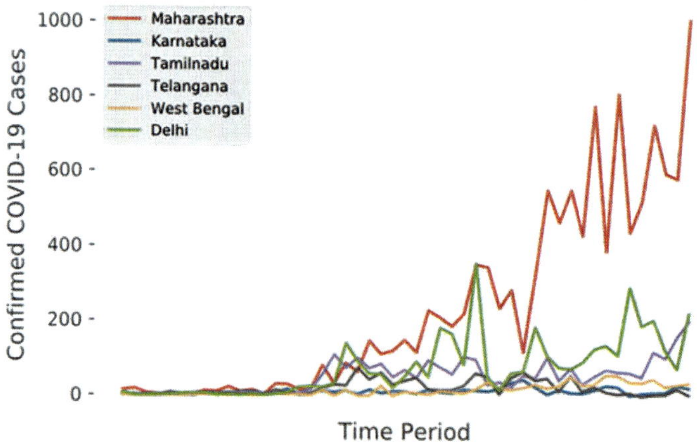

Fig. 11 Number of confirmed cased in the Major States of India

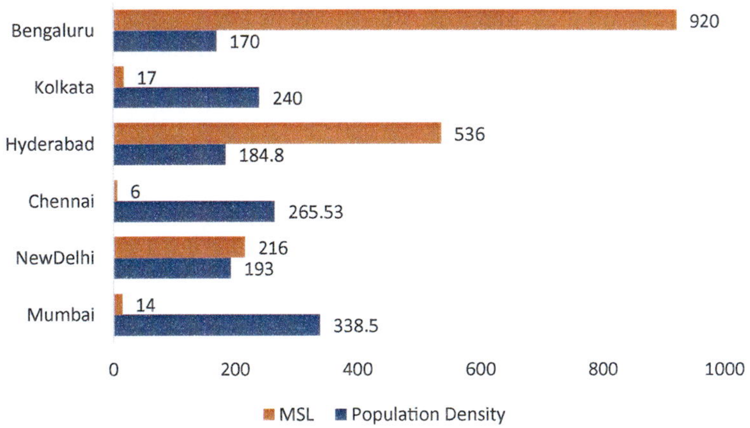

Fig. 12 Population density and MSL of identified red-zone districts in India

3 Results and Discussion

Table 2 represents the result of Spearman's correlation analysis. As mentioned earlier, air pollutants such as $PM_{2.5}$, PM_{10}, Ozone, CO, SO_X, and NO_X and weather parameters such as average temperature, average humidity, and dew point are considered for this study. All these parameters are evaluated for the duration of lockdown between 24th March till 4th May 2020.

In Table 2, the positive correlation coefficients are in italic and the negative correlation coefficients are in bold. By comparing the Fig. 12 and Table 2, expect Kolkata, all the other red-zone districts are having positive temperature correlation with COVID-19 cases. From the results, it is evident that for the Mumbai district, all the air pollutants are having negative correlation coefficient with a maximum for SO_X which is – 0.73162. The maximum value of the positive coefficient is for the dew point which is

Table 2 Results from Spearman's Correlation Analysis

	Mumbai	Bengaluru	Chennai	Hyderabad	Kolkata	New Delhi
$PM_{2.5}$	**– 0.53081**	**– 0.30462**	**– 0.52129**	*0.139179*	**– 0.78619**	**– 0.23106**
PM_{10}	**– 0.5335**	**– 0.41196**	**– 0.07703**	**– 0.21973**	**– 0.78224**	**– 0.14839**
O_3	**– 0.53388**	**– 0.05906**	*0.274844*	*0.111364*	**– 0.28197**	*0.694413*
CO	**– 0.69787**	**– 0.55868**	**– 0.07287**	**– 0.11002**	**– 0.72236**	**– 0.19363**
NO_X	**– 0.71124**	**– 0.17373**	0.064664	**– 0.05174**	**– 0.74995**	**– 0.4349**
SO_X	**– 0.73162**	**– 0.3357**	**– 0.53746**	**– 0.0486**	**– 0.78756**	**– 0.27337**
Avg. temperature	*0.587189*	*0.231033*	*0.630225*	*0.382372*	**– 0.13478**	*0.479886*
Avg. humidity	*0.320102*	**– 0.03885**	*0.187166*	**– 0.28948**	*0.713122*	**– 0.57176**
Dew point	*0.775292*	**– 0.00647**	*0.537491*	**– 0.13348**	*0.578741*	**– 0.17347**

0.775292. In other words, in Mumbai, as the dew point increases the confirmed cases are also increasing and the SO_X is decreasing with an increase in confirmed cases. For the Bengaluru district, all the air pollutants are having negative correlation coefficient along with humidity and dew point. Here the maximum negative correlation coefficient is obtained for PM_{10} which is − 0.41196. In other words, in Bengaluru, as the confirmed cases increase the level of PM_{10} is decreasing, and as the temperature increases the number of COVID-19 cases increases. For the Chennai district, air pollutants such as $PM_{2.5}$, PM_{10}, CO, and SO_X are having negative correlation coefficient out of which the maximum is for SO_X (− 0.53746). The other parameters such as O_3, NO_X, average temperature, average humidity, and dew point are having positive correlation coefficients out of which the maximum is for temperature. For the Hyderabad district, except for parameters such as $PM_{2.5}$, O_3, and average temperature, all the other parameters are having negative correlation coefficient out of which the maximum is for humidity which is − 0.28948. For the Kolkata district, all the parameters are having a negative correlation coefficient except for humidity and dew point. The maximum value of negative correlation is for $PM_{2.5}$. For New Delhi district, expect O_3 and average temperature all the other values are having negative correlation coefficient out of which humidity is having maximum value (− 0.57176).

The average temperature was found to be highly correlated with COVID-19 cases in Turkey [8]. Whereas from this study, it is clear that, there is no single parameter which is having a negative correlation coefficient for all the districts. A comparison of correlation results for temperature between [8] and the current study is shown Fig. 13. From Fig. 13, it is clear that the temperature is not having negative correlation in majority of the major cities of India. From this study, it is clear that, the guidelines to be formulated must be regional-specific rather than a nationwide framework under the current situation to convert these red-zones to orange and then to green.

This study may have a few limitations. The study results may not reflect the actual situation in the selected red-zones. In recent reports, it was found that people are traveling between districts due to various reasons. One among those is intra-state

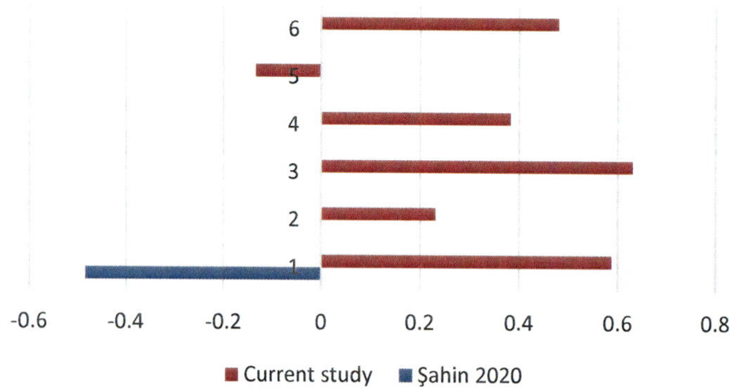

Fig. 13 Comparison of temperature correlation coefficient of major cities of Turkey and India

transportation to carry the essentials goods to maintain the supply chain. Therefore, without the knowledge of people traveling between these major cities, converting the red-zone to orange and then green is challenging.

4 Conclusion

Compared to other developed and developing countries, India, having the largest population has controlled the spread of the virus more effectively. However, it is still challenging for those red-zone districts which have a positive trend in confirmed cases in the current scenario which provides an alarming sign. In this study, the impact of air pollutants (such as $PM_{2.5}$, PM_{10}, Ozone, CO, SO_X and NO_X) and weather parameters (such as average temperature, average humidity, and dew point) from 14th March 2020 to 4th May 2020 on confirmed COVID-19 cased using data in India is presented. The results obtained show that the correlation coefficient is diverse for all the regions of India. Therefore, in the current situation, it is essential to have a decentralized control and guideline mechanism among the red-zone districts identified in India to convert those into orange and then green.

References

1. Ahmadi M, Sharifi A, Dorosti S, et al (2020) Investigation of effective climatology parameters on COVID-19 outbreak in Iran. Sci Total Environ 729:1–7. https://doi.org/https://doi.org/10.1016/j.scitotenv.2020.138705
2. Bashir MF, Ma B, Bilal, et al (2020) Correlation between climate indicators and COVID-19 pandemic in New York, USA. Sci Total Environ 728:138835. https://doi.org/10.1016/j.scitotenv.2020.138835
3. Gautam S, Hens L (2020) SARS-CoV-2 pandemic in India: what might we expect? Environ Dev Sustain 22:3867–3869. https://doi.org/https://doi.org/10.1007/s10668-020-00739-5
4. https://www.mygov.in/covid-19 (2020) Mygov. https://www.mygov.in/covid-19
5. International Energy Agency (2020) Global Energy Review 2020
6. National Air Monitoring Programme Central Pollution Control Board of India. http://cpcb.nic.in/
7. National Health Mission (2020) Identified Hotspot Districts in India to fight Covid19. National Health Mission
8. Şahin M (2020) Impact of weather on COVID-19 pandemic in Turkey. Sci Total Environ 728. https://doi.org/10.1016/j.scitotenv.2020.138810
9. WAQI Project World's Air Pollution : Real time Air Quality Index. https://waqi.info/
10. World Health Organization Coronavirus (COVID19). https://covid19.who.int/
11. Cássaro FAM, Pires LF (2020) Can we predict the occurrence of COVID-19 cases? Considerations using a simple model of growth. Sci Total Environ 728:138834. https://doi.org/10.1016/j.scitotenv.2020.138834
12. Worldatlas.com (2011) The 150 Largest Cities in the World. https://www.worldatlas.com/citypops.htm
13. Yunus AP, Masago Y, Hijioka Y (2020) COVID-19 and surface water quality: Improved lake water quality during the lockdown. Sci Total Environ 139012. https://doi.org/10.1016/j.scitotenv.2020.139012

14. Zambrano-Monserrate MA, Ruano MA, Sanchez-Alcalde L (2020) Indirect effects of COVID-19 on the environment. Sci Total Environ 728:138813. https://doi.org/https://doi.org/10.1016/j.scitotenv.2020.138813
15. U.S. Embassy and Consulates' Air Quality Monitor in India. https://in.usembassy.gov/embassy-consulates/new-delhi/air-quality-data/

Impact of Lockdown Due to COVID 19 Pandemic on Air Quality of Global Environment

Madhav Saket, Kanchan Deoli Bahukhandi, Siddiqui Nihal Anwar, Koranga Nirmala, Agrawal Shilpi, Singhal Shailey, and Vohra Shalini

Abstract In present time countries all over the world are facing lockdown to slow down the transmission of COVID 19, an infectious respiratory disease with human to human transmission. The name COVID 19 represents the disease Corona Virus Disease 2019. The outbreak took place in December 2019 in Wuhan, Hubei province, China. Severe Acute Respiratory Syndrome Corona Virus 2 (SARS-CoV-2) is known to be accountable for occurrence of this disease. As per the latest report of World Health Organization, a total of 3,588,773 cases has been reported worldwide. It is believed that the number of total cases will increase in coming days which will force countries worldwide to extend lockdown. Due to this throughout the world are following work from home culture, schools and colleges have been shut down. Malls, shopping complexes, amusement parks, multiplexes and all other places of social gatherings are facing closure; peoples worldwide are not allowed to come out of their homes. On road vehicular movements, flights, cruises are at halt. In this research paper an attempt has been made to compare the air quality index during lock down and prelock down period.

Graphical Abstract

HIGHLIGHTS:

- COVID 19 has been declared Global Pandemic by WHO.
- Virus spreads through human to human transmission.
- Lockdown has been taken as preventive measure throughout world to check its spread.
- This resulted in reduction in vehicular movements.
- Reduction in emission of air pollutants was observed.

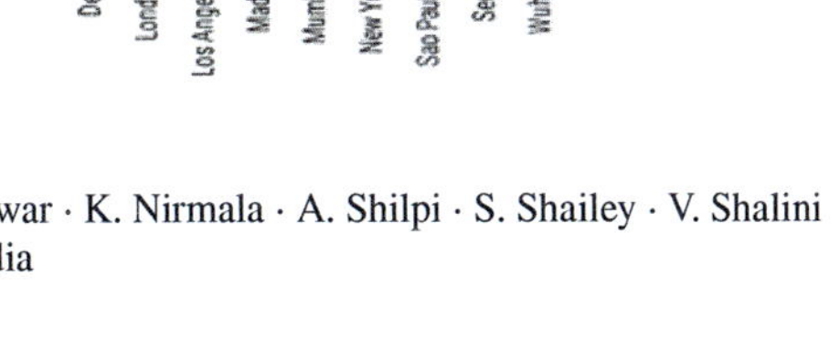

M. Saket · K. D. Bahukhandi (✉) · S. N. Anwar · K. Nirmala · A. Shilpi · S. Shailey · V. Shalini
School of Engineering, UPES, Dehradun, India
e-mail: kanchan@ddn.upes.ac.in

N. A. Siddiqui et al. (eds.), *Advances in Environment Engineering and Management*, Springer Proceedings in Earth and Environmental Sciences,
https://doi.org/10.1007/978-3-030-79065-3_6

Keywords Covid 19 · Global pandemic · Ambient air quality

1 Introduction

In late December 2019 COVID 19 pandemic originated in Wuhan, China and spread globally [1]. COVID 19 spreads through human to human transmission, till date it has spread worldwide. Until now there is no vaccine, the only way to put an end on the spread is to control human to human transmission. Countries worldwide are following lockdown to do so which has led to shut down of industrial activities globally. All types of transport road, air as well as sea came to halt.

The restrictions on these activities due to lockdown has led to reduction in release of air pollutants. National Aeronautics and Space Administration (NASA), European Space Agency (ESA) released data indicating that the pollution level globally has reduced [2]. On one hand the COVID 19 pandemic has resulted in worldwide destruction by taking many lives, causing lockdown and demobilizing global economy but on the other hand it has put a positive impact on air quality globally [3]. The objective of this paper is to study the changes in quality of air before and during COVID 19 pandemic worldwide.

The quality of air worldwide is degrading rapidly, which has both acute and chronic effects on human health. 91% of the total world population in 2016 was living in the areas where air quality parameters did not met WHO guidelines [4]. The disease caused by air pollution is irritation in respiratory tract, heart disease, lung cancer, bronchitis, asthmatic attacks and several others [5]. According to World Health Organization the estimated death caused by air pollution is 7 million per year [6]. The major pollutants contributing to air pollution are PM_{10}, $PM_{2.5}$, CO, NO_2, SO_2 and O_3 [5] with mean value set by WHO are 20 $\mu g/m^3$ (annual mean), 10 $\mu g/m^3$ (annual mean), 30 ppm, 40 $\mu g/m^3$ (annual mean), 20 $\mu g/m^3$ (24 h mean) and 100 $\mu g/m^3$ (8 h mean) respectively [4]. Air Quality Index is defined as a number for reporting quality of air with respect to its effects on human health and is calculated by using concentrations of pollutant in some mathematical formulae, different countries have different air quality indices [7]. The WHO air quality guidelines indicates that air pollution related deaths can be checked by 15% by reducing the PM_{10} concentration from 70 to 20 μg per m^3 [4]. Nitrates, ammonia, sodium chloride, mineral dusts, water, black carbon are major components of Particulate Matter (PM).

2 Result and Discussion

The countries which are most affected by COVID 19 globally and facing lockdown are India, China, United States of America and European countries. The effect of lockdown on air pollution in each of the mentioned country is discussed below.

2.1 Air Pollution in India During Lockdown

To prevent human to human transmission of COVID 19, restrictions on human activities were imposed on public by initiating a 21 days lockdown from 24th of March which was followed by lockdown of another 21 days from 14th of April continuing till 4th of May. The lockdown resulted in reduced emission of pollutants from transportation and other industries. The process industry, power plant, construction activities, automobile emissions are the major sources of emission i.e. SO_2, NO_2, CO, RSPM (Respirbale particulate matter), SPM (Suspended Particulate matter), PM 2.5. Nation wide lockdown significantly improved the quality of air. The AQI in Delhi and NCR region was largely under moderate category in first week of lockdown 1 [8]. The emission level of PM2.5, PM10 and NO_2 were reduced significantly during lockdown period. The PM 2.5 was reduced to 46%, PM 10 reduced to 50% in Delhi. The NO_2 and CO were significantly reduced up to 56% and 37%. The reduction of Benzene was 47% and SO2 was 17%. The hourly trend of PM2.5 ranged from 127 to 244 $\mu g/m^3$. In Gujarat the AQI (Air Quality Index) was reduced up to 37%. The AQI reading was 220 in the month of January before lock down and after lockdown it was 98 (on 31st March) showing good improvement in air quality index of Gujarat (Energy World.Com April 2020).

The Odisha state which is famous for its coal reserve and number of underground and open caste mine where the pollution level was reduced up to 52% during lockdown period. The AQI reading was ranged from 96 to 204 on 30th March while in January it ranged from 166 to 416 and came under severe category of AQI Index. Talchar situated in eastern coast of Vishakhapatnam is a major industrial center where the large scale establishment of producing steel, metal, petroleum and Heavy engineering equipment. The AQI before lockdown was 118 considered under poor category and during lockdown it was 63, considered under satisfactory category of AQI (Table 1).

The AQI further improved to good AQI. In March the AQI were ranged from 45 to 192, during April 73 to 155 [8] in Delhi.

The Kanpur which is famous for leather industry has improved air quality up to 60% during lockdown (March and April 2020) when compared with the data of previous year (2019). The AQI was 163 during March 2019 which further improved and its value was 64 during (31st March 2019) (CPCB http://moef.nic.in, http://www.

Table 1 Air quality index (AQI)

S.No.	Characteristics	AQI Range
1	Good	0–50
2	Satisfactory	51–100
3	Poor	101–200
4	Very poor	301–400
5	Severe	401–500

cpcb.nic.in). The PM 10 and PM 255 were reduced by about 51.84% and 53.11% respectively during lockdown period in Kanpur. The NO_2, CO, SO_2 and NH_3 were decreased around 52.8%, 30.35%, 12.33% respectively during lockdown at Kanpur state of UP.

Air Quality Index in north, south, east, central and western India reduced by 44, 33, 29, 15 and 32% respectively. A reduction of 43, 31, 10, 18% were observed in quantity of $PM_{2.5}$, PM_{10}, CO and NO_2 throughout India [9]. According to Central Pollution Control Board (CPCB) the air quality index has significantly reduced throughout India. The AQI in India is on a scale of 0–500 with 0–50 representing good air quality which have minimal impact on humans, 51–100 shows that air quality is satisfactory in which sensitive people may experience minor breathing discomfort, 101–200 shows that air quality is moderate and people with asthma, heart and lung disease will face breathing discomfort, 201–300 shows that air quality is poor and its prolonged exposure will cause discomfort to most people, 301–400 shows air quality is very poor and prolonged exposure will cause respiratory illness, 401–500 shows that air quality is severe and effects both healthy and unhealthy people [10]; It is calculated by using concentrations of 3 pollutants out of which 1 is either $PM_{2.5}$ or PM_{10}, the remaining are any 2 of CO, NO_2, SO_2, Pb, O_3, NH_3 of which minimum concentrations are available [9]. The Chart shown below compares air quality index in prominent cities of India like Delhi in North, Mumbai in West, Kolkata in East, Indore in Centre and Chennai in South on 30/04/2019 to 30/04/2020 [10] (Fig. 1).

The satellite image of India published by NASA shown below clearly displays that Aerosol level in terms of Aerosol Optical Depth in India are at 20 year low. Tiny liquid and solid particles suspended in the air are termed as Aerosols; they reduce visibility and can damage human lungs and heart [12]. While on the other hand Aerosol Optical Depth (AOD) is measurement of absorption and reflection of light by airborne particles in the atmosphere. AOD of 1 or more indicates hazy condition where as AOD less than 0.1 indicates clean and clear condition [12] (Fig. 2).

The first 5 maps show AOD in India during March 31 to April 5 periods starting from 2016 till 2020. The 6th map (Anomaly) show AOD in 2020 compared to previous years.

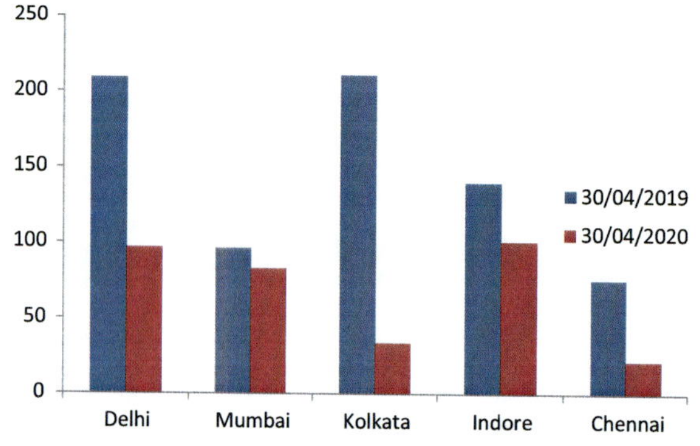

Fig. 1 Air quality index in different cities of India on 30/04/2019 [11] and 30/04/2020 [8]

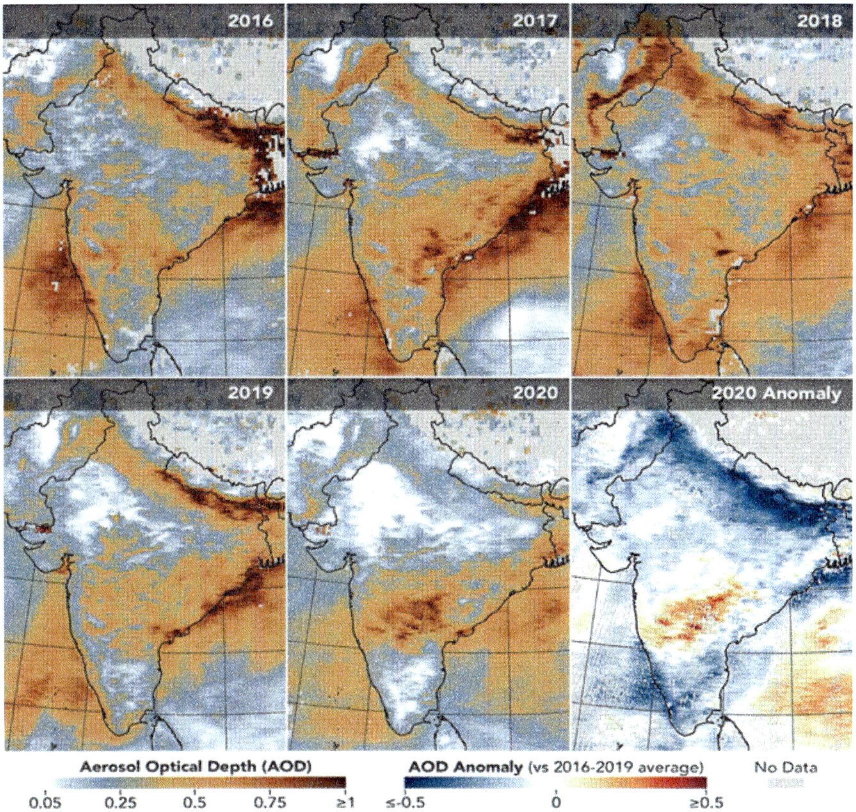

Fig. 2 Aerosol optical depth of India [12]

2.2 *Air Pollution in China During Lockdown*

Initially the COVID 19 outbreak took place was in China. In late December 2019 cases of fever, sore throat, dry cough, chest pain, rashes caused by COVID 19 were reported in Wuhan. Later spread of COVID 19 was confirmed throughout the country [13]. In January 2020 World Health Organization confirmed human to human transmission. To slow down the transmission of COVID 19 pandemic the Wuhan was placed under lockdown followed by Hubei province and subsequently the whole China on January 23rd [2]. There was a huge reduction in urban transportation and industrial activities throughout China which resulted in decline of coal and energy consumption. Since burning of coal produces major air pollutants like CO, SO_2, NO_2, PM and heavy metals [13], a decline in consumption of coal resulted in reduction in air pollutants. The Air Quality Index was better than previous years in China during January 2020 and February 2020. The China included $PM_{2.5}$ and O_3 concentrations as two new indicators to the National Ambient Air Quality Standards in the year 2012s [14]. The Air Quality Index of China reduced from 71.6 (lockdown) to

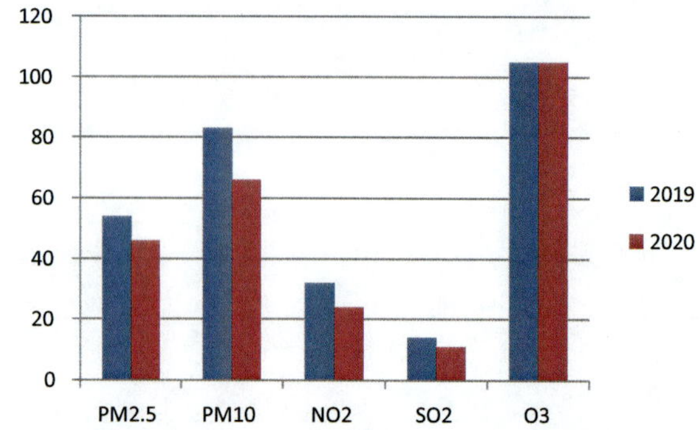

Fig. 3 Concentration of air pollutants in $\mu g/m^3$ in China [13]

89.6 (before lockdown) [15]. The satellite image published by NASA have shown decrease in air pollution level (Figs. 3, 4 and 5), [2, 13].

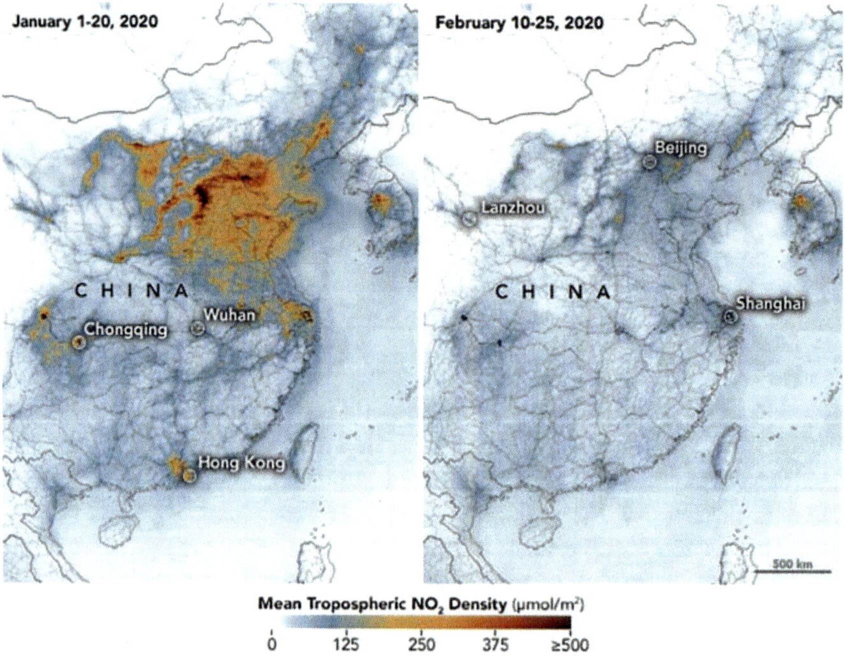

Fig. 4 Concentration of NO_2 in China in January and February 2020 [16]

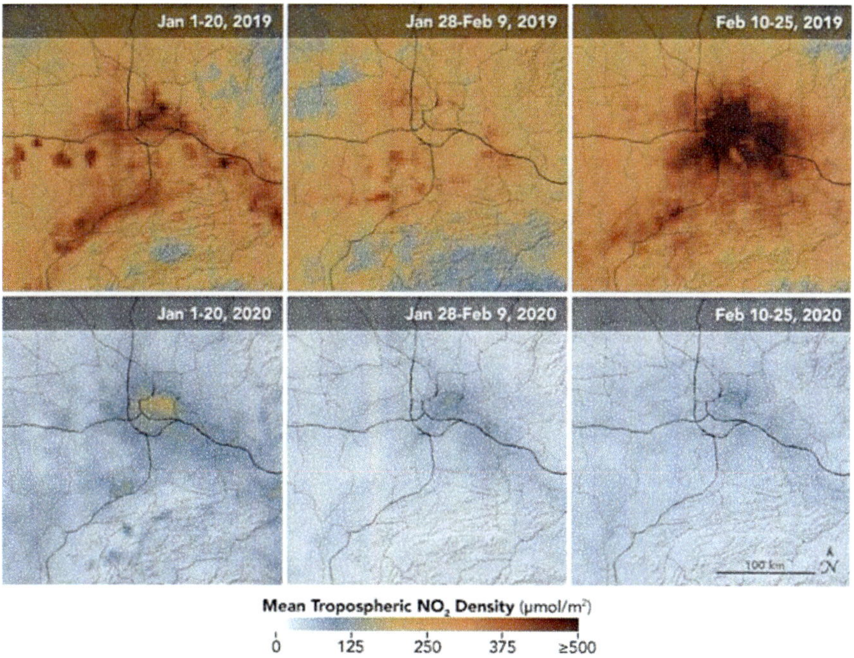

Fig. 5 Concentration of NO_2 in Wuhan [12]

2.3 Air Pollution in United States of America During Lockdown

The first COVID 19 case reported in USA on January 15, 2020 and within in short period of time USA became the epicenter of COVID 19 Pandemic [17]. To keep a check on increasing number of cases restrictions were laid down on the public USA which affected its economy in a negative manner and substantially improved air quality of ambient air. In USA Air Quality Index uses six major pollutants PM_{10}, $PM_{2.5}$, CO, NO_2, SO_2, O_3 [18]. But the most hazardous pollutant among all of them is $PM_{2.5}$ which concentration was reduced during lockdown period (March 23–April 13, 2020). Los Angeles saw a drop of 31% in concentration of $PM_{2.5,}$ its average value came down to 5.5 $\mu g/m^3$ from 7.97 $\mu g/m^3$. New City saw a drop of 25% in concentration of $PM_{2.5,}$ its average value came down to 4.4 $\mu g/m^3$ from 5.86 $\mu g/m^3$ [19]. The satellite image captured by NASA showed a reduction of 30% in concentration of NO_2 during March in north eastern part of United States as compared to average concentration in March 2015–2019 (Fig. 6) [2].

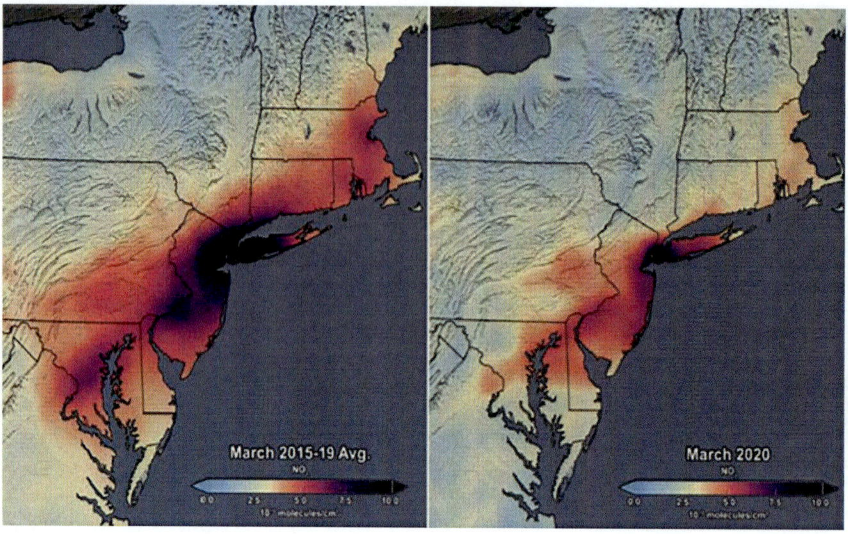

Fig. 6 Concentration of NO_2 in northeastern America before and after lockdown [12]

2.4 Air Pollution in Italy

Italy is one of the worst affected countries and to keep a check on the rising number of COVID 19 cases Italy went into lock down from early March. Restrictions on human activities resulted in less vehicular movements which in turn resulted in decline of NO_2 concentrations [20]. The reduction in NO_2 emissions across Italy was 20 to 30% [2]. The satellite image published by ESA has shown reduction in NO_2 concentration in March 2020 when compared to March 2019 (Fig. 7).

France implemented lockdown from mid of March; it resulted in drop of NO_2 concentration. The satellite image of France captured by ESA showing reduction in

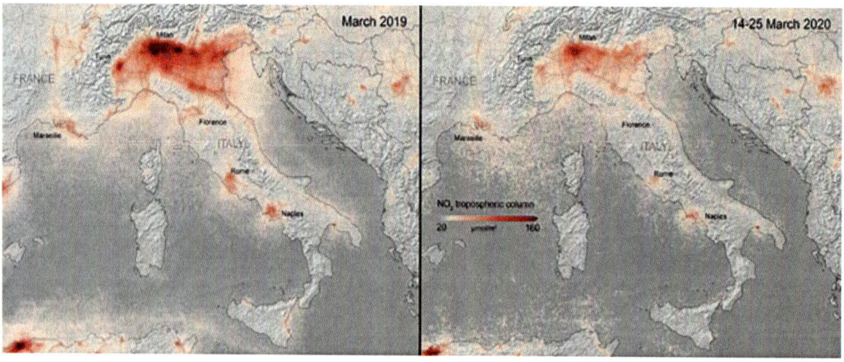

Fig. 7 Reduction in NO_2 in Italy [16]

emissions of NO_2 concentration in March 2020 compared to March 2019 (Fig. 8) [2]. Spain went into lockdown from March 23 [19], during lockdown the reduction in emissions of NO_2 observed compared to March 2019 (Fig. 9) [2]. There was reduction of 11% in concentration of $PM_{2.5}$ in comparison to last year, its value reduced to 6.4 $\mu g/m^3$ from 7.19 $\mu g/m^3$ [19] in Madrid, Spain.

UK implemented lockdown from March 23 to curb rising number of corona virus cases, the reduction in human activities improved the quality of air. A reduction of 9% in the concentration of $PM_{2.5}$ was noticed in London during the lockdown period, the concentration of $PM_{2.5}$ reduced to 16.2 $\mu g/m^3$ from 17.80 $\mu g/m^3$ [19].

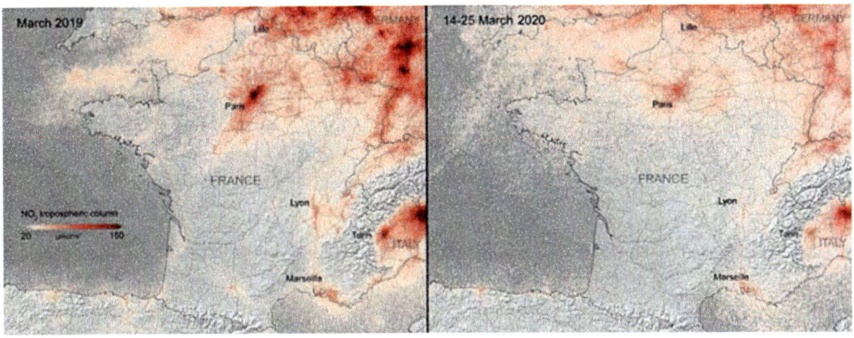

Fig. 8 Reduction in NO_2 in France [16]

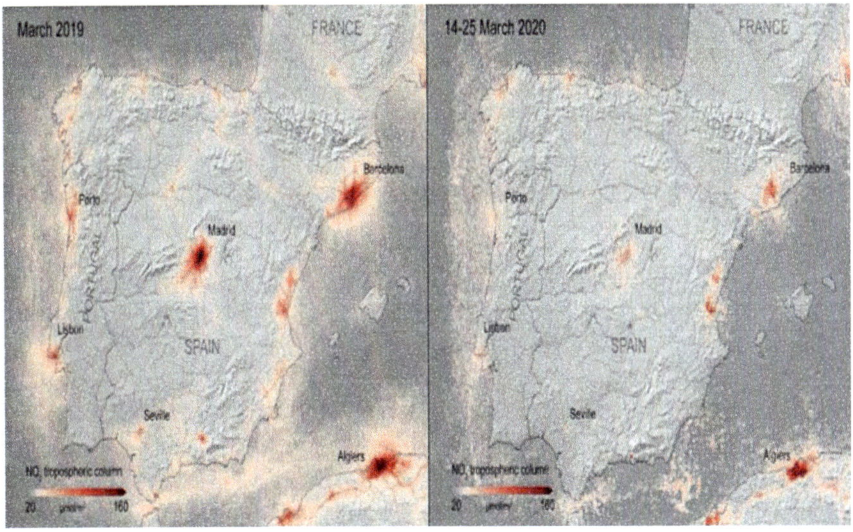

Fig. 9 Reduction in NO_2 in Spain [16]

3 Conclusion

The quality of air have been improved due to Covid 19 Pandemic lockdown in various part of the world. The major air pollutants like PM_{10}, $PM_{2.5}$, CO, NO_2, SO_2, O_3 were found to be decreased during lockdown period.

References

1. S.A. Meo, T. Al-Khlaiwi, A.M. Usmani, A.S. Meo, D.C. Klonoff, T.D. Hoang, Biological and Epidemiological Trends in the Prevalence and Mortality due to Outbreaks of Novel Coronavirus COVID-19, J. King Saud Univ. Sci. (2020) https://doi.org/10.1016/j.jksus.2020.04.004.
2. S. Muhammad, X. Long, M. Salman, COVID-19 pandemic and environmental pollution: A blessing in disguise?, Sci. Total Environ. 728 (2020) 138820. https://doi.org/10.1016/j.scitotenv.2020.138820.
3. I. Chakraborty, P. Maity, COVID-19 outbreak: Migration, effects on society, global environment and prevention, Sci. Total Environ. 728 (2020) 138882. https://doi.org/10.1016/j.scitotenv.2020.138882.
4. World Health Organization, 2018. Ambient (outdoor) air pollution [WWW Document]. Who.
5. M. Kampa, E. Castanas, Human health effects of air pollution, Environ. Pollut. 151 (2008) 362–367. https://doi.org/10.1016/j.envpol.2007.06.012.
6. WHO | CLEAN AIR FOR HEALTH: Geneva Action Agenda [WWW Document], 2018. . Who. https://www.who.int/phe/news/clean-air-for-health/en/
7. B. Bishoi, A. Prakash, V.K. Jain, A Comparative Study of Air Quality Index Based on Factor Analysis and US-EPA Methods for an Urban Environment, 9 (2009) 1–17.
8. CPCB, 2020. Central Pollution Control Board LIST OF AQI STATIONS.(b)
9. S. Sharma, M. Zhang, Anshika, J. Gao, H. Zhang, S.H. Kota, Effect of restricted emissions during COVID-19 on air quality in India, Sci. Total Environ. 728 (2020) 138878. https://doi.org/10.1016/j.scitotenv.2020.138878.
10. CPCB, 2014. National Air Quality Index. Cent. Pollut. Control Board 1–44.
11. CPCB, 2019. Central Pollution Control Board LIST OF AQI STATIONS.(a)
12. "Live: Launch America — NASA's Spacex Demo-2 Mission To The International Space Station". 2020. *NASA*. https://www.nasa.gov/content/live-launch-america-nasas-spacex-demo-2-mission-to-the-international-space-station.
13. Q. Wang, M. Su, A preliminary assessment of the impact of COVID-19 on environment – A case study of China, Sci. Total Environ. 728 (2020) 138915. https://doi.org/10.1016/j.scitotenv.2020.138915.
14. Zhan, D., Kwan, M.P., Zhang, W., Yu, X., Meng, B., Liu, Q., 2018. The driving factors of air quality index in China. J. Clean. Prod. 197, 1342–1351. https://doi.org/10.1016/j.jclepro.2018.06.108
15. Y. Wang, Y. Yuan, Q. Wang, C. Liu, Q. Zhi, J. Cao, Changes in air quality related to the control of coronavirus in China: Implications for traffic and industrial emissions, Sci. Total Environ. 731 (2020) 139133. https://doi.org/10.1016/j.scitotenv.2020.139133.
16. European Space Agency [WWW Document], 2006. . Choice Rev. Online.
17. M.F. Bashir, B. Ma, Bilal, B. Komal, M.A. Bashir, D. Tan, M. Bashir, Correlation between climate indicators and COVID-19 pandemic in New York, USA, Sci. Total Environ. 728 (2020) 138835. https://doi.org/10.1016/j.scitotenv.2020.138835.

18. M. Mirabelli, S. Ebelt, S. Damon, Air Quality Index and air quality awareness among adults in the United States, Environ. Res. 183 (2020) 109185. https://doi.org/10.1016/j.envres.2020.109185.
19. IQAir, 2020. Covid-19 Air Quality report 1–14.
20. Anjum, N., Good in The Worst: COVID-19 Restrictions and Ease in Global Air Pollution Naser A. Anjum, (2020). https://doi.org/10.20944/preprints202004.0069.v1.

Stress Condition Modeling to Optimize Mud Weight Window for Efficient Drilling

Dharmendra Kumar Gupta, Kamal Chandra Dani, and Pushpa Sharma

Abstract Geomechanical instability refers most significant incidents during drilling operations such as wellbore collapse or failure. In general, wellbore instability is related to drill pipe sticking, tight spots, caving production, wellbore collapse and unscheduled sidetracks. These mechanical conditions are mostly cause by unknown rock stress conditions and lead to increased costs during drilling and completion operations. Wellbore instability issues are important in the success of drilling operations. One of the main goals of any drilling mission is to drill the well economically viable. Wellbore instability can be detrimental to this goal and it is included during well planning stage. Therefore, It is quite difficult to measure the state of stress acting on a borehole wall. The uses of inclined and horizontal wells in the exploitation of natural resources have increased considerably and so the need of wellbore instability while drilling arises. This is particularly true for long reach, highly deviated and horizontal wells where the cost of downtime is very high. During the development phase engineers seek to optimize wellbore instability through determination of optimal well trajectories and safe mud weight window. The response of the wellbore to the drilling process can monitored and compared to the rock failure criterion. Therefore, wellbore instability is a function of several factors such as well bore inclination and azimuth, in situ stresses, mud weight, rock strength properties etc. Some of these factors are controllable and some are not. Among the controllable factors are inclination, azimuth and mud weight. By changing these parameters, one can reduce instability problems significantly. Theoretically, it is possible to design the well trajectory in a way to face least instability problems. Present paper describes that the linear elastic model along with Mohr's-Coulomb failure criterion can be utilize to calculate optimum mud weight window to avoid wellbore instability problems.

D. K. Gupta (✉) · P. Sharma
School of Engineering, University of Petroleum and Energy Studies, Dehradun, Uttarakhand 248007, India
e-mail: dkgupta@ddn.upes.ac.in

K. C. Dani
Drilling and Completion, Division of Exploration and Production, Reliance Industries Limited, Mumbai 400021, India

N. A. Siddiqui et al. (eds.), *Advances in Environment Engineering and Management*, Springer Proceedings in Earth and Environmental Sciences,
https://doi.org/10.1007/978-3-030-79065-3_7

Keywords Drilling · In-situ stress · Mohr's coulomb failure criterion · Wellbore instability · And mud weight

1 Introduction

Wellbore stability modeling is relevant to the full lifecycle of oil exploration and production. Wellbore stability analysis has therefore been included at the well planning stage and been studied extensively. Increasingly, drilling programs are incorporating a more proactive approach to mitigate this frequent and expensive source of drilling cost over-runs [1]. Therefore, to maintain wellbore integrity, geomechanical modeling techniques are using available data to provide a solid technical foundation from which to manage the drilling process. Mostly stability problems occur in shale formations. The main causes are high original pore pressure and drilling-induced pore pressure perturbations. In addition to well-known shear failure, tensile failure of shales is one of the most troublesome and hard-to-predict forms of wellbore destabilization. Modeling of the near-wellbore stress environment requires consideration of the complex interaction of wellbore mechanical and hydraulic forces (Fig. 1).

Number of platforms can be minimize just by implementing deviated production wells. The directional and horizontal wells expose more area for flow of crude oil, result increase in production. However, if azimuth and inclination angle are not analyzed properly, other problem like transportation of cutting, well bore stability becomes the key challenges.

Initially, the subsurface formation is in equilibrium condition under in situ stress balance. After drilling, remove cuttings that create mechanical disturbance to the

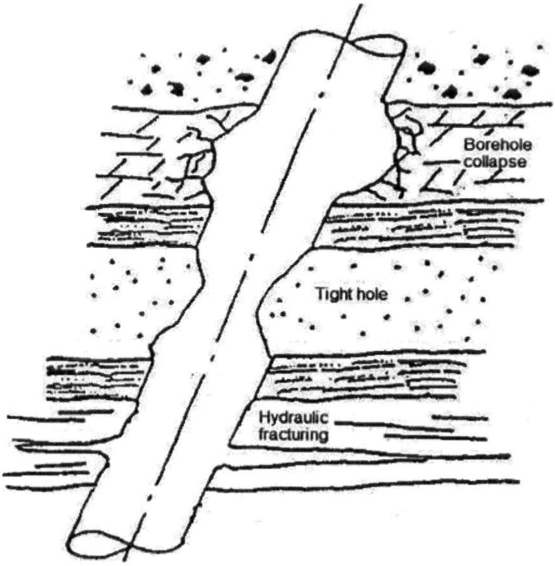

Fig. 1 Typical stability problems during drilling

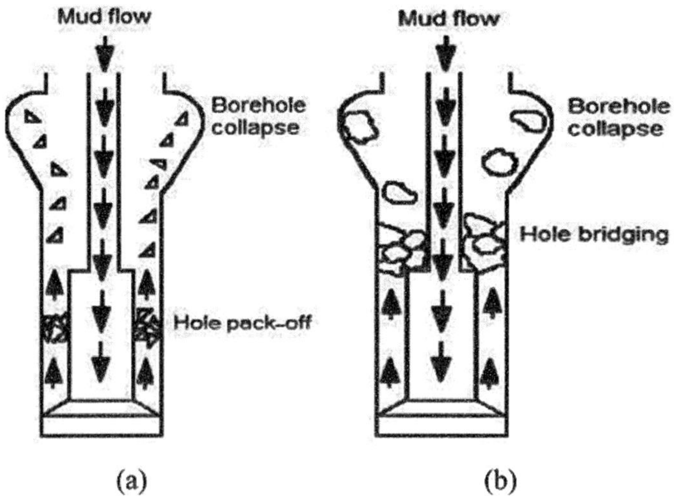

Fig. 2 Pipe sticking problem i.e. pack-off/hole bridge

equilibrium condition. Hence to overcome the new situation, stresses around bore well are developed and are known as induced stresses. If induced stresses are not managed properly than well may collapse or tensile failure (Fig. 2).

2 Model Structure and Methodology

The overriding objective of present research work is to develop conceptual learning of most confusing and complicated area of Geomechanics and to investigate the effect of stresses on wellbore stability under varying borehole orientation. The work is classify into three stages as describe below.

A. **Model Development**
It emphasizes the estimation techniques of in-situ earth stresses, transposition of in-situ earth stresses from global earth co-ordinates to wellbore co-ordinates and the variation of induced wellbore stresses along the periphery of the wellbore. Wellbore stability analysis is carry based on present field procedures and techniques involved in the geomechanical model development.

B. **Programming**
Since manual calculation to solve wellbore stability analysis takes so much time because of the very lengthy calculation procedure, a suitable C-Program was develop for various output conditions to simplify the calculation work to determine the induced wellbore stresses for various borehole inclination and azimuth angles.

C. **Computational Analysis**

With the aid of C-Programs, the induced wellbore stresses for various well profiles namely vertical, directional and horizontal wellbore were determined. Based on the calculated stress values, various wellbore stress plots were generate with the help of Matlab software to understand the stress concentration variation around the wellbore and can easily inferred from the stress plots that stress concentration varies strongly as a function of position around the wellbore.

3 Stresses Around Borehole

Before drilling, rock stress can describe by the in-situ stresses and effective stresses. As the hole was drill, the support provided by the rock is remove and replaced by hydrostatic pressure. This change alters the stresses. The stress at any point on or near the wellbore can now be described in terms of

1. Radial stress σ_{rr} acting along the radius of the wellbore
2. Hoop stress $\sigma_{\theta\theta}$ acting around the circumference of the wellbore (tangential)
3. Axial stress σ_{zz} acting parallel to the well path and
4. The additional shear stress components designated by ($\tau_{r\theta}$, $\tau_{\theta z}$, $\tau_{z\theta}$).

These stresses are perpendicular to each other and for mathematical convenience is use as a borehole coordinate system.

Considering Cartesian (x, y, z) and cylindrical (r, θ, z) co-ordinate systems for the analysis of stresses around wellbore as shown in figure. Following equations are useful for the modeling of the problem (Figs. 3, 4 and 5):

The most complicated stress solution around well bore are given as

$$\sigma_r = \left(\frac{\sigma_x^o + \sigma_y^0}{2}\right)\left(1 - \frac{a^2}{r^2}\right) + \left(\frac{\sigma_x^o + \sigma_y^0}{2}\right)\left(1 + 3\frac{a^4}{r^4} - 4\frac{a^2}{r^2}\right)\cos 2\theta$$
$$+ \sigma_{xy}^0\left(1 + 3\frac{a^4}{r^4} - 4\frac{a^2}{r^2}\right)\sin 2\theta + P_w\frac{a^2}{r^2}.$$

$$\sigma_\theta = \left(\frac{\sigma_x^o + \sigma_y^0}{2}\right)\left(1 + \frac{a^2}{r^2}\right) - \left(\frac{\sigma_x^o + \sigma_y^0}{2}\right)\left(1 + 3\frac{a^4}{r^4}\right)\cos 2\theta$$
$$- \sigma_{xy}^0\left(1 + 3\frac{a^4}{r^4}\right)\sin 2\theta - P_w\frac{a^2}{r^2}.$$

$$\sigma_s = \sigma_s^0 - v\left[2(\sigma_x^o - \sigma_y^o)\frac{a^2}{r^2}\cos 2\theta + 4\sigma_{xy}^o\frac{a^2}{r^2}\sin 2\theta\right].$$

$$\sigma_{r\theta} = \left[-\left(\frac{\sigma_x^o - \sigma_y^o}{2}\right)\left(1 - 3\frac{a^4}{r^4} + 2\frac{a^2}{r^2}\right)\sin 2\theta\right] + \sigma_{xy}^0\left(1 - 3\frac{a^4}{r^4} + 2\frac{a^2}{r^2}\right)\cos 2\theta.$$

$$\sigma_{\theta z} = \left(-\sigma_{xz}^0\sin\theta + \sigma_{yz}^0\cos\theta\right)\left(1 + \frac{a^2}{r^2}\right),$$

$$r = (x^2 + y^2)^{1/2}, \quad \theta = \arctan(y/x).$$

$$x = r\cos\theta, \quad y = r\sin\theta.$$

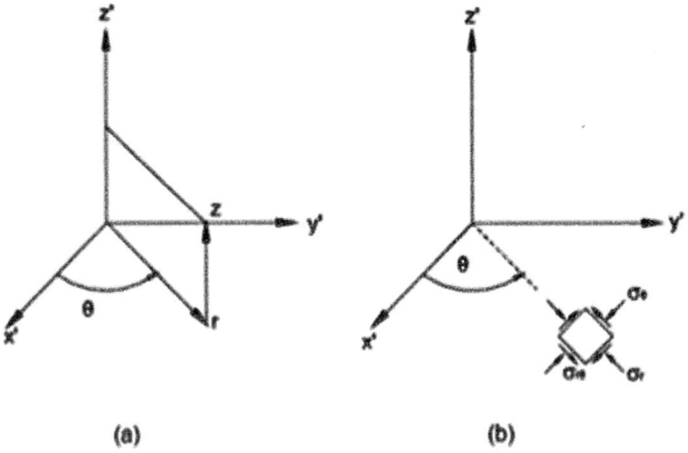

Fig. 3 Stress transformation from Cartesian to cylindrical system

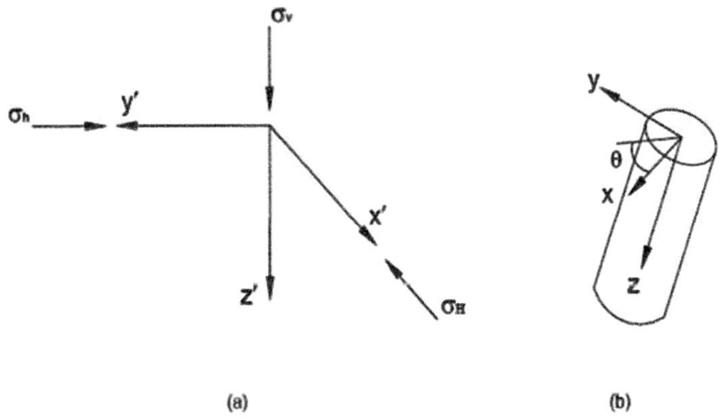

Fig. 4 Far field stress co-ordinate system

$$\sigma_{rz} = \left(\sigma_{xz}^0 \cos\theta + \sigma_{yz}^0 \sin\theta\right)\left(1 - \frac{a^2}{r^2}\right),$$

Note as,
a—radius of the wellbore
P_w—internal wellbore pressure
ν — Poisson's ratio

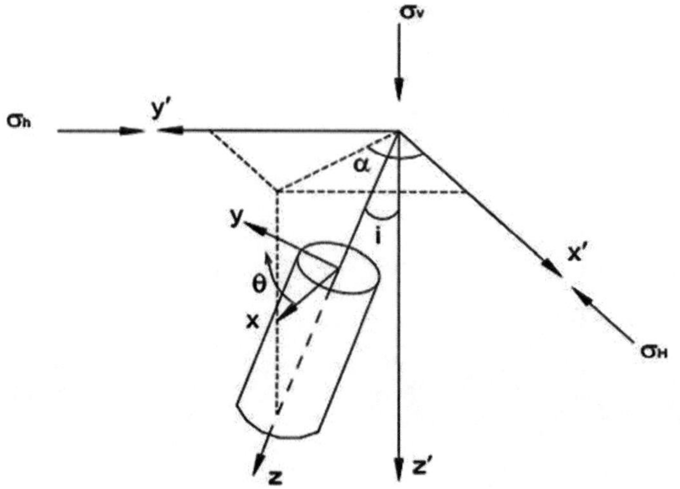

Fig. 5 Stress transformation system for arbitrary borehole

θ—Angle measured clockwise from the *x*-axis.

Moher's Coulomb Criterion

This rock failure criterion states that the material resists the shear stress across a plane Cohesion (C) and normal stress (σ′) such that (Fig. 6):

$$|\tau| = C + \sigma' \tan \emptyset$$

Fig. 6 Coulomb strength envelopes in terms of shear and normal stresses

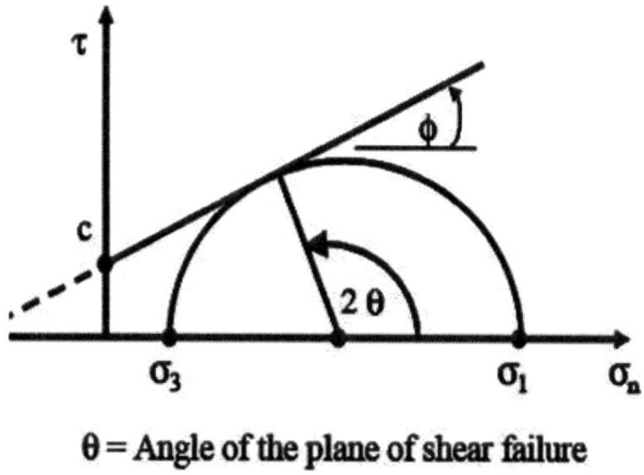

4 Results and Discussion

Results

Various wellbore stress plots were draw for different well conditions to understand the stress concentration around the wellbore.

Reservoir Data for Stress Plots (Vertical and Horizontal Well)

Considering following wellbore data:
Wellbore Pressure Pw = 5000psi.
Poisson Ratio v = 0.2
Vertical In-situ Stress = 12,000 psi.
Max Horizontal Stress = 10,000 psi.
Min Horizontal Stress = 9000 psi.
Using this field data, we plotted Induced Tangential stress, Axial Stress versus Circumferential.
Angle (θ) for a vertical and horizontal wellbore.

4.1 *Stress Plot for Vertical Wellbore*

Discussions

- Referring to Fig. 7 for a **vertical wellbore orientation**, the induced wellbore hoop stress is always greater than the induced wellbore axial stress
- The induced wellbore stresses namely hoop stress and axial stress are independent of transposed wellbore in-situ stresses and are depending only on wellbore circumferential wall angle.

4.2 *Stress Plot for Horizontal Wellbore*

Discussions

- Referring to Fig. 8, for a **Horizontal wellbore orientation**, the induced wellbore hoop stress is always greater than the induced wellbore axial stress
- The induced wellbore stresses namely hoop stress and axial stress are dependent on transposed wellbore in-situ stresses and are depending on wellbore circumferential wall angle.

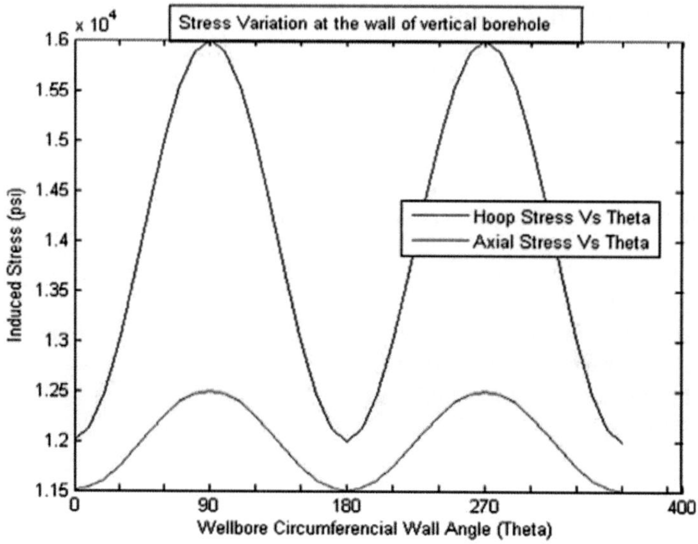

Fig. 7 Stress plot for vertical wellbore

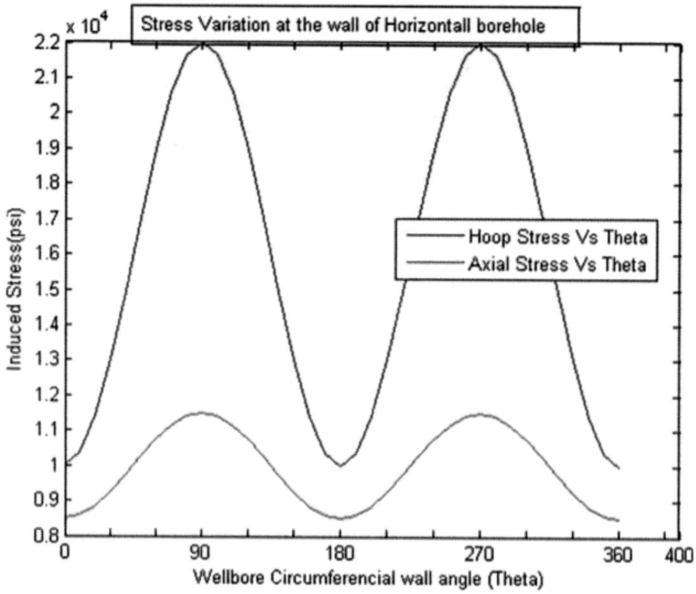

Fig. 8 Stress plot for horizontal wellbore

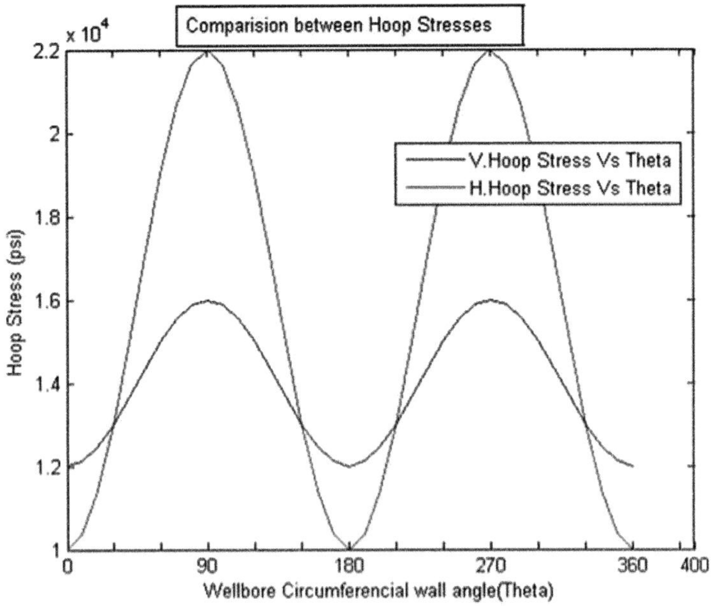

Fig. 9 Comparision of stresses-vertical and horizontal wellbore

4.3 Comparision of Stresses-Vertical and Horizontal Wellbore

Discussions

- Referring to Fig. 9, when we compare a vertical wellbore orientation and horizontal wellbore orientation, the induced wellbore hoop stress in a vertical well is always greater than the induced wellbore hoop stress in a horizontal well.
- The in-situ stress acting along the horizontal well bore axis greatly influences the magnitude of induced hoop stress
- Referring to Fig. 10, when we compare a vertical wellbore orientation and horizontal wellbore orientation, the induced wellbore axial stress in a vertical well is always greater than the induced wellbore axial stress in a horizontal well.
- The in-situ stress acting along the horizontal well bore axis greatly influences the magnitude of induced axial stress.

4.4 Variation of Hoop Stress with Mud Weight and Circumferential Wall Angle

Discussions.

- Referring to Fig. 11, Decrease in mud weight increases Hoop Stresses.

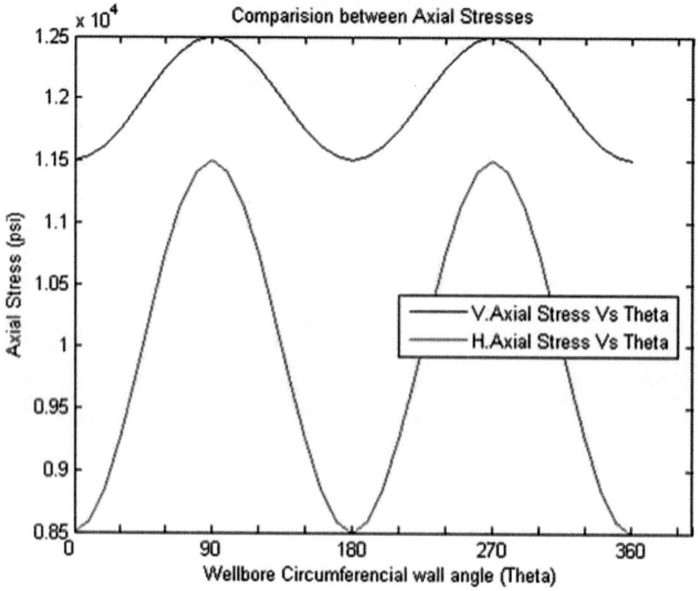

Fig. 10 Comparison between axial stresses

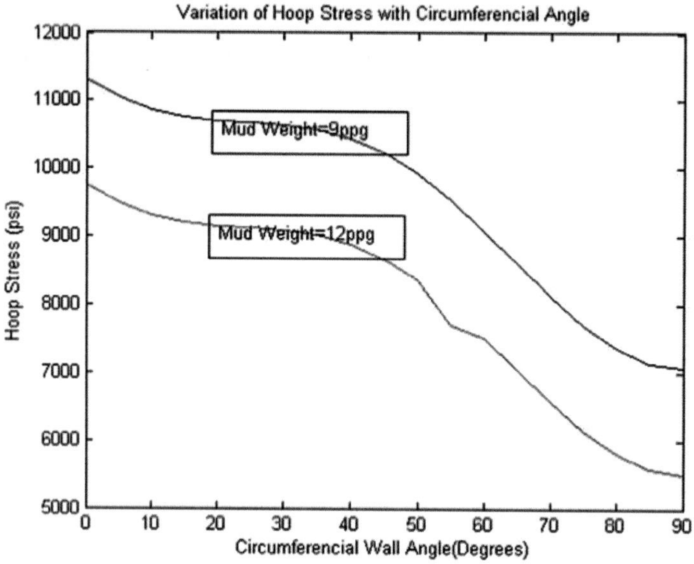

Fig. 11 Variation of hoop stress with circumferential angle

- Increase in mud weight increases Radial Stresses as radial stress is directly proportional to Mud Weight as seen from the previous mathematical expressions.

4.5 Validity of Results

- The discussions in above sections matches with Sect. 3 of Amoco's Wellbore Stability-Drilling Handbook.
- The pattern of various stress concentration plots around the wellbore wall we obtained using Matlab is matching with pattern of stress concentration plots around the wellbore wall of Sects. 6.3 and 6.5 of Reservoir Geomechanics by M. D. Zoback.
- The results we got in above section is matching with the published SPE Paper 20405.

5 Safe Mud Weight Determinations

By computing the stresses at points around the circumference of the wellbore using equations and comparing them with various failure criterions, we calculated at what mud weights either tensile or compressive failure is initiated. The critical stress points are manually calculated first around the periphery of the wellbore wall by varying angle θ from 0 to 90. C-Programming was also developed to make complicated calculations as simple as possible.

Field Data Assumptions

When no direct measurements of the horizontal stresses (minimum and maximum principal horizontal in-situ stresses) in the reservoir were available, an arbitrary value equivalent to 0.75 psi/ft was choose for both horizontal stresses. The vertical stress is assume as 1 psi/ft. The formations were normally pressured equivalent to 0.45 psi/ft.

Minimum Mud Weight Comparison

Well inclination (°)	Collapse pressure (psi)	Minimum mud weight (ppg) manual	SPE 20405 (ppg)	Error (%)
0	3992	8.999	8.9964	0.3
30	4250	9.5815	9.8294	2.5
50	4422.6	9.9707	10.6624	6.48
70	4594.8	10.359	11.0789	6.5
90	4767.07	10.747	11.4121	5.8

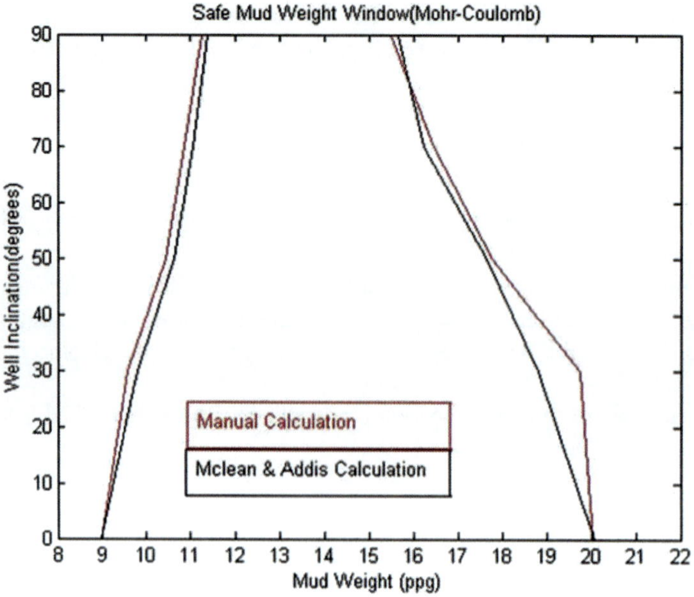

Fig. 12 Minimum mud weight required

Wellbore Stability Plot (Mohr Coulomb)

The calculated optimum mud weights were draw against well inclination to obtain the wellbore stability plot (Fig. 12).

Discussions

- The Minimum mud weight required is directly proportional to the well inclination and Maximum mud weight required is inversely proportional to the well inclination.
- Hoop stress contributes to collapse failure (ie, stress induced failure) whereas Radial stress contributes to tensile failure (ie, drilling induced failure).

6 Conclusion

i. Rapid or frequent changes in wellbore pressure at the wellbore wall can have an effect on the effective radial stress and these changes can lead to circumferential tensile failure.

ii. When radial stress increases hoop stress decreases, under this condition the wellbore wall can go into tension and contribute to the occurrence of Drilling-Induced Tensile Fractures.
iii. The stress concentration varies rapidly with respect to position around the wellbore and distance from the wellbore wall.
iv. Inclined wells lose their relative stability very quickly with inclination. This is due to the fairly high overburden gradient.
v. Depending upon the stress regime, the minimum and maximum horizontal principal In-Situ Stress may exceed the Overburden In-Situ Stress.
vi. In conclusion stability model should never be considered as a purely deterministic trial.

References

1. McLennan, J. D., & Abou-Sayed, A. S. (2002, October). Some advances in near wellbore geomechanics. In SPE/ISRM Rock Mechanics Conference, Irving, Texas, October 2002. https://doi.org/10.2118/78194-MS

COVID-19 Fallout: India, Power Sector and Lockdown

Deepali Yadav and V. Balaji Venkateswaran

Abstract The corona virus disease 2019 has affected the life of almost every individual and the impact on various sectors of any country is still ongoing. The leaders across the globe are working towards containing the disease by enforcing lockdown measures and hence struggling to stabilise economy. India saw a little more than two months of lockdown, which tremendously affected manufacturing, transport, change in working pattern, supply chain and other industrial activities. This has directly affected the electrical power sector of the country, with change in power consumption and generation pattern during lockdown and delay in upcoming/ongoing projects post lockdown. This paper presents a detailed discussion on power generation and demand trend with change in energy mix during the lockdown period in the country. Subsequent impact on already stressed distribution utilities, power market, government measures and contributing factors are presented. At the end, prospective techno-economic recommendations are discussed to overhaul the power sector to make a resilient system for future.

1 Introduction

1.1 COVID-19 Fallout

The Corona Virus Disease 2019 (COVID-19) broke out in December 2019 in the Wuhan City of Hubei province in China. Since then, the deadly virus has taken a toll on human life across the world causing an unprecedented impact on global health and economy. By end of January 2020, first few cases of COVID-19 infected individuals were reported in India and by February 2020, the World Health Organization

D. Yadav (✉)
Electrical and Electronics Engineering Department, University of Petroleum and Energy Studies, Dehradun 248001, India

V. Balaji Venkateswaran
School of Technology, Woxsen University, Hyderabad, Telangana, India

N. A. Siddiqui et al. (eds.), *Advances in Environment Engineering and Management*, Springer Proceedings in Earth and Environmental Sciences,
https://doi.org/10.1007/978-3-030-79065-3_8

(WHO) declared it a pandemic [1]. Initially, the count of infected individuals was low, but being the second highest populated country in the world, risk for India was very high. In response, the Government of India (GoI) declared a nation-wide lockdown on March 23 with an aim to save as many lives as possible [2]. Between March 24 and May 31, lockdown in India saw varying levels of restrictions and hence varied degree in opening country's economy. Necessary sectors and services like power, healthcare, and daily essentials came under the must run status during the lockdown period. This nation-wide lockdown called for social-distancing, hence imposing several restrictions on individual movements and economic activities. The transport services, imports, exports, educational institutions, industries and manufacturing all came to a halt, except essential services and goods (must run status) including the electrical power generation and supply [3]. This paper presents the impact of covid-19 on Indian power sector and variation in power generation and consumption during lockdown. The time frame selected is from March 2020 to May 2020.

1.2 Power Sector in India—Overview

As of April 30, 2020 the total installed capacity of India totals 370.348 GW including fuel, hydro, nuclear and renewable energy sources [4]. Globally, India ranks third in terms of both overall generation and consumption and in top five for RES installed power capacity [5]. Factors like increasing demand, policy and reform support, attractive prospects and opportunities and increasing investments favour the power sector in India. Hence, this sector has evolved from its introductory stage before 1956 to nationalisation and liberalisation stage between 1956 and 2003 and currently stands in its growth era [5]. From 2009 to 2019, the average annual percentage growth in generation from conventional energy sources is 5.7% and electricity demand growth rate was 7% during 2007–17 [4, 6]. With this India has set attractive and ambitious power targets like its goal of achieving 175 GW RE by 2022 [7]. As per the 19th Electric Power Survey Committee (January 2018), the 2021–22 projections for energy requirement and peak demand are 1566 Billion Units (BU) and 225.751 MW respectively. The committee also projects an All-India installed capacity of 479.418 GW by the end of 2020–21 [8]. A 2 GW solar power project has been delayed to come online in 2022 owing to constraints in equipment and labour availability. However, despite a growing trend and a favourable scenario and having surplus capacity, India faces numerous challenges like weak infrastructure and long-term outstanding debt of numerous generation units, poor financial status of distribution companies (DISCOMs) and high aggregate technical and commercial (AT&C) losses [6]. Subsequently, by the end of 2019, the power sector in India owed an outstanding debt of more than Rs. 80,000 crores (due to non-performing/stressed thermal generation units and poor cash flow of DISCOMs) [9]. The pandemic breakout in first quarter of 2020 has worsened this situation due to various factors arising because of nation-wide lockdown, moratoriums and social distancing measures, especially for upcoming projects and DISCOMS.

2 Government's Response and Its Impact on Power Sector

With an aim to contain the virus, the GoI enforced nation-wide lockdown from March 25 to May 31 in four phases. Each phase saw a varying degree of restrictions, for instance the phase 1 of lockdown (March 25 to April 14) restricted public from stepping out of homes, halted the activities of educational and industrial establishments, suspended hospitality, transport and other services including banks, petrol pumps, ATMs and food outlets [10]. Manufacturing and transport of essential goods and services were permitted [10]. Phase 2 (April 15 to May 3) brought an ease in restrictions for areas where spread had been contained (called green zones) and allowed for resumption of agriculture and horticulture sector [11]. Similarly, subsequent phase 3 and 4 further eased the extent of restrictions and by May 31 allowed for normal movement of people, private offices with 33% staff, inter-state movement of goods and e-commerce services for green zones [12].

During lockdown, with work from home pattern and halted industrial activities, the industrial and commercial load reduced considerably and domestic load contributed to the major share of power consumption. Between March 24 and April 19, all India power supply reduced by 25% year-on-year and where there was an increase of 7% in power supplied in February 2020, between March 1 and March 24 saw a decrease of 3% [13]. This scenario also affected the energy mix in generation adversely affecting the thermal units as RES units were accorded the must run status by GoI for lockdown period [14]. Subsequent sections present a graphical depiction of variation in energy mix, demand and consumption during the lockdown period in India. In this study non-covid factors like temperature, season, economic factors and meter issues, etc. have not been considered.

In addition, the enforcement of lockdown measures bought with it adverse impact due to factors like—supply chain disruption due to halted transport and reduced or no labour for all sectors including the power sector—in terms of both operation and finances.

2.1 Electrical Power Demand and Energy Consumption

About 1.3 billion people in the country obliged the rules laid down by the government as part of COVID-19 lockdown by staying at home and in most cases working from home. The lockdown also suspended industrial activities and non-essential commercial services. Consequently, the power demand from these customers considerably reduced and on the other hand, domestic demand expectedly increased. As per Central

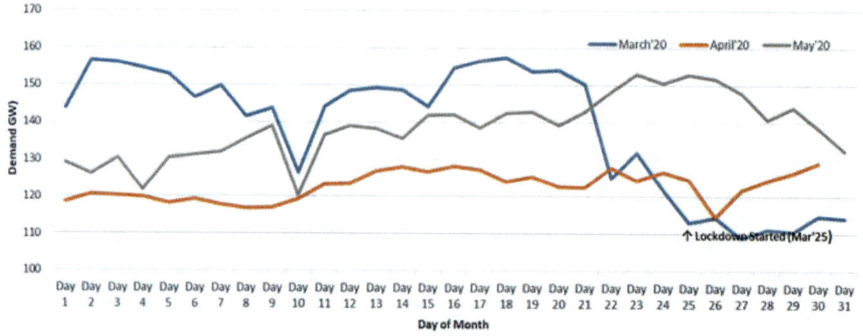

Fig. 1 All India demand (evening peak hour, GW)

Electricity Authority (CEA) report 2018–19, of the total all-India demand- industrial, commercial, agricultural and domestic load consumes approximately 41.2%, 8.2%, 17.7% and 24.8% respectively [15]. Figure 1 depicts variation in all-India peak hour demand during lockdown months from March 2020 to May 2020 [16]. It can be fairly concluded that for phase 1 (March 25 to April 14) of lockdown peak demand reduced considerably falling to as low as 108.95 GW on March 27. Phase 2 of lockdown (April 15 to May 3) depicts slight increase in demand, as certain restrictions were eased and agriculture and horticulture services were resumed. With subsequent phases of lockdown (Phase 3 from May 4 to May 17 & phase 4 from May 18 to May 31) restrictions were further relieved and thus a smooth increase in peak power demand is noticed [12, 16]. However, the pattern still does not reflect the pre-covid trend. Figure 3 suggests an average reduction of approximately 17.14% (y-o-y) in peak power demand during the lockdown with maximum reduction of 24.44% occurring in April 2020 compared to April 2019 and 11.38% compared to March 2020. This can be attributed to the fact that most part of April saw strict lockdown rules when compared to relatively eased restrictions in May and only last quarter of March facing lockdown. Figure 2 represents a similar pattern showing variation in all-India energy consumption during lockdown. Figure 4 represents total energy consumption from March to May 2020, depicting a comparison with corresponding non-covid months of year 2019 and with previous month of 2020 having varied degree of lockdown restrictions. Later part of lockdown saw resumption in most industrial and commercial activities as is evident from Figs. 3 and 4, peak power demand and total energy consumption in May 2020 increased by 12.72% and 23.9% respectively relative to April 2020.

2.2 *Source Wise Generation—Change in Energy Mix*

With lockdown restrictions across the country, the energy mix shows an increasing trend in share of RES and hydro generation (Fig. 5). The key factors for this include

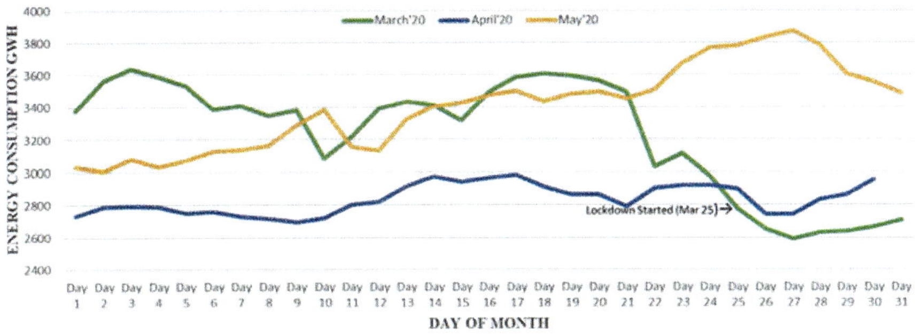

Fig. 2 All India energy consumption (GWh)

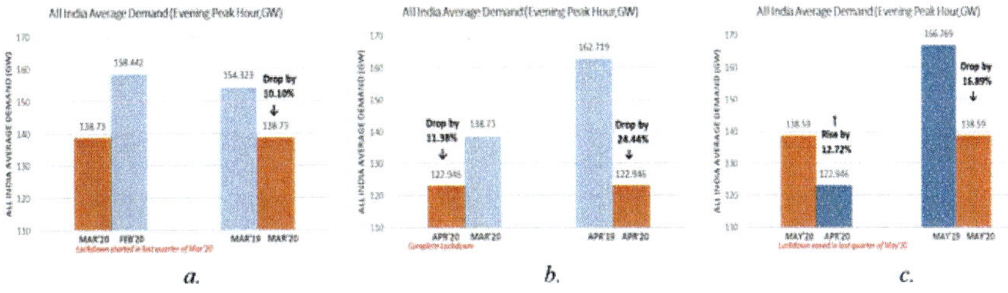

Fig. 3 All India average demand (evening peak, GW) **a** MAR'20. **b** APR'20. **c** MAY'20

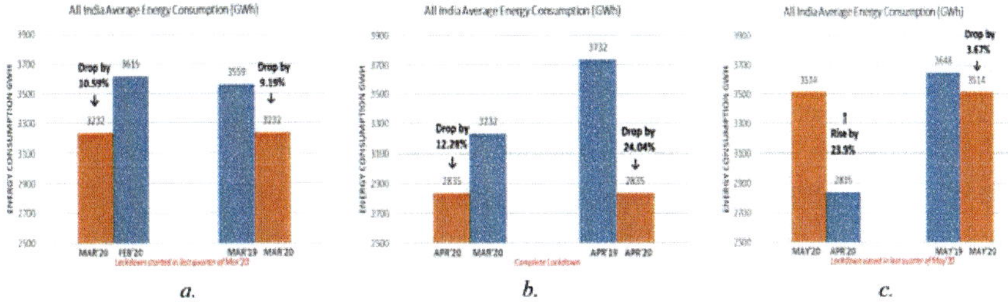

Fig. 4 All India average energy consumption GWh. **a** MAR'20. **b** APR'20. **c** MAY'20

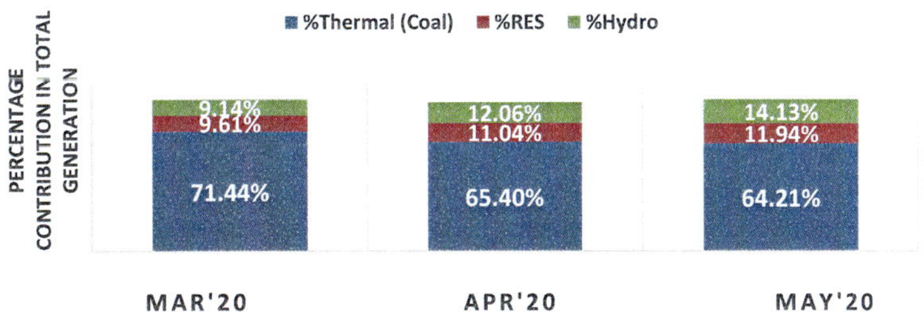

Fig. 5 Source wise contribution in total generation (%)

Table 1 Percent source/fuel wise contribution in total monthly generation

Source/fuel	March 2020 (%)	April 2020 (%)	May 2020 (%)
Coal	71.44	65.40	64.21
Lignite	2.42	2.65	2.20
Hydro	9.14	12.06	14.14
Nuclear	3.43	3.89	3.14
Gas, naphtha and diesel (GND)	3.94	4.88	4.36
RES (wind, solar, biomass, other)	9.61	11.04	11.94

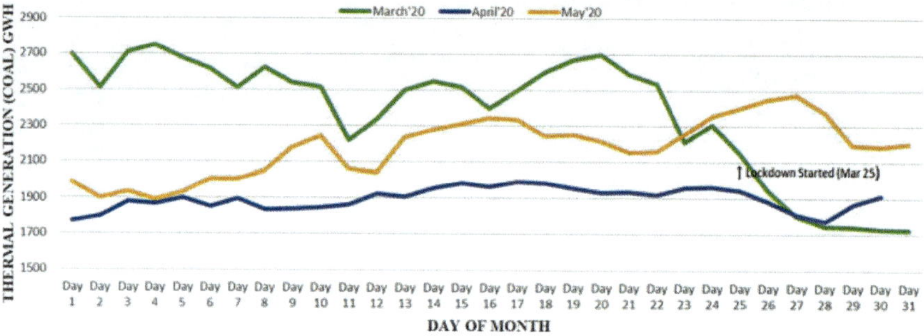

Fig. 6 All India thermal generation (coal, GWh)

reduced power demand, must run status to RES units, lower operating cost and priority access to grid through government framework and regulations [17, 18]. Table 1 shows significant reduction in the gap between coal based thermal and renewable generation [16]. Percentage share of coal consistently remains below 70% (Fig. 6), recovering only in later half of May when lockdown restrictions soften and power demand recovers. The pre-lockdown months recorded average 73% share in generation from coal, which reduced to 64% during lockdown [16]. This suggests an adverse impact on coal based power plants compared to other energy sources. Already the thermal power plants have seen a drastic fall in Plant Load Factor (PLF) from 77.5% in 2009–10 to 42.4% in April 2020 [4]. Such scenario could subsequently lead to these plants lying idle and their fixed cost would be a burden. In addition, increase in coal stocks and its reduced demand could be a concern too, as power sector is major consumer (87%) of domestic coal [19]. Such factors would further challenge the financial viability of these plants due to decreased utilisation of capacity.

On other hand, RES generation sees a rising trend with approximately 24.44% average month-on-month rise in the total generation share from March to May 2020 (Fig. 7). In addition, the share of hydro generation also increased from March to May 2020 (Table 1). On bright side, although not for long-term, this has resulted in increased use of clean energy and hence reduced CO_2 emissions [20].

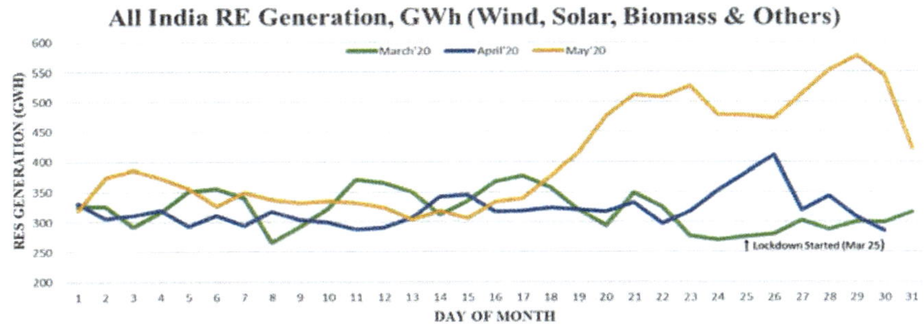

Fig. 7 All India RE generation (GWh)

2.3 *Impact on Grid, Reforms and Power Market*

The domestic and agriculture load tariff in India is low when compared to industrial and commercial loads. It is lower than the average supply cost. With people confining to their homes and industries shutdown, industrial and commercial demand reduced and domestic expectedly increased. This has coincided with a dip in clearing volume and market clearing price [20]. As illustrated in Fig. 2 on March 16 (pre covid lockdown) the energy demand was 3494 GWh and in last week of March (lockdown imposed) energy consumption varied between 2600 and 2800 GWh. Hence, the already stressed DISCOM utilities suffered a major revenue loss due to reduced demand and incapability to ensure cross subsidy to domestic consumers. Additionally, compliance of long-term power purchase agreements (PPAs) with generator units would be a challenge too [21]. Also, owing to the financial challenges faced by the citizens due to pandemic, the government provided moratorium to consumers in electricity bill payments and simultaneously the utilities have maintained continuity of supply. Subsequently, with delay in bill payment cycle the cash flow status of DISCOMS would be compromised further [17].

India ambitiously aims for 176 GW capacity addition by 2020, which includes 118 GW form RES, 6.8 GW from hydro, and 6.4 GW from thermal (coal based) [8]. The impact of pandemic lockdown has adversely affected these targets halted construction activities, global supply-chain disruption leading to difficulty in availability of components that are imported from China or Europe and poor revenue due to reduced demand that financially incapacitates the companies [22]. Moreover, if the government wants to achieve its high set RES future targets it must take measures to ensure sufficient capital, as renewable projects require significant capital investment and has long payback time which can be a challenge due to current economy of the country. The GoI has always been pro-active for reforms and policies in power sector. The implementation of real time market for electricity was scheduled on April 1.2020 which faced a delayed of two months [23]. There are numerous policy measures in pipeline for 2020–21 including UDAY 2.0, a scheme aiming to provide installation of smart meters among other things and coal block auction for commercial mining [24, 25]. However, both plans seem to be delayed owing to the current situation.

3 Government Measures

In May 2020, the government announced special relief package to revive economy, the *Atmanirbhar Bharat* (self-reliant India) plan. For power sector, it provided key reforms in tariff policy, privatization of utilities in Union Territories, and a liquidity infusion of Rs. 90,000 crore by Power Finance Corporation (PFC) and Rural Electrification Corporation (REC) to aid the hard hit DISCOMs stabilize their financial status [25]. With this discom utilities could clear their outstanding dues with generation and transmission units which owing to default had to stop supplies to distribution companies. Additionally, Ministry of New & Renewable Energy (MNRE) has presented the start of campaign to foster domestic manufacturing for RE sector [26].

4 Recommendations

The COVID-19 affected the interconnected power system network with major demand variations and variable and unpredictable RE generation, subsequently, the grid operator must take suitable measures to manage the demand and at the same time ensure grid stability and reliability [27]. Various options explored in different power markets to manage the large variations in demand provide recommendations to mitigate the risks covid has posed to the power sector and a holistic approach to modify business with innovations in technology and digitization.

4.1 Use of Existing Hydro and Thermal Fleet

The hydro generation is either reservoir based or run-of-river type. The reservoir-based plants can be used to manage system variability due to the fast response from these machines. The hydro machine has capability to reach full load within 5 min so these units provide the most ideal solution for managing large variations. However, their availability depends on monsoon and so this may limit their actual contribution. Most of the existing thermal fleet comprises of coal and gas-based power generations. Coal plants while provide about 1% of ramp up/down capability, the gas turbines are much faster. A combined cycle gas turbine would provide about 3–5% ramp up/down performance depending on the class of machine whereas a simple cycle operation could provide a ramp rate of between 8 and 10%. A judicious use of gas turbines could help manage the system variability.

4.2 *Modifying the Existing Thermal Fleet*

There is an ongoing effort to reduce the emission from the thermal units that requires retrofit of suitable emission reduction equipment in the installed units. It may also be noted that the performance of the existing units is most optimal at the rated capacity. These units may be suitably modified to provide near optimal performance at part loads, improved ramp rate and reduced emission. All these modifications would ensure better performance of these units and the grid.

4.3 *Possible Future Measures*

Energy storage solutions have been increasingly perceived as fast solution for frequency control. The duration of these solutions could vary from a shorter duration of few seconds to few hours/days. A proper grid study can help establish the need for the right solution and it would be worth exploring the possible solutions. Flywheel and ultra-capacitors can add smaller watts in the system and provide quick frequency response. Pumped hydro would be able to provide higher capacity for longer durations. Batteries additions into the network in several grids can provide for absorbing variability of renewable power. These can cover timeframes from very short duration to large duration and rating varies from few kilowatts (KW) to megawatts (MW). Demand Response is another solution that may be worth considering by the distribution companies. This approach can help in optimizing the demand and manage the demand variability. Although this is an emerging solution, it would entail implementing smart metering infrastructure and smart devices to be able to participate.

5 Conclusion

In the lockdown period, it can be observed that the collapse in electrical power consumption has risen from 9.6% in the month of June to 2.6% in July This can be accounted for unlock activities owing to presumption of major commercial and industrial units nationwide. In addition, the peak demand has shown a considerable rise from complete lockdown months of April–May to June–July 2020, when unlock was initiated. The peak demand in power in year 2019 for months April to July has shown a dip in its respective values in year 2020, with slump ranging from 25% in April to 9.6% in June 2020 relative to corresponding values in 2019. Although minor relaxations in lockdown can be one reason for rise in demand from May 2020 onwards but the fact that increasing temperatures of summer months with extreme heat can also be accounted for a contributing factor for demand rise. The leaders and specialists in the power and energy field across the globe project normal levels of demand and power consumption in the coming months, owing to government's planned stages of unlock activities.

After reviewing the present status of power sector and from the expert opinion, recommendations to revive power sector are derived in this article.

References

1. Catrin Sohrabia, Zaid Alsafib, Niamh O'Neilla, Mehdi Khanb, Ahmed Kerwanc, Ahmed Al-Jabirc, Christos Iosifidisa, Riaz Agh, "World Health Organization declares global emergency: A review of the 2019 novel corona virus (COVID-19)", International Journal of Surgery 76 (2020) 71–76.
2. Khanna, Utpal Bhaskar, Anuja, Pretika (11 April 2020). "Modi says 'jaan bhi jahaan bhi', signals a shift in India's coronavirus playbook". Livemint. Retrieved 16 April 2020.
3. "Modi's 21-day India lockdown: Here's the list of essential services that will remain operational", The Economic Times, English Edition, E-paper, March 25, 2020.
4. "Power Sector at a Glance" , Government of India, Ministry of Power, [Online]
5. "Power (PDF)", Indian Brand Equity Foundation (IBEF), April 2020. Retrieved 6 May 2020.
6. IEA (2020), India 2020, IEA, Paris. https://www.iea.org/reports/india-2020.
7. "Report of the Expert Group on 175 GW RE by 2020 (PDF)", National Institution for Transforming India (NITI). Retrieved 20 April 2020.
8. "National Electricity Plan, Volume 1, Generation (PDF)", Government of India, Ministry of Power, Central Electricity Authority, January, 2018, Retrieved 20 April 2020.
9. "The burden of Non-Performing Assets in India-Thermal Power Sector (PDF)", IEEFA, December 2019, Retrieved 6 May 2020.
10. "Guidelines.pdf" (PDF). *Ministry of Home Affairs*, Retrieved 6 May 2020
11. "India to allow farmers back to work amid lockdown". *BBC News*. 15 April 2020.
12. "MHA extend lockdown period" (PDF). mha.gov.in. Retrieved 8 May 2020.
13. Surya, Saket (23 April 2020), "Impact of COVID-19 on the power sector", PRS Legislative Research. Retrieved 8 May 2020.
14. Ministry of New and Renewable Energy, "Office Orders & Memorandums." [Online]. Available: https://mnre.gov.in/public-information/office-orders.
15. Bhagwat, Pradyumna, "Identifying impacts of COVID-19 on the Indian Power Sector", FSR Global, 01 April 2020.
16. "Daily Reports", POSOCO. [Online], Available: https://posoco.in/reports/daily-reports/.
17. "Office Memorandum (PDF)", Ministry of New & Renewable Energy (MNRE), April 2020. Retrieved 10 May 2020.
18. IEA (2020), Covid-19 impact on electricity, IEA, Paris https://www.iea.org/reports/covid-19-impact-on-electricity.
19. "Provisional Coal Statistics 2018–19 (PDF)", Government of India, Ministry of Coal, Coal Controller's Organisation, Kolkata. Retrieved 15 May 2020.
20. "Market Snapshot", Indian Energy Exchange (IEX), [Online]. Available: https://www.iexindia.com/marketdata/market_snapshot.aspx.
21. "Office Memorandum (PDF)", Ministry of New & Renewable Energy (MNRE), March 2020. Retrieved 10 May 2020.
22. Current Regulations, Central Electricity Regulatory Commission, [Online]. Available: http://www.cercind.gov.in/Current_reg.html.
23. "UDAY 2.0 Scheme", Ministry of Power, 17 March 2020, [Online]. Available: https://pib.gov.in/PressReleaseIframePage.aspx?PRID=1606775.
24. "Auction of Coal Mines for Sale of Coal (PDF)" (Under Mines and Minerals (Development and Regulation) Act, 1957 & Coal Mines (Special Provisions) Act, 2015), 14 January 2020, Ministry of Coal, Government of India. Retrieved 28 May 2020.

25. "Covid-19 Economic Relief Package—8 Significant Announcements", BTG Legal, Mondaq, 10 June 2020.
26. Verma, Kanika, "India's renewable energy sector powers the nation during COVID-19 pandemic", Invest India (National Investment Promotion & Facilitation Agency), 24 April 2020.
27. Hall, William Patrick, "Impact of Coronavirus on the Indian Energy Sector", 07 May 2020, TerraGreen, The Energy and Resources Institute.

Assessing the Impact of Novel Coronavirus (COVID-19) on Renewable Energy Sector of India During the Lockdown

Deepali Yadav and V. Balaji Venkateswaran

Abstract This paper presents the impact of coronavirus disease on the Renewable Energy (solar and wind) market of India. This study describes the impact on RE business over its five dimensions—continuity of operation, year on year (YOY) growth, securing financials, client and employee, and social responsibility. This commentary presents the effect on the implementation of upcoming projects, policies, and frameworks in 2020–21 are presented in detail with special emphasis on challenges faced by RE developers and off-grid companies, economic slowdown and government response and actions to mitigate the same. In the end, the paper presents possible measures and recommendations with a short-term and long-term vision, immediately aiming for safeguarding public health and gradually for the country's economic revival and developing shock resilient RE sector.

Keywords Novel Coronavirus (COVID-19) · Solar power · Wind power

1 Introduction

Novel Coronavirus or more commonly called as Corona Virus Disease-2019 (COVID-19) originated in late December 2019, with first human cases reported officially in Wuhan City, China [1, 2]. Since, then the disease has taken a toll on human life all over the world.

By March 2020, COVID 19 was declared a global pandemic by WHO (World Health Organisation) causing a public health threat and thus triggering a crunch in the global economy at an unprecedented level. In India (and other countries too), the respective governments have taken suitable measures to contain the disease and lessen its impact at a relatively initial stage [3]. With lockdown measures, suspended

D. Yadav (✉)
Department of Electrical and Electronics Engineering, University of Petroleum and Energy Studies, Dehradun, Uttarakhand, India

V. Balaji Venkateswaran
School of Technology, Woxsen University, Hyderabad, Telangana, India

N. A. Siddiqui et al. (eds.), *Advances in Environment Engineering and Management*, Springer Proceedings in Earth and Environmental Sciences,
https://doi.org/10.1007/978-3-030-79065-3_9

import/export and rate cuts there has been a significant impact on all industry sectors, including the RE market of India. The pandemic timing has been such that the RE sector in India was at its boom, ambitiously aiming to achieve the 450 GW target by 2030 [4]. But with nationwide lockdown, all plans and developments remain suspended [5].

Hence, there is a need to take rightfully immediate and adequate measures to contain the crisis and subsequently develop a framework and policies to revive economic growth and make headway for the future RE sector with increased shock resilience.

2 RE Scenario in India Before COVID-19

Being the first country in the world to set up a ministry (Ministry of New and Renewable Energy, MNRE) for its RE sector, the total Renewable Energy Sources (RES) installed capacity (utilities) of India as of March 31, 2020, is 87.027 GW (including state, private and central ownerships), which accounts for almost 23.5% of the total installed capacity of Power Stations [6].

Out of this total RES capacity, solar and wind power accounts for 34.627 GW and 37.693 GW respectively [6]. The pace of RE sector growth has been more than satisfactory for India as its target for 2022 of installing 20 GW of solar power had been achieved four years in advance of its deadline [7]. With this, the new targets aim at 113 GW and 66GW from solar and wind power respectively by 2022. The government's renewable power framework and policies would speed up this rapid expansion in 2020–21.

Out of the top five largest solar parks in the world, three are in India, with Bhadla Solar Park being the largest in the world as of March 2020 with a capacity of 2.245 GW [8]. Even in the category of largest wind power installed capacity, India is amongst the top five nations in the world [9]. With such a strong base and attractive targets for the RE market, India has certainly established itself to become one of the top nations for clean energy and sustainable power projects.

3 Impact of COVID 19 on Upcoming and Operational RE Projects

In an attempt to contain COVID 19, the Indian government responded with measures such as nationwide lockdown and restricted human activities. Although, on March 24, 2020, MNRE designated declared electrical power as an 'essential service' and on April 1, 2020, designated a 'must run' status to RE generating stations [10]. Given this scenario, the major challenges faced by operational and upcoming RE projects are suspended manufacturing, restricted imports, workforce impact, restricted transports,

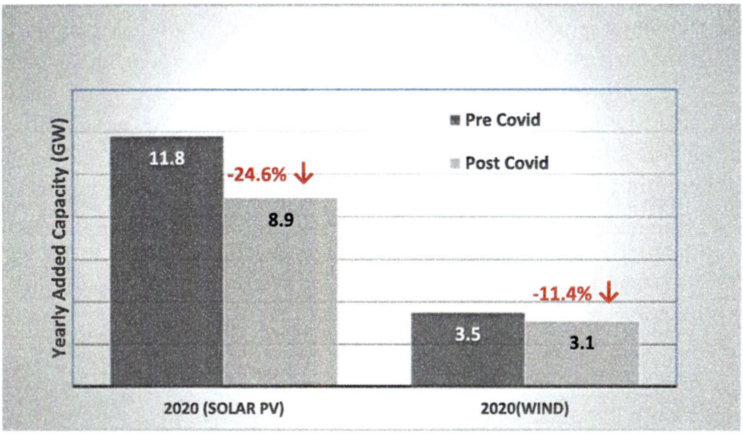

Fig. 1 Projected solar and wind power installations (GW) for the year 2020 (showing impact due to COVID 19)

trade, and financial market impact [11]. For the Atal Jyoti Yojana (AJAY) Program, the MNRE issued an order on April 23, 2020, stating the shutting down of phase II of the project due to pandemic outbreak [12]. Thus, it is expected that approximately 3 GW of solar and wind power projects would face delay with nearly 24.6 and 11.4% drop in projected installed capacity of solar and wind projects (Fig. 1) [13].

The major obstacles for the pipelined projects are supply chain and labour disruptions. Solar manufacturing units in India produce PV cells and modules with 1.1 GW cells and 1.8 GW of modules and some limited accessories, however, its solar market mostly remains dependent on imports, mainly from China [14]. Manufacturing units in China supply approximately 70% of global solar panels and 10–15% of supplies come from Chinese firms based in South East Asia [15].

On contrary to solar, the Indian wind market has been dependent on its wind turbine and wind gearbox manufacturers, but with shutdown and suspension in production, the execution of under implementation solar and wind power projects has come to a standstill. Consequently, India's roadmap to achieve 2022 targets face a crisis, as the financing and deployment of RE ventures need to accelerate this year.

Meanwhile, with social distancing, it is difficult to have multiple meetings and human interaction, which play a significant role in renewable projects, as social acceptance of these projects has been a challenge and requires interaction with the local community at all levels. This again directly leads to the delaying of upcoming RE projects [15].

The operational projects and off-grid companies are at a lower risk due to supply chain disruption as compared to the upcoming ventures. The existing RE developers face challenges in production facilities, supplies, project site, and limited workforce, but disruption for these projects remain minimal as they are dependent on natural resources (solar and wind) [16].

Since the major power consumption facilities (malls, complexes, university campuses, metro trains, etc.) are shut down, the all India power demand has reduced (Figs. 2 and 3) (from 154 GW on March 20, 2020, to 114.9 GW on March 30, 2020) [17] and for the existing demand (mainly domestic) few developers are facing problems in generating monthly power sale invoices [16]. With this, the DISCOMs face disruption in cash flow and may be forced to call for RE curtailment. Also in an attempt to lessen the pressure on DISCOMs, the Central Electricity Regulatory Commission (CERC) has reduced the surcharge payable due to late payment, by DISCOMs to generation units [16]. This eventually may lead to DISCOMs delaying payments to renewable projects.

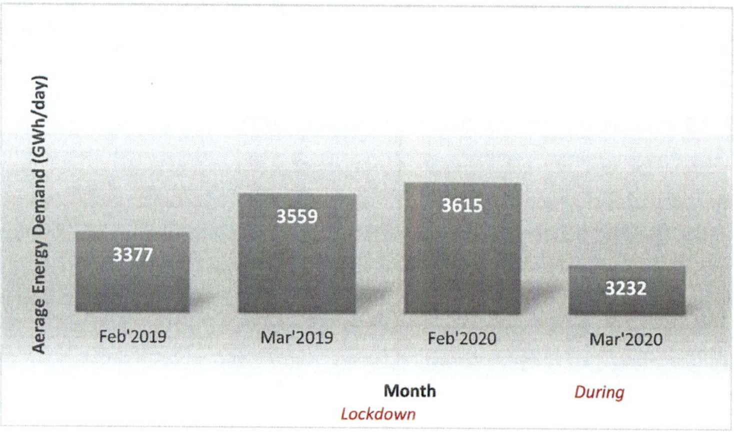

Fig. 2 All India average energy demand (GWh/day)

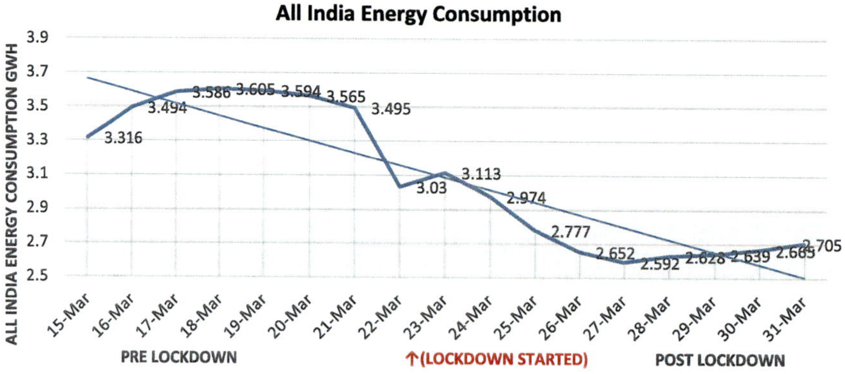

Fig. 3 Variation in all India energy consumption (GWh) pre and post lockdown due to COVID 19

4 Impact on Off-Grid RE Companies

The effect on off-grid companies are more adverse than their grid-connected counterparts do, as most of them are small or mid-sized companies. Their major challenges to deal with include, sustaining operation, service and maintenance, employee and customer satisfaction, securing financials, growth, and revival strategy. Due to lockdown, the installation of solar rooftops has stopped, and businesses or households planning to install solar PVs may altogether postpone or cancel their plans owing to financial burden and economic crunch [15]. With declining investment and suspended supplies, these companies have to work hard to maintain cash, focus on existing customers, ensure essential service status to off-grid supply and at the same time come up with revival plans for sustaining business growth.

5 Economic Impact and Subsequent Government Measures

The COVID 19 crisis has had an unparalleled impact on the economy. By mid-April, the power demand in India had fallen by 25–30% [18]. Due to supply chain disruption, production slowdown, and no or reduced workforce the upcoming solar and wind projects will face delay, this may most likely lead to missing deadlines. This puts the developers at risk of penalties; bank guarantees encashment and lower credit ratings. As already stated, the DISCOMs are facing collection issues (difficulty in raising physical invoices during lockdown), which in turn affects their cash liquidity status and leads to delay in payments made by DISOCOMs to generation and transmission companies.

In wake of this and RE developers' representation, the MNRE notified on Apri' 17 2020, that supply chain disruption due to COVID-19 would be a case of natural calamity and invoked Force Majeure Clause. This has granted the developers a suitable time extension in completion and normalization post lockdown [10]. The 'must-run' status had already been given to RE generation by MNRE, but on April 1, 2020, it issued an office memorandum directing DISCOMs to make payments to RE Generating Stations as they form a small part of the total power generation in the country [10].

In this pandemic situation, the government has been more than active in providing relief for the country's RE sector from time to time. Even before the corona crisis hit the country, MNRE had granted relief to developers in view of supply disruptions due to the lockdown situation in other countries (China, USA, Europe) [10]. The main directions by MNRE in response to the corona crisis include—granting 'must-run' status, invoking force majeure, extension in the deadline for project developers without the requirement of submitting any proof or evidence to avail same, normalization period, and directing DISCOMs for payment to RE generation units.

6 Recommendations by Global Energy Leaders

The leaders of the RE sector across the globe recommend strategy and plan that concurrently meet the objective of managing the crisis, recovery, and securing future growth [19]. The main dimensions of this journey may include—shifting mindset from present to future, identify uncertainties and their implications, clarity in the objective, and recovery plan on persistent issues [20]. Practical steps for renewable energy sector would require—development of contingency plans, re-assessment of the dynamics in supply chain and reflect on within-country manufacturing of equipment to minimize sourcing from other countries, and as learning from this crisis, accelerate the implementation of digital and automation technology for future projects [15].

7 Conclusion

The COVID 19 crisis for the solar and wind power market of India implies a major economic let-down, not only in the current situation of lockdown but also post lockdown due to fall in Indian rupee which subsequently may lead to short term rise in equipment price. However, it may also be fair to assume that the same will be stabilized by excess supply from China due to reduced demand [16]. Although the government has taken required measures and these may seem fit in the short-term, however, policymakers need to take actions, aligned with short-term measures, for the country's economic growth and sustainable energy development in long run. There is a need to reduce dependency on a single country for supplies, and plan for our supply unit, provide stimulus packages to channelize funds to new RE technologies that are not yet commercialized and are cost-effective (e.g. floating off-shore wind and marine technology), incentives and energy policies providing tax credits, investment grants and specific loan schemes for technologies that are highly vulnerable like solar PV [15].

While now it is right for policymakers and leaders to focus on public health and contain the disease, it is also important, in the long-run, to carve our way out of this pandemic in reviving the economy, developing new and restructuring existing policies and framework for a more adaptive, shock resistant, cleaner and sustainable renewable energy sector.

References

1. World Health Organization, "Coronavirus Disease (COVID-19) Situation Report—103," 2020.
2. World Health Organization, "Coronavirus Disease 2019 Situation Report—94," 2020.
3. World Health Organization, "Coronavirus (COVID19)." [Online]. Available: https://covid19.who.int/.

4. Central Electricity Authority, "Overview of Renewable Power Generation," 2018.
5. https://www.mygov.in/covid-19, "Mygov," 2020. [Online]. Available: https://www.mygov.in/covid-19.
6. Central Electricity Authority, "All India Installed Capacity of Utility Power Stations," 2020. [Online]. Available: http://www.cea.nic.in/monthlyinstalledcapacity.html.
7. Ministry of New and Renewable Energy, "Year End Review 2017," 2017.
8. I. MERCOM, "With 2,245 MW of Commissioned Solar Projects, World's Largest Solar Park is Now at Bhadla," 2020. [Online]. Available: https://mercomindia.com/world-largest-solar-park-bhadla/.
9. Global Wind Energy Council, "Wind Industry COVID-19 Response Hub | Global Wind Energy Council," 2020. [Online]. Available: https://gwec.net/wind-industry-covid-19-response-hub/.
10. Ministry of New and Renewable Energy, "Office Orders & Memorandums." [Online]. Available: https://mnre.gov.in/public-information/office-orders.
11. MONDAQ Limited, "Impact Of COVID-19 On Energy Sector—Recent Government Notifications And Initiatives," 2020. [Online]. Available: https://www.mondaq.com/india/renewables/928680/impact-of-covid-19-on-energy-sector-recent-government-notifications-and-initiatives.
12. Ministry of New and Renewable Energy, "Closing of Atal Jyoti Yojana (AJAY): Phase II." 2020.
13. Wood Mackenzie, "India's renewables installation could fall by a fifth due to lockdown," 2020. [Online]. Available: https://www.woodmac.com/press-releases/indias-renewables-installation-could-fall-by-a-fifth-due-to-lockdown/.
14. FICCI, "Securing the Supply Chain for Solar in India," 2013.
15. Heymi Bahar, "The coronavirus pandemic could derail renewable energy's progress. Governments can help," *IEA*, 2020. [Online]. Available: https://www.iea.org/commentaries/the-coronavirus-pandemic-could-derail-renewable-energy-s-progress-governments-can-help.
16. A. Khanorkar, H. Shah, K. Arora, and S. Manager, "COVID-19 Impact: Indian Renewable Sector," 2020.
17. POSOCO, "Daily Reports—2019–20." [Online]. Available: https://posoco.in/reports/daily-reports/daily-reports-2019-20/.
18. Ashish Khanna, "Renewable Energy 'New Normal' and Impact of Covid-19," *Energy-World.com*, 2020. [Online]. Available: https://energy.economictimes.indiatimes.com/energy-speak/renewable-energy-new-normal-and-impact-of-covid-19/4167.
19. Felipe Requejo, "COVID-19's impact on power, utilities & renewables companies," *Deloitee*, 2020. [Online]. Available: https://www2.deloitte.com/global/en/pages/about-deloitte/articles/covid-19/covid-19-s-impact-on-power--utilities---renewables-companies--de.html.
20. PWC, "Energy industry and COVID-19 (coronavirus): strategising for the 'new normal,'" 2020. [Online]. Available: https://www.pwc.com/gx/en/issues/crisis-solutions/covid-19/energy-utilities-resources-coronavirus.html.

Study of the Ecological Effect of Bismuth Layered Structured Ferroelectric Materials Over Lead Zirconate Titanate

Gagan Anand

Abstract Lead is extensively used in many electronic components, and various types of sensors used to measure temperature, pressure, etc. lead zirconate titanate (PZT) is extensively expended, because of its high dielectric constant. Lead is an environmentally hazardous and long term impact on humankind due to its high toxicity. Since the last five decades, People have been working on developing lead-free materials that possess better properties than PZT Martials. It is always necessary to check the impact on the environment from developing the content, production, consumption, and disposing of it. Comparing PZT and Bismuth layered structured ferroelectric (BLSF) materials like $SrBi_4Ti_4O_{15}$, $BaBi_4Ti_4O_{15}$ and $CaBi_4Ti_4O_{15}$ have been extensively studied with various dopants, these materials pose high dielectric constant and have a higher value of transition temperature making these materials a first choice for high-temperature sensors. These BLSF materials utilize lower energy for production in comparison to PZT ceramics.

Keywords Lead free · Bismuth · Ecology · Sensors

1 Introduction

The growing need for materials and manufacture techniques that are ecologically benevolent united with universal strategy initiatives such as European Union advice on waste electrical and electronic equipment (WEEE) and Restriction of Hazardous Substances (RoHS) [1–6] has stimulated the expansion of bismuth layered structured ferroelectrics (BLSF) ceramics for electronics, High-temperature sensor applications. Lead is used extensively in various appliances such as solder, paints, high-temperature sensors, piezoelectric transducers, and a host of other electronic modules [7]. On the contrary, lead is considered as environmentally hazardous and non-biodegradable material having extreme effects on living organisms.

G. Anand (✉)
Department of Physics, SOES, University of Petroleum and Energy Studies, Dehradun 248007, India
e-mail: ganand@ddn.upes.ac.in

N. A. Siddiqui et al. (eds.), *Advances in Environment Engineering and Management*, Springer Proceedings in Earth and Environmental Sciences,
https://doi.org/10.1007/978-3-030-79065-3_10

The general formula represents the Aurivillius family of compounds. $(Bi_2O_2)^{2+}$ $(P_{n-1}Q_nO_{3n+1})^{2-}$ the constants are, P can be a tria, mono, divalent ion, or a combination of all, e.g. Q = Sr, Ba, Bi, Ca, Na, Pb, and numerous rare-earth ions, where B = Nb, Fe, Ti, Ta, W, Cr, and Mo, these are exceedingly charged ions.

In the last eight decades, over 100+ crystals have been identified as ferroelectricity in nature. Ferroelectric oxides, in general, may be classified into two groups:

1. Materials which exhibits spontaneous polarization laterally one crystallographic axis in the unpolarized condition, this group comprises potassium, di-hydrogen phosphate, and related compounds, tartrates, etc.
2. Ferroelectrics having polar axis in more than 1 direction may additionally be segmented into 4 edifice types: Pyrochlores (e.g.$Cd_2Nb_2O_7$); bismuth layered compounds (lead-free) (e.g.$SrBi_4Ti_4O_{15}$); Perovskites (e.g. Barium titanate); and Tungsten bronze (e.g. Lead niobium oxide);

All the above structures have a common attribute to the comportment of small dimension with an enormous charge (e.g. Tantalum, titanium etc.) in the oxygen octahedral, which are connected over angles arising due to an incessant chain of oxygen-metal-oxygen Such structural arrangements are favourable for the occurrence of ferroelectricity in metal-oxides. The addition of Lanthanum, Niobium, etc. to barium titanate oxide has produced the positive temperature coefficient of resistivity (PTCR) effect, which has resulted in its efficacy in thermistor formulation [8, 9]. The layered structure of the BLSF materials is shown in Fig. 1.

Piezoelectric transducers have a global market of approximately \$10 billion and growing yearly at a pace of 15%. Piezoelectric sensors are used in various industries:

- Automobile: Audible alarms, fuel vaporizers, air drift sensors, and seat belt indicators.
- Computer: Printers, temperature, and pressure sensors.

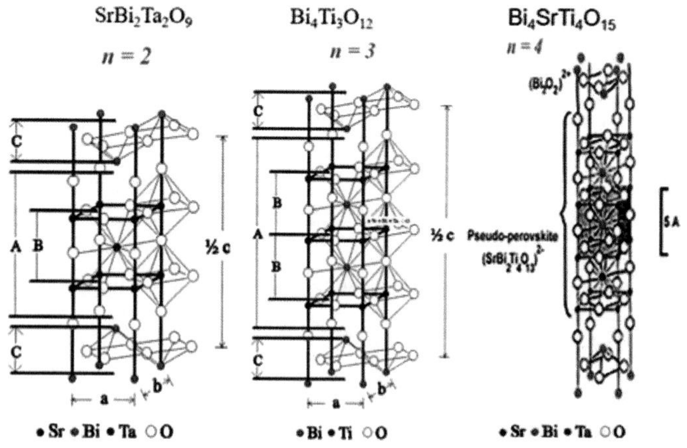

Fig. 1 Bismuth layered structured ferroelectric compounds (n indicates the number of levels)

- Medical diagnostics and Cure
- Energy production and storage.

For the successful implementation of eco-friendly piezoelectric materials better than PZT, it is necessary to understand the preparation technique, production, distribution, consumption, and disposing of the material. A substantial role is played by Life cycle assessment (LCA) in understanding the impact of the materials on the environment. LCA helps us in ascertaining the result of material production and developing strategies with the government authorities, which can be used to develop environmental regulations. In this work, we would like to compare BLSF materials properties with PZT and show that BLSF materials are eco-friendly compare to PZT.

2 Experimental

2.1 Fabrication Techniques (PZT or BLSF)

The piezoelectric materials are produced using a solid-state reactive (SSR) technique, and a detailed process is shown in Fig. 2.

Initially, the chemicals are taken in proper ratio as per the stoichiometry. The sample is mixed with water and subjected to a high energy mill for 18 h. As shown in the flow chart during the milling process, the triturate grains get ensnared between the balls and undergo buckle repeatedly. The slurry is taken in a glass crucible and placed on the magnetic stirrer for uniform distribution. The process is supported by heating the liquid at 800 °C to reduce the slurry. Calcination is a process to eliminate

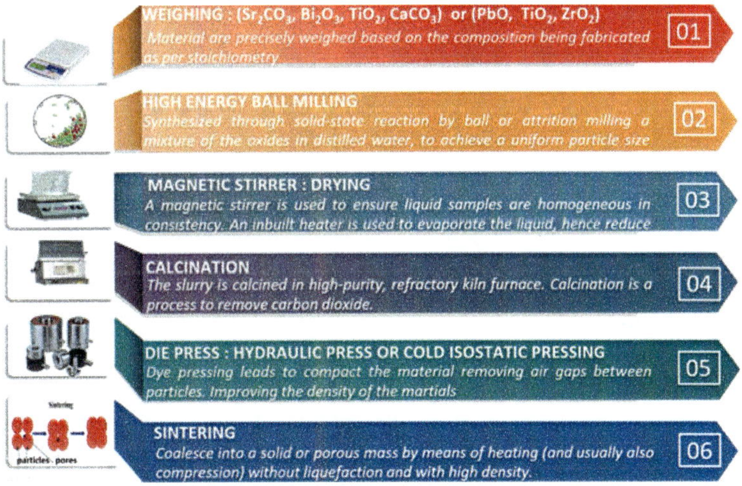

Fig. 2 Flow chart explaining the production of PZT or BLSF materials

CO_2 from the composition. Now at this stage, the residual is the final composition in powdered form. The powder is taken in a die of predefine shape and size to compress the pellets. The pellets are lastly sintered at elevated temperature to obtain high-density samples. The density of samples is measured using the water displacement technique developed by Archimedes as reported by Anand et al. [10].

3 Results and Discussion

3.1 Life Cycle Assessment (LCA)

LCA is built on the subsequent four crucial phases:

1. *Purpose and opportunity*: Aim is to define, and LCA systems limitations are fixed.
2. *Analysis*: Evaluation of the roaster in which inputs and outputs at every step of manufacturing are systematically collected and assimilated throughout the whole structure.
3. *Evaluation*: evaluation of the effects on the environment, itemizing LCA calculations leading to the same unit for relative study.
4. *Impact Assessment*: The elucidation of the LCI and compelling valuation of outcomes, hence ascertaining the environmental high impact zones [11–13].

LCA involves the various steps in the manufacturing, production, distribution of the inventions. It assesses the aggregate of supply chain contributions essential to manufacture a given operative unit (i.e. 1 kg of PZT/BLSF). Using lifecycle portfolios, the process LCA can be conveyed arithmetically as:

$$= \sum_{j=1}^{n} K_{p(j)} X E_{p(j)} \tag{1}$$

where: K_p is the contributions (j) into an inventions (i.e. PZT vs. BLSF piezoelectric materials) sequence of processes involved in the production and distribution etc.; n is the entire count of bustle aids (j) into the inventions allocation manacle, and Ep is the output concentration across several ecological and workable measurements (e.g. Greenhouse gases emissions, the terrestrial area used, etc.), for each contribution (j) into the inventions supply chain emanations.

Figure 3a emphasizes total energy utilization throughout the production accomplishments (i.e. Sintering, Calcination and pressing). PZT ingests supplementary electrical and thermal energy across the significant production deeds except in crushing where it utilizes circa identical to PZT. It is evident that the disproportion in energy utilization between BLSF and PZT exists; this is due to PZT has a higher specific heat capacity than BLSF materials. PZT devours more considerable thermal energy through the heat up successions tangled in production [14].

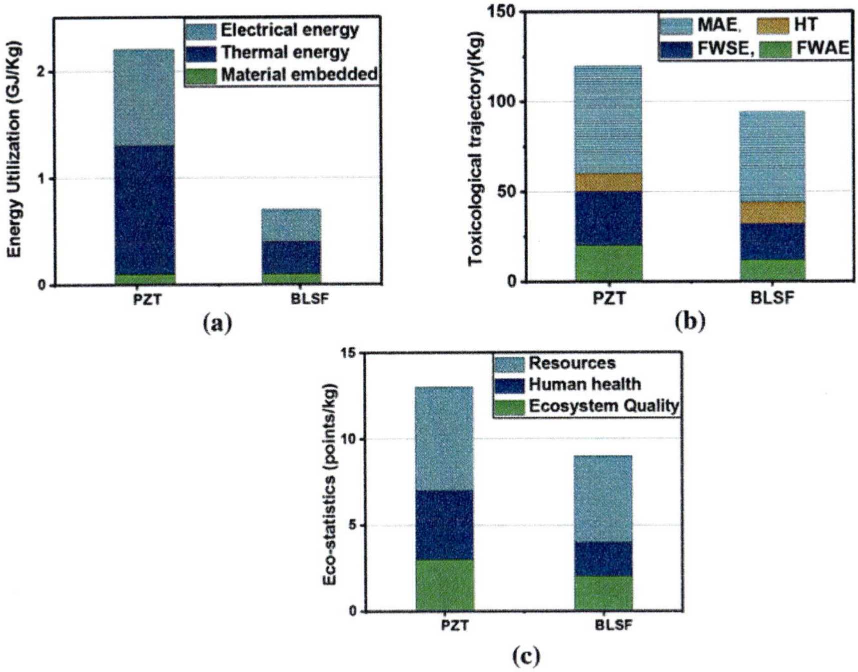

Fig. 3 Evaluation of BLSF versus PZT. **a** Total energy utilization, **b** toxicological trajectory, **c** eco-statistics

Figure 3b focuses on the toxicological trajectory. Our calculation of the toxicological path enmeshes Freshwater aquatic eco-toxicity (FWAE), marine residue eco-noxiousness (FWSE), human toxicity (HT), and sea aquatic eco-toxicity (MAE), it discloses frighteningly extraordinary toxicological trajectory of PZT when associated with BLSF. The main factor effecting is lead (Pb), which is ecologically hazardous. Figure 3c indicates the eco-statistics, which also confirms the same as determined in Fig. 3a, b.

3.2 Properties

The expected properties of piezoelectric materials are high dielectric constant, high transition temperature for sensor applications. BLSF materials show high transition temperature from 500 to 800 °C in comparison to PZT. The PZT materials have a high d_{33} than BLSF materials. Both materials possess higher remnant polarization and lower coercive field. Hence BLSF materials are a viable substitute for PZT ceramics.

4 Conclusion

A large scale comparative LCA, the ecologically sustainable capabilities of BLSF and PZT were studied over an eclectic array of statistics. The statistics reveal that BLSF is more environmentally friendly than materials as compare to PZT. Above all, BLSF materials have high transition temperature, which makes them a deserving candidate for high-temperature sensors applications.

References

1. Cucchiella F, D'Adamo I, Koh SL, Rosa P. Recycling of WEEEs: An economic assessment of present and future e-waste streams. Renewable and Sustainable Energy Reviews. 51. (2015) 263–72.
2. Lusiola T, Bortolani F, Zhang Q, Dorey R. Molten hydroxide synthesis as an alternative to molten salt synthesis for producing K0. 5Na0. 5NbO3 lead-free ceramics. Journal of Materials Science. 47. (2012) 193842.
3. Koruza J, Rozic B, Cordoyiannis G, Malič B, Kutnjak Z. The large electrocaloric effect in lead-free K0. 5Na0. 5NbO3-SrTiO3 ceramics. Applied Physics Letters. 106. (2015) 202905.
4. Saito Y, Takao H, Tani T, Nonoyama T, Takatori K, Homma T, et al. Lead-free piezoceramics. Nature. 432. (2004) 847.
5. Nour Mohammadi A, Bahrevar M, Schulze S, Hietschold M. Electrodeposition of lead zirconate titanate nanotubes. Journal of Materials Science. 43. (2008) 4753–9.
6. He C, Li X, Wang Z, Liu Y, Shen D, Li T, Growth of Pb (Fe 1/2 Nb 1/2) O 3–Pb (Yb 1/2 Nb 1/2) O 3–PbTiO 3 piezo-ferroelectric crystals for high power and high-temperature applications. Cryst Eng Comm. 14. (2012) 4407–13.
7. Reichmann K, Feteira A, Li M. Bismuth sodium titanate based materials for piezoelectric actuators. Materials. 8. (2015) 8467–95.
8. Anand G, James A., and Sarah P. Dielectric and Polarization Studies of Ca Substituted in Bi4SrTi4O15 Ceramics 116 (2010) 137–144.
9. Anand, G., James, A., Krishna, T. R., & Sarah, P. Synthesis and Characterization of Strontium Bismuth Titanate Ceramics via High-energy Mechanical Milling. Defence Science Journal, 57. (2007), 29–34.
10. Anand, G., Kuchhal P., & Sarah, P. AC and DC Conductivity Studies on Lead-Free Ceramics: Sr1−x Ca x Bi4Ti4O15 (x= 0, 0.2, 0.4, 0.6, 0.8). Particulate Science and Technology 33 (2015), 41–46
11. T. Ibn-Mohammed, S. Koh, I. Reaney, A. Acquaye, D. Wang, S. Taylor, and A. Genovese: Integrated hybrid life cycle assessment and supply chain environmental profile evaluations of lead-based (lead zirconate titanate) versus lead-free (potassium sodium niobate) piezoelectric ceramics. Energy Environ. Sci. 9. (2016) 3495.
12. Ibn-Mohammed T, Greenough R, Taylor S, Ozawa-Meida L, Acquaye A. Operational vs embodied emissions in buildings—A review of current trends. Energy and Buildings. 66. (2013) 232–45.
13. Acquaye AA, Wiedmann T, Feng K, Crawford RH, Barrett J, Kuylenstierna J, Identification of 'carbon hotspots' and quantification of GHG intensities in the biodiesel supply chain using hybrid LCA and structural path analysis. Environmental science & technology. 45. (2011) 2471–8.
14. Collado-Ruiz D, Ostad-Ahmad-Ghorabi H. Comparing LCA results out of competing products: developing reference ranges from a product family approach. Journal of Cleaner Production. 18. (2010) 35564.

Convalescent the Ecosystem by Enhancing the Life Cycle of Lubricants with Blended Nano Oxides

Harsh Gupta, Santosh Kumar Rai, Nippaani Satya Krishna, and Gagan Anand

Abstract Engine oil or lubricating oil is the lifeblood of any engine or machinery. Engine oil is basically a selected petroleum oil fraction that has been suitably refined and compounded with various types of additive agents. The primary assignment of lubricating oil is to lubricate the metal parts of the engine or machinery. Present paper stresses upon impacts of used oil on the environment, and improving life of used lubricating oils to minimize its adverse effects on surroundings. Worn lubricating oil is extremely hazardous contaminating outcome as well as these oils are costly enough and intention of preserving precious overseas trade has emphasized the awareness for re-refining used/worn oil. Used lubricating oils (ULOs) represent a serious problem for environment and human health due to the presence of highly harmful contaminants. Negligence in managing the waste/worn lubricating oil is a severe ecological concern. Nearly every sorts of used oil comprise of the prospective to be re-refined unharmed, preserving an expensive conventional or non-renewable source and simultaneously reducing environmental contamination. Unluckily, a large amount of waste/used oil is tackled inappropriately. Some of it is spilled openly into the sewers, which mixed with wastes of water and hence harmfully upsetting the industrial plants meant for water treatment. Some oil is discarded openly onto the land to destroy wild plants or is spilled over soiled roads or is thrown in desert area, where it can pollute surface water as well as ground water. Discarding worn oils into the bionetwork gives birth to serious ecological problems. Stringent laws and rules are being imposed all over the globe for the dumping of used petro-products

H. Gupta
Department of Mechanical Engineering, I.T.S. Engineering College, Greater Noida, Uttar Pradesh, India

S. K. Rai
Petrology and Geochemistry Group, Wadia Institute of Himalayan Geology, 33, GMS Road, Dehradun, Uttarakhand, India

H. Gupta · N. S. Krishna · G. Anand (✉)
SOES, University of Petroleum and Energy Studies, Bidholi, Via-Premnagar, Dehradun, Uttarakhand 248007, India
e-mail: ganand@ddn.upes.ac.in

N. A. Siddiqui et al. (eds.), *Advances in Environment Engineering and Management*, Springer Proceedings in Earth and Environmental Sciences,
https://doi.org/10.1007/978-3-030-79065-3_11

and all the authentic attempts should be done for its recycling, re-refining and re-use. In majority of cases, used/waste oil can be used again post treatment with or without additives ensuing in large saving and preservation of expensive oil. Hence renewal, recovery or recycling of exhausted lube oils has developed into a significant processing plant which is exercising different methodologies for purification of used oil like wiped thin film evaporation technique.

Keywords Used oil · Environment · Lubricating oil · Re-refining

1 Introduction

Lubricating or engine oils are the liquids having required viscosity and they are used to lubricate rotating or sliding parts of an engine or a machine, whereas used lubricating oil is a petroleum derived oil or synthetic oil which left after usage in the lubrication, cutting operations, etc. Lubricating oils (LOs) are conventionally obtained from crude oil, so that, chemical composition of LOs consists on average of about 80–90% base oil and about 10–20% chemical additives and other compounds [1]. During operation time, LOs deteriorate, as well as their additives, and their physical and chemical properties become unsuitable for further use [2]. Increasing temperature deteriorate the lubricating oils, hence it leads to declining in various properties like; viscosity, density, specific gravity etc. Through petroleum refining processes, fuels and lube base oils are produced. There are several types of lubricating oils as; engine oil, gear oil, turbine oil, grease, cutting oil, heat transfer oil, hydraulic oil, industrial transmission oil, heat treatment oil, electrical oil, rolling oil, motor oil etc. Used or exhausted lube oil is actually toxic and useless. A single drop of used oil has the capability to pollute thousands gallon of normal water. Problems associated with used oils are; slow to degrade, attaches to metallic surfaces and due to this oil contamination, the waterways are polluted and finally it becomes the one of the major sources of pollution of drinking water [3]. Used lubricating oil is an extremely hazardous pollutant. It consists of PAH (Polynuclear Aromatic Hydrocarbons) and also heavy metals in large amounts. These heavy metals include Zn, Cr, Pb, Cu, Cd and Ni, and these metals are having extreme toxicity which is injurious to human, wildlife and ecosystem. Pollution through lubricating oil can spoil the earth (soil) as well as the aqueous environment, and it also affect the atmosphere by means of burning of waste oil. That's why it is essential to think about worn lubricating oil contamination seriously. Unfortunately, the required attention has not given to the contamination/ pollution created due to used lubricating oil. Waste/used oil can be refined again to produce the virgin oil or base oil which is processed into fuels and can be utilized as raw materials for making various petroleum products. To prevent polluting and contaminating the environment, used oil must be handled wisely by local municipal authorities or auto service or repair shops.

There are several reasons to re-refine used oil and re-use it:

Fig. 1 Generation and composition of used oil [3]

- Recycling used oil and making it fit for re-use prevents the environmental pollution because waste or used oil pollutes soil and water.
- Used oil does not wear out, so it is very precious resource. Used oil only becomes dirty and contaminated while its usage, so it's recycling preserves an expensive resource.
- Producing re-refined base oil consumes quite lesser energy and resources than that of producing virgin base oil directly from crude oil which requires more energy and resources.
- The re-refined oil is as good as virgin base oil.
- One gallon (4.5 L) of used oil produces the same 2.5 quarts (2.85 L) of new lubricating oil as produced by 42 gallons (189 L) of crude oil.

2 Generation and Composition of Used Lubricating Oil

It is observed from Fig. 1 that fresh lubricating oil consists of 80–90% base lube oil and 10–20% additives. When this oil has been used in a vehicle for a particular time or stipulated mileage covering, we obtain used lubricating oil. This used oil consists of 70–75% base lube oil, 5–10% fuel, 10–20% water, 10–20% used additives and 1% sludge. The important thing to notice here is that this used oil still have good amount of lube oil and so it has the potential to re-refine to become reusable again and protect the environment.

Various major contaminants or impurities in waste oil are antimony trichloro trifluoro ethane, aluminium dichloro difluoro methane, benzene, toluene, arsenic, 1-1-1 trichloro ethane xylene, barium trichloro ethylene, chromium polychlorinated biphenyls other PAHs, nickel, phosphorus, cobalt, copper. nitrogen, lead, magnesium, manganese, mercury, silicon, sulphur and zinc [4, 5].

3 Environmental Impacts of Used Oil

How does used lubricating oil harm our environment?

When used lube oil is exhausted into environment from hydraulic machines, engine, turbine, gearbox, etc.:

- This oil consists of wear debris contamination.
- This used oil has now deteriorated and it has degraded due to acids content.

- Other chemical species are obtained due to the decomposition of additives mixed in oil.
- Different solvents, degreasers and fluids from processes have now mixed with used oil.

3.1 Spilling of Used/Worn Oil on Land/Soil

There are four diverse behaviors, used oil can be isolated into the earth or soil: loss of oil during operation of vehicle/machinery; for dust control on the rural road; asphalt which contains used engine oil used for asphalting and last but not the least it is spilled openly in a landfill.

One-third of lube oil come in practice is vanished when in use; a few of it is vanished on asphalt road surfaces, on the surfaces of streets, highways and in the parking. This spilled oil leftovers on the various earth surfaces as long as rain water or cleaning service network of municipal board clean off this oil and then in urban areas it is collected through a pipe in a container mounted on a truck, or on the outer roads it is rejected to nearby land or soil.

Used oil spilling on the soil/land creates some significant alterations in micro-organism community participating in soil's nitrogen cycle, and also in metabolic action of aerobic micro-organisms having ability to oxidise hydrocarbons. Many pores between particles of soil are filled by petroleum products and due to this oxygen access is hampered. A considerable number of micro-zones come up, which have an oxygen deficiency, in the soil aggregates. Other factor of promoting the development of anaerobic micro-zones here in this polluted soil/land is nothing but the growth of aerobic hydrocarbon oxidizing micro-organisms. Various nitrogen fixing, de-nitrifying and ammonifying micro-organisms present in the polluted earth are higher as compared to non-contaminated earth [6].

Majorly slowing down occurs in mineralisation of carbon, transformations of nitrogen, chief production, and mineralisation of phosphorus and sulphur due to an availability of heavy metals into the exhausted oil [7]. In used oil, biodegradation of hydrocarbon can be diminished due to the presence of lead, as it has been revealed to take place while lead is mixed to contaminated/polluted soil along with petrol [8]. For one year after the spillage, used oil alongwith its lead content stay upto 20 cm depth of soil, means it stays upto this depth where most of the microbial activity occurs [9].

3.2 Spilling of Used Oil in the Aqueous Environment

As mentioned earlier, used oil is usually spilled over earth or soil but it can be moved quickly to aqueous environment by running and rain waters. This kind of movement is significant in the streets and roads of urban areas close to canals, rivers, lakes and seas. The waste water in urban area incoming at a water treatment plant, the hydrocarbons present in it are various organic matters which are mainly opposing to the bio-degradation process, in contrast to another organic matter [10]. The most important outcome of the existence of hydrocarbon in aqueous environment is undoubtedly an alteration, like we observe in land/soil contamination, in the compositions of micro-organisms communities [11]. Though this outcome can be a rise in the number of entire microorganisms able to biodegrade hydrocarbons [12], typically these microorganisms are firmly inhibited. Various species of microbes in sea and fresh water environment have been examined with respect to toxicity of several petroleum products [13–15].

3.3 *Used Oil Burning*

Mostly used/worn oil is working like a supporting petro-fuel in unique stoves/burning devices made in support of such kind of worn oil used for gas cleaning systems. This kind of burning of used oil is carried out in insufficient installations. This burning emits huge amount of heavy metals. Particles of emission and residue yields are the most toxic in compare to other particles of environment. Particulate matters obtained through burning worn/used oil can magnify and becomes a source of respiratory troubles, and can consequence in the failure of lung functionality and failing in ability to resist the infection. Serious respiratory problems can happen due to sulphur dioxide and nitrogen dioxide produced by burning used oil.

4 Methods of Improving Life of Lubricating Oils

The researchers examined various effects of unprocessed used oil as a fuel to distillation and re-refining [16]. In this comparison, the prospective of toxicity of unprocessed used oil was about 150 times larger for earthly eco-systems, and five times larger for human being. Zinc and cadmium exhibited the supreme warning to eco-system; human health effects are also influenced by lead and chromium.

The results encourage the substitute 'used-oil' managing strategies, by giving incentives for treatment of used oil and by supporting markets for processed oil products, as per Boughton, the best choice among all choices—use less oil primarily [16].

- Oil filters better designing and changing of oil less frequently, are the two methods to trim down the health and environmental challenges of used engine/motor oil.
- One way to achieve this would be enhancing oil changing intervals from the national average 7000 km to more than 14,000 km or from a period of 6 months to 12 months. The most important point is that engine oil quality is improving continuously, but oil filters have remained same basically for the last 50 years.

Mostly the filters are prepared by element of paper that becomes choked after 7000 km, hence engine oil required to be refilled as it has become contaminated and dirty. However filters having high effectiveness are accessible now to common public. If the automotive manufacturing companies install these effective filters on new cars and made mandatory these filters at every oil change, the volume of used oil could be decreased by half at least from the automobile sector.

4.1 Re-refining/Recycling Treatment Process of Used Oil

Re-refining Process Flow

Used lube oils to light base oil having SN-150 grade and heavy base oil having SN-500 grade through distillation and wiped film evaporator for refining process contain four processes as mentioned below (Figs. 2 and 3):

1. Prior treatment—Ash removal and demetallization
2. Distillation—Separating the light ends
3. Thin wiped film evaporation—Separating the oil fractions
4. Solvent/hydrofining treatment option—finishing state.

1. **Segregation of the light ends through column tower**

Priory treated used lube oil (i.e. dehydrated and filtered) came from store is pumped by feed pump to the distillation column (Fig. 4) passing through filters. To remove any solids from feed, filters are installed additionally. Light fractions from used oil are segregated by distillation column which is worked under vacuum. Diesel and water is received as the products on top (which is condensed), in distillation column and by gravity both are collected in receivers. From reboiler the bottom product is supplied to a thin wiped film evaporation system so that further segregation takes place through WFE-I feed pump. For condensation of top vapors from column, cooled water is used in condenser and as a heating medium, the thermo fluid (hot oil) is used in reboiler shell.

Stage I: Dehydration and Defueling:

2. **Segregation of oil fractions through thin wiped film evaporators**

Thin wiped film evaporation system comprises of two thin wiped film evaporators, WFE-1 and WFE-2 to obtain light base oil (SN-150) and heavy base oil (SN-500) grades respectively. These evaporators consist of a vertical cylindrical surface which is enclosed into a heating jacket having an internal rotor. Rotor divides the feed in the shape of a layer of thin film onto the heated wall by using a wiper blade. To condense the vapors produced by the hot surface, internal condenser is installed. WFE-1 (Fig. 5) is worked at a pressure of around 1–2 torr vacuum and a heating temperature of 250–280 °C by thermo fluid contained in jacket. From the column bottom, the feed free from diesel is fed to WFE-1, where the light base oil SN-150 is produced as a product and it is collected in receivers and its residue is collected

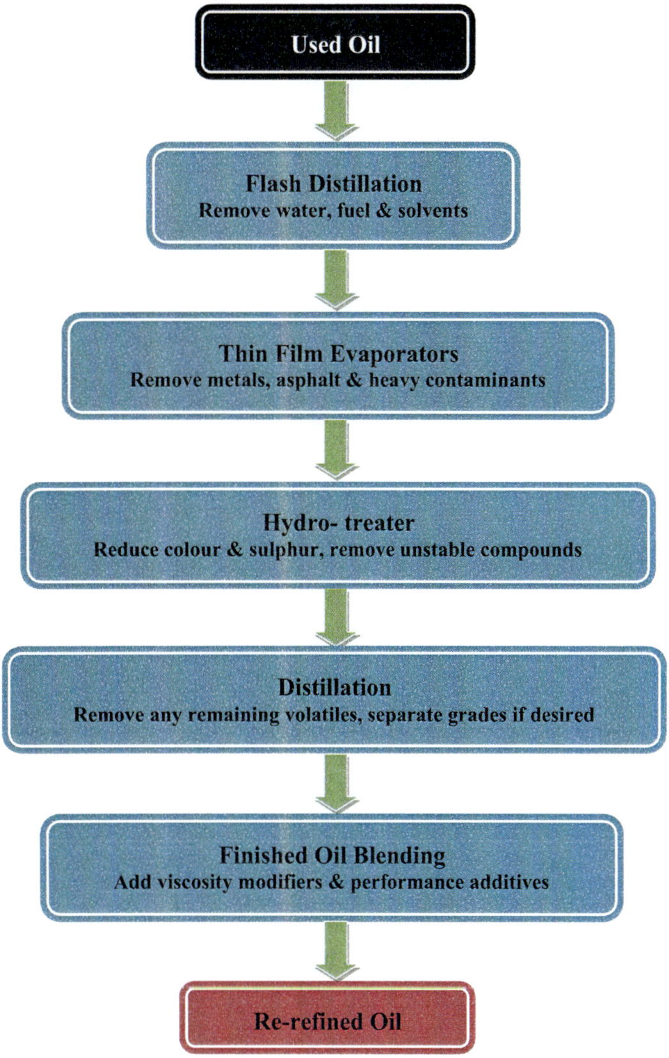

Fig. 2 Re-refining process of used oil

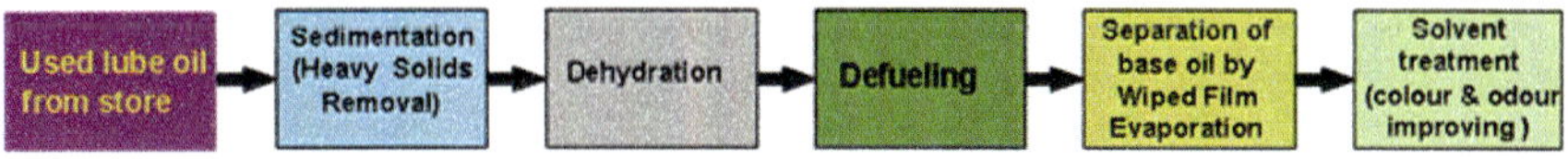

Fig. 3 Re-refining process flow

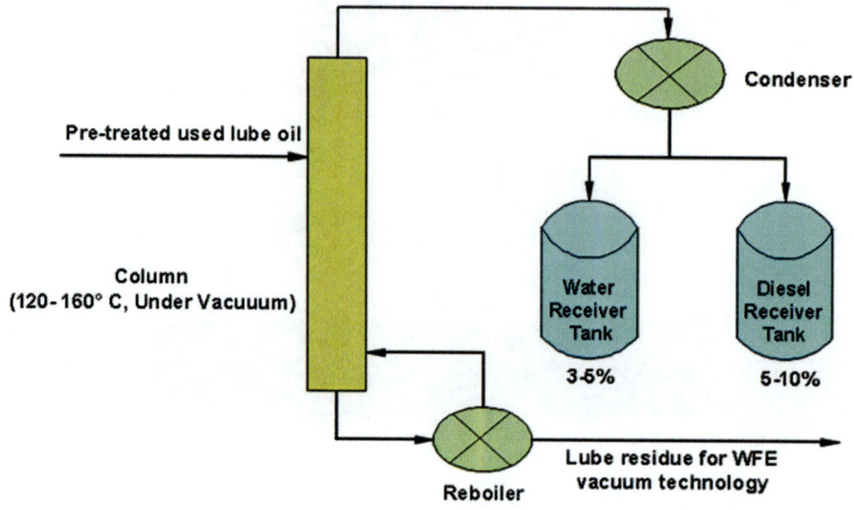

Fig. 4 Dehydration and defueling of used oil

in residue receivers through gravity. Through pump, light base oil SN-150 is sent to main storage tank. The residue obtained from WFE-1 is then pumped and fed to the WFE-2 through rotary gear pump. In WFE-2 (Fig. 6), heavy base oil SN-500 is produced as a product and it is collected in product receiver tank and its residue is collected in residue tank through gravity. WFE-2 is worked at the pressure of 0.1 torr vacuum and a heating temperature of 315 °C by a thermo fluid contained in jacket.

Stage II: Light base oil (SN-150) segregation through thin wiped film evaporator (WFE-1).

Stage III: Heavy base oil (SN-500) segregation through thin wiped film evaporator (WFE-2).

3. **Solvent/hydrofining treatment option—finishing state**

The final filtration process occurs by the mixture of solvents in this stage and colour and smell of base oil is improved. The solvent utilized are reused for 300 cycles and hence the cost is cheaper than regular solvents. If we want the best quality of base oil group 2+ oil, we have to exercise a hydro-fining (finishing) process and for this a precise equipment is used for filtration process which comes with a renew process.

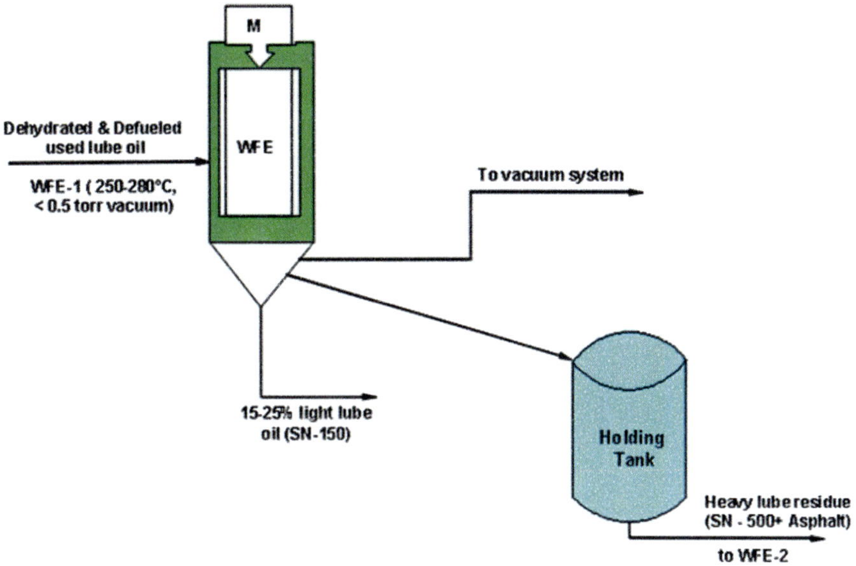

Fig. 5 SN-150 light base oil separation by thin wiped film evaporator-I

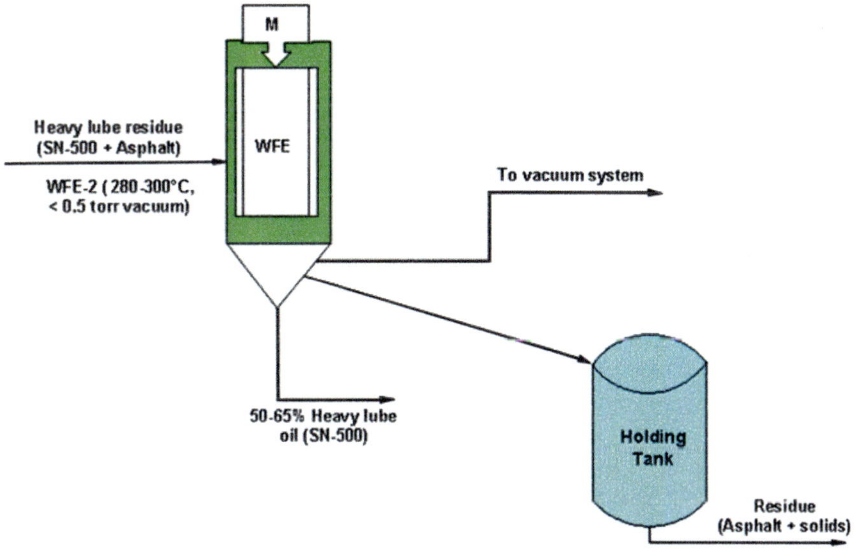

Fig. 6 SN-500 heavy base oil separation by thin wiped film evaporator-II

5 Conclusion

Used or waste oil is useless, toxic and polluting, its polluting nature is reduced when we re-refine it. The worth of recycled and re-refined oil is as good as virgin lube oil. The repeated use of these used oils are enabled through re-refining of lubricating oil, hence enhancing the buy and sell balance through domestic oil supplementing and thus strengthening the native oil. Hence, the most significant course to preserve lubricating oils is that through recycling/reprocessing. Recycling is a general word for processing meant for lubricating oils to recover valuable substance through re-refining and recovery. Oil filters better designing and changing of oil less often by improving the quality of various oils, are the two ways to downsize the health and environmental challenges of used engine/motor oil. Used oil must be wisely discarded on account of its more injurious impacts than utilities, when untreated. Fortunately used oil can come into use again by re-refining, recycling, and hence its re-use is possible nowadays, one of the best method available for this is thin wiped film evaporation technique.

References

1. Rincón, J., Cañizares, P., García, M.T., Gracia, I., Regeneration of used lubricant oil by propane extraction. Ind. Eng. Chem. Res. 42 (2003) 4867–4873.
2. Tsai, W.T., An analysis of used lubricant recycling, energy utilization and its environmental benefit in Taiwan. Energy 36, (2011) 4333–4339.
3. Arslan Enginery Pvt. Ltd, Waste and used oil recycling plant, arslanenginery.com/used-oil-recycling/
4. Rec 93 Proceeding (volume III); International Recycling Congress, Geneva, Switzerland, (1993) 19–23.
5. U.K. The Hazardous Waste Inspectorate, Third Report-London (1988).
6. Ismailov, N.M., Effect of oil population on the nitrogen cycle in soil, Mikrobiologiya, 52 (1983)1003–1007.
7. Babich, H. and Stotzky, G., Heavy metal toxicity to microbe-mediated ecologic processes; a review and potential application to regulatory policies, Environ. Res. 36 (1985) 11–137.
8. Jensen, V., Effects of lead on the biodegradation of hydrocarbons in soil, Oikos, 28 (1977) 220–224.
9. Raymond , R.L., Hudson, J.O. and Jamison , V.W., Oil degradation in soil, Appl. Environ. Microbiol., 31 (1976) 522–535.
10. Farrington , J. and Quinn , J., Petroleum hydrocarbon and fatty acids in wastewater effluents, J. Water Pollut. Control Fed., 45 (1973) 704–712.
11. Atlas, R.M., Microbial degradation of petroleum hydrocarbons; an environmental perspective, Microbiol. Rev., 45 (1981) 180–209.
12. Horowitz, A. and Atlas, R.M., Response of microorganism to an accidental spillage in arctic freshwater ecosystem, Appl. Environ. Microbiol., 33 (1977) 1252–1258.
13. Griffin, L.F. and Calder, J.A., Toxic effect of water-soluble fractions of crude, refined and weathered oils on the growth of marine bacterium, Appl. Environ. Microbiol., 33 (1977) 1092–1096.
14. Hsieh, Y.P., Tomson, M.B. and Ward, C.H., Toxicity of water-soluble extracts of No. 2 fuel oil to the freshwater alga *Selenastrum capricornutum* Dev. Ind. Microbiol., 21 (1980) 401–409.

15. London, S.A., Mantel, C.R. and Robinson, J.D., Microbial growth effect of petroleum and shale-derived fuels, Bull. Environ. Contam. Toxicol., 32 (1984) 602–12.
16. Waste Motor Oil Management May Pose Threat To Health And The Environment, Environmental Science & Technology, a peer-reviewed journal of American Chemical Society. ScienceDaily (15 January 2004).

Traditional and Modern Agricultural Implements Used in Agro Sciences – A Case Study in the Part of National Capital Region of Uttar Pradesh, India

Ganesh Datt Bhatt, Mahesh Singh, Archna Bharti, and Deepali Rana

Abstract The research work was undertake to discover a variety of conventional and modern crops growing; implements used for agriculture operations by the rural farmers of Dankaur and adjoining villages of National Capital Regions (NCR), a part of Gautam Buddha Nagar, Uttar Pradesh, India. The traditional and modern agriculture implements are very economical with respect to man power, finance and economy as well as saving of valuable time. The traditional and modern agriculture implements are prepared of the vicinity accessible resources which are: iron, firewood along with its substitute products. These agriculture implements are easily too operate without any special skills and training. The information related to agricultural implements was documented through the interactions with local villagers, personal observations and existing data are also used. The detailed information about each agriculture implements was collected and informative notes were taken during interview with the rural farmers. The information were collected during the field work from July 2018 to March 2020 and found approximately fifty three (53) traditional and modern agriculture implements were identified and briefly explained with scratched photographs prepared by a research team of School of Agriculture. These agriculture implements are: bill hook, hand trowel, secateurs, sickles, girdle, pruning knife, shovel, spud, knap-sack sprayer, pick axe, garden hoe, weeder etc. The traditional economy of farmers is for subsistence only not to profit. The farmers are characterizing by undersized and patchy land property, small yield of crop and domestic animals, disguised joblessness, reduced profits and short hazard manner skill.

Keywords Agriculture · Implements · Traditional and modern implement · GB nagar · NCR

G. D. Bhatt (✉) · M. Singh · A. Bharti
School of Agriculture, Galgotias University, Greater Noida, Uttar Pradesh 203201, India

D. Rana
Department of Zoology, Dolphin (P.G.) Institute of Biomedical and Natural Sciences, Manduwala, Dehradun, Uttarakhand 248007, India

N. A. Siddiqui et al. (eds.), *Advances in Environment Engineering and Management*, Springer Proceedings in Earth and Environmental Sciences,
https://doi.org/10.1007/978-3-030-79065-3_12

1 Introduction

Agriculture is the main spinal column of Indian financial system [1]. In India, agriculture is the life line of the financial system, which provides the foundation for foodstuff and source of revenue safety and sustain for the financial enlargement and community conversion of the Nation [2]. Cultivation in India is sole within its uniqueness, where more than 250 diverse crops be cultivate within its wide-ranging agro-climatic region, disparate 25–30 crops grown inside numerous of the urbanized Nation of the Globe [3]. India with its constructive agro-climatic state of associations and prosperous natural reserve bottom has grown to be the Globe main manufacturer diagonally a variety of supplies [4]. India is the principal creator of coconut, mango, banana, milk and dairy foodstuffs, cashew nuts, pulses, ginger, turmeric and black pepper and so on. It be too the second major creator of rice, wheat, sugar, cotton, fruits and vegetables and so forth [5]. The growth of farming computerization has been strongly connected through in general progress in invention crop growing [6]. The farm in 1950 incredibly hardly any farmers overcome main movers similar to tractors, engines and motors and the like, the weighty farming tractors and equipment be import by administration organization largely for soil recovery and growth of huge administration farm [7, 8].

The image is changed speedily throughout the early on sixties with the foreword of far above the ground yielding variety of wheat and additional crop growing which desirable irrigation conveniences [9]. The going up creation of food stuff grain consequential as of the extend part below high yielding variety could not be handle inside the regular harvest and threshing period [10]. The farmers in North India suffer profound fatalities as a consequence of spoil to harvest wheat throughout the behind schedule sixties and early on seventies since the threshing of greater than before wheat assembly could not be accomplished prior to the inception of pre-monsoon rain [11, 12]. Out sized extent taking up of threshers operates by electric motors, engines and tractors that follow in near the commencement seventies beyond was an effect to do with to supreme threshing process rapidly. Agro-processing includes farm-level dispensation to develop superiority of construct and knowledge for thrashing preclusion in storage space, usage and bring [13, 14]. Most important equipment which have been developed and adapted for farm level dealing out take account of cleaners, graders, dryers, shellers decorticators and store room structure mill implement [15]. The key rationale of this study is to gather reliable in relation to the traditional and modern agriculture implements which is extensively used by local peoples in and around of Dankaur villages of NCR of Uttar Pradesh, India.

2 Material and Methods

2.1 *Study Area*

The study area in the Dankaur village and adjoin villages is situated in Gautam Buddha Nagar, Tehsil of Gautam Buddha Nagar district of Uttar Pradesh, India. The Dankaur village is situated on 28.35° N 77.55° E. It has an average elevation of 194 m. It is positioned around 55 km East of Delhi alongside the bank of river Yamuna. As per [16], the Dankaur had an inhabitant of 13,520; males represent 54% of the inhabitants and females 46%. The Dankaur villages' have standard literacy rate of 57%, lower than the countrywide average of 59.5%: male literacy is 67% and, female literacy is 47% correspondingly.

2.2 *Climate*

The study area has vigorous climatic conditions. It's very nearby to New Delhi; so the temperature is parallel to that of New Delhi. In the summer, the climate is quite scorching and similarly the winter is also quite frosty. The rains are an average as it's located in the semi arid region and part of Upper Indo Gangetic plains of Uttar Pradesh. The monsoon start in the study area is mid June to mid September. The winter season in the study area starts from October to February and in the summer from March to June correspondingly. Approximately, 90% of the annual rainfall occurs in monsoon season, variation of rainfall from one to another [17].

2.3 *Flora and Fauna*

The study area has sub-tropical deciduous forest and it does not have expansion of natural forests. The fragmented forest is maintained by Uttar Pradesh Forest Department through plantation and management practices [18]. The dominant trees species are: Shishum (*Dalbergia sissoo* DC.), Mango (*Mangifera indica* L.), Jamun (*Syzygium cumini* (L.) Skeels), Imli (*Tamarindus indica* L.) and Babool (*Vachellia nilotica* (L.) P.J.H. Hurter & Mabb.), *Azadirachta indica* A. Juss., *Butea monosperma* (Lam.) Taub., *Cordia dichotoma* G. Forst., *Ficus religiosa* L., *Holoptelea integrifolia* Planch., *Mitragyna parvifolia* (Roxb.) Korth., *Neolamarckia cadamba* (Roxb.) Bosser., *Prosopis cineraria* (L.) Druce, *P. juliflora* (Sw.) DC., *Phoenix sylvestris* (L.) Roxb., *Sesbania sesban* (L.) Merr., *Syzygium cumini* (L.) Skeels, *Terminalia arjuna* (Roxb. ex DC.) etc. The shrubs and bushes primarily found in the study area are: ber (*Ziziphus jujuba* Mill.), madar (*Calotropis procera* (Aiton) Dryand.), (*Calotropis gigantea* (L.) Dryand.), *Ricinus communis* L., and *Tamarix ramosissima* Ledeb. etc.

The study area is rich in herbal diversity which includes economically and medicinally important plant species which are: *Amaranthus viridis* L., *Achyranthes aspera* L., *Cannabis sativa* L., *Chenopodium murale* L., *Citrullus colocynthis* (L.) Schrad., *Convolvulus prostrates* Forssk., *Croton bonplandianus* Baill., *Digera muricata* (L.) Mart., Mosyakin and Clemants, *Eclipta prostrate* (L.) L., *Heliotropium ellipticum* Ledeb., *Oxystelma esculentum* (L.f.) Sm., *Phyllanthus niruri* L., *P. maderaspatensis* L., *Phyla nodiflora* (L.) Greene, *Physalis peruviana* L., *Stellaria media* (L.)Vill., *Tribulus terrestris* L., and *Withania somnifera* (L.) Dunal etc., [18]. The study area is also rich in faunal diversity and a few fauna are: Wolves (*Canis lupus*), Monkey (*Simiiformes catarrhini*), Nilgai (*Boselaphus tragocamelus*), Indian Grey Mongoose (*Herpestes edwardsii*), Indian Hare (*Lepus nigricollis*), black buck deer (*Antilope cervicapra*), Leopard (*Panthera pardus*) and Golden Jackal (*Canis aureus*), Wild boar (*Sus scrofa*) etc., are found. The common birds are also found in the study area which are: Peacock (*Pavo cristatus*), owl (*Strigiformes*) Partridge (*Perdix perdix*), Snipe (*Gallinago gallinago*), The Lesser Florican (*Sypheotides indicus*) and Sarus cranes (*Grus antigone*) etc. The trees accessible in the study area are providing firewood and wood for furniture. The ornamental trees are planted in the road side as part of social forestry under Uttar Pradesh State Forest Department scheme.

2.4 Soil

On the banks of the river which is mostly low land, different types of soil are found. In the Khadar, the soil composition is sandy, clay and alkaline. The chief soil of the district is rich loam. A distinctive feature of this soil is that it dries white or a very light gray and becomes dark of colour when manure and moisture.

2.5 Agriculture and Crop Pattern

Agriculture is acting as a major role in the study area economy, growth and development. The crop growing system in the district is well developed. The scientific methods of agriculture cultivation and availability of elevated docile diversity of seed, knowledge about balanced utilize of fertilizers and the rest, are the main contribution factors which have enhanced the productivity of different crops. The Rabi and Kharif crops are two main harvests in the district. Wheat and Gram in Rabi and Maize and Paddy in Kharif are main crops of the district. The enhanced undeveloped practice, such as use of far above the ground yielding diversity of seed, fertilizers, superior farming equipment, plant safety trial have exposed expectant fallout and gross and net produce of many crops have improved. Out of the total gross area sown the main crops of the district are: wheat and sugarcane followed by paddy, bajra and maize. The District Agriculture Department looks after the problems of the farmers. They provide training on improved seeds, modern method of cultivation and modern technologies as well. The major agricultural crops of the study area are: wheat (*Triticum aestivum* L.), maize (*Zea mays* L.), barley (*Hordeum vulgare* L.), rice (*Oryza sativa* L.), pigeon pea (*Cajanus cajan* (L.) Millsp.), pearl millet (*Pennisetum glaucum* (L.) R.Br.), potato (*Solanum tuberosum* L.) and sugarcane (*Saccharum officinarum* L.) etc.

3 Results and Discussion

The universally used traditional and modern agricultural implements in the various small and large scale farmers dominating villages are illustrated in the study, which includes english and local name, description and their usages. A qualitative research approach was adopted in this study. In India, the poor farmers have been using a variety of traditional and modern agriculture implements for agricultural practices, which are identified and discussed. In the northern India the traditional and modern agricultural implements are used by small scaled farmers. Generally, small scale farmers and farm owner of spread land are not capable to use tractors and in that casing ox-ploughs are extremely favored. The traditional and modern agriculture implement is made up for reducing human efforts. The farmers are using implements, equipment and machinery to improve farms production and reducing their efforts as well. In this research paper we have briefly explained the use of traditional and modern agriculture implements which is frequently used by small and large scaled farmers in and around the Dankaur village area. The brief descriptions of agriculture implements are below as:

1. **Bill Hook** (**बिल्हुक**)–Bill hook is a conventional hurtful implement which is used commonly in crop growing and forestry for cutting slighter wooded bits and pieces i.e., bushes and twigs. The School of Agriculture has two models of bill hook namely FBH-705 and FBH-706 (Fig. 1) [Plate I].
2. **Hand Trowel** (**खुरपी**)–Hand Trowel is recognized as garden trowel. It is an implement with a sharp, scoop-shaped, made up of metal sharp edge and wood. It has metal with synthetic handle. It is second-hand for contravention up earth, dig small hole, particularly for plant and weed, mixing in manure, other additives, and transfer flora to pot. The School of Agriculture has four models of bill hook viz., FSS-4002, FWT-2003, FWT-204 and FPKM-200 (Fig. 2) [Plate I].
3. **Hand Pruners** (**दस्ती कैंची**)–It is also call hand pruners, secateurs and it is a type of scissors for utilize on flora. It is physically powerful to prune rigid brushwood of trees and shrubs; from time to time; it's equal to two centimeters bulky. It is use in garden, dendrology, plant garden center, agriculture, flower arrange, and scenery management, wherever fine-scale habitation administration is essential (Fig. 3a, b) [Plate I].

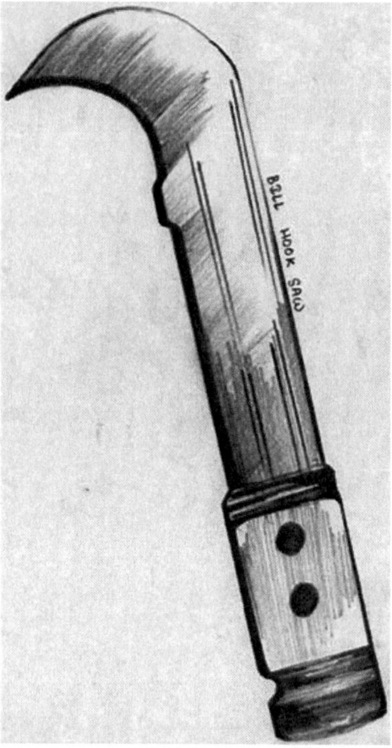

Fig. 1 Bill hook

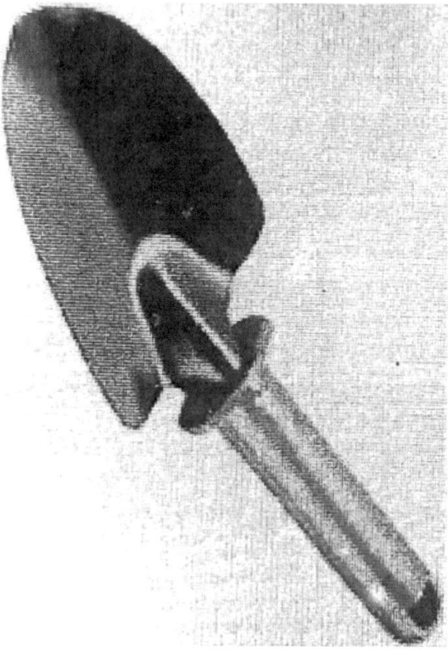

Fig. 2 Hand trowel

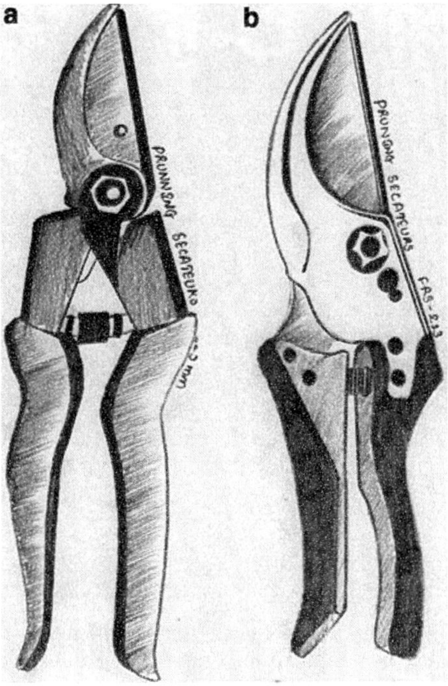

Fig. 3 **a**, **b** Hand pruners

4. **Weeding Trowel** (खरपतवार करणी)–Weeding trowel is an implement through a sharp, scoop-shaped metal blade and wood, metal and plastic lever. It is used for breach up ground, dig undersized hole, particularly for plant and weed, integration in manure with additional additives, and transfer plants to pot. It is second-hand to take away weed from crops sticks harvest them so that they can be used to make bio chaff. The School of Agriculture has two models of weeding trowel FW-900 and FW-9000 (Fig. 4) [Plate I].
5. **Sickle** (हँसिया) A sickle is bag hook is also known as reaping-hook. It is a hand-held farming implement which is premeditated with various bent blade and characteristically use for harvest, reap, granule crop, cutting succulent forage primarily for feed livestock, moreover recently cut and dried up as feed (Fig. 5a–c) [Plate I].

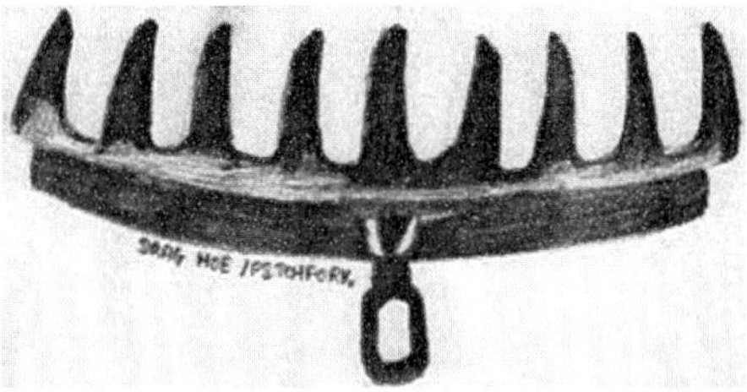

Fig. 4 Weeding trowel

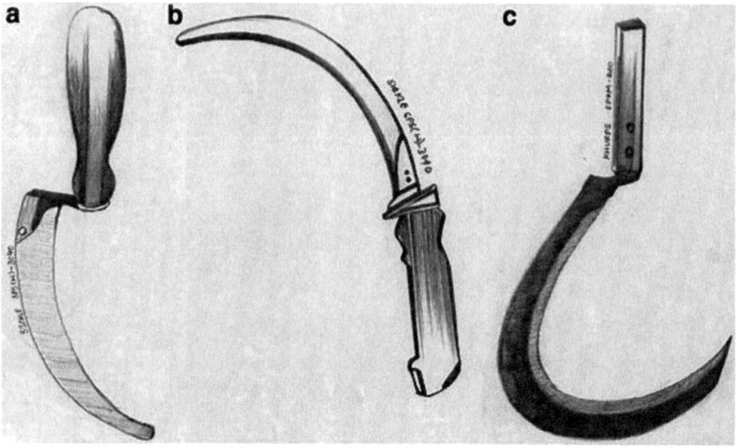

Fig. 5 **a–c** Sickles

6. **Girdle Knife** (करधनी)–Its use as a technique to strength a fruit-bearing plant to allow better outgrowth. A planter would place a girdle-bark removal at its base of a bulky bough at the stalk. In the grapes girdling and cincturing; its use to make the grapes large and sweeter on the grape canopy. The School of Agriculture has two models of Girdle SPG-98 and SPG-99 (Fig. 6) [Plate I].
7. **Hand Sickle** (दरांती)–A sickle is an arched, hand-held agriculture implement which is normally use for harvest puffed rice crop and spiteful grass for hay. Inside of the arc, it has acerbic edge and it's notched. The farm-hand swings the razor blade next to the base of the crop cutting through the stem with a saw

Fig. 6 Girdle knife

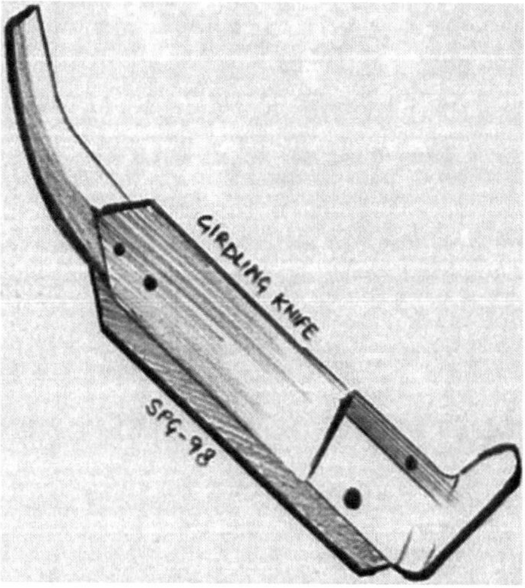

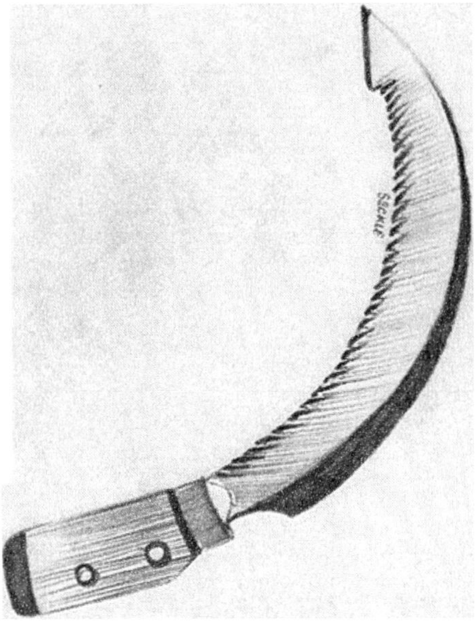

Fig. 7 Hand sickle

action. The School of Agriculture has one model of hand sickle SPS (W)-3040 (Fig. 7) [Plate I].

8. **Prunning Knife** (पोधो को छॉटने की छुरी)–Make use of pruning knife is to slice over-turn pointless growth in a irregular plant which is set pressure on the shoot and unkind by means of a pointed slice progress absent beginning the stiff. It keeps the plant vigorous by acerbic stem at least ½ in. absent from the stem from which they are upward. The School of Agriculture has model of punning knife FPTS-001 (Fig. 8) [Plate I].
9. **Shovel** (बेलचा)–It is use for soil integration, soil compost level of soil exterior for little area, level soil in pot. It can be use also for heavy pot with pot mix, insertion manure dispersal manure and many other uses. The School of Agriculture has three models of Shovel; FPKM-100, FPKM-200 and FPKM-400 (Fig. 9) [Plate I].
10. **Kinnow Cutter** (किन्नो कटर)–It is use for plug for kinno fruits. The fruit is mature in January and February month. Their peel has high juice content. The School of Agriculture has model of kinnow cutter PKC-91 (Fig. 10) [Plate II].
11. **Grafting Knife** (कलम बांधने का चाकू)–Grafting Knife implements can be use for together bud and graft on the plants. It gives fine cuts and make budding and grafting easy with this right implement in hand. The metal parts of the implements are encrusted for guard from oxidation and handle of the implements is especially considered for enhanced grasp. It is extensively use for grafting grape vines and for roses. This model of grafting knife FPTM-16 is available in the School of Agriculture (Fig. 11a, b) [Plate II].

Fig. 8 Prunning knife

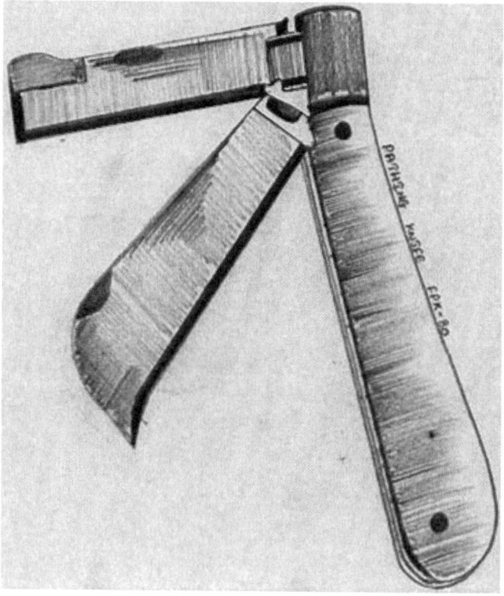

Fig. 9 Showel

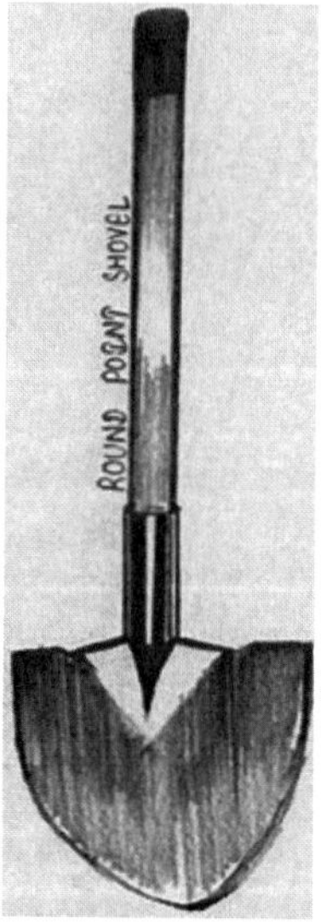

12. **Fruit Shear** (**फल की बड़ी कैंची**)–The fruit shears is also call hand pruners and secateurs be a variety of shears for use on flora. They are physically powerful sufficient to trim solid brushwood of tree and shrub, on occasion up to implement usually used to cut fruits and vegetables (Fig. 12) [Plate II].
13. **Prunning Saw** (**छंटाई**)–Pruning saw is ergonomic deliberate, deliver the excellence require to get behind twigs and limb. Strong blade prepared out of carbon steel with three edges sharpens. The blades are specifically considered to cut on

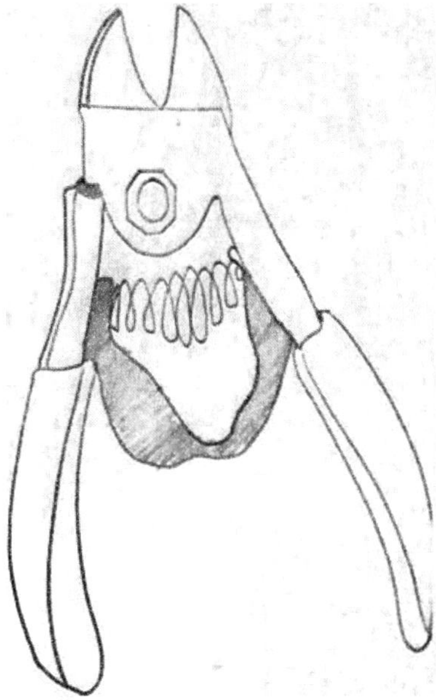

Fig. 10 Kinnow cutter

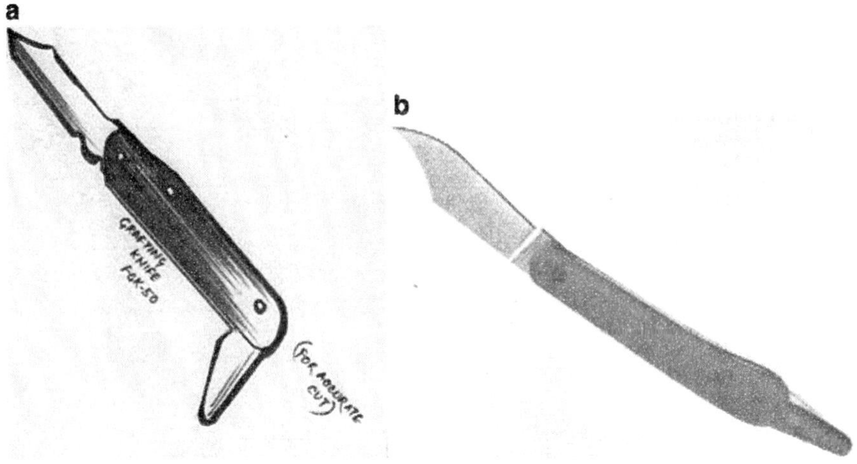

Fig. 11 **a**, **b** Grafting knives

the drag knock. It consists of a tough blade chain with a hand notched border. It is use to scratch flog substance and wood. The cut is ready by insertion the notched edge beside the stuff and affecting if powerfully back; otherwise incessantly frontward. The School of Agriculture have many type models of pruning saw FPS-18, FPS-21, FPS-100, FPS-30 and FPS-333 (Fig. 13) [Plate II].

Fig. 12 Fruit shear

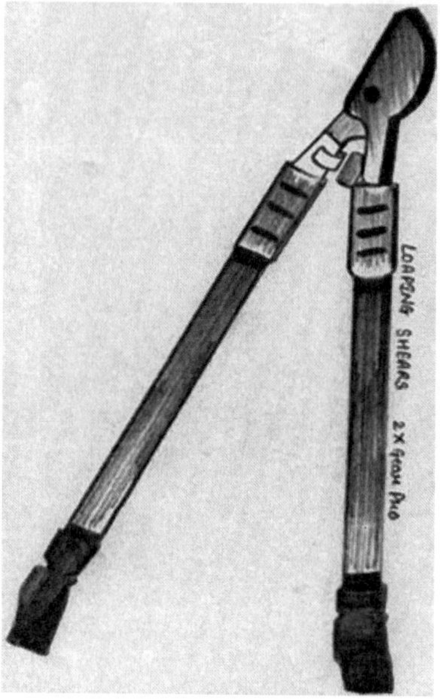

Fig. 13 Prunning saw

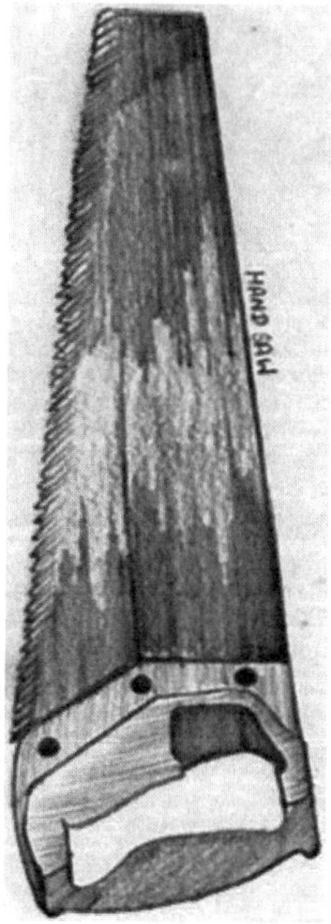

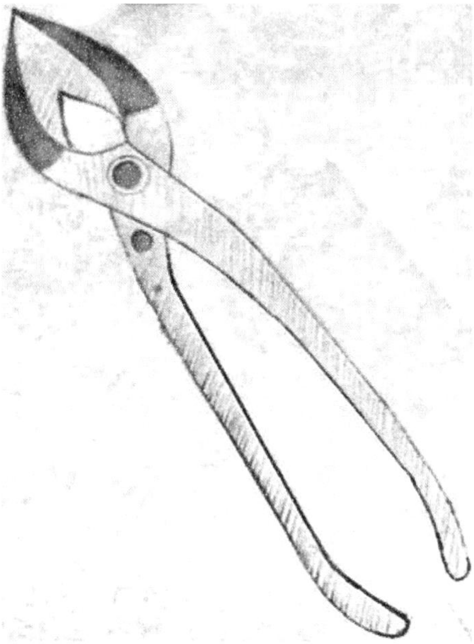

Fig. 14 Bonsai cutter

14. **Bonsai Cutter** (बोनसाई कटर)–It is a procedure to construct small trees in confiners that impersonate the form and range of full-size trees considered for spiteful twigs glow to the stem. The ensuing wood can then cure more rapidly with very slight scar the curved in cutter is the single the majority significant implement for Bonsai. The School of Agriculture has two type of model bonsai cutter FBT-50 and FBT-60 (Fig. 14) [Plate II].
15. **Hedge Shear** (बाड़ा कैंची)–It is physically operate hand implement for pruning, trimming and cutting of hedge and bushes. The implements are basically consists of two blades with tangs. The tangs are inserting in the wooden handle and protected by ferrule. The cutting action takes place between two blades. The School of Agriculture have two three types of models hedge shear FSH-888, FSH-999W and FSH-999P (Fig. 15) [Plate II].
16. **Spud** (खुरपा)–It is a small handle cutting implement by a flat blade use for dig soil and weed in small gardens and vegetables farms. It is use to cut branches of vegetation. The School of Agriculture has three types of models SPS-100, SPS-2000 and PK-1100 (Fig. 16) [Plate II].
17. **Maize Sheller** (मकई शेलर)–It is considered for moderate partition of Maize kernel from the cobs and elimination of rachis. They locate function mostly in the seed dealing out subdivision, where maize crop are harvest as a complete part. In the dispensation plant, the mechanism is placed after husking and cob ventilation. The School of Agriculture has model of Maize Sheller FPMP-1990 (Fig. 17) [Plate II].
18. **Garden Leaf Rake** (उद्यान में पत्ते जमा करने का औजार)–The garden leaf rake is an implement consisting of a notched bar set diagonally to a grip and use to

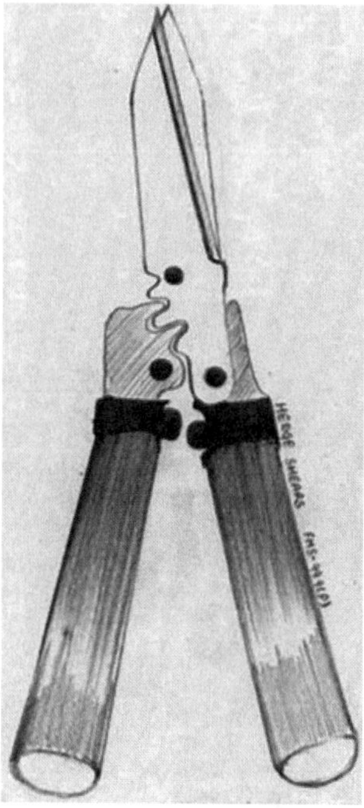

Fig. 15 Hedge shear

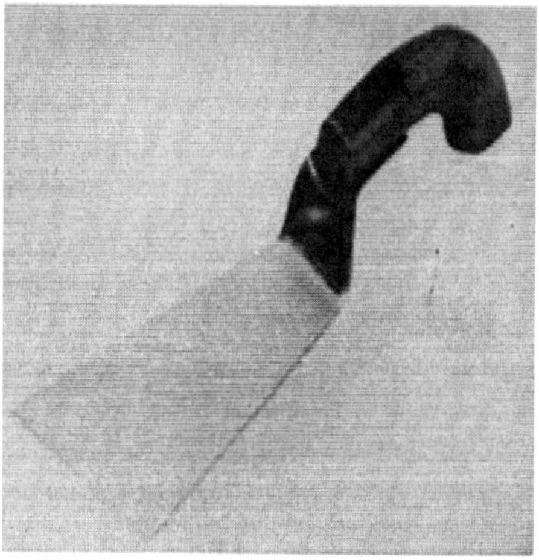
Fig. 16 Spud

accumulate foliage, hey and grass. It is also used in agriculture, for loosen the soil, weed, leveling, remove grass from lawns to perform in agriculture by the harrow. The School of Agriculture has three types of model garden leaf rake FGR-8, FGR-12 and FGR-16 (Fig. 18) [Plate II].

19. **Wire Hand Rake** (**पत्ते जमा करने का औजार**)–It is a faultless implement which is use to clean tiny flower buds and ventilate moss enclosed area. It is ergonomically

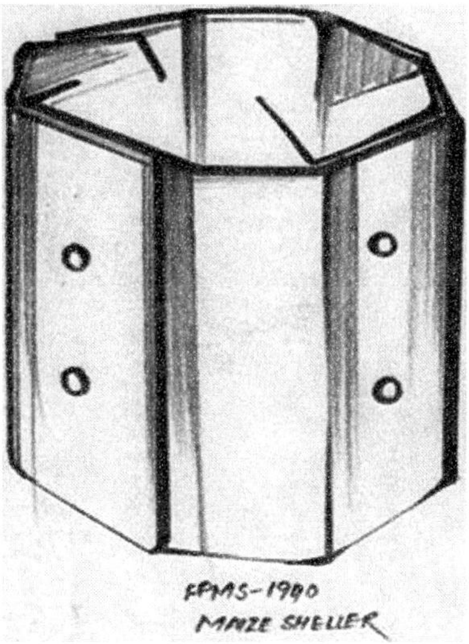

Fig. 17 Maize sheller

Fig. 18 Garden leaf rake

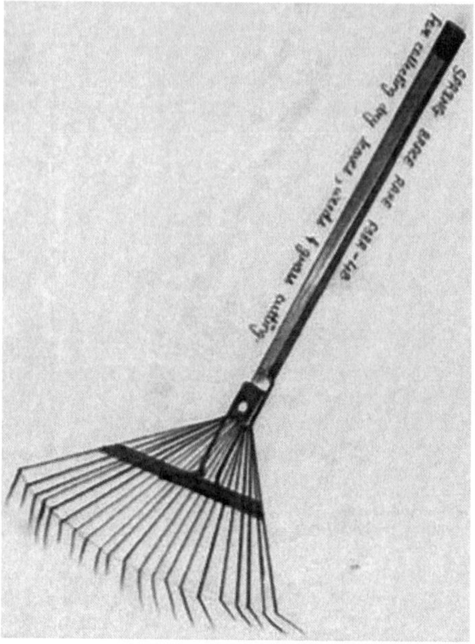

Fig. 19 Wire hand rake

shaped grip with its incorporated spongy piece, hysterics flawlessly in the hand over and is inclined at the end to avoid slip. It is prepared of high feature steel. The School of Agriculture has model of wire hand rake FWHR-9, SBR-48 and FPLR-50 (Fig. 19) [Plate II].

20. **Hand Sprayer** (**फुहारा**)–An attachment to a shower, sink, unit etc., which is not fixed in place but can instead be moved in order to aid washing. It is a piece of equipment for spraying water and another liquid over growing plants (Fig. 20) [Plate III].
21. **Knap-Sack Sprayer**(**नैपसैक फुहारा**)–It gives gardeners the ability to bring large amounts of liquid to other raised inaccessible areas of their property. Knap-sack sprayer resemble hand sprayer in terms of functionality and use, but they

Fig. 20 Hand sprayer

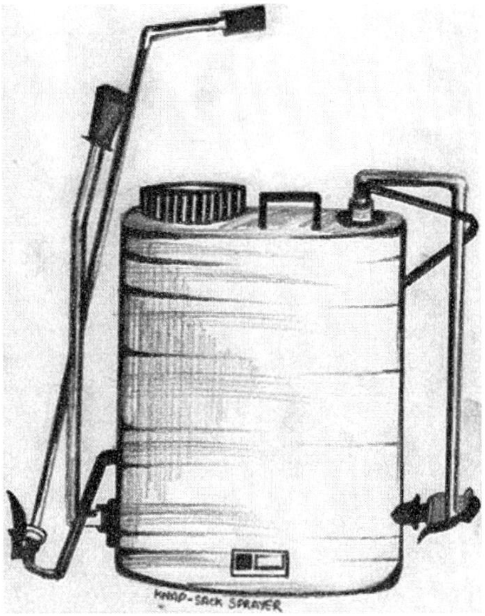

Fig. 21 Knap-sack sprayer

tend to hold more liquid plus you carry the device comfortably on your back (Fig. 21) [Plate III].

22. **Folding Spade** (**मुड़ जाने वाली फावड़ा**)–The folding spade is a collapsible dig implement that has a sward sharp edge for chop and saw roots to build in glass-filled nylon lever with a fine particles layered steel spade head. It's premeditated so that it workings both as a spade and as a hoe. The School of Agriculture has three types of models folding spade SPKW-25, SPKW-50, and SPKW-1000 (Fig. 22) [Plate III].
23. **Digging Shovels** (**कुदाल**)–It is an implement which is use for dig, elating and pitiful mass material i.e., soil, coal, snow, sand and majority shovel are hand implement consisting of a wide blade set to a middle span handle. The School of Agriculture has digging shovel model of FSS-4002 (Fig. 23a, b) [Plate III].
24. **Garden Hoe** (**खुदाई बेलचा**)–It is a very old and adaptable farming and horticulture hand implement which is use to shape soil, take out weed, apparent soil about thc bottom of plants, dig slight furrow and petty trench for plant seed and bulb. The School of Agriculture has two types of models garden hoe FGWH-200 and FGWH-100 (Fig. 24) [Plate III].
25. **Spade** (**फावड़ा**)–Spade is an implement primarily used for digging. A spade can both break and move the soil in most situations by escalating the effectiveness of implement with a sharp edge, characteristically rectangular, metal blade and a long handle, used for dig, cutting soil, and sand turf respectively. The School of Agriculture has four types of models SPKW-25, SPKW-50, SPK-25 and SPKW-2000 (Fig. 25a, b) [Plate III].
26. **Dutch Hoe** (**डच फावड़ा**)–It works similar to straight blade hand hoe. It quickly removes weeds with the help of the grant pro series. This shank type implement

Fig. 22 Folding spade

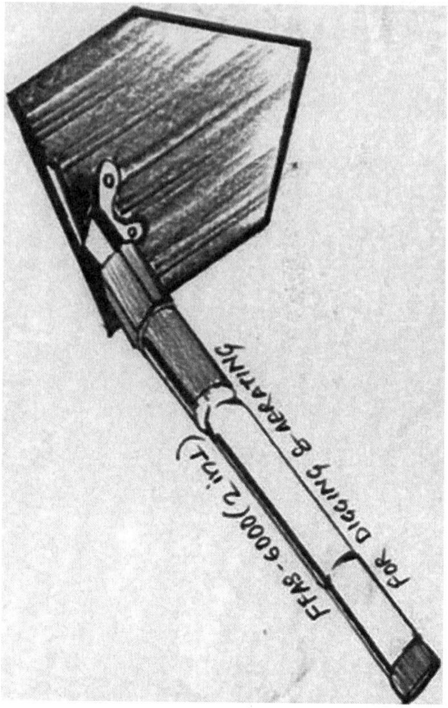

is equipped with a 20.3 cm thin, ultra-robust steel blade. It is reliable and efficient implement. The School of Agriculture has two types of models, which are: Dutch hoe FWH-5140 and FWH 5150 (Fig. 26) [Plate III].

27. **Snake Catcher** (साँप पकड़ने वाला उपकरण)–It is suitable for gardeners and mountaineers to catch snakes. It is made up of high quality steel blade to ensure its durability, high strength and solid structure. Highly polished and nicked blade is uneasy to get rusty. The steel teeth help to fix stably without hurting. The School of Agriculture has two types of models FPSC-66 and FPSC-44 (Fig. 27) [Plate III].
28. **Pick Axe** (गैंती)–It is made up of carbon steel. The pick axe has two edges with provision of axial hole for attachment with handle. One edge of pick axe is pointed and another is broadened. It is used for digging hard, compact and stony soils (Fig. 28a, b) [Plate III].
29. **Sky-Bird Sprayer** (स्काई - बर्ड फुहारा)–It is an implement which is used to scatter the liquor. In gardening, a sprayer is a part of apparatus that is use to concern herbicide, pesticides, insecticides, nematicide, bactericides, and manure on crop growing (Fig. 29) [Plate IV].
30. **Water Sprinkler Can** (फव्वारा)–An irrigation sprinkler; is also known as water sprinkler is an apparatus use to water in crop growing, lawns, landscape and other areas as well. The water sprinkler is also use for cool and manages airborne dust. It is a transferable urn, typically through a lever and a spout use

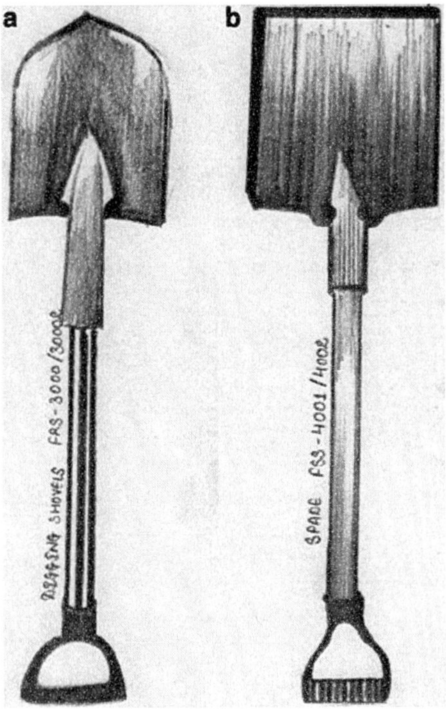

Fig. 23 **a**, **b** Digging shovels

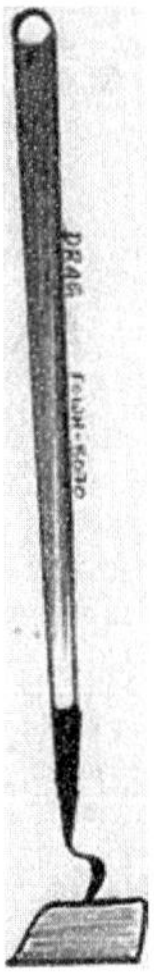

Fig. 24 Garden hoe

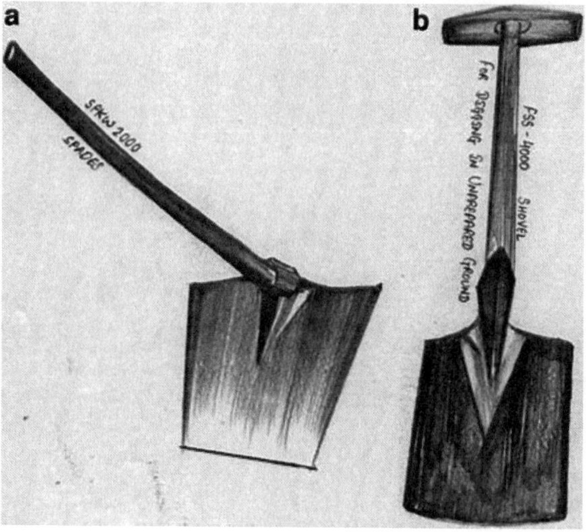

Fig. 25 **a**, **b** Spades

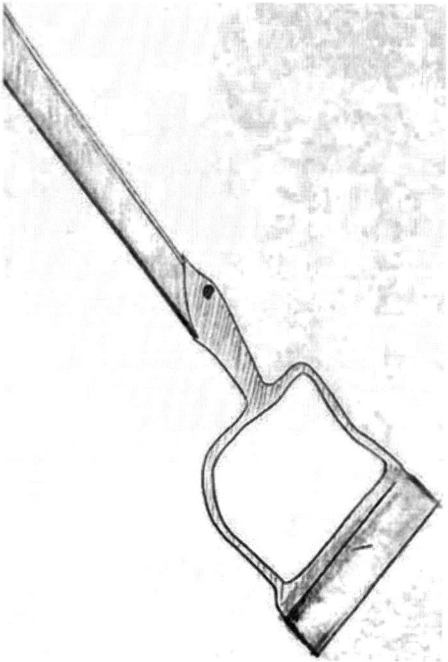

Fig. 26 Dutch hoe

to water plants by hand. The capacity of urn can be everywhere approximately from 0.5 to 1.0 L. It is made up of metal ceramic and plastic as well. The water sprinkler irrigation is the process of apply water in a prescribed approach in the mode alike to rainfall. The water is dispersed throughout a system that may consist of pumps, valves, pipes, and sprinklers (Fig. 30) [Plate IV].

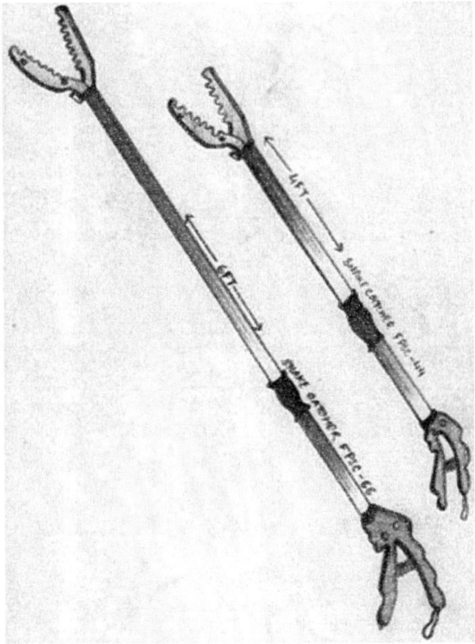

Fig. 27 Snake catcher

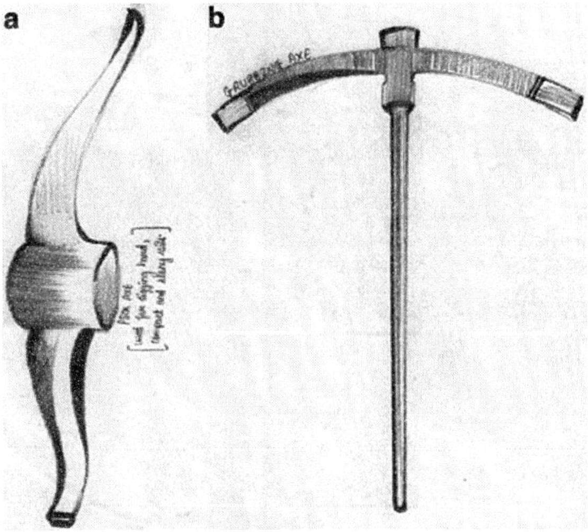

Fig. 28 **a**, **b** Pick axes

31. **Pressure Sprayer** (छिड़कने वाला यंत्र दाब)–A pressure sprayer is an apparatus use to spray a liquor, where spray can be usually use for outcrop of water, weed killer, crop performance material, pest maintenance chemical, as well as manufacturing and production line ingredients. It is far above the ground worth, suitable versatile sprayer that can be use for numerous applications key reimbursement comprise—modifiable cone nozzle from jet to mist (Fig. 31) [Plate IV].

Fig. 29 Sky-bird sprayer

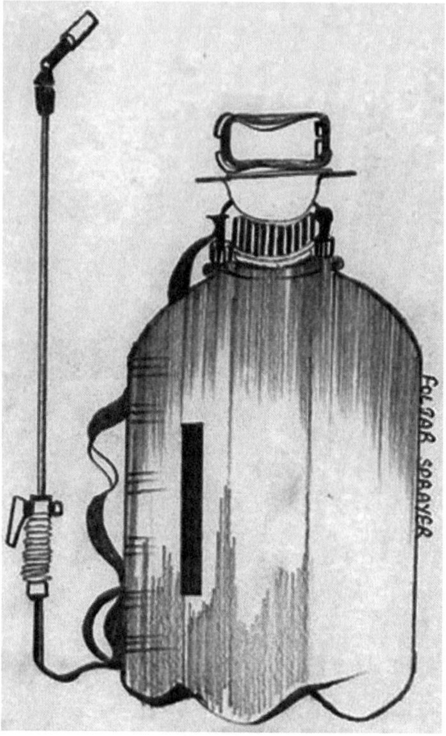

Fig. 30 Water sprinkler can

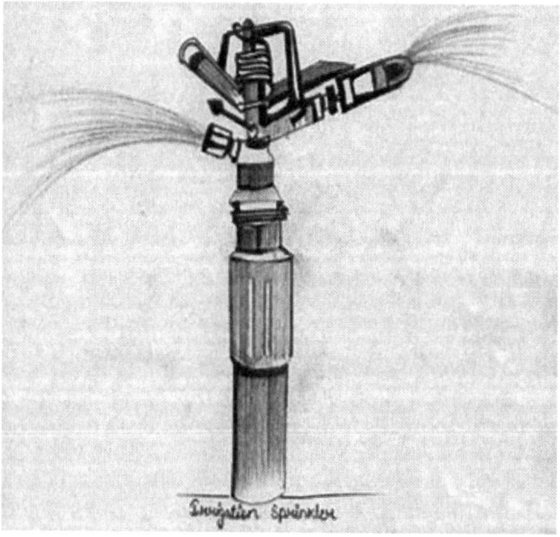

32. **Vaccum Pressure Sprayer** (**वैक्यूम दबाव स्प्रेयर**)–It is a high class, expedient multi-use sprayer that can be use for manifold application. The key profit contain—variable cone nozzle from jet to value (Fig. 32) [Plate IV].

Fig. 31 Pressure sprayer

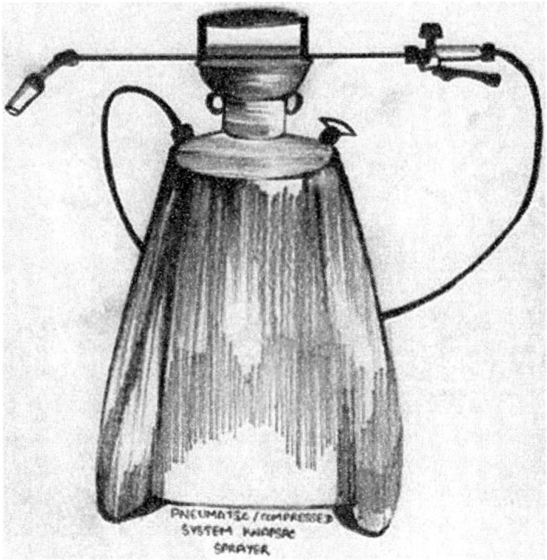

Fig. 32 Vaccum pressure sprayer

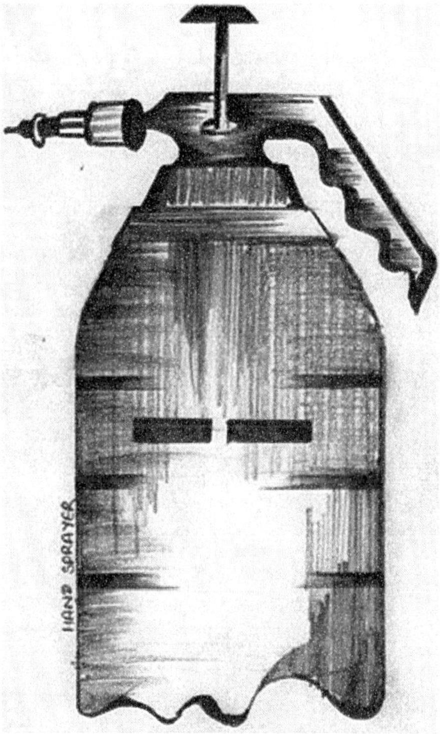

Fig. 33 Home garden gloves

33. **Home Garden Gloves** (बगीचे के दस्ताने)–It is necessary gloves for each and every gardener for horticulture as well as agriculture work. The hand gloves are fit to all size of hand (Fig. 33) [Plate IV].
34. **Grass Sword** (घास काटने की तलवार)–Some species of grasses have blades that are sharp enough to cut human skin, so grass sword is used for their removal. It's useful for jungle grass and brushes. The School of Agriculture has model of grass sward SGS-2008 (Fig. 34) [Plate IV].

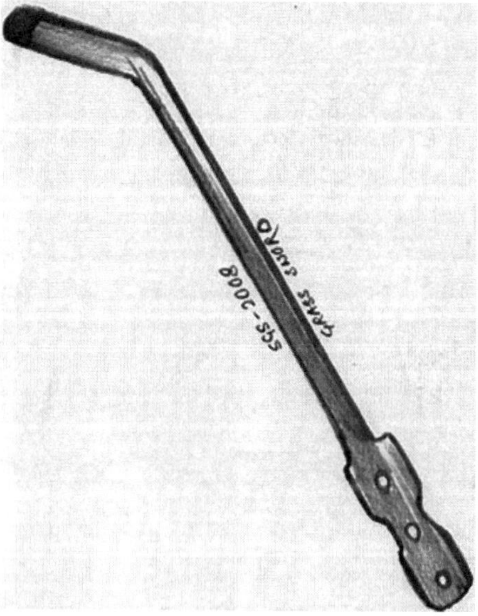

Fig. 34 Grass sword

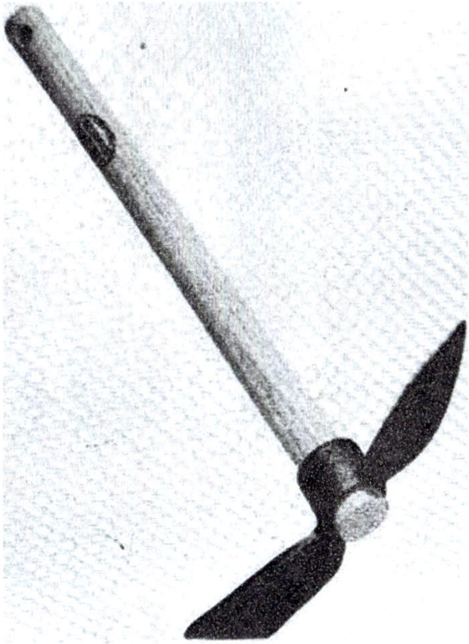

Fig. 35 Garden hoe

35. **Garden Hoe** (कुदाली)–Garden hoe is useful in a lettuce field. It is a slight metal knife blade so as to it's frequently use to disintegrate dust. The School of Agriculture has model of garden hoe FGWH-200 (Fig. 35) [Plate IV].
36. **Weeder** (निराने वाला)–A weeder implement is intended to easiness the task of remove weed from garden, lawns and agriculture field as well. It is premeditated to assist and eliminate unnecessary weeds with a top root (Fig. 36a, b) [Plate IV].
37. **Edging Knife** (चाकू)–An edge knife is also known as lawn edger in a garden implement. It is moreover instruction manual, mechanical to discrete limitations among a lawn, characteristically consisting of a grass of ground cover.

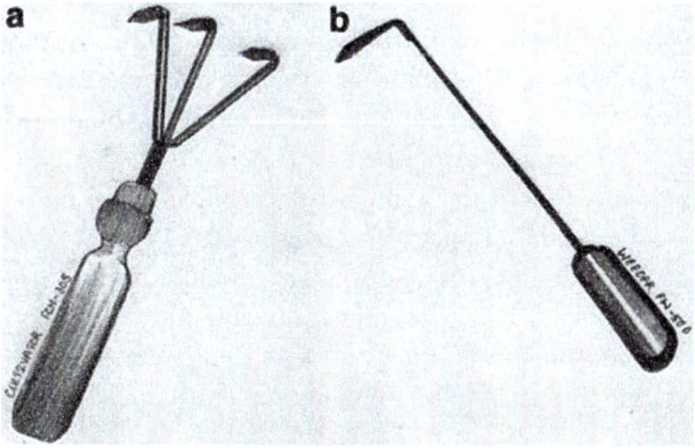

Fig. 36 **a**, **b** Weeders

Fig. 37 Edging knife

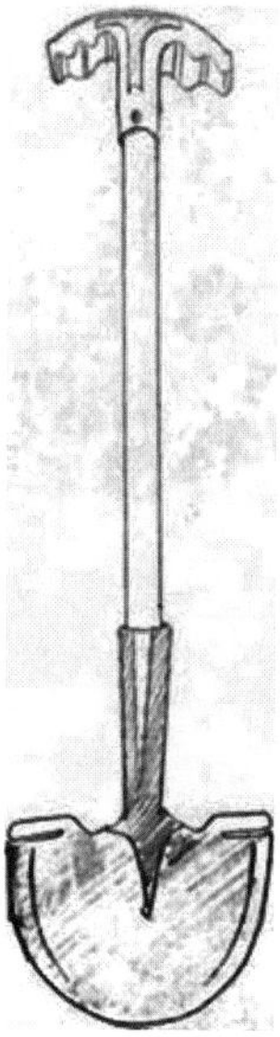

It is another ground surface feature i.e., cemented, concreted, and asphalted area, rough material such as sand, gravel, simply uncovered soil, i.e., limitless garden. A pointed edge hold of solid material which is the ray of a balance pivot about which a pendulum is balanced (Fig. 37) [Plate V].

38. **Thorn Remover** (शूल हटाने वाला)–It is the idea to eliminate a thorn from a rose plant. It is awfully practical piece for individuals who contain their personal garden and purchase flowers frequently (Fig. 38) [Plate V].
39. **Manual Aerating Rake**–Aeration breaks up compacted by soil and also by creating holes in it, helping grass grow vigorously. The School of Agriculture has three type models of manual aerating rakes FGR-8, FGR-12 and FGR-16 (Fig. 39) [Plate V].
40. **Long Reach Pruners** (रीच प्रूनर)–This implement help to prune plants that are out of normal reach and will help with standard pruning jobs, if you have reduced reach or mobility (Fig. 40) [Plate V].
41. **Manual Lawn Mower** (लान मौवेर)–In this implement, the blade rotate upright and employ scissoring stroke to hack out the blade of grass. It is hard-pressed by public and cut grassland incredibly by far (Fig. 41) [Plate V].
42. **Electric Chain Saw** (चेन सॉ मशीन)–It is a form of a saw which has a set of teeth attach to a revolving sequence. It is serve up numerous purpose i.e., tree fell and buck prune (Fig. 42) [Plate V].
43. **Rake** (रेक)–A rake is a garden and arable farm implement. The farmers are using it to get together movable stuff i.e., hay and grass. In farming, rakes loosen soil and remove wild plant and deceased grass from lawns. These are indispensable implements for a variety of purposes, a bow rake is great to use

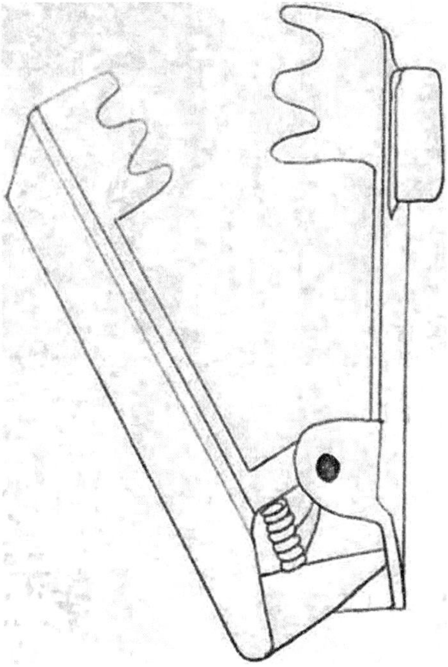

Fig. 38 Thorn remover

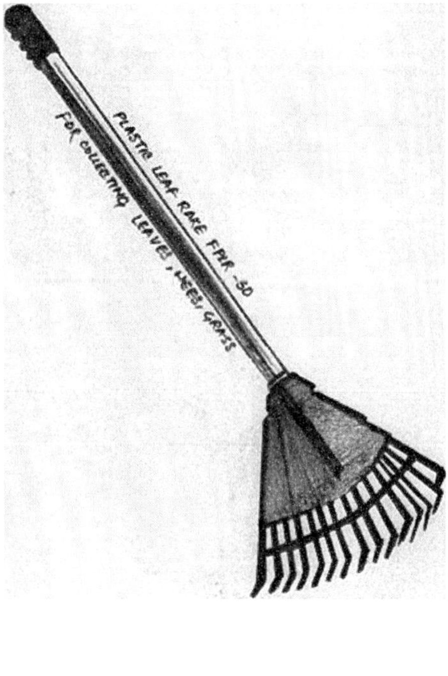

Fig. 39 Manual aerating rake

Fig. 40 Long reach pruners

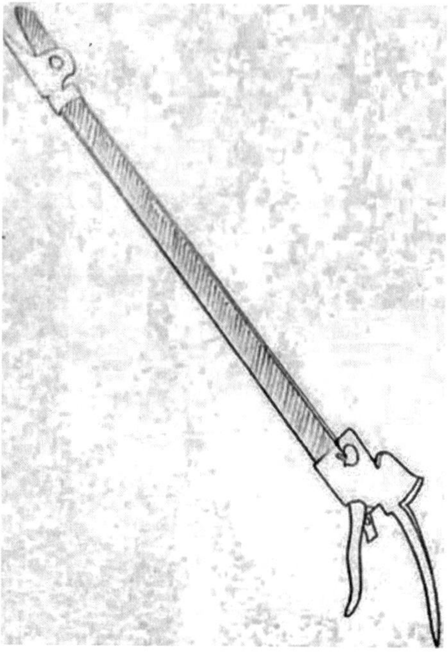

Fig. 41 Manual lawn mower

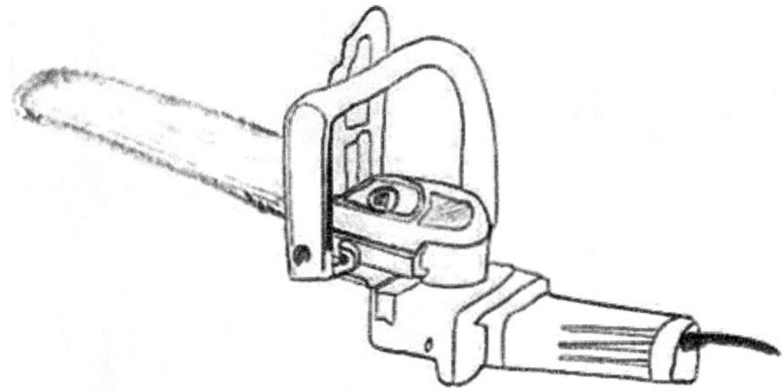

Fig. 42 Electric chain saw

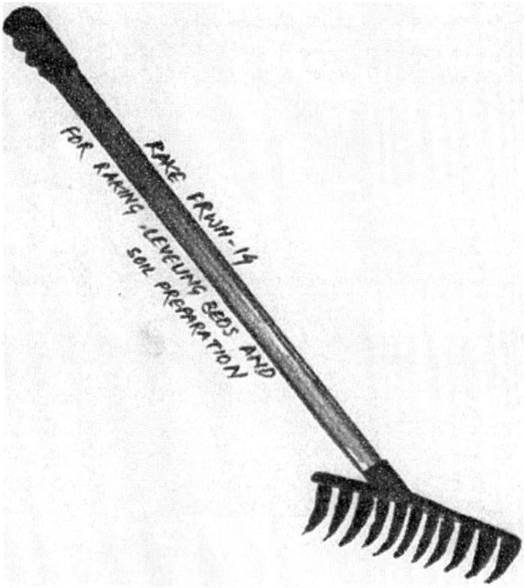

Fig. 43 Rake

in the garden, used to clear leaves and spread much. The School of Agriculture has FRWH-14 (Fig. 43) [Plate V].

44. **Wheelbarrow** (बैरो)–A wheelbarrow is a miniature hand-propelled means of transportation, generally with just one wheel, consider to be pressed and guide by a only one individual by two handle at the back and by a flow to drive the prehistoric wheelbarrow by wind. The use of wheelbarrows is ordinary in the manufacture engineering and in agriculture (Fig. 44) [Plate V].
45. **Loopers** (लूपर)–The loppers might be taking place at several times throughout the increasing time of year. Infestations early in the period on sapling plants may consequence in injure to cotyledons, fresh leaves, and fatal buds. The lopper's eggs are deposit individually, generally on the underneath of younger leaves. The larva provide for superior, additional full-grown leaves (Fig. 45) [Plate V].
46. **Lawn Edger** (एजर)–A lawn edger is used to cut a clean line in the soil between gross and a sidewalk, driveway or a garden bed. They are generally designed

Fig. 44 Wheel barrow

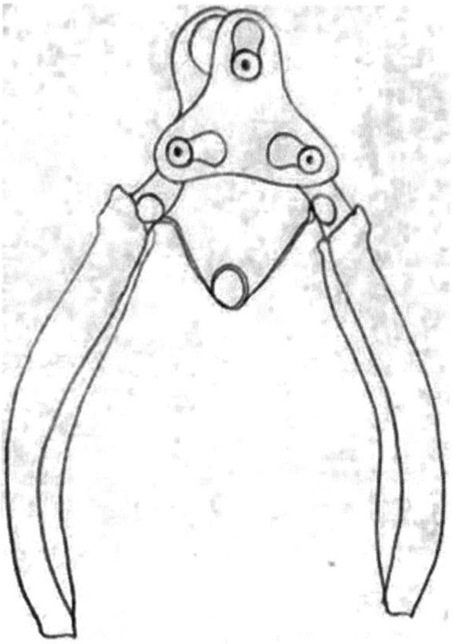

Fig. 45 Loopers

in half circle shape with a lip on the top where you can press the implement (Fig. 46) [Plate V].

47. **Spading Fork** (**कुदाल कांटा**)–It is also called digging fork, garden fork, a grip, similar in appearance to a pitch fork. Usually, it has four study tines perfect for loosing hand dirt and lifting soil. It is also used to mix fresh compost in to established beds (Fig. 47) [Plate VI].
48. **Digging Fork** (**खुदाई का कांटा**)–A digging fork is also known as spading fork and dig fork, is a farming implement, with a grip and numerous small and

Fig. 46 Lawn edger

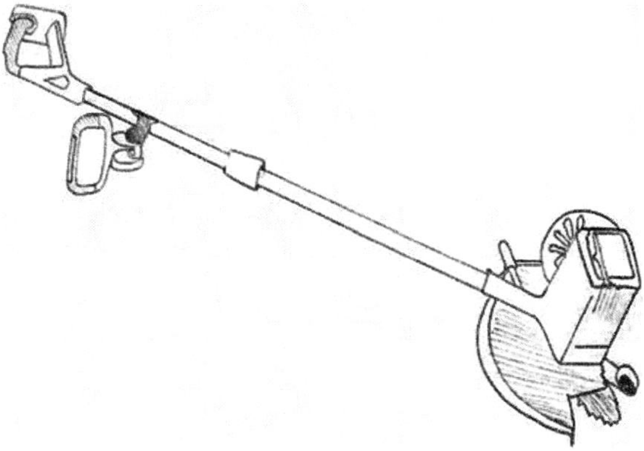

Fig. 47 Spading fork

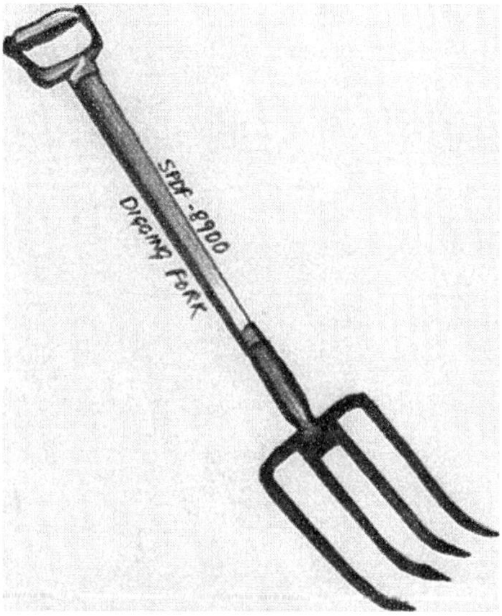

strong tines. It is use for loosen, elating and rotating above soil in farming and agriculture (Fig. 48a, b) [Plate VI].

49. **Digging Bar** (**खुदाई बार**)–A digging bar is a extensive, straight metal bar use for diverse purpose, with as a post gap not to be puzzled with a bent crowbar, which is considered to give influence relatively to dig. The frequent use of digging bars includes observance clay, solid, ice-covered land, and additional rigid material (Fig. 49) [Plate VI].

50. **Flail** (**मूसल**)–A flail is a farming implement use for threshing, the practice of sorting out grain from their husks. It is generally finished from two or additional huge brushwood attaches by a small sequence; single stick is detained and swings, because the extra to hit a heap of granule, loosen the husks (Fig. 50) [Plate VI].

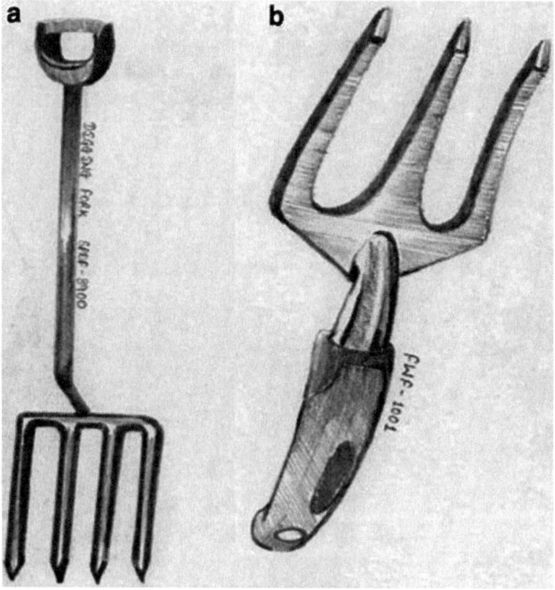

Fig. 48 **a**, **b** Digging forks

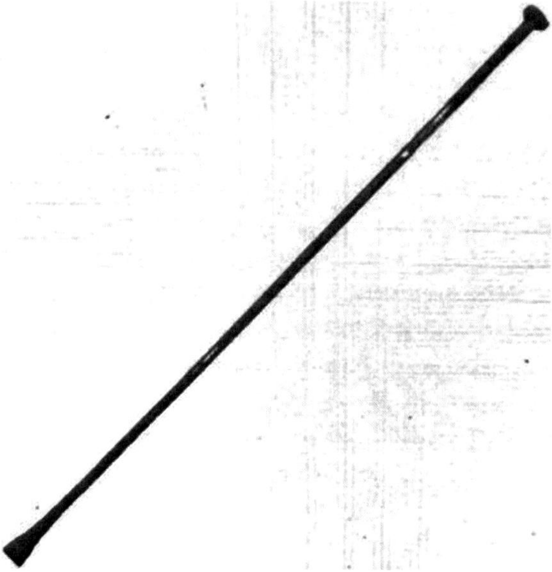

Fig. 49 Digging bar

51. **Foot Sprayer** (**फुट स्प्रेयर**)–The foot sprayer is one of the perfect and adaptable sprayers use for flexible spraying job. The sprayer consists of a pump operate by the foot knob, suction hosepipe with filter, delivery hose, spray lance fixed

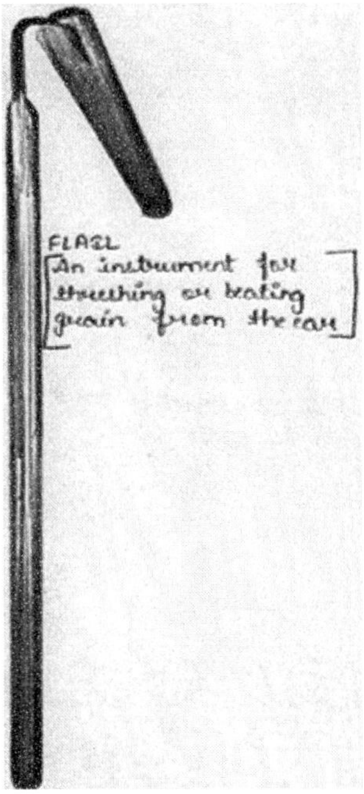

Fig. 50 Flail

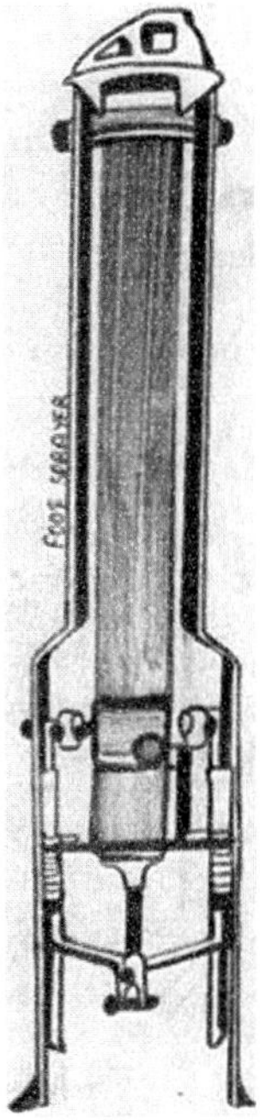

Fig. 51 Foot sprayer

Fig. 52 Rotavator

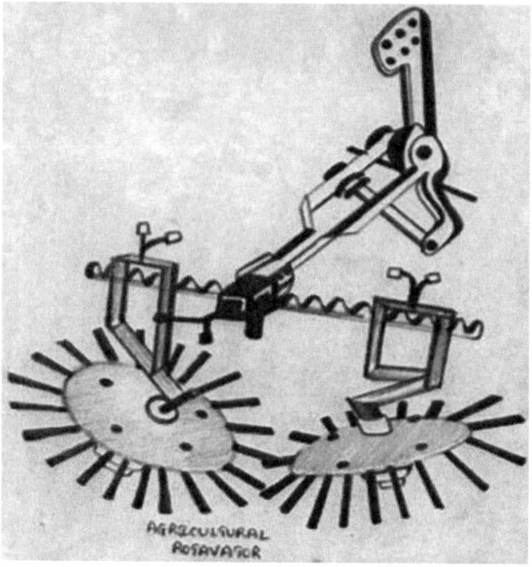

with close off pistol valve, gooseneck curve and modifiable nozzles (Fig. 51) [Plate VI].

52. **Rotavator** (**रोटा वेटर**)–It be capable of assist to accumulate instance and attempt with their revolving blade. It works efficiently and rotates the ground in excess of and breaks it downward. The rotavator support with aerating this help to perk up soil drainage in the garden, vegetable scrap, allocation allows plant living, vegetables to mature faster (Fig. 52) [Plate VI].
53. **Rubber Tapping Implement** (**रबड़ टैपिंग टूल**)–It is the implement by which is composed commencing a rubber tree. The latex is harvest by slice a furrow into the growl of the tree at the deepness of a quarter inch with a captivated knife and cracking back the bark (Fig. 53) [Plate VI].

4 Conclusion

The traditional and modern agriculture implements are well thought-out victorious since these are cost-effective, viable and sustainable. It can extend rapidly and effortlessly from single province to an additional. Still these implements are universal in use normally un-preferred since they minor the effectiveness and boost exhaustion of the operative. As a result of by the contemporary perception, these traditional and modern implements desirable to be consistent observance in mind the financial system of rural meager. Appropriate, scheming in harmony with the farmer's necessities certainly popularizes these implement in near future.

Fig. 53 Rubber tapping tool

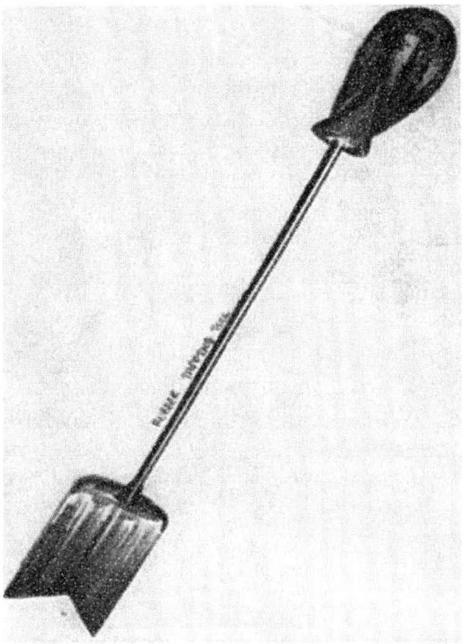

Acknowledgements The authors are acknowledged to the local peoples of Dankaur and adjoining villages for shared the information of traditional and modern agriculture implements which they are using in daily agriculture practices. The authors are also gratefully acknowledged the help, support and encouragement received from Honorable Pro-Vice-Chancellor (Academics) Prof. Dr. Pradeep Kumar and Honorable Chancellor Shri Sunil Galgotia Ji.

References

1. Madhusudhan, L.: Agriculture Role on Indian Economy. Journal of Business and Economics. Volume 6:(4). (2015) 1000176.
2. HLPE: Sustainable Agricultural Development for Food Security and Nutrition: What Roles for Livestock? A Report by the High Level Panel of Experts on Food Security and Nutrition of the Committee on World Food Security, Rome. (2016).
3. Steiner A., Aguilar G., Bomba K., Bonilla J.P., Campbell A., Echeverria R., Gandhi R., Hedegaard C., Holdorf D., Ishii N., Quinn K., Ruter B., Sunga I., Sukhdev P., Verghese S., Voegele J., Winters P., Campbell B., Dinesh D., Huyer S., Jarvis A., Lobo Guerrero RAM, Millan A., Thornton P., Wollenberg L. and Zebiak S.: Actions to transform food systems under climate change, Wageningen, Netherlands, CGIAR Research Program on Climate Change, Agriculture and Food Security (CCAFS). (2020).
4. FAO: Territorial Tools for Agro-Industry Development - A Source book by Eva Gálvez Nogales and Martin Webber (Eds.), Rome, Italy. (2017).
5. OECD/ICRIER: Agricultural Policies in India, OECD. Food and Agricultural Reviews, OECD Publishing, Paris (2018) https://doi.org/10.1787/9789264302334-en.
6. Emami M., Almassi M., Bakhoda H., and Issa K.: Agricultural Mechanization, A Key to Food Security in Developing Countries: Strategy Formulating for Iran. Agriculture & Food Security. Volume 7(24). (2018).
7. Government of India: Indian Agriculture in Brief, 44th Edition, Directorate of Economics and Statistics Department of Agriculture & Cooperation, Government of India, New Delhi. (2012).

8. Anonymous,: Statistical Hand Book of Nagaland, Directorate of Economics and Statistics, Government of Nagaland, Kohima. (2008)
9. Kathryn S.: The Green Revolution of the 1960's and Its Impact on Small Farmers in India. Environmental Studies Undergraduate Student Theses. (2010).
10. Basavaraja H., Mahajanashetti S.B. and Udagatti N.C.: Economic Analysis of Post-harvest Losses in Food Grains in India, a Case Study of Karnataka, Agricultural Economics Research Review. Agricultural Economics Research Association (India). Volume 20(1). (2007).
11. Kumar D., and Kalita P.: Reducing Postharvest Losses during Storage of Grain Crops to Strengthen Food Security in Developing Countries. Foods. Volume 6(1):8. (2017)
12. Das P.K., and Nag D.: Traditional Agricultural Tools: A Review. Indian Journal of Traditional Knowledge. Volume 5(1): 41-46. (2006)
13. Manchikanti, P. and Sengupta, M,,: Agricultural Machinery in India: IPR Perspective. Journal of Intellectual Property Rights. Volume 16. (2011).
14. Elzubeir A.S.: Traditional Agricultural Tools and Implements Used In Sudan, International Journal of Agriscience. Volume 4 :(2). (2014) 140–146.
15. Karthikeyan C., Veeraragavathatham D., Karpagam D., and Firdouse A.: Traditional Tools in Agricultural Practices. Indian Journal of Traditional Knowledge. Volume 8(2). (2008)
16. Census of India,: Data from the Census, Including Cities, Villages and Towns (Provisional). Census Commission of India. (2001).
17. Joshi B.C.: District Brochure of Gautam Buddha Nagar, UP, India. 1–22 (2008–2009).
18. Shrama J.K., Tripathi A.K. and Ahmad M.: Illustrated Flora, Part of Western Uttar Pradesh and Delhi NCR, India, Siya Publishing House and Shiv Nadar University, India. 606. (2018).

Synthesis, Characterization and Evaluation of Iron-Molybdenum Oxide Catalyst for the Hydrothermal Liquefaction of Wastewater to Bio-oil

Bhavish Kumar Reddy Koteru, B. Praveen Kumar, and Gopalakrishnan Govindasamy

Abstract Handling of ever increasing municipal sewage containing organics has become a challenge in one hand whereas getting the supply of cheaper and assured supply of biomass to produce biofuels, a replacement for fossil fuels is a challenge on the other hand. Thus developing suitable technologies to process municipal sewage to biofuels is of huge interest to address both the UN sustainable development goals (SDG) 11.6 and 7. Hydrothermal liquefaction (HTL) is the only thermochemical conversion that can process the municipal liquid sewage to bio-oil, which upon hydrotreating could be the substitute for gasoline, kerosene and diesel. The development of appropriate recyclable heterogeneous catalyst with high deoxygenation activity is crucial for increasing the yield and quality of bio-oil to make the process commercially viable. Iron-molybdenum oxide was prepared from iron and molybdenum precursors of Mo/Fe mole ratio 1.5 and 2.2 by co-precipitation method, calcined, characterized by XRD and FTIR. The XRD and FTIR characteristics indicated the presence of Fe (III) molybdate and molybdenum oxide (MoO_3) phases in both the Fe–Mo Oxide (1.5) and (2.2) catalysts but the Fe (III) molybdate phase was the predominant phase in the former, since the Mo/Fe mole ratio of Fe (III) molybdate phase is also1.5. The catalysts were evaluated for hydrothermal liquefaction of synthetic wastewater to bio-oil under subcritical conditions and Fe–Mo Oxide (1.5) gave Total Oil Yield of 36.5%, which was more than double that of Fe–Mo Oxide (2.2). The higher bio-oil yield of Fe–Mo Oxide (1.5) was attributed to the deep deoxygenation activity of Fe (III) molybdate phase predominantly present.

Keywords Bio-oil · Hydrothermal liquefaction · Wastewater · Catalyst · Sustainable development

B. K. R. Koteru · B. Praveen Kumar · G. Govindasamy (✉)
Department of Chemical Engineering, School of Engineering, University of Petroleum and Energy Studies (UPES), Dehradun, India

N. A. Siddiqui et al. (eds.), *Advances in Environment Engineering and Management*, Springer Proceedings in Earth and Environmental Sciences,
https://doi.org/10.1007/978-3-030-79065-3_13

1 Introduction

Rapid urbanization, increase in per capita income and consumption by urban population have resulted in an enormous increase in municipal solid and liquid waste generation which is having huge impact on the habitants and environment of the cities [1]. The sustainable development goals SDG 11.6 of United Nations aims at reducing the per capita environmental impact of cities through better way of municipal and other waste management and SDG-7 aims at providing access to reliable, sustainable, affordable, and modern energy for all by 2030 [2]. The two SDGs are seemingly different and independent goals but can be attained together, thus there is a paradigm shift in municipal solid and liquid waste treatment from waste reduction to waste to energy (WTE). WTE technologies for municipal solid waste (MSW) such as incineration, pyrolysis, gasification and anaerobic digestion, their comparison in terms of efficiency of energy recovery, impact on environment and suitability for varying MSW characteristics been reported [1, 3–5]. Gas Technology Institute, USA invented integrated hydropyrolysis and hydroconversion ($IH^{2®}$) process and further developed it with CRI/Criterion Inc., a subsidiary of Shell to convert cellulosic part of MSW to gasoline, jet and diesel range hydrocarbon fuels [6]. Anaerobic digestion (AD) is the most commonly used technology for the treatment of sewage sludge (SS) generated in the domestic wastewater treatment plants (WWTP) and biogas produced is upgraded to bio-methane, the energy product [7]. Potential of anaerobic co-digestion of the organic fraction of MSW and sewage sludge to produce bio-methane was reported [8]. Interest on faster and efficient thermochemical conversion of sewage sludge to energy products gathered interest as AD is inherently slow.

Pacific Northwest National Laboratory (PNNL), USA converted SS by hydrothermal liquefaction (HTL) to bio-crude of 44% yield as energy product along with bio-char, aqueous phase containing significant amount of water soluble organics (WSO) and gases and as coproducts [9]. The bio-crude was upgraded to naphtha, diesel and heavy oil range hydrocarbons by hydrotreating. Roy Posmanik et al. [10] reported that processing the aqueous phase of HTL by AD, increased the energy return by converting the WSO to bio-methane. HTL is performed with hot compressed water which has low dielectric constant of 19.66 compared to 78.85 for ambient water and thus enhanced solvency for biomass. Further, its ionization constant is 500 times as that of water under ambient conditions. This enhances its ability as an acid–base catalyst to several fold for hydrolyzing the biopolymers present in the wastewater like starch, cellulose and hemicellulose to simple sugars, fats to fatty acids, lignin to phenols and proteins to amino acids [11], constituting WSO. However, the deoxygenation of WSO formed to water immiscible bio-oil yield depends on the deoxygenation activity of catalyst [12]. HTL of SS was studied by Yokoyama et al. [13] using sodium carbonate as catalyst and reported that the bio-oil yield was 50% having calorific value of 33 MJ/kg. However, alkali was reported to be difficult to recover and recycle besides corrosive [6]. An overview of catalysts studied for HTL of SS to bio-oil, emphasized on the importance of further research for furthering the understanding of the catalyst's effect on the bio-oil yield and its characteristics

[14]. Maag et al. studied the HTL of food waste and reported that the bio-oil yield was higher over heterogeneous $CeZrO_x$ catalyst in comparison with homogeneous Na_2CO_3 catalyst, which was attributed to the reduction in the WSO yield by 50% [15]. Iron-molybdenum oxide catalyst was studied by Gopalakrishnan et al. [16] for the HTL of sugarcane bagasse and found to yield 52% bio-oil. Hence in the present work, iron-molybdenum catalyst was studied for the HTL of synthetic wastewater.

2 Materials and Methods

2.1 *Materials*

Ferric nitrate nonohydrate (98%), ammonium molybdate tetrahydrate (98%) of SRL make and nitric acid (AR) of SD's fine chemicals make were used for preparing the iron-molybdenum oxide. For the preparation of synthetic wastewater, potato starch (Pure), casein (fat free purified) of SRL make and sun flower edible oil (Fortune) were used. Diethyl ether (99%), ethyl acetate (99%) and acetone (99%) of SRL were used for the extraction of bio-oil. Hydrogen used as the process gas was of UHP grade, supplied by Gupta Industrial Gases.

2.2 *Synthesis and Characterization of Iron-Molybdenum Oxide Catalyst*

Ammonium molybdate solution was acidified to pH of 2 with concentrated nitric acid. To this ferric nitrate solution was added under stirring so that Mo/Fe of the mixture after addition would be 1.5 (Fe–Mo Oxide 1.5) and 2.2 (Fe–Mo Oxide 2.2). Yellow precipitates formed were ripened at 373 K for 3 h under stirring and separated by suction filtration. The precipitate was air-oven dried at 120 °C for 5 h and calcined at 773 K for 6 h.

XRD of the synthesized catalysts were studied using Bruker D8 advance diffractometer with nickel-filtered Cu Kα radiation ($\lambda = 1.54$ Å) and liquid nitrogen cooled germanium solid-state detector having crystallography open database (REV89244 2013.10.11). The FTIR spectra of the catalysts made into pellets with KBr were recorded using Frontier model of Perkin Elmer.

2.3 HTL Studies

Preparation of Synthetic Wastewater

100 g of synthetic wastewater containing 15 wt% of organic wastes was prepared by mixing 4.5 g of casein, 8.25 g of potato starch, and 2.25 g of edible oil in 85 g of distilled water.

Catalyst Evaluation for HTL

High-pressure batch reactor (BR 100, Berghof) made up of stainless steel having a tubular furnace around it with a magnetic stirrer cum heater and digital temperature controller cum indicator was used for studying the catalytic HTL of synthetic wastewater. The schematic diagram of the reactor assembly used in the present study was already reported [16]. Synthesized wastewater was transferred BR 100 reactor and 0.5 g of catalyst was added under magnetic stirring. The reactor was closed, purged with hydrogen and then filled with hydrogen until the pressure reached the initial pressure of 40 bar. Then the reactor was heated to 250 °C and kept for 2 h. After 2 h, heating was stopped and the reactor was cooled to ambient temperature and the products of HTL were analyzed by the reported procedure [17] with few modifications and the same is depicted in Fig. 1. The gases were vented out and the remaining

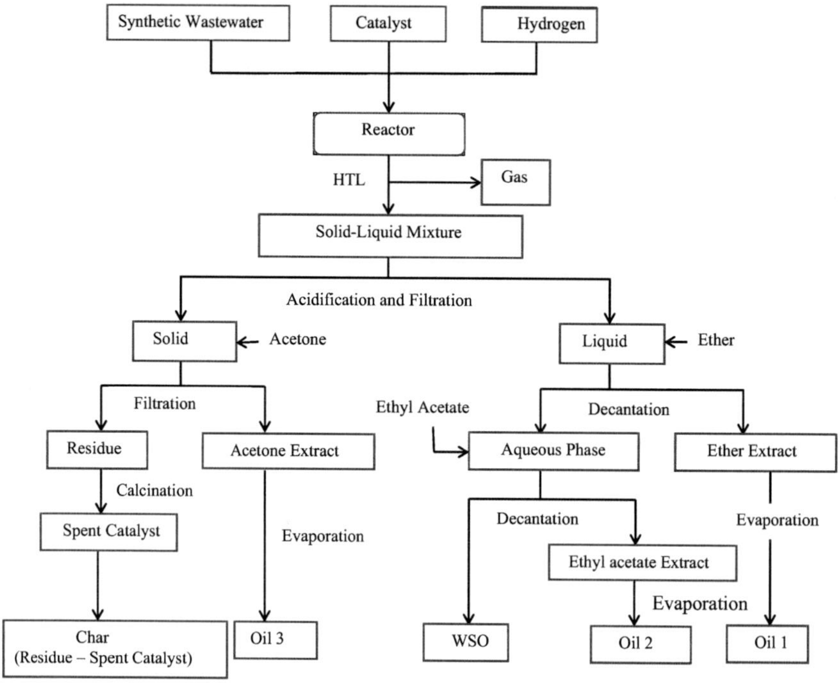

Fig. 1 Process flow diagram

solid–liquid product mixture was acidified with HCl to pH around 1–2, refrigerated overnight and separated by suction filtration. The filtrate obtained was extracted with diethyl ether first, then the extract was dehydrated with sodium sulfate, filtered and rotary evaporated to get Oil 1. Raffinate was extracted with ethyl acetate and the extract was dehydrated with sodium sulfate, filtered and rotary evaporated at 50 °C to get Oil 2. The raffinate obtained was the aqueous phase containing water-soluble organics (WSO). The solid phase obtained during suction filtration was extracted with acetone in Soxhlet apparatus until the solvent in the thimble became colorless. The acetone extract was rotary evaporated to get Oil 3. The acetone insoluble solid was dried at 100 °C and then weighed to get the amount of residue containing char formed and the catalyst used in the HTL. The same was heated in a silica crucible at 650 °C for ½ h, cooled and weighed to find the weight of Ash. The weight of char was determined by subtracting the weight of ash from the weight of residue. The % Conversion, % Yield of the products and % Yield of Total Oil were calculated using the equations reported earlier [16].

3 Results and Discussion

3.1 Catalyst Characterization

The XRD pattern of the catalysts are presented in Fig. 2. The peaks at 2Θ position of 22.89, 25.73 and 27.36 indicated the presence of Fe (III) molybdate [$Fe_2(MoO_4)_3$] phase while the peaks at 23.35, 30.17 and 33.81indicated the presence of MoO_3 phase in both the iron-molybdenum oxide catalysts with Mo/Fe mole ratio of 1.5 and 2.2 [18]. Further, from the analysis of XRD data by the inbuilt software, it was found that amount of MoO_3 phase in the iron-molybdenum catalyst prepared from iron

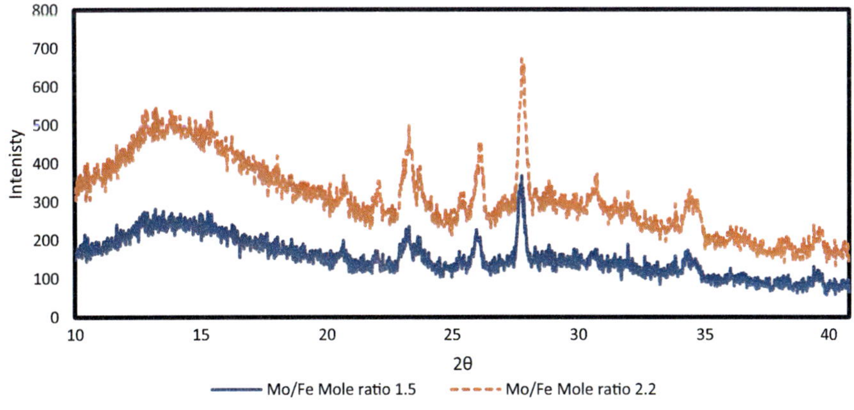

Fig. 2 X-ray diffraction pattern of the catalysts

and molybdenum precursors with the starting Mo/Fe mole ratio of 2.2 was higher compared to the same catalyst prepared from starting Mo/Fe mole ratio of 1.5. It is because Mo/Fe mole ratio of Fe(III) molybdate [$Fe_2(MoO_4)_3$] phase is 1.5 and hence it was the major phase present when the starting Mo/Fe mole ratio was also 1.5. But with an increase in starting Mo/Fe mole ratio to 2.2, excess MoO_3 phase present in the catalyst increased. The FTIR spectra of the iron-molybdenum oxide catalyst prepared with starting Mo/Fe ratio 1.5 and 2.2 are given in the Fig. 3a and b respectively. The weak narrow band around 960 and 990 cm^{-1} observed in both the catalysts were due to Fe–O–Mo vibration of the Fe(III) molybdate [$Fe_2(MoO_4)_3$] phase and Mo–O vibration of MoO_3 phase respectively reaffirming the inferences made from XRD result.

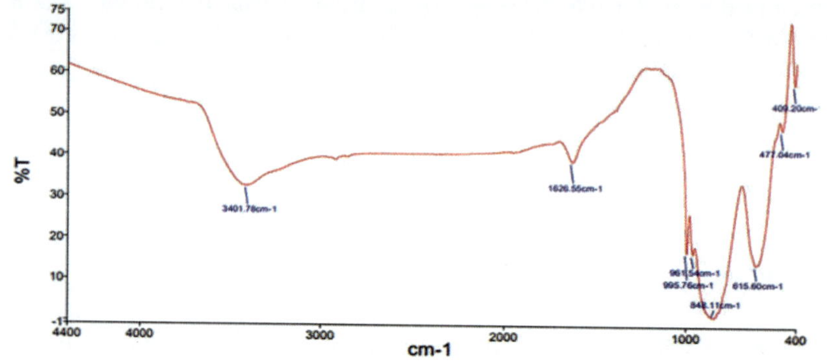

(a)

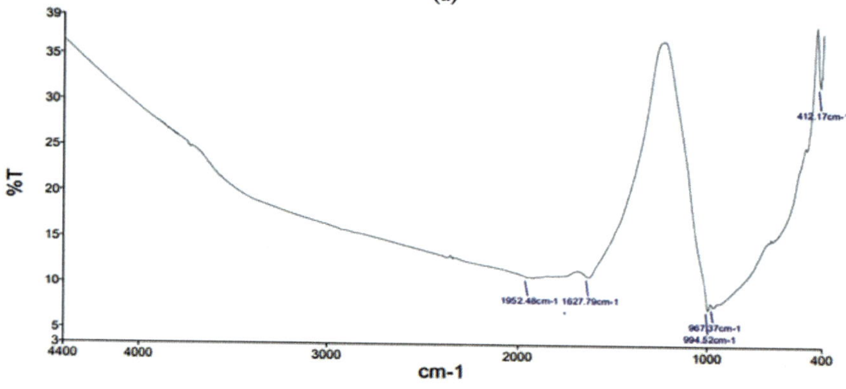

(b)

Fig. 3 FTIR spectra of Iron-molybdenum oxide catalyst **a** Mo/Fe mole ratio 1.5 **b** Mo/Fe mole ratio 2.2

Table 1 HTL of synthetic wastewater

Catalyst (Mo/Fe Mole Ratio)	Yield of products (g)					Percentage%		
	Oil 1	Oil 2	Oil 3	Char	WSO and gas	Conversion	Total oil yield	WSO and gas yield
Fe–Mo oxide (1.5)	0.09	2.81	2.57	3.04	6.49	79.1	36.5	43.3
Fe–Mo oxide (2.2)	0.27	1.36	0.94	2.61	9.82	82.6	17.1	65.5
Without catalyst	0.91	1.13	2.11	3.31	7.54	77.9	27.7	50.3

3.2 *HTL of Synthetic Wastewater*

The yield of products, % Conversion%, % Total Oil Yield and % WSO and Gas Yield for the HTL of synthetic wastewater both in the presence and absence of catalyst are presented in Table 1.

The % Conversion obtained in the presence of catalyst was higher than the non-catalytic hydrothermal liquefaction of synthetic wastewater. The % Total Oil Yield in the presence of Fe–Mo Oxide (1.5) was higher whereas lower for Fe–Mo Oxide (2.2) in comparison with non-catalytic HTL. The % Total Oil Yield of 36.5% observed for Fe–Mo Oxide (1.5) was higher compared to 34.7% shown by hierarchical Fe–Co–ZSM-5 [19]. % Total Oil Yield was reported to be dependent on the catalyst's deoxygenation activity [12], which deoxygenates the WSO containing the hydrolysates of biopolymers present in the wastewater such as starch, proteins etc., to bio-oil. Hence it could be concluded that the Fe (III) molybdate [$Fe_2(MoO_4)_3$], the major phase present in the Fe–Mo Oxide (1.5) catalyst might be the most active for deoxygenation compared to MoO_3 phase.

4 Conclusion

Iron-molybdenum oxide catalysts were prepared from iron and molybdenum precursors of Mo/Fe mole ratio 1.5 and 2.2 by co-precipitation method of which the one obtained from Mo/Fe mole ratio of 1.5 predominantly contained Fe (III) molybdate [$Fe_2(MoO_4)_3$] phase. As the same gave the higher bio-oil yield, it was concluded that Fe (III) molybdate [$Fe_2(MoO_4)_3$] phase was the most active for deoxygenation of WSO to bio-oil.

Acknowledgements The authors gratefully acknowledge UPES R&D for providing the financial support for this work and Mr. Charu Chandra Pant, Centralized Instrumentation Center for XRD and FTIR characterization.

References

1. 1. Liu, J., Zhan, X.: Overview of Bio-Oil from Sewage Sludge by Direct Thermochemical Liquefaction Technology. In: Journal of Sustainable Bioenergy Systems. Volume 2. (2012) 112–116.
2. 2. Yong, Z.J., Bashir, M.J., Ng, C.A., Sethupathi, S., Lim, J.W., Show, P.L.: Sustainable Waste-to-Energy Development in Malaysia: Appraisal of Environmental, Financial, and Public Issues Related with Energy Recovery from Municipal Solid Waste. Processes. Volume 7. (2019) 676–704.
3. 3. Bridgwater, A.V.: Renewable fuels and chemicals by thermal processing of biomass. In: Chemical Engineering Journal. Volume 91. (2003) 87–102.
4. Sipra, A.T., Gao, N., Sarwar, H.: Municipal solid waste (MSW) pyrolysis for bio-fuel production: A review of effects of MSW components and catalysts. Volume 175. (2018) 131–147.
5. 5. Van, D.P., Fujiwara, T., Tho, B.L., Toan, P.P.S., Minh, G.H.: A review of anaerobic digestion systems for biodegradable waste: Configurations, operating parameters, and current trends. In: Environmental Engineering Research. Volume 25(1). (2020) 1–17.
6. 6. Perkins, G., Batalha, N., Kumar, A., Bhaskar, T., Konarova, M.: Recent advances in liquefaction technologies for production of liquid hydrocarbon fuels from biomass and carbonaceous wastes. In: Renewable and Sustainable Energy Reviews. Volume 115. (2019) 109400–109423.
7. Liu, X., Han, Z., Yang, J., Ye, T., Yang, F., Wu, N., Bao, Z.: Review of enhanced processes for anaerobic digestion treatment of sewage sludge. In: IOP Conf. Series: Earth and sEnvironmental Science 113 (2018) 012039 doi: https://doi.org/10.1088/1755-1315/113/1/012039.
8. 8. Sosnowski, P., Wieczorek, A., Ledakowicz, S.: Anaerobic co-digestion of sewage sludge and organic fraction of municipal solid wastes. In: Advances in Environmental Research. Volume 7. (2003) 609–616.
9. 9. Seiple, T.E., Coleman, A.M., Skaggs, R.L.: Municipal wastewater sludge as a sustainable bioresource in the United States. In: Journal of Environmental Management. Volume 197. (2017) 673–680.
10. 10. Posmanik, R., Labatut, R.A., Kim, A.H., Usack, J.G., Tester, J.W.: Coupling hydrothermal liquefaction and anaerobic digestion for energy valorization from model biomass feedstocks. In: Bioresource Technology. Volume 233. (2017) 134–143.
11. 11. Toor, S.S., Rosendahl, L., Rudolf, A.: Hydrothermal liquefaction of biomass: a review of subcritical water technologies. In: Energy. Volume 36. (2011) 2328–2342.
12. 12. Duan, P., Savage, P.E.: Hydrothermal liquefaction of a microalga with heterogeneous catalysts. In: Ind Eng Chem Res. Volume 50. (2011) 52–61.
13. 13. Yokoyama, S., Suzuki, A., Murakami, M., Ogi, T., Koguchi, K., Nakamura, E.: Liquid fuel production from sewage sludge by catalytic conversion using sodium carbonate. In: Fuel. Volume 66(8). (1987) 1150–1155.
14. 14. Castello, D., Pedersen, T.H., Rosendahl, L.A.: Continuous Hydrothermal Liquefaction of Biomass: A Critical Review. In: Energies. Volume 11. (2018) 3165–3199.
15. 15. Maag, A., Paulsen, A., Amundsen, T., Yelvington, P., Tompsett, G., Timko, M.: Catalytic hydrothermal liquefaction of food waste using CeZrOx. In: Energies Volume 11. (2018) 564–577.
16. Gopalakrishnan Govindasamy., Rohit Sharma., Sunu Subramanian.: Studies on the effect of heterogeneous catalysts on the hydrothermal liquefaction of sugarcane bagasse to low-oxygen-containing bio-oil. In: Biofuels. Volume 10(5). (2019) 665–675.

17. 17. Karagoz, S., Bhaskar, T., Muto, A., Sakata, Y., Oshiki, T., Kishimoto, T.: Low temperature catalytic hydrothermal treatment of wood biomass: analysis of liquid products. In: Chemical Engineering Journal. Volume 108. (2005) 127–137.
18. Liu, X., Kong, L., Liu, C., Xu, S., Zhang, D., Ma, F., Zheng-ping Lu, Z., Sun, J., Chen, J.: Study on the formation process of MoO3/Fe2(MoO4)3 by mechanochemical synthesis and their catalytic performance in methanol to formaldehyde. In: J Therm Anal Calorim (2020). https://doi.org/10.1007/s10973-020-09483-4
19. Synthesis and evaluation of hierarchical Fe-Co-ZSM-5 for catalytic hydrothermal liquefaction of waste water to bio-oil.
20. Rama Tejaswi Pasupula, Sourodipto Modak and Govindasamy Gopalakrishnan
21. International Conference on Advances in Chemical Engineering (AdChE-2020), February 5–7, 2020, UPES, Dehradun.

Synthesis, Characterization and Biological Aspects of Novel Indole Derivatives

Indu Singh and Sanjeev Kumar Bhatt

Abstract A series of novel ethyl 2(5-methoxy-1H-indolyl)acetate (1), ethyl 2(5-methoxy-1H-indolyl)acetohydrazide (2), N-(substitutedbenzylidene-2-(5-methoxy-1H-indolyl)acetohydrazide (3a-f) and N-(-2-(substitutedphenyl)-3-chloro-4-oxoazetidinyl)-2-(5-methoxy-1H-indol-1-yl)acetamide (4a–f) derivatives have been synthesized. Compound 2 was obtained by reaction of compound 1 with hydrazine. Compound 2 treated with different aromatic aldehydes gives arylidene derivatives 3(a–f). Compounds 3(a–f) were cyclized with chloroacetylchloride in presence of triethylamine to yield N-(-2-(substitutedphenyl)-3-chloro-4-oxoazetidinyl)-2-(5-methoxy-1H-indol-1-yl) acetamide (4a–f). The recently prepared compounds were verified by spectral analysis (IR as well as ^{1}HNMR) and elemental analysis. The biological activity i.e. antibacterial activities of these synthesized compounds have been illustrated. Among of these compounds, some compounds have shown better inhibition against different bacterial growth. This activity was performed with help of cup plate method against staphylococcus aureus, proteus vulgaris and *Escherichia coli*. The antibacterial activity of these compounds was tested in form zone of inhibition in mm and this activity was compared with standard drug ampecillin.

Keywords Indole · Azetidinone · Antibacterial activity · Ampecilline

1 Introduction

Heterocyclic compounds play an important role in our life because they are necessary to performed metabolic process in all living things. Heterocyclic compounds and medicines are interrelated because humans are totally dependent on drugs and these drugs are derived from heterocyclic compound. Indole is a branch of heterocyclic compound with chemical formula C_8H_7N. It contains six-member benzene ring with five member nitrogen containing pyrrole ring. Many researchers have reported

I. Singh (✉) · S. K. Bhatt
Department of Chemistry, Meerut College Meerut, Meerut, Uttar Pradesh, India

N. A. Siddiqui et al. (eds.), *Advances in Environment Engineering and Management*, Springer Proceedings in Earth and Environmental Sciences,
https://doi.org/10.1007/978-3-030-79065-3_14

that indole derivatives possessing a wide variety of biological activity like antimicrobial [1, 2], antibacterial [3, 4], anti-inflammatory [5–9], analgesic [10, 11] and anticonvulsant [12–16] activities etc. Moreover, large number of indole derivatives having azetidinone moieties increases the antibacterial activity. From the literature it has been found that an azetidinone derivative shows biological activities such as antimicrobial [17–19], antibacterial [20–23], antifungal [24–26], antitubercular [27] and anticonvulsant [28] activities. Therefore, it was thought worthwhile to synthesize severalindole derivativesincluding azetidinone moiety with hope to get better antibacterial molecule with improved antibacterial activities.

2 Methods

Melting points and purity of new compounds were confirmed by ordinary methods. Perkin-Elmer 2400 analyzer, Beckman Acculab-10 spectrometer and BruckerDPX-300 MHz were used for elemental (C, H and N), IR spectra and ^{1}H NMR spectra respectively. Diagrammatical structure of target compounds has displayed in Scheme 1.

2.1 Biological Studies

Target compounds were tested for their antibacterial property. This activity was performed with help of agar cup plate method [29] against different type of bacteria like staphylococcus aureus, proteus vulgaris and *Escherichia coli.* The antibacterial activity of these compounds was recorded in form zone of inhibition in mm and compared this activity with standard drug ampecillin.

3 Results and Discussion

3.1 Chemistry

Ethyl 2(5-methoxy-1H-indolyl) acetate (1).

Indole (0.01 mol) with ethyl chloroacetate (0.01 mol) was taken in acetone (12 ml) then stirred with anhydrous potassium carbonate (2.2 g) on magnetic stirrer for 20 h. This mixture was poured in beaker with 100 ml ice water and separated by ether. After removing of ether, the product was obtained. The physical, other properties and spectral records are represented in Tables 1 and 2.

Ethyl 2-(5-methoxy-1H-indole-1-yl)acetohydrazide (2).

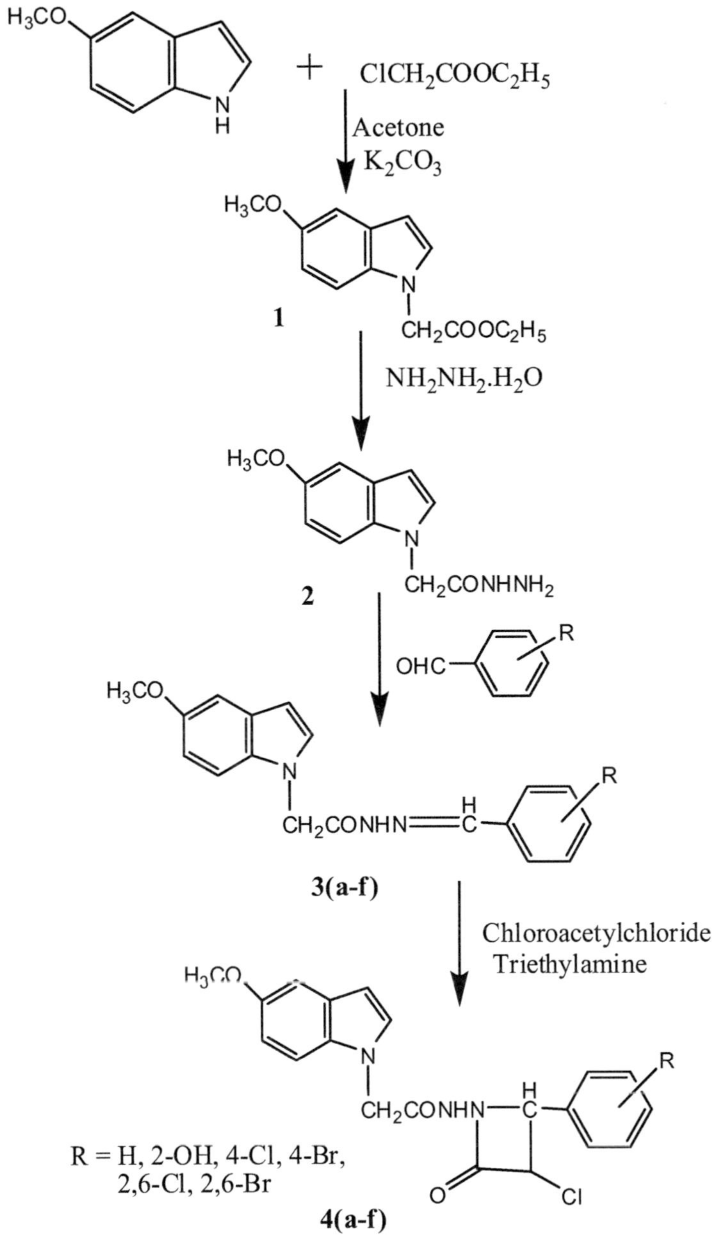

Scheme I

Scheme 1 Compound 1 Ethyl-2-(5-methoxy-1H-indolyl)acetate
Compound 2 Ethyl-2-(5-methoxy-1Hindole-1-yl)acetohydrazide
Compounds 3a-3f N-Substituted benzylidene-2-(5-methoxy-1H-indole-1-yl)acetohydrazide
Compounds 4a-4f N(3-chloro-2-oxo-4-substituted phenylazetidinyl)2-(5- methoxy-1Hindole-1-yl)acetamide

Table 1 Physical and analytical data of compounds 1-4f

Compound No.	R group and position	Molecular formula	mp in °C	Yield	Recrystalised solvent	Elemental analysis					
						%C		%H		%N	
						Calcd	Found	Calcd	Found	Calcd	Found
1		$C_{13}H_{15}NO_3$	154	67	Ethanol	66.94	66.96	6.48	6.45	6.00	6.02
2		$C_{11}H_{13}N_3O_2$	160	65	Methanol	60.26	60.27	5.98	5.96	19.17	19.18
3a	H	$C_{18}H_{17}N_3O_2$	165	62	Ethanol	70.34	70.37	5.58	5.59	13.67	13.64
3b	2-OH	$C_{18}H_{17}N_3O_3$	168	60	Acetone	66.86	66.87	5.30	5.33	13.00	13.03
3c	4-Cl	$C_{18}H_{16}ClN_3O_2$	173	61	Ethanol	63.25	63.27	4.72	4.74	12.29	12.31
3d	4- Br	$C_{18}H_{16}BrN_3O_2$	169	59	Methanol	55.97	55.94	4.18	4.14	10.88	10.86
3e	2,6-diCl	$C_{18}H_{15}Cl_2N_3O_2$	171	57	Acetone	57.46	57.48	4.02	4.04	11.17	11.15
3f	2,6-diBr	$C_{18}H_{15}Br_2N_3O_2$	176	54	Ethanol	46.48	46.46	3.25	3.24	9.03	9.04
4a	H	$C_{20}H_{18}ClN_3O_3$	189	52	Acetone	62.58	62.56	4.73	4.75	10.95	10.98
4b	2-OH	$C_{20}H_{18}ClN_3O_4$	185	49	Benzene	60.08	60.06	4.54	4.55	10.51	10.54
4c	4-Cl	$C_{20}H_{17}Cl_2N_3O_3$	190	47	Acetone	57.43	57.46	4.10	4.12	10.05	10.04
4d	4-Br	$C_{20}H_{17}Br_2ClN_3O_3$	192	44	Toluene	51.91	51.94	3.70	3.72	9.08	9.09
4e	2,6-diCl	$C_{20}H_{16}Cl_3N_3O_3$	196	40	Acetone	53.06	53.08	3.56	3.52	9.28	9.27
4f	2,6-diBr	$C_{20}H_{16}Br_2ClN_3O_3$	198	38	Benzene	44.35	44.38	2.98	2.96	7.76	7.78

Table 2 Spectral data of compounds 1-4f

Comp. No.	IR (KBr) ν_{max} in cm^{-1}	^{1}HNMR($CDCl_3$ + DMSO-d_6) δ in ppm
1	1347 (C–N–), 1567 (–C=C–), 1610 (C–C of phenyl), 1683 (–C = O),3045 (C–H ring)	2.48 (s, 3H, –CH_3), 2.84 (s, 2H, COO–CH_2–), 4.23 (s, 2H, –N–CH_2–), 4.58 (d, 3H, OCH_3), 6.01–6.25 (d, 2H, pyrrole ring), 7.01–8.02 (m, 3H, Ar–H)
2	1289 (–N–N–), 1346 (–C–N), 1560 (–C=C–), 1619 (–C–C– of phenyl),1682 (–C=O), 3041 (–C–H– ring), 3420 (N–H)	4.25 (s, 2H, –N–CH_2), 4.56 (d, 3H, OCH_3), 6.03–6.27 (d, 2H, pyrrole ring), 6.57 (s, 1H, –NH–), 6.74 (s, 2H, NH_2), 7.21–8.02 (m, 3H, –Ar–H)
3a	1285 (–N–N–), 1343 (–C–N–), 1562 (–C=C–), 1612 (C–C of phenyl), 1636 (–C=N–),1685 (–C=O), 3044 (C–H ring), 3423 (N–H)	4.19 (s, 2H, –N–CH_2), 4.52 (s, 3H, OCH_3), 6.05–6.26 (d, 2H, pyrrole ring), 6.55 (s, 1H, NH), 6.87 (d, 1H, = CH–Ar), 7.52–8.52 (m, 8H, –Ar–H)
3b	1283 (–N–N–), 1347 (–C–N–), 1568 (–C=C–), 1614 (C–C of phenyl), 1630 (–C=N–),1686 (–C=O), 3045 (C–H ring), 3421 (N–H), 3475 (C–OH)	4.24 (s, 2H, –N–CH_2–), 4.54 (s, 3H, OCH_3), 6.03–6.22 (d, 2H, pyrrole ring), 6.56 (s, 1H–NH), 6.98 (d, 1H, CH–Ar), 7.56–8.53 (m, 7H, ArH),11.89 (s, 1H, OH)
3c	761 (–C–Cl), 1287 (–N–N), 1345 (–C–N), 1562 (–C=C–), 1619 (–C–C– of phenyl), 1636 (C=N–),1682 (–C=O), 3048 (C–H ring), 3424 (N–H)	4.25 (s, 2H, –N–CH_2), 4.56 (s, 3H, OCH_3), 6.06–6.27 (d, 2H, pyrrole ring), 6.59 (s 1H NH), 6.92 (d, 1H, CH–Ar), 7.52–8.56 (m, 7H, ArH)
3d	615 (–C–Br), 1280 (–N–N), 1346 (–C–N), 1569 (–C=C–), 1612 (–C–C– of phenyl), 1633 (C = N–),1680 (–C = O), 3047 (–C–H ring), 3426 (N–H)	4.30 (s, 2H, –N–CH_2), 4.52 (s, 3H, OCH_3), 6.02–6.25 (d, 2H, pyrrole ring), 6.55 (s, 1H NH), 6.94 (d, 1H, = CH–Ar), 7.50–8.51 (m, 7H, Ar–H)
3e	764 (–C–Cl), 1286 (–N–N–), 1342 (–C–N), 1564 (–C = C–), 1617 (–C–C– phenyl), 1631 (C = N–),1680 (–C = O), 3048 (–C–H ring), 3427 (N–H)	4.24 (s, 2H, –N–CH_2–), 4.53 (s, 3H, OCH_3), 6.03–6.20 (d, 2H, pyrrole ring), 6.56 (s, 1HNH), 6.98 (d, 1H, CH–Ar), 7.54–8.53 (m, 6H, Ar–H)
3f	613 (–C–Br), 1282 (–N–N–), 1345 (–C–N), 1567 (–C = C–), 1610 (–C–C– phenyl), 1630 (C = N–),1683 (–C = O), 3042 (–C–H ring), 3424 (N–H)	4.26 (s, 2H, –N–CH_2–), 4.57 (s, 3H, OCH_3), 6.01–6.25 (d, 2H, pyrrole ring), 6.51 (s, 1H NH), 6.91 (d, 1H, CHAr), 7.51–8.56 (m, 6H, ArH)
4a	768 (–C–Cl), 1288 (–N–N), 1340 (–C–N), 1563 (–C = C–), 1615 (–C–C phenyl), 1681 (–C = O), 3049 (–C–H ring), 3422 (N–H)	4.19 (s, 2H, –N–CH_2), 4.53 (s, 3H, OCH_3), 4.80 (d, 1H, CH–Cl of azetidinone), 6.00–6.21 (d, 2H, pyrrole ring), 6.49 (s, 1H–NH), 6.99 (d, 1H, CH–Ar), 7.57–8.56 (m, 8H, Ar–H)
4b	760 (–C–Cl), 1289 (–N–N), 1345 (–C–N), 1569 (–C = C–), 1610 (–C–C phenyl), 1683 (–C = O), 3045 (–C–H ring), 3424 (N–H), 3468 (C–OH)	4.25 (s, 2H, –N–CH_2–), 4.59 (s, 3H, OCH_3), 4.83 (d, 1H, CH–Cl of azetidinone), 6.03–6.24 (d, 2H, pyrrole ring), 6.54 (s, 1H NH), 6.95 (d, 1H, CHAr), 7.56–8.58 (m, 7H, ArH),11.93 (s, 1H, OH)

(continued)

Table 2 (continued)

Comp. No.	IR (KBr) ν_{max} in cm^{-1}	^{1}HNMR($CDCl_3$ + DMSO-d_6) δ in ppm
4c	767 (–C–Cl), 1285 (–N–N), 1343 (–C–N), 1562 (—C = C–), 1612 (–C–C phenyl), 1680 (–C = O), 3048 (–C–H ring), 3420 (N–H)	4.23 (s, 2H, –N–CH_2–), 4.54 (s, 3H, OCH_3), 4.84 (d, 1H, CH–Cl of azetidinone), 6.01–6.23 (d, 2H, pyrrole ring), 6.50 (s, 1H, -NH), 6.96 (d, 1H, CHAr), 7.59–8.54 (m, 7H, Ar–H)
4d	611 (–C–Br),763 (–C–Cl), 1283 (–N–N), 1347 (–C–N), 1565 (–C = C), 1613 (–C–C phenyl), 1686 (–C = O), 3047 (–C–H ring), 3428 (N–H)	4.22 (s, 2H, –N–CH_2–), 4.55 (s, 3H, OCH_3), 4.86 (d, 1H, CH–Cl of azetidinone), 6.04–6.24 (d, 2H, pyrrole ring), 6.59 (s, 1H–NH), 6.98 (d, 1H, CHAr), 7.56–8.57 (m, 7H, Ar–H)
4e	765 (–C–Cl), 1281 (–N–N), 1340 (–C–N), 1565 (–C = C–), 1614 (–C–C phenyl), 1680 (–C = O), 3042 (–C–H ring), 3421 (N–H)	4.29 (s, 2H, –N–CH_2), 4.54 (s, 3H, OCH_3), 4.86 (d, 1H, CH–Cl of azetidinone), 6.05–6.21 (d, 2H, pyrrole ring), 6.58 (s 1H NH), 6.97 (d, 1H, CHAr), 7.51–8.56 (m, 6H, ArH)
4f	613 (–C–Br),760 (–C–Cl), 1288 (–N–N), 1345 (–C–N), 1567 (–C = C–), 1610 (–C–C phenyl), 1683 (–C = O), 3045 (–C–H ring), 3424 (N–H)	4.24 (s, 2H, –N–CH_2–), 4.51 (s, 3H, OCH_3), 4.85 (d, 1H, CH–Cl of azetidinone), 6.05–6.28 (d, 2H, pyrrole ring), 6.50 (s 1H–NH), 6.95 (d, 1H, CHAr), 7.55–8.59 (m, 6H, ArH)

An equal amount of indole ester (0.01 mol) and NH_2NH_2 (0.01 mol) were kept in appropriate solvent and refluxed it for 9 h. This mixture was transfered in ice water and kept it overnight. The product of this reaction was obtained by filtered and washed using cold water. The physical, other properties and spectral records are represented in Tables 1 and 2.

N-benzylidene- 2-(5-methoxy-1H-indole-1-yl)acetohydrazide (3a).

Compound 2 (0.01 mol) with benzaldehyde (0.01 mol) was taken in C_2H_5OH (55 ml) and refluxed it about 10 h. This mixture was cooled at normal temp and transfered it into a beaker with ice water containing triturated sodium bisulfate solution and stirred for 5 h. The product was separated from ethanol. The final product was recrystallized by using of appropriated solvent to give compound 3a. The physical, other properties and spectral records are represented in Tables 1 and 2.

Compounds 3(b–f) were manufactured by using same method, explained for compound 3a. The physical and spectral data of compounds 3(b–f) are giving in Tables 1 and 2.

N(3-chloro-2-oxo-4-phenylazetidinyl)-2-(5-methoxy-1H-indol-1-yl) acetamide (4a).

Taken the solution of chloroacetyl chloride (0.01 mol) indioxane (53 ml), was added in compound 3a (0.01 mol). This mixture was stirred continuously 4 h, cooled andtransfered it into a beaker with water. A solid product was obtained, filtered and then washed by using of water and recystallized by acetone to yield compound 4a.

Table 3 Details of antibacterial activity

Compounds	R	Bacterial inhibition zone in mm		
		S. aureus	*E. coli*	*P. vulgaris*
3a	H	9	6	8
3b	2-OH	11	–	9
3c	4-Cl	14	10	–
3d	4-Br	13	11	11
3e	2,6-Cl	15	13	14
3f	2,6-Br	11	12	11
4a	H	15	13	12
4b	2-OH	17	15	14
4c	4-Cl	20	16	15
4d	4-Br	16	14	16
4e	2,6-Cl	22	18	19
4f	2,6-Br	21	16	17
Ampecillin		20	18	18

The physical, other properties and spectral records are represented in Tables 1 and 2.

Compounds 4(b–f) were manufactured by using same method explained for compound 4a. The physical and spectral data of compounds 4(b–f) are giving in Tables 1 and 2.

Antibacterial Activity

Compounds 3a–3f and 4a–4f were tested for antibacterial activity and pharmacological details of these compounds have mentioned in the Table 3. From the table it was cleared that these compounds have shown significance inhibition against microorganism. Methoxy group on the indole nucleus increase the antibacterial activity. Compounds having azetidinone moiety exhibited better antibacterial activities. Compound 4e was found more potent antibacterial agents against *S. aureus* and *P. vulgaris* with standard drug ampecillin. Compounds 3e, 4b, 4c, 4d and 4f have shown moderate activity while rest compounds showed mild activity.

4 Conclusion

We have synthesized indole derivatives containing azetidinone moiety using easily available starting materials. These compounds were confirmed by spectral and elemental details and antibacterial activity of these compounds was also investigated against different bacteria i.e. staphylococcus aureus, proteus vulgaris and *Escherichia coli* using cup plate method. Biological activity was recorded in form zone of inhibition in mm and compared with standard drug ampecillin. Mostly compounds have shown moderate degree of activity. Compounds 4a–4f containing azetidinone ring exhibited better antibacterial activity than compounds 3a–3f. 2,6-dichloro substituted indole derivatives have shown more efficient because it contain more electronegative atom than other substituents. Compounds 4e, 4f against *S. aureus* and compound 4e against *P. vulgaris* have shown good antibacterial activity than standard drug.

References

1. 1. Quazi, I., Sastry, V.G., Ansari, J.A.: Synthesis and antimicrobial activity of indole derivative bearing the pyrazole moiety. International Journal of Pharmaceutical Sciences and Research. 8 (3). (2017) 1145–1152.
2. Singh, A.K., Prasad, R.K., Singh, C.S.: Synthesis, characterization and pharmacological evaluation of some novel 3-indole derivatives. 5 (2). (2013) 311–119.
3. 3. Nataraj, K.S., Rao, J.V., Jayaveera, K.N.: Synthesis and antimicrobial activity of new indole derivatives. Int. Chem. Sci. 8 (1). (2010) 609–616.
4. Hishmat, O.H., Rahman, A.H.A., Nasef, M.H., Abdel-Hamid.: Synthesis and antimicrobial activity of some indole derivatives. Archives of Pharmacal Research. 11 (4). (1988) 266–269.
5. 5. Chandra, T., Garg, N., Kumar A.: Synthesis and anti-inflammatory activity of indole derivatives. International Journal of Chem Tech Research. 2 (2), (2010) 762–773.
6. 6. Liu, Z., Tang, L., Zhu, H., Xu, T., Qiu, C., Zheng, S., Gu, Y., Feng, J., Zhang, Y., Liang, G.: Design, synthesis and structure activity relationship study of novel indole-2-carboxamide derivatives as anti-inflammatory agents for the treatment of sepsis. Journal of medicinal Chemistry. 59 (10). (2016) 4637–4650.
7. 7. Georgel, M., Joseph, L., Thomas, A.M.: Synthesis, characterization and biological screening of novel derivatives for certain pharmacological activities. International Journal of Academic Research and Development. 2 (4). (2017) 240–243.
8. Jaishree, B., Manjulatha, K., Girish, M., Adil, S., Purohit, M.G.: Synthesis and biological evaluation of some N-substituted indoles. ARKIVOC. (xii). (2009) 217–231.
9. 9. Swathi, K., Sarangapani, M.: Synthesis and screening of anti-inflammatory activity of indole derivatives. International Journal of Pharmaceutical Research and Biomedical Analysis. 1 (3). (2012) 10–14.
10. 10. Priyanka, B., Pogula, M., Suresh, S., Thirupathi, K., Sammaiah, G.: Synthesis and characterization of new indolederivatives for analgesic activity. International Journal of Pharmacy and Pharmaceutical Sciences. 4. (2012) 231–233.
11. 11. Abdellatif, K.R.A., Lamie, P.F., Omar, H.A.: 3-Methyl-2-phenyl-1-substituted-indole derivatives as indomethacin analogs: design, synthesis and biological evaluation as potential anti-inflammatory and analgesic agents. Journal of Enzyme Inhibition and Medicinal Chemistry. 31 (2). (2016) 318–324.
12. 12. Kulkarni, S.D., Tankar, A.N., Patwardhan, K.B., Govindwar, R.B., Anandrajgopal, K. and Tiwari R.V.: Evaluation of anticonvulsant activity of novel indole derivatives. Int. J. Chem. Sci. 6 (2). (2008) 926–932.
13. 13. Patil, P.O., Bari, S.B.: Synthesis, characterization and screening for antidepressant and anticonvulsant activity of 4,5-dihydropyrazole bearing indole derivatives. Arabian Journal of Chemistry. 9 (4). (2016) 588–595.
14. 14. Rohini, R.M., Manjunath, M.: Synthesis and anti-convulsant activity of triazothiole/thiazolylthiazolidinone derivatives of indole. Der Pharma Chemica. 4 (6). (2012) 2438–2441.

15. 15. Kaur, H., Kumar, S., Kumar, A.: Synthesis, antipsychotic and anticonvulsant activity of some new pyrazolinyl/isoxazolinylindol-2-ones. International Journal of Chem Tech Research. 2 (2). (2010) 1010–1019.
16. 16. Swathi, K., Sarangapani, M.: Synthesis and screening of biologically significant indole derivatives for anticonvulsant activity. International Journal of Pharmacy and Biological Sciences. 3 (1). (2013) 628–635.
17. 17. Patel, T.M., Patel, A.M.: Synthesis and antimicrobial activity of newly azetidinone derivatives. Der Pharmacia Lettre. 4 (2) (2012) 579–583
18. 18. Mehta, P., Davadra, P., Pandya, J.R., Joshi, H.S.: Synthesis, characterization and antimicrobial activity of 2-azetidinone derivatives of benzimidazoles. International Letters of Chemistry, Physics and Astronomy. 11 (2). (2014) 81–88.
19. 19. Deep, A., Kumar, P., Narasimhan, B., Lim, S.M., Ramasamy, K., Mishra, R.K., Mani V.: 2-Azetidinone derivatives: synthesis, antimicrobial, anticancer evaluation and qsar studies. ActaPoloniaePharmaceutica- Drug Research. 73 (1). (2016) 65–78.
20. Gawande, S.K., Khadsan, R.E.: Synthesis and biological evaluation of azetidinone derivatives from 2-α (phenylacetyl) benzohydrazide moiety by microwave method. Der Pharma Chemica.); 6 (2): (2014) 70–74.
21. 21. Patel, D.B., Desai, V.: Synthesis and biological evaluation of azetidinone derivatives as antibacterial and antifungal agents. International Journal of Pharm Bio. Sci., 8(1). (2017) 67–173.
22. 22. Jambu, S.P., Patel, Y.S.: Synthesis, characterization and biological aspects of novel azetidinone derivatives, Journal of chemical and Pharmaceutical Research. 8 (1). (2016) 745–749.
23. 23. Ali, A.T., Mosa, M.N., Alshaheen, Z.G., Ali, M.A.M.: Sythesis, characterization ad antibacterial evaluation of oxaazetidin- benzene sulfonamide derivatives as a hybrid antimicrobial agents. Sys Rev Pharm. 11 (2): (2020) 487–494.
24. 24. Toraskar, M.P., Kadam, V.J., Kulkarni, V.M.: Synthesis and antifungal activity of some azetidinones. International Journal of Chem Tech Research. 1 (4). (2009) 1194–1199.
25. 25. Sulthana, R.M., Quine, S.D.: Synthesis, characterization and biological activities of some azetidin-2-ones as potential antifungal agents. World Scientific News. 92 (2). (2018) 247–259.
26. 26. Kumbhar, S.P., Mayur, Y.C., Poul, B.N., Waghmare, P.V.: Synthesis of some new azetidinones derived from heterocyclics and their antimicrobial activity. International Journal of Pharmaceutical and Clinical Research. 7 (1). (2015) 84–91.
27. 27. Sharma, R., Samadhiya, P., Srivastava, S.D., Srivastava, S.K.: Synthesis and biological significance of some 2-azetidinone derivatives. Journal of Sciences, Islamic Republic of Iran. 23 (2). (2012) 139–146.
28. Archana.: Synthesis of newer substituted azetidinone and thiazolidinone derivatives as potent anticonvulsant agents. International Journal of Techno Chem. Research. 2 (2). (2016) 121–126.
29. 29. Smith, Q. E.: In pharmacological screening tests progress in medicinal chemistry. Butterworths London. 1 (1960) 1–33.

A Review of Challenges and Safety Measures in Petroleum Industry During COVID-19

Siddharth Sharma, Surajbhan Singh, Kanchan Deoli Bahukhandi, and S. M. Tauseef

Abstract The whole world is going through the Coronavirus pandemic. At this time of COVID-19, Indian authority has also chosen the path of lockdown and everything has come to a halt. Each and everything has shut off except essential commodities. Transportation, construction, small and big industries, and factories have also been closed. This pandemic has affected the economy of the whole world. The petroleum industry is the key industry that contributes to the economy of the country and it has also been shut down because of this pandemic. Now the whole world is working on reconstructing their economy in the middle of this pandemic. The Government has allowed factories, industries (including the petroleum industry) to restart their work. This study will provide overview of the safety measures and occupational safety which need to be incorporated after the restarting of the petroleum industry.

1 Introduction

In the petroleum industry the safety is always at the forefront. The workers working in the petroleum industry are always exposed to high risk [1]. At this time of the Corona pandemic, the risk in the petroleum industry has increased, so the workers need to be given careful attention to safety in the petroleum industry and the culture of preventive safety should be encouraged. As the whole world is facing unprecedented lockdown, everything has come to halt including the petroleum industry. COVID-19 overshadows all human evolution, intelligence, and technology. Today pandemic is making roots in our life and giving us mighty harm. Now the world is restarting and because of the reopening of industries, markets, and public places the severity of COVID-19 has also increased. So, at this time more awareness is needed and the applications of more preventive measures are required [2]. The pandemic causing huge impact on petroleum industry. The less demand for petroleum products the

S. Sharma (✉) · S. Singh · K. D. Bahukhandi (✉) · S. M. Tauseef
Student of Department of Health Safety and Environment, University of Petroleum and Energy Studies, Dehradun, India
e-mail: kanchan@ddn.upes.ac.in

N. A. Siddiqui et al. (eds.), *Advances in Environment Engineering and Management*, Springer Proceedings in Earth and Environmental Sciences,
https://doi.org/10.1007/978-3-030-79065-3_15

petrochemical industries are going face huge challenges. We have also witnessed the lowest oil price in the history of the US. The oil price was decreased to mines 32.63$ [3]. The main problem which will arise during this COVID-19 pandemic is the maintenance of the machines and sanitizing of the workplace including the machinery [4–6].

2 Problems in Restarting the Petroleum Industry

Due to this COVID-19, In India, a lockdown was announced by Honorable Prime Minister Shri Narendra Modi India. Many problems were going to arise for the restarting of the petroleum industry. Some of them are discussed below.

i. Lack of human resource: In India more than 1 Million Migrant workers who were working in the petroleum industry are moving to their respective native places [7]. So for work, 85% of the workers are not available. As to fulfill their needs they are not getting enough wages and by this, the transportation of raw material has also been adversely affected. Some of them are still living nearby industries, in camps, run by the government or industry itself. Now the Government has allowed opening the industry. At the time when COVID-19 was expanding its roots in society, most of them moved to their native places. Because of this, it's impossible for them to come back and start working again. There is no transportation facility for workers available this time and they don't want to risk their life too. COVID-19 affects them economically as well as mentally and mental illness is a very severe issue while working in the petroleum industry [8].
ii. Impact on Cost: Restarting of work is a costly process, for any industries especially in an oil refinery. The reopening of oil refineries needed maintenance and inspection of the heavy machinery [4, 9]. The risk assessment also plays a very important role here [10]. It can be understood by a recent accident which was happened at Vishakhapatnam. In that accident, Styrene Gas was leaked due to cooling system failure. More than 1000 people were affected and more than 11 people lost their life [11]. Petroleum Industry needs to take more precautions due to increased chances of accidents due to shut down of work. The transportation expenditure will also increase. Suppose in an industry there were 3 buses required picking and dropping their employees but now they required 7 seven buses because they have to maintain social distancing for avoiding the spread of the COVID19 virus. Before COVID19 they didn't have to spend anything on sanitization and hygiene [2, 12]. Now they have to spend money on the sanitization of the whole premises including buses, machinery, and all other equipment. Petroleum Industry has to maintain social distancing while working like this can also reduce the speed of the work. Industries have to spend more on the PPE and medical facilities also. PPE plays an important role to avoid not only occupational hazards, but it can also help in stopping

the spread of COVID-19. Storing and cleaning of PPE kits contribute to cost increment and also require special maintenance and care [13, 14].

iii. Lack of experienced workers: As migrants have moved to their native places, now the petroleum industry is facing a new problem which is a lack of workers who are experienced. Many experienced, employees have been moved on, and now the industry is unwillingly forced to work with inexperienced workers. Working hours all increased by the government for taking the economy on track [7]. With increased working hours, chances of accidents also increased and it will have a negative effect on the health of the workers. It has been seen that increased working hours increased cardiovascular diseases among the workers [15]. Now working speed has been increased and also, it's an open invitation to an accident. Inexperienced new workers required training and time to replace the experienced workers. So, industries have to go through the whole procedure again and it requires money and time as well.

As discussed above in restarting the petroleum industry various problems like Lack of human resource, Impact on Cost and Lack of experienced worker has occurred due to which hazards have increased and have led to many accidents. Some of the hazards are discussed below.

3 Hazards Present in the Industry

There are many hazards that are present in the petroleum industry. Some of them are

i. Physical health or wellbeing hazard
ii. The risk of chemical health hazard
iii. The risk to Biological Safety
iv. Risk to ergonomic health
v. Risk to psychosocial wellbeing [16, 17, 10, 18].

At this time of COVID-19 pandemic, not only in the petroleum industry but in all industry the risk of these hazards has increased.

It is well known that sanitization has become an essential requirement in the petroleum industry and most of the sanitizers which are used to sanitize are alcohol-based. As it is known that alcohol is a combustible material so it can ignite into flames and can result in a severe accident. Employee health and safety is paramount. Due to the properties of petroleum products, operations tasks, and processes in the petroleum industries are dangerous, an occupational hazard in the oil industry is a major issue [19, 20].

4 Accidents Occurred in the Petroleum Industry During COVID-19

COVID-19 period has surely increased the number of accidents that happened in the Petroleum, Construction, and Process Industries. Accidents which are happened during COVID-19

1. Visakhapatnam Gas Leakage—On the morning of 7th of May 2020, at the LG polymers plant located, Visakhapatnam, Andhra Pradesh, India, an accident happened. Styrene gas leaked from the storage tanks because of improper maintenance. According to the primary investigation, storage tanks had unattended since March 2020 due to COVID-19. The failure of the refrigerating system of the storage tanks leads to the evaporation of chemicals. Styrene gas spread over 5 km around the plant. More than 1000 people were affected 11 people had died in this accident. Because of this accident people have to leave their homes and move to other places [11].
2. Siberia Oil leak—On 29 MAY 2020, the oil storage tank started leaking from the bottom holes of the tanks. More than 21,000 Ton oil leaked and emerged in the Ambarnaya River. The Cause of the leakage was holed which was made due to corrosion. This is the second-highest oil leakage accident in the history of Russia. Russia declares this accident is a state of emergency. Environmentalists are saying it will take decades in restoring the ecological balance of the river [21].
3. Assam's Tinsukia Oil Well- On the early morning of 27 MAY 2020 an oil well of Tinsukia, Assam, India was a blowout. Gas had been leaking continuously and after 13 days of continuous flow out well caught fire and while the sweeping accident happened and 2 firemen lost their life. From the primary investigation, machine failure is the cause of the blowout of the oil well. More than 1600 families evacuated from the hazardous area [22].
4. Texas Eastern Gas Pipeline Explodes in Kentucky—On 4 MAY 2020 at around 5 PM, a natural gas pipeline was blown out in Fleming County, Kentucky, USA. There was no injury reported. Line no 10 of natural gas was blowout Enbridge company's crew members responded quickly and secured the incident area [23].
5. Explosion Valero Refinery: On the early morning of 10/4/2020 explosion occurred at Valero Energy Corp Meraux, Louisiana, and refinery on the east side of New Orleans. One person had injured in the fire. The cause of the explosion is unclear till now [24].
6. AL-Sabriya Refinery in Kuwait—On 11 JUN 2020 fire accident happened et al. Sabriya Refinery in Kuwait. One person died in that fire accident and one worker injured. An accident occurred at dawn during the drilling process. The investigation is under process [25].
7. Telangana Hydroelectric Power Plant Fire Accident—On Aug 20, an accident happened at a Hydroelectric Power Plant located in Telangana India. Around 10:30 PM fire stated due to the short circuit, at the time 20 workers were on the sift 11 were managed to escape and 9 people lost their life. Due to heavy

smoke on the escape tunnel, those 9 workers can't manage to escape. After the lockdown and the heavy rain, the power plant was in full swinging. An investigation is in under process of finding the exact cause of the short circuit and till then power generation has been suspended [26].

8. Bharuch tank explosion—On 3 June, 12 PM an explosion happened at a tank of Argo-chemical factory called Yashashvi Rasayan Private Ltd in Dahej Industrial Estate of Gujarat's Bharuch district. 8 people lost their lives and 51 people were injured. According to the primary investigation, the explosion happened due to an unsafe working act. Minister said the whole factory has to go under safety audit and it's charged 250 million as plenty [27].

Summary of the accidents are shown below in the form of Table 1.

As we can see there are many accidents which happened in the petroleum industry during COVID-19, each and every accident has different causes but the one thing which is common in all these accidents was the lack of attention to the machinery which should have been paid by the industry workers. As in this time of COVID-19, the industry has been closed for a while and the workers have also moved to their places. So the maintenance has been totally ignored by them which have led to these accidents. As per the above-mentioned accidents, the causes for these misfortunes were maintenance failure or any other reasons which are related to this pandemic. So, the indirect cause was COVID-19.

In this COVID-19, everybody realizes that cleansing is a must, so in the wake of opening all industries, everybody should play it safe while doing sanitizing. So, because of this pandemic, it has become a necessity of sanitizing the industry. Generally, the alcohol-based sanitizer contains a mix of isopropyl liquor, ethanol (ethyl liquor), or n-propanol, containing 60–95% liquor by volume. As all these are combustible and effectively can ignite into flames with a modest quantity of start source so precautions have to be taken by keeping these points in mind. Similarly, the alcohols which are alike to ethanol and methanol can burn effectively because they form a vapor cloud of ethanol and methanol if they spill off at room temperature, on the grounds during the process of sanitization and making the threat of fume start increasingly huge whenever spilled or spread during the procedure of cleansing in the petroleum industry which can lead to a terrible accident [12, 28, 29].

5 Measures Taken for Safety and the Maintenance of the Machinery

To ensure the protection of the workers against COVID-19 there is a need to set up SOP, i.e., Standard Operating Procedure. There are numerous estimates which must be adopted by an employer at the time of working in the industry, which is as follows [30]:

Table 1 Summary of the accidents

S. No.	Date of the accident	Location of the accident	Cause of the accident
1	7 MAY 2020	At LG polymers chemical plant located at R. R. Venkatapuram village, Visakhapatnam, Andhra Pradesh, India	Styrene gas leaked from the storage tanks because of improper maintenance. The failure of the refrigerating system of the storage tanks leads to the evaporation of chemicals According to the primary investigation, storage tanks had unattended since March 2020 due to COVID-19 [11]
2	29 MAY 2020	At Norilsk-Taimyr Energy (NTEK), Siberia, Russia	Oil leaked from Tank 5 of the Norilsk-Taimyr Energy (NTEK) which is a coal-fired combined heat and power plant. The Cause of the leakage was holes that were made due to corrosion. So improper maintenance and inspection leads them to 2nd largest oil spillage accident in the history of Russia [21]
3	27 MAY 2020	At Tinsukia, Assam, India	From the primary investigation machine failure is the cause of the blowout of the oil well [22]
4	4 MAY 2020	At Fleming County, Kentucky, USA	The investigation is under process still reason for the fire explosion is unknown [23]
5	10 APR 2020	At Valero Energy Corp Meraux, Louisiana	The investigator said that a pressure relief valve blew up with a spark and that sparkled to an explosion [24]
6	11 JUN 2020	et al.-SabriyaRefinery, Kuwait	According to the primary investigation the cause of the explosion was the sudden ignition of accompanying gases during a drilling operation at the tower [25]
7	20 AUG 2020	At Telangana, India, Hydroelectric Power Plant	According to primary investigation the cause of the fire was a short circuit [26]
8	3 JUN 2020	At Bharuch, Gujarat, India, Yashashvi Rasayan Private Ltd	According to primary investigation unsafe act was the main cause of the explosion. The Factory had been not following safe working acts [27]

- Maintenance and safe work practices are the keys in the petroleum industry in this COVID-19 pandemic and that could be achieved by good leadership. In restarting, the phase value of inspection and maintenance will be increased.
- Training is also an important aspect in the restart phase, many experienced workers absquatulated with what they have because of no job security. It should be given to both old and newly recruited employees.
- All parts of the petroleum industry working environment including the industry's entrance gate, washroom, cafeterias, lift, wall and other various surfaces must be fully clean daily.
- Administrators need to coordinate transportation for all company workers who don't depend on public transportation and make sure that each and every vehicle is only loaded up to a maximum capacity of 25–35%.
- Both automobiles and equipment which are coming in this sector should be thoroughly cleaned and disinfected.
- Compulsory thermal inspection of each person entering the industry and leaving the industry and proper PPE kit should be provided to the workers to avoid the spread of COVID-19.
- Administrators must provide medical care to all workers in the industry.
- The Administrators must provide a sufficient amount of hand sanitizer, if possible, with the touch-free machinery on all passage and exit areas and at all mutual zones of the working environment.
- Working environments should have a break of at least sixty minutes between each and every move.
- Gatherings in social events exceeding ten individuals are disheartened and there must be a distance of 6 feet separation between every worker while seated.
- Special sessions should be introduced for providing knowledge about personal hygiene and social distancing.
- The limit of employees who are allowed in a lift at once should be two and use of the stairs should be followed.
- A severe restriction should be there on tobacco-based products and expelling in the premises should be banned and there must be fines for this.
- All unnecessary visitors should be strictly prohibited from the industry.
- All employees in the industry exceeding the age of 65 must be urged to work, from their home itself.
- All Employees in the industry must present concentrated training and communication skill on good cleanliness, cleanliness practices, particularly in the working environment [9, 10, 31].

All the preventive measures which should be taken by the workers after entering the petroleum industry (Fig. 1).

- Cleanliness—Cleanness is very important in this COVID19 period. After every shift and before starting the shift work area should be washed or sanitized properly.
- Latrines and urinals: Latrines and urinals washed properly frequently.
- Spittoons: No one should be allowed to spit inside the premises except spittoons and whoever spits in contravention shall be punishable with a fine.

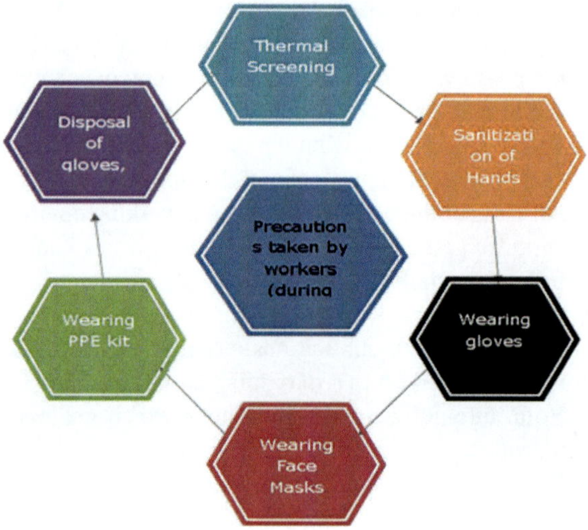

Fig. 1 Flow chart of preventive measures

- Washing facilities: There should be washing facilities inside the refinery separate for males and females. Workers should be properly screened and employers should make sure no one should be contaminated.
- Disposal: Everyone should get their own PPEs and those PPEs will be properly disposed of or sanitized. Disinfection and reuse of the mask and other PPE facility should be present on the premises. If not then disposable masks should be provided. Mainly radioactive waste, oily hazardous waste is generated by the Oil refineries so it is very essential to handle waste carefully [7, 32–35].

As discussed above, if someone not follows the guideline provided by the safety committee then there should be a penalty for offenses. Occupational health safety is every employee's right. So, by taking into the consideration of COVID19 special safety officers should be appointed who will take care of employees from COVID19 and also provide guidelines [1, 2, 16, 36]. The operational planning and control measures are needed for working in the petroleum industry during COVID-19 (Fig. 2).

6 Conclusion

The chances of accidents have been increased because of improper maintenance which is due to COVID-19 and the lack of experienced employees. The industry should follow different safety practices at this time of the COVID-19 pandemic. Cleanliness, use of PPE kit, thermal scanning, avoid gathering, washing facilities, and how to sanitize the industry, how much alcohol should be present in the sanitizer or disinfectant should be the main practices that industry should follow. With the help of good leadership maintenance and healthy work activities in the industry can be

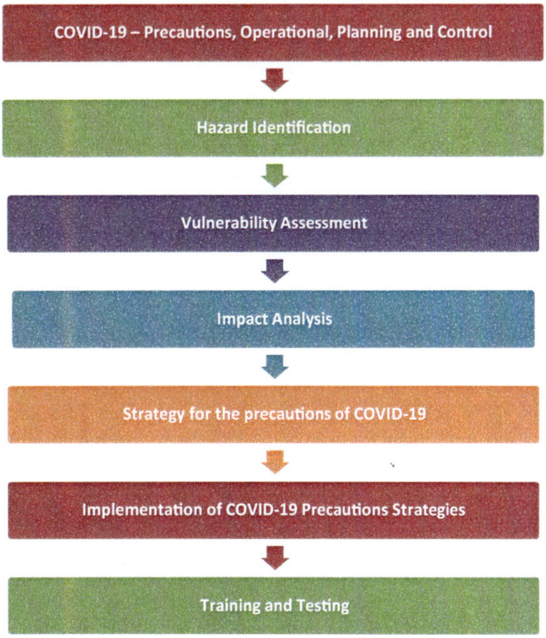

Fig. 2 Flowchart of operational planing

achieved. Training of each and every employee in the industry is also important at this time of COVID-19 pandemic. Special Sessions should be introduced to the workers for the training. This pandemic and lockdown have also shown many positive aspects as working from home was not normal in the industry however this New Normal working from the offsite location has become the new standard.

References

1. Pintelon, Liliane, and Peter N. Muchiri. 2009. "Safety and Maintenance. Pp. 613–48 in *Handbook of Maintenance Management and Engineering*. London: Springer London.
2. "WHO India | World Health Organization." https://www.who.int/india (accessed Jun. 22, 2020).
3. "Explained: Why Did Oil Prices Turn Negative in the US and What Does It Mean for Indian Consumers?, Energy News, ET EnergyWorld." Retrieved November 24, 2020.
4. A. L. Kosta and K. Kishore, "Refinery Inspection and Maintenance," in Petroleum Refining and Natural Gas Processing, ASTM International, 2013, pp. 393–435.
5. The Institution of Occupational Safety and Health (IOSH). (2015) "Systems in focus - Guidance on occupational safety and health management systems." http://www.iosh.co.uk/systems (Erişim tarihi: 27.02.2019).
6. Li, Y.,Guldenmund F.W. (2018) " Safety management systems: A broad overview of the literature", Safety Science, p. 100, 2018.
7. "MINISTRY OF LABOUR AND EMPLOYMENT Government of India."

8. A. McCloughen, K. Foster, M. Huws-Thomas, and C. Delgado, "Physical health and wellbeing of emerging and young adults with mental illness: An integrative review of international literature," International Journal of Mental Health Nursing, vol. 21, no. 3. Int J Ment Health Nurs, pp. 274–288, Jun. 2012, doi: https://doi.org/10.1111/j.1447-0349.2011.00796.x.
9. 11. S. Çaliş and B. Y. Buÿükakinci, "Occupational Health and Safety Management Systems Applications and A System Planning Model," in Procedia Computer Science, 2019, vol. 158, pp. 1058–1066, doi: https://doi.org/10.1016/j.procs.2019.09.147.
10. EzejioforJIN, IwualaMOE, Osuala FOU, NwigweH.C.Risk assessment: Prevalent Occupational hazards in Nigeria Petroleum Industry. Jour Med Invest and Pract. 2012; 8(1): 1–3.
11. "India gas leak: At least 11 dead after Visakhapatnam incident - BBC News." https://www.bbc.com/news/world-asia-india-52569636 (accessed Jun. 22, 2020).
12. 14. White, C., et al., "The Effect of Hand Hygiene on Illness Rate Among Students in University Residence Halls," American Journal of Infection Control, October 2003, Vol. 31, No. 6, pp. 364–70.
13. 15. N. J. Rowan and J. G. Laffey, "Challenges and solutions for addressing critical shortage of supply chain for personal and protective equipment (PPE) arising from Coronavirus disease (COVID19) pandemic – Case study from the Republic of Ireland," Sci. Total Environ., vol. 725, p. 138532, Jul. 2020, doi: https://doi.org/10.1016/j.scitotenv.2020.138532.
14. 16. A. Tabah et al., "Personal protective equipment and intensive care unit healthcare worker safety in the COVID-19 era (PPE-SAFE): An international survey," J. Crit. Care, vol. 59, pp. 70–75, Jun. 2020, doi: https://doi.org/10.1016/j.jcrc.2020.06.005.
15. J. Li et al., "The effect of exposure to long working hours on ischaemic heart disease: A systematic review and meta-analysis from the WHO/ILO Joint Estimates of the Work-related Burden of Disease and Injury," Environment International, vol. 142. Elsevier Ltd, p. 105739, Sep. 01, 2020, doi: https://doi.org/10.1016/j.envint.2020.105739.
16. F. Eyayo, "Evaluation of Occupational Health Hazards among Oil Industry Workers: A Case Study of Refinery Workers," 2014. Accessed: Jun. 17, 2020. [Online]. Available: www.iosrjournals.org.
17. S. and E. D. I. (SEDInst), "PATTERN OF OCCUPATIONAL HAZARDS AND PROVISIONS OF OCCUPATIONAL HEALTH SERVICES AND SAFETY AMONG WORKERS OF KADUNA REFINERY AND PETROCHEMICAL COMPANY LTD (KRPC), KADUNA, NIGERIA - A.A. Aliyu and S. Saidu".
18. 18. A. O. Eguvbe, N. Akpede, and J. E. Egbagba, "Occupational health hazards in small and medium-scale manufacturing industries in Anambra State, South East, Nigeria," World J. Prev. Med., vol. 5, no. 1, pp. 26–32, 2017, doi: https://doi.org/10.12691/jpm-5-1-4.
19. 19. B. Krishnakumar and S. Rana, "COVID 19 in INDIA: Strategies to combat from combination threat of life and livelihood," J. Microbiol. Immunol. Infect., vol. 53, no. 3, pp. 389–391, Jun. 2020, doi: https://doi.org/10.1016/j.jmii.2020.03.024
20. 20. A. Senthil et al., "Perception and prevalence of work-related health hazards among health care workers in public health facilities in Southern India," Int. J. Occup. Environ. Health, vol. 21, no. 1, pp. 74–81, Jan. 2015, doi: https://doi.org/10.1179/2049396714Y.0000000096.
21. "Arctic river will take decades to recover from fuel spill: Russian official, Energy News, ET EnergyWorld." https://energy.economictimes.indiatimes.com/news/oil-and-gas/arctic-river-will-take-decades-to-recover-from-fuel-spill-russian-official/76166505 (accessed Jun. 22, 2020).
22. "Widespread Anger Against Company, Govt After Two Die in Oil Field Fire. https://thewire.in/environment/assam-oil-gas-well-fire-two-killed. (accessed June 25, 2020).
23. "Texas Eastern Gas Pipeline Explodes in Kentucky | Pipeline & Gas Journal." https://pgjonline.com/news/2020/05-may/texas-eastern-gas-pipeline-explodes-in-kentucky (accessed Jun. 22, 2020).
24. "Explosion rocks Valero refinery." https://www.hydrocarbonprocessing.com/news/2020/04/explosion-rocks-valero-refinery (accessed Jun. 22, 2020).

25. "Oil worker killed, another injured in Al-Sabriya refinery - KOC - TimesKuwait." https://www.timeskuwait.com/news/oil-worker-killed-another-injured-in-al-sabriya-refinery-koc/ (accessed Jun. 22, 2020).
26. "Telengana Hydroelectric Power Plant Fire Accident: At Least 9 People Killed In a Massive Fire https://energy.economictimes.indiatimes.com/news/power/nine-killed-in-fire-at-telanganas-srisailam-hydroelectric-power-plant/77686331 (accessed Aug. 20, 2020).
27. Bharuch tank explosion - At Least 5 People killed and 32 Injured In Massive Tank Explosion https://timesofindia.indiatimes.com/city/surat/gujarat-5-dead-32-injured-in-dahej-pesticide-company-explosion/articleshow/76175766.cms (accessed Jun. 5, 2020).
28. W. M. Cavage, "Flammability Test of Alcohol-Based Hand Sanitizer," 2010.
29. Marker, T. and Do, D., "Fire Testing of Ethanol-Based Hand Cleaner," DOT/FAA/ARTN98/15, April 1998.
30. "Standard Operating Procedures Introduced For Social Distancing In Workplaces In India - Coronavirus (COVID-19) - India." Retrieved November 24, 2020.
31. "FUNDAMENTAL PRINCIPLES OF OCCUPATIONAL HEALTH AND SAFETY."
32. "Factories Act Government Of India".
33. "Recycling and Recovery Routes of Plastic Solid Waste (PSW): A Review." *Waste Management* 29(10):2625–43.
34. "ENVIRONMENTAL IMPACT ASSESSMENT OF OIL&GAS INDUSTRY | Mamdouh F. Abdel-Sabour | Research Project." Retrieved November 24, 2020.
35. *MINISTRY OF LABOUR AND EMPLOYMENT Government of India.*
36. 37. R. E. Witter, "Guidelines for hazard evaluation procedures. Second edition," Plant/operations Prog., vol. 11, no. 2, pp. 50–52, Apr. 1992, doi: https://doi.org/10.1002/prsb.720110207.

Sustainable City Planning Strategy Review for Next-Level Technology, Earthquake-Resistant Buildings, Dehradun City

Geetanjli Rani, P. A. Arun, Nitty Ann Abraham, Shariq Ansari, and Umar Muktar

Abstract The Indian state of Uttarakhand was formed in 2000 to transform the life and economy of people in hilly region. The state government aims to achieve the target set by United Nation's Sustainable Development Goals framework by 2030. The state's vulnerability to various natural disasters can be potential challenge in achieving these targets. The capital city of Uttarakhand, Dehradun is situated in seismic zone IV and in the last few decades Dehradun has witnessed rapid urbanization causing construction of buildings compromising on safety and people occupying vulnerable location. This paper focus on spatial and social vulnerability analysis of Dehradun city suing GIS mapping. The study classifies wards of Dehradun city into different categorizes based on risks such as slope, built up area, population density and identifies Spatial Hazard Risk Zones (SHRZ) of Dehradun city. Based on the study recommendations are made for effective management of earthquake risk in Dehradun city. The study proposes few techniques that city planners may consider for earthquake resistant construction in future. Outcomes and recommendations of the study can be handy to city planners and administration team to enable them to plan for these critical risks and achieve state's sustainability targets.

1 Introduction

The Indian state of Uttarakhand was formed on 9th of November, 2000 to transform the life and economy of people in hilly region. The state government aims to achieve the target set by United Nation's Sustainable Development Goals framework by 2030

G. Rani (✉) · P. A. Arun · N. A. Abraham · S. Ansari · U. Muktar
School of Engineering, UPES, Dehradun, India
e-mail: grani@ddn.upes.ac.in

P. A. Arun
e-mail: arun.pa@ddn.upes.ac.in

S. Ansari
e-mail: 500051292@stu.upes.ac.in

N. A. Siddiqui et al. (eds.), *Advances in Environment Engineering and Management*, Springer Proceedings in Earth and Environmental Sciences,
https://doi.org/10.1007/978-3-030-79065-3_16

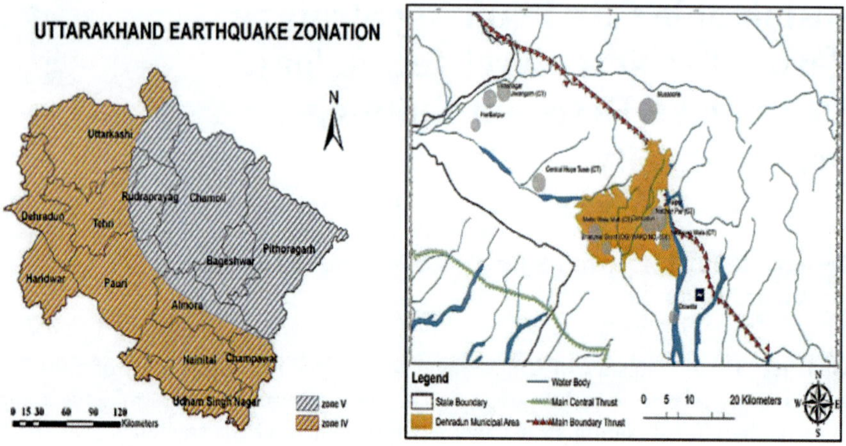

Fig. 1 Dehradun earthquake zonation and location between MCT and MBT

[1]. The state's vulnerability to various natural disasters can be potential challenge in achieving these targets. Livelihood of people are being severely affected by frequent occurrence of earthquakes, landslides, flashfloods, cloudbursts, droughts and cold waves in this region. The entire state falls under Zone IV and Zone V of earthquake zonation, and state capital, Dehradun is located in Zone IV [2]. After becoming the state capital, Dehradun has witnessed unpreceded growth in the number of industries, infrastructure and different types of residential and industrial structures being constructed in this area. It is estimated that the urban built-up has grown from 982.64 Ha in 2003 to 7162 Ha in 2015. This has led to overcrowding, mass encroachments and constructing buildings in vulnerable areas compromising on building regulation [3, 4]. Figure 1 shows the earthquake zonation maps for Dehradun city.

The impacts of tremors on a poorly arranged community have demonstrated to be destroying with long-term effects. Seismic risk reduction (SRR) looks into the measures, which can be taken to forestall or decrease impact of an earthquake, essentially before one occurs. Preparedness is a significant segment of risk reduction, involving taking measures beforehand to caution and educate communities on ways to be prepared in the event of an earthquake, and teaching people to react properly.

It is well documented that usage of technology can help the city planners in mitigating the associated with earthquakes. Aim of this research work is to prepare the earthquake hazard vulnerability maps of Dehradun city, strategize the safety of buildings in high-risk areas based on spatial and social vulnerability analysis through GIS mapping, and recommend devices suitable for technology implementation. This work prepares ward wise risk maps of Dehradun municipal area and suggests various Non-Destructive Testing (NDT) that can be utilized for assessing the structural integrity of buildings in vulnerable areas. Recommendations are also made to strengthen earthquake disaster preparedness to create earthquake resilient communities.

2 Methodology

The methodology adopted for this work is depicted in Fig. 2. Extensive literature review was carried out to comprehend the earthquake disaster management framework in Dehradun city, these includes review of disaster management plans and guidelines at national, state and district levels, city development plans, Building byelaws and details pertaining to geographical and social livelihood of Dehradun city. Seismic risk mitigation strategies adopted by countries which are highly vulnerable to earthquake risks were also reviewed. The data collected are fed into GIS software and created spatial, social vulnerability maps of Dehradun city. Based on national guidelines and standards the city map is categorized into different risk zones. From these risk maps highly vulnerable zones and wards are identified. Based on the review of best practices adopted by developed countries, techniques are suggested to assess the structural integrity of buildings in vulnerable areas and suggestion are made to improve the earthquake disaster management preparedness in Dehradun city.

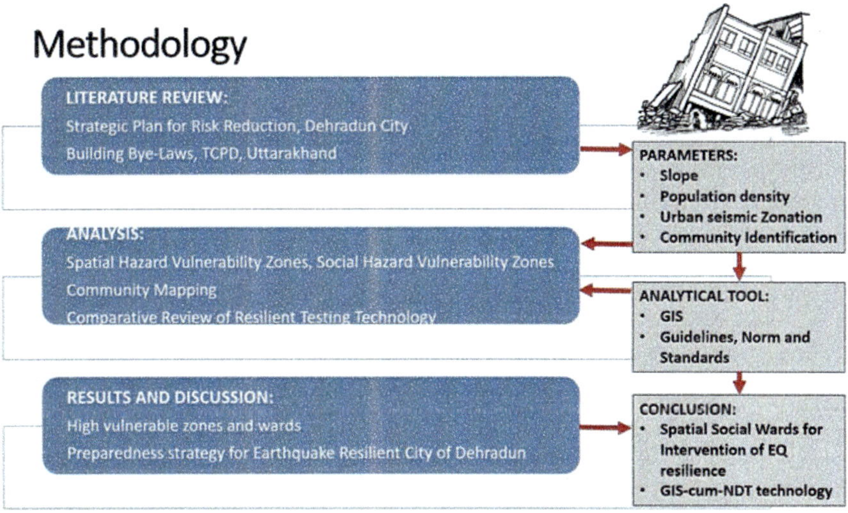

Fig. 2 Methodology

Table 1 History of earthquake in Karakoram fault and injuries in earthquake of Uttarkashi (1991) and Chamoli (1999)

S. No	Year	Location	Magnitude
1	1803	Upper Ganga Valley	6.5
2	1816	Near Gangotri	6.5
3	**1916**	**Near Dharchula**	**7.5**
4	1958	Kapkot	6
5	1980	Dharchula-Bajang	6.1
6	**1991**	**Uttarkashi**	**6.6**
7	**1999**	**Chamoli**	**6.8**

3 Analysis

3.1 *Earthquakes of Karakoram Fault and Indus Suture in the Tibetan Plateau*

There have been several earthquakes in the Karakoram Fault and Indus Suture in Tibetan Plateau. This is an oblique-slip fault system in the Himalayan region across India and Asia. There has been series of earthquakes heavy loss of life and property since recorded 1803–1999. The most sever one occured near Dharachula (Pithoragarh district of Uttarakhand) of 7.5 magnitude. Most recent being in Chamoli, having 6.6 magnitude of earthquake with 21 km focal depth (Table 1). In 1991, Uttarkashi experienced an earthquake of 6.8 magnitude and focal depth of 13 km. About 1.87 lakh buildings were damaged with 106 fatalities and 453 injuries in case of Chamoli whereas much destruction was also seen during Uttarkashi earthquake (Table 1 and Fig. 2). The history of earthquake has been devastating. Dehradun City lies in between two of Karakoram fault called the Main Boundary Thrust (MBT) and Main Central Thrust (MCT) (refer Fig. 1). Earlier researches indicate that Dehradun might experience Peak Ground Acceleration shakings of about 2.2 m/s^2 for 225 years return period and around 4.6 m/s^2 for 2500 years return period [5] (Fig. 3).

3.2 *Sustainable Land Resources Management of Built-Environment and City Planning Strategies*

Ground water changes qualified to meddle with building's strength because of fault rupture during earthquake [6, 7]. In seismic areas, the level of groundwater frequently changes due to earthquakes in addition to tide and barometric pressure [8]. Occurrence of slope failure was experienced when the magnitude of earthquake is above six. Frequency of 5–10 Hz ground motion is very capable of moving solid structure at ground surface. An active fault line in vicinity attributes to quick landslip in areas of

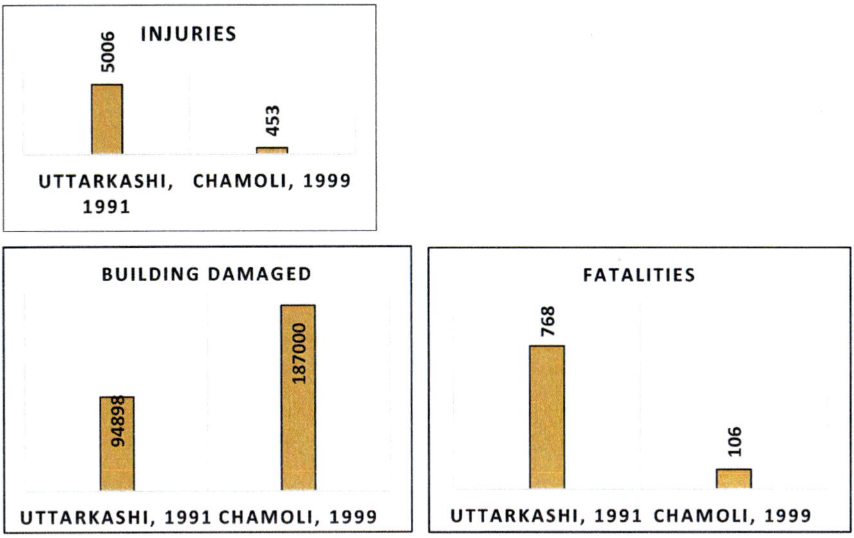

Fig. 3 Buildings damaged and fatalities in earthquake of Uttarkashi and Chamoli [11]

steep slope [9]. As per Building Materials and Construction Technologies (BMTPC 2016), earthquake hazard zoning atlas (2016) showed housing category A (clay and stone) and B (brick) are prominent in Uttarakhand State creates highly vulnerable built-environment to life and property during an event of earthquake [10]. Local engineers and masons ought to be imparted with training of earthquake resilient building construction [11].

3.3 Provisions and Technicalities of Uttarakhand Government

The technical section of strategy implementation, monitoring and evaluation provided for GIS aided spatial approach. Additionally, its section on integrated mitigation, planning and preparation offers for risk management capable community/stakeholder lessening the level of probable loss of life and assets [12]. The building bye-laws guides for parameters of low risk, medium risk and high-risk topographical conditions of slope (refer Table 2) [13].

Table 2 Building bye-laws, TCPD, Uttarakhand

PARAMETERS		RISK LEVEL		
		LOW RISK	MEDIUM RISK	HIGH RISK
BUILDING CLASSIFICATION		Low hazard occupancies as defined in NBC-2005 (non-assembly buildings)	Moderate hazard occupancies in NBC-2005 (open assembly buildings)	High hazard occupancies in NBC-2005 (closed assembly buildings)
HEIGHT	PLAINS	Up to 9.0 m	> 9.0 m up to 21.0 m	>21.0 m
	HILLS	Up to 7.5 m	> 7.5 m up to 9.0 m	> 9.0 m
FLOOR AREA		Covered area on each floor <= 350 sq m	Covered area on each floor >350 sq m and <=500 sq m	Covered area on each floor > 500 sq m
SLOPE		<= 10 degrees	> 10 degrees and <= 26.5 degrees	> 26.5 degrees
EXPERIENCE OF THE DESIGN AND BUILDING TEAM		More than 3 (three) buildings	Fewer than 3 (three) buildings	no previous experience

4 Results

4.1 *Urban Sprawl and Built-Up*

Study area witnessed rapid urbanization soon after its formation as the active capital city of State of Uttarakhand in year 2000. Increase growth rate of housing in the state was by 1.5% per year, which increased to 3.4% in 2019–2020. These numbers continue to increase the urban sprawl shown during year 2011–2020 (map). The built-up growth remains tremendously expanding in south-west direction towards Selaqui and south-east towards Haridwar. Population is dwelling upon unsafe and highly vulnerable land. Houses with high number of storeys get located in river flood plains, steep slopes and on soft soil (Map 1).

4.2 *Spatial Hazard Risk Zones*

Slope Analysis is an important factor during the study of earthquakes. In places where ground is sloping, a static shear stress develops in the soil. These stresses cause deformation to accumulate towards downhill direction. To ensure disaster resilient built form for mitigation of structural loss owing to vulnerability of the urban form, the building bye-laws has incorporated 'risk-based classification' of buildings with the slope angle for construction activities for building classification on the basis of risk level. As per the building Byelaws, those structures which are on

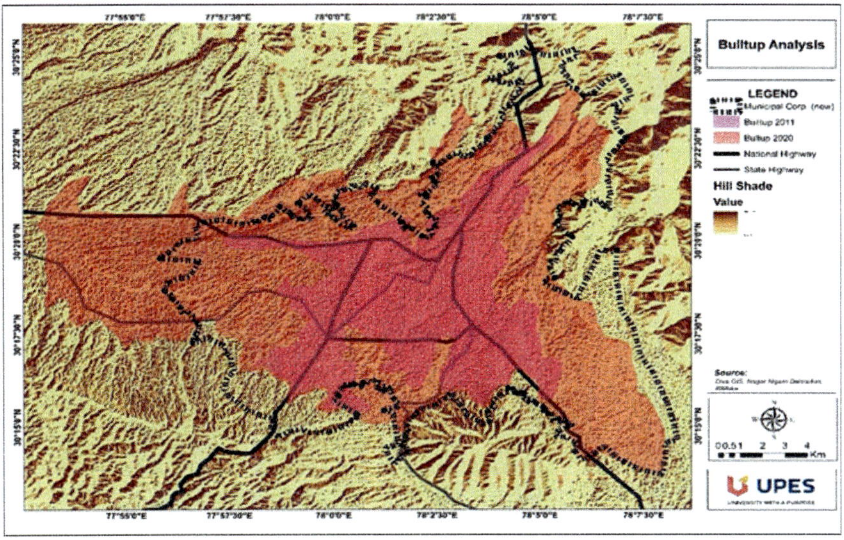

Map 1 Urban sprawl and built-up change from 2011 to 2020, Dehradun City. *Source* https://earth.google.com/web/@0,0,0a,22251752.77375655d,35y,0h,0t,0r, https://bhuvan.nrsc.gov.in/home/index.php

the slope less than 10° are at low risk while those buildings which are between 10° and 26.5° are at medium risk and those buildings which lies on the slope more than 26.5° are at high risk. Majorly the study area lies in the medium risk zone which is between 10 and 26.5° of slope. The figure below overlays the buildings on the slope map. All these buildings in low, moderate and high-risk regions have 2–3 storeys of buildings. Quite a number of buildings were found when zoomed into the red code of high vulnerability. The buildings in high risk zones must be further studied and strengthened for making them resilient towards earthquake (Map 2).

4.3 Social Hazard Risk Zones

Seismic micro-zonation maps from the study by [14] (evaluating the shear-wave velocity (Vs) using geophysical technique) has been overlaid on the base map to indicate the vulnerable zones. Central city wards were found to be densely populated with high impact of spectral acceleration (Map 3).

Community Mapping

The communities were identified for interventions of community strengthening and preparedness. These were Prem Nagar, Uttranchal University, Kheri Gaon on west of the city. Mahesh Indresh area (Nanak Vihar, Derakhas), Irrigation Colony (between

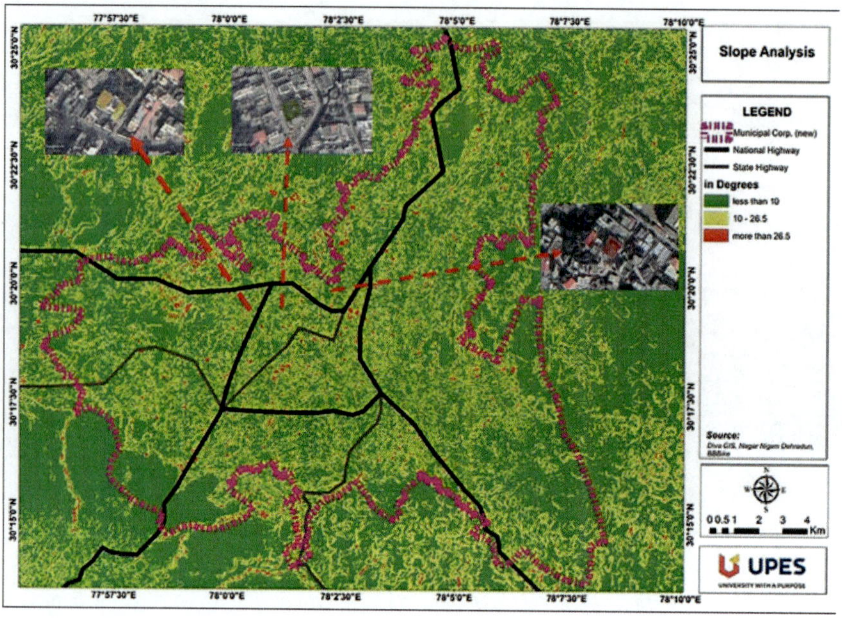

Map 2 Slope analysis. *Source* https://earth.google.com/web/@0,0,0a,22251752.77375655d,35y,0h,0t,0r, https://bhuvan.nrsc.gov.in/home/index.php

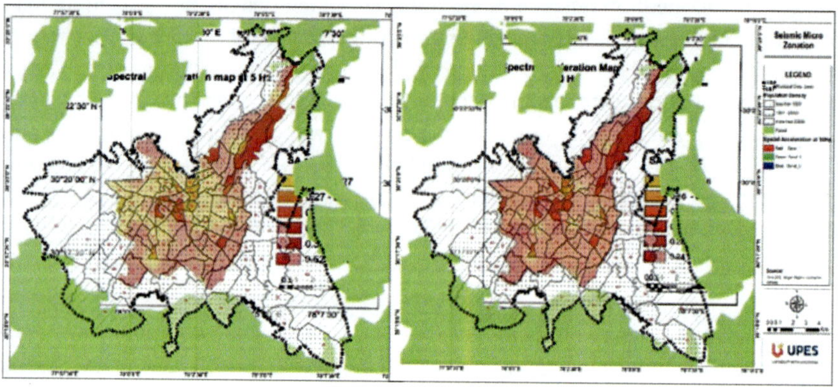

Map 3 Micro zonation map overlaid on base map (spectral acceleration 5 and 10 Hz) *Source* [14], Census of India 2011, https://earth.google.com/web/@0,0,0a,22251752.77375655d,35y,0h,0t,0r, https://bhuvan.nrsc.gov.in/home/index.php

Rajpur road and E. C. Road) and Sahastradhara colonies lies on the north-east side. These colonies are highly vulnerable to disaster of earthquake.

4.4 Comparative Review of Earthquake Resilient Structure Technology

Since much unplanned development or built-up has already come in the city, the impact of earthquake would be devastating. Redevelopment of the entire colonies would also be difficult. It was wise to check of best practices of other countries dealing with similar threat. Case of Turkey, Japan and Bangladesh were compared to find that they have experienced more number of earthquakes and are investing in massive retrofitting of identified buildings through Non-Destructive Testing (NDT) (Table 3).

Non-Destructive Testing

NDT is a reasonable method to discover the properties of a structure without demolishing it. Evaluation of a prevailing structure is tough since construction does not always take place according to specifications as designed and various deformities or, errors occur during construction, which goes unnoticed. Also, the natures of materials deteriorate over time. For structures that may not be strong against earthquakes, an NDT evaluation serves as a decent method for estimating the structural safety

Table 3 Review of NDT of different countries [11]

Region	Turkey	Japan	Bangladesh
Seismicity	High	High	Moderate
No. of Earthquakes (1900–2016)	77	61	
Action	Large scale remedial initiative—retrofitting/renewal of all high-risk buildings within 20 yrs	Contributing US$ 725,200 for using nuclear technology to verity the structural integrity of buildings after earthquakes or any other disasters	NDT to evaluate the condition of existing structures against different loading patterns
Common NDT Methods Used	For on-site strength test of concrete: UPV—Pull out method—Drilling resistance method—Rebound hammer	Seismic vulnerability assessment using NDT, Determining seismic performance indices	Ferroscan—Rebar detector—Ultrasonic device—Rebound hammer—Microtremor—Ground Penetrating Radar

at low budget. Usage of NDT during the development of structures and testing all through their life expectancy can fundamentally limit collapses during earthquakes and spare lives. Various NDT tools that can be used are Rebound hammer (Schmidt hammer), Ultrasonic Pulse Velocity (UPV) instrument, Ground Penetrating Radar (GPR), rebar detector, Ferro scanner etc. Penetration resistance, Pull-off testing and drilling core tests are partially destructive methods adopted.

5 Discussion

Spatial hazard risk zone (SHRZ) in conjunction with the slope analysis for seismic risk mentioned in the byelaws has been prepared indicating areas of low, moderate and high risk. These to be overlaid with building footprint and building height, check areas of violation of building byelaws, earthquake risk and vulnerability to life and property. Further, check there building layouts for non-resilient construction as it amplifies the susceptibility if buildings are in the high risk of SHRZ. Thus, further suggests screening these buildings through Non-Destructive Testing (NDT) in areas of high-risk zone for the once that are not in compliance with building-bye laws, construction codes and standards. Social Vulnerability Hazard zones and wards has been identified of highly vulnerable population to carry community-strengthening programme for integrated mitigation, planning and preparedness.

6 Conclusion

Study carried socio-physical integrated spatial strategy with GIS-Techno aided seismic risk reduction strategy in Dehradun city and community strengthening. Strategic location of high-risk areas has been identified for further building details, testing and carry programs of community strengthening. Morphology of city, urban sprawl and narrow road network fears higher chance for road blockage and cut-off access of rescue and relief to affected area and would make dreadful emergency period. Comparative structural resilience technologies were described in the study, which can be, incorporated appropriately in retrofit (sync Smart City Mission), seismic strengthening and construction.

This study also highlights, authorities and organizations to ensure the compliance to building bye-laws for further construction works. Engineers and contactors must be licensed to do works in these areas, which are under seismic risk. Highly vulnerable communities identified through social hazard risk zone for earthquake preparedness and strengthening. GIS and NDT technology provides cutting edge to save existing buildings that are highly earthquake vulnerable.

Acknowledgements We express our sincere thanks to the state Government of Uttarakhand for providing sufficient information for completing this report. We thank Mr. Ranjan Bhora, Disaster

Management Specialist in UNDP India, and Centre for Public Policy and Good Governance (CPPGG), the division under the department of Uttarakhand Government, Mr. Manas Ranjan Mishra (Program Manager, DM) for coordination. Gratitude to Dr. Nihal Siddiqui (Program Director), Dr. S. M. Tauseef (Head of Department) and faculties of Department of HSE and Civil for giving the opportunity. The Dean-School of Engineering Dr. Kamal Bansal of University of Petroleum and Energy Studies (UPES) who played an integral part in providing support and relevant guidance, a special thanks goes to him.

References

1. T. Endow et al., "Uttarakhand Vision 2030," Department of Planning Government of Uttarakhand, 2018.
2. Uttarakhand State Disaster Management Authtority, "State Management Plan 2020–21." 2020.
3. K. Gupta, "Unprecedented Growth of Dehradun Urban Area: a Spatio-Temporal Analysis", *Int. J. Adv. Remote Sensing, GIS Geogr.,* pp. vol. 1, no. 2, pp. 47–56, 2013.
4. G. o. U. Dehradun Municipal Corporation, ""Dehradun Smart City Proposal", pp. 1–92," 2016.
5. C. L. Mukat Sharma, "Earthquake Hazard Assessment for Dehradun, Uttarakhand, India, Including a Characteristic Earthquake Recurrence Model for the Himalaya Frontal Fault (HFF)," *Research Gate,* pp. 1–5, 2011.
6. H. Wakita, "Water wells as possible indicators of tectonic strain." *Science,* vol. https://doi.org/10.1126/science.189.4202.553, p. 189(4202):553–555, 2003.
7. E. A. R. S. S. B. F. S. R. A. W. Records, "Hydrologic effects on water level changes associated with episodic fault creep near Parkfield, California," *Advanced Earth and Space Science,* vol. J Geophys Res, p. 94(B9):12387–12402, 1989.
8. Y. C. P.-Y. C. Y.-C. C. &. T.-L. T. Ching-Yi Liu, "Impacts of hydrogeological characteristics on groundwater-level changes induced by earthquakes," *Springer Link,* pp. https://link.springer.com/article/10.1007/s10040-017-1684-z#:~:text=Sustained%20groundwater%20level%20changes%20are,to%20tide%20and%20barometric%20pressure. 2017.
9. M. A. T. I. S. U. G. Y. H. Noda, "Earthquake Induced Slope Failure," in *Ninth World Conference on Earthquake Engineering, Aug 2–9,* Tokyo-Kyoto, Japan Vol. III, 1988.
10. N. D. M. A. NDMA, "Design for National Seismic Risk Management Programme, Incception Report," NDMA, Government of India, 2019.
11. P. Pande, *Seismic Hazard of Uttarakhand: Lessons Learnt from the Study of Damaging Earthquake,* 2017.
12. D. M. S. DMS, *Strategic Plan for Risk Reduction, increasing resilience through effective responce, recovery, mitigation and preparedness, Urban Risk Hotspot, Dehradun City,* Government of Uttarakhand, 2018.
13. U. H. a. D. Authority, *Building Bye laws,* Town and Country Planning Department, TCPD.
14. S. S. R. R. R. S. P. C. r. C. W. A.K. Mahajan, "Seismic microzonation of Dehradun City using geophysical and geotechnical characteristics in the upper 30 m of soil column," vol. J. Seismol. 11, no. https://doi.org/10.1007/s10950-007-9055-1, p. 355–370, 2007.

Investigation of Energy Generation on Large Rotary Wing Unmanned Aerial Vehicle's Propeller Using Coupled Engineering Approaches

K. Naveen Kumar, S. Meenakshi, K. Deviparameswari, R. Vaidegi, R. Nandhagopal, M. Ramesh, and R. Vijayanandh

Abstract Low endurance is the predominant issue in the multi-rotor UAVs, which has been solved through various inventions such as advanced cum integrated propulsive systems and implementation of aerodynamic efficient designs. Similarly, this work deals with the extraction of energy through the components of UAV and its working principle to tackle energy insufficient issue, in which the methodologies are used for this extraction, is coupled engineering approaches. The fundamental platform of this work is a 20-inch two-bladed propeller, which can be suitable for both Rotary wing UAVs and Multi-rotor UAVs. Two primary material families are used in this comprehensive investigation, which is Glass Fiber Reinforced Polymer (GFRP) and Carbon Fiber Reinforced Polymer (CFRP). The subordinate material types used in this simulation are prepreg, wet, woven, and uni-directional for both the classifications. Because of the work nature of UAV's propeller, the probability of non-linear in vibrations may happen in numerous manners. Therefore the study about vibration relayed energy harvesting is imposed on the short-listed propeller and thereby energy obtained through Piezoelectric layers. In this regard, the natural frequencies, aerodynamic pressure estimations on composite UAV's propeller are estimated through advanced computational tools and thus the outcomes are coupled with an analytical energy estimator. At last, the energy extracted by the piezoelectric layer for different composite materials are determined, which guided the material optimization.

Keywords Composite materials · Energy · Piezoelectric · Propeller · UAV · Vibration

K. Naveen Kumar · S. Meenakshi · K. Deviparameswari · R. Vaidegi · R. Nandhagopal · M. Ramesh · R. Vijayanandh (✉)
Aeronautical Engineering, Kumaraguru College of Technology, Coimbatore, Tamil Nadu, India

N. A. Siddiqui et al. (eds.), *Advances in Environment Engineering and Management*, Springer Proceedings in Earth and Environmental Sciences,
https://doi.org/10.1007/978-3-030-79065-3_17

1 Introduction

In today's world, UAVs are taking over many things like surveillance, animal detection, rescue operation, etc. To complete the above operations effectively, UAVs should have acceptable endurance. Instead of increasing battery size, which will decrease endurance, energy can be extracted from UAV's components using piezoelectric materials to elevate endurance. By converting rotational motion of motor to linear motion and by driving air down, propellers are producing low amount of pressure in top and high amount of pressure in bottom which results in lift of UAV and here, electrical energy which is supplied by battery to motor is converted into kinetic energy. During flight, more amount of vibration occurs at propeller than other UAV components. To utilize this vibration in effective way and to choose appropriate material for propeller which can be used to handle low endurance of UAV, this study performed coupled engineering approach on 20-inch propeller with ten materials, for this purpose piezoelectric vibration energy harvesters are installed in propeller's top surface of both blades to generate electrical energy from kinetic energy [1].

2 Literature Survey

A theoretical calculation and experimental verification of a cantilevered Piezoelectric Vibration Energy Harvester (PVEH) were conducted. Using Euler–Bernoulli beam theory, the theoretical calculation was calculated and to verify it experimentally, a MEMS-scale cantilevered harvester without tip mass is fabricated. The paper concluded that maximizing active area does not always increase output power and the active electrode area should be present in cantilever structure where there is more than 31% of maximum bending stress [1].

A feasibility study of the Quadcopter propeller at different rotational speeds was demonstrated to find the energy produced from the propeller vibrations. From the SolidWorks simulation results and experimental results, it was observed that the Quadcopter CCW propeller can be deflected without a tip mass also the power extracted through PVEH is suitable to implement in the UAV's propeller [2].

A study by Steven R. Anton, deals with investigation of piezoelectric energy harvesting of an RC aircraft during its flight condition. Based on two energy harvesting device i.e., piezoelectric devices and photovoltaic panels, the test was performed and it is concluded that both the devices have the capability of charging energy storage devices. The piezoelectric patches were having some advantages over solar panels, since solar panels are dependent on environmental condition [3].

The PVEH was analysed using strain energy method. In order to solve differential equations, MATLAB software had been used. The Piezoelectric layers were calculated in both series and parallel connections and the results shown that both are producing equal amount of power, but the voltage and current are different at optimum resistive load [4].

A review on mechanisms of piezoelectric energy harvesting (PEH) based on fluid–structure interaction, human movement, and vibration were studied. Qualitative and quantitative analysis of existing PEH mechanisms had been analysed and recommending to use piezoelectric materials with a higher coupling coefficient, to gain high energy and to reduce the need of batteries and also proved that the efficiency of vibration-based PEH can be improved by optimizing the geometry of the system [5].

A theoretical and experimental study on an UAV during its flight was tested using solar and piezoelectric vibration energy harvesting devices. The results of both the test conclude that the flight endurance of an UAV can be increased, when the thin film solar panels and flexible piezoelectric devices were added to an aircraft [6].

The piezo aeroelastic analysis of a cantilevered plate like a wing with embedded piezo ceramics energy harvesting was investigated using the Doublet-Lattice Method. The resistive and resistive-inductive circuit in series are considered in the electrical domain. Among the two circuits, the resistive inductive circuit improves the performance of piezo aeroelastic and the power output is 20 times larger than the resistive case [7].

2.1 *Summary*

The relevant literature surveys are collected and studied for perfect guidance. From this survey, the following factors are noted for further processing; (1) The electrical energy generation methodology i.e., PVEH is finalized. (2) The base platform for PVEH is selected, i.e., UAV's propeller based on the high probability of vibrations generation while rotating. (3) The coupled engineering approaches such as analytical and computational investigations are finalized. (4) In the computational simulations, the vibrational procedures and MRF-CFD procedures are collected with proper conditions. (5) The conceptual design of UAV's propeller and the clear view about on PVEH's design are obtained.

3 Methodology Used

The coupled engineering approaches are implemented in this Vibration based energy extraction, which is predominantly depended upon analytical estimation and computational simulations. In addition to that, another coupled approaches also involved in the aerodynamic pressure estimations on the UAV's propeller, which is MRF (Moving Reference Frame)-CFD (Computational Fluid Dynamics). The alternate energy extraction through UAV propeller's vibration is estimated based on analytical formulae, which is provided in the Eqs. (1) and (2). In which, the predominant inputs are natural frequency and aerodynamic load. In this regard, the vibrational

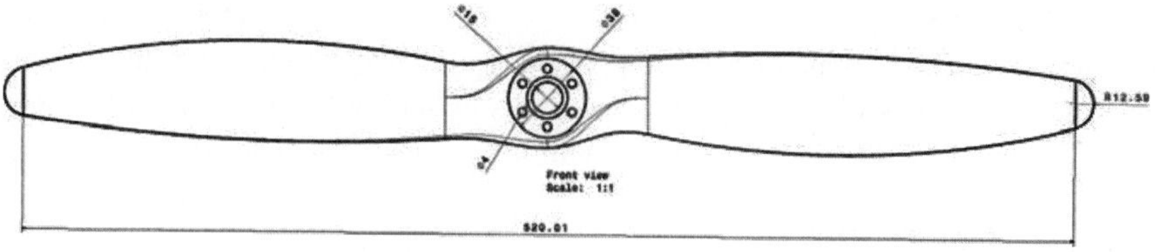

Fig. 1 Conceptual design of 20 inch UAV's propeller

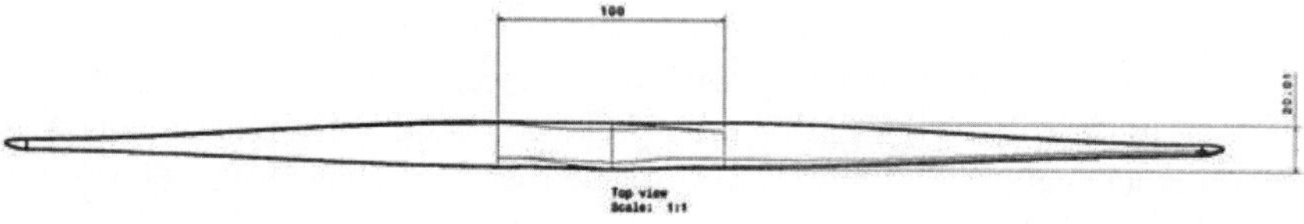

Fig. 2 A typical view of 20-inch UAV's propeller

analyses are executed through FEA tool and aerodynamic analyses are executed through CFD tool [8].

3.1 *Conceptual Design of Propeller*

The focused application of this work is high payload large UAVs so the 20-inch modelled propeller is chosen as platform of this work, which is revealed in Figs. 1 and 2. The pitch of the shortlisted rotor is 4.5 inch.

3.2 *Discretization*

Two kinds of discretizations are used in this work, which are structural based grid generation and aerodynamic fluid analyses based grid generation. Figure 3 is revealed the grid generation on the UAV's propeller, which is used for vibrational analyses. The facilities used for this grid construction are refinement on propeller surface, curvature cum proximity based grid set-up also jointly implemented in this grid formation [9].

Apart from this vibrational grid, the aerodynamic computations also important for energy extraction so the external control volumes are formed and constructed through un-structural elements. Both of the grids are generated through fine un-structural set-up with refinement so the grids are achieved the quality of 0.9862.

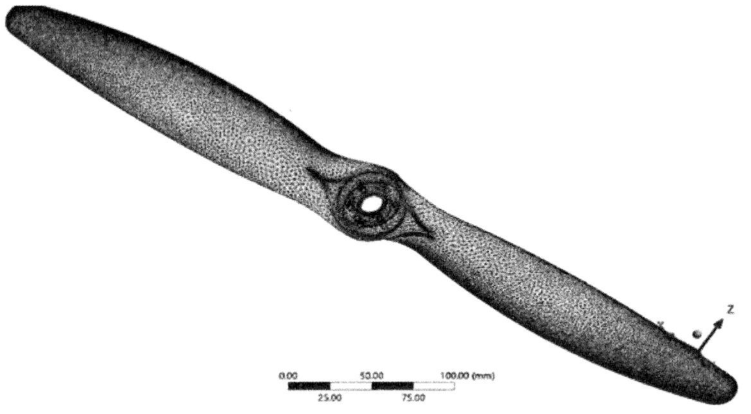

Fig. 3 Discretization of 20 inch UAV's propeller

3.3 Boundary Conditions

At the vibrational analyses, two boundary conditions are contributed principally, which are rotational remote displacement at the inner hub of the propeller and the composite mechanical properties over the entire propeller. The comprehensive material properties are listed in the Table 1. At the CFD analyses, the inlet fluid velocity and rotational velocity of the propeller are major contributors, which were estimated through literature survey. The focused UAV's weight is approximately 15 kg and the success configuration planned is Hexacopter. The Thrust to weigh ratio is assumed as 2 so thrust requirement by the individual propeller is 5 kg. In the electronics perspective, the 4S Li–Po battery with 280 kV BLDC Motor and 30A ESC are finalized for future process. Thus the maximum rpm is obtained as 4144, 75% of RPM is obtained as 3108, and 50% average RPM is determined as 2072. Through the proper instrument, the following velocities are measured: minimum input velocity is 1 m/s, average velocity is between 5 and 10 m/s, and maximum velocity is 20 m/s. Finally, the inlet velocity is given as 5 m/s and RPM is given as 3108. Apart from the boundary conditions, the control volumes and its dimensional parameters are also important in the attainment of convergence. To represent the rotating nature of the UAV's propeller, rotating control volume is formed in the cylindrical shape with the dimensions 0.6 m as diameter and 0.25 m as thickness [10, 11]. In addition to the rotating domain, the fixed domain is also constructed for the purpose of control volume formation with the cylindrical shape of diameter is given as 2.54 m and thickness is given as 10.16 m.

Table 1 The mechanical properties of all the materials

Material properties	Material name									
	CFRP-UD-prepreg	CFRP-UD-wet	CFRP-UD-prepreg-395	CFRP-woven-prepreg	CFRP-woven-wet	CFRP-woven-prepreg-395	GFRP-E-UD	GFRP-E-Wet	GFRP-E-woven	Al alloy
ρ (kg/m^3)	1490	1518	1540	1420	1451	1480	2000	1850	1646	2770
E_1 (GPa)	121	123.34	209	61.34	59.16	91.82	45	35	5.802	71
E_2 (GPa)	8.6	7.78	9.45	61.34	59.16	91.82	10	9	4.966	
E_3 (GPa)	8.6	7.78	9.45	6.9	7.5	9	10	9	4.966	
G_{12} (GPa)	4.7	5	5.5	19.5	17.5	19.5	5	4.7	1.898	0.33
G_{23} (GPa)	3.1	3.08	3.9	2.7	2.7	3	3.85	3.5	0.949	
G_{13} GPa)	4.7	5	5.5	2.7	2.7	3	5	4.700	1.898	
υ_{12}	0.27	0.27	0.27	0.04	0.04	0.05	0.3	0.28	0.17	69.61
υ_{23}	0.4	0.42	0.4	0.3	0.3	0.3	0.4	0.4	0.17	
υ_{13}	0.27	0.27	0.27	0.3	0.3	0.3	0.3	0.28	0.17	

3.4 Explanation of Computational Analysis

Both of the analyses used in this work are comes under non-linear category. Natural frequency estimation and its outcomes are non-linear in the structural displacements with respect to applied load and or supports so the modal analysis based numerical tool is implemented in all the advanced materials. The transient conditional based solver is used in the CFD analyses because of the rotodynamic nature of the propeller. The second order based derivates are implemented in this complicated analysis in order to capture the flow behaviour perfectly.

3.5 PVEH (Piezoelectric Vibrational Energy Harvesting)

The UAV's propeller with piezoelectric is revealed in Fig. 4, in which the piezoelectric layer length is used as 0.472011 m (100%) and the thickness is used as 0.0003846 m (100%).

4 Results and Discussion

Figures 5, 6 and 7 are revealed the structurally displaced propellers with GFRP, Aluminium alloy and CFRP materials respectively. The predominant initial conditions of this analysis are rotational remote displacement, which is acted about Y-axis. Totally six degree of freedoms are available for rotational displacement, wherein rotation is made free and other freedoms such as rotation about X-axis, rotation about Z-axis, displacement in X-axis, displacement in Y-axis and displacement in Z-axis are made as fixed one through zero deformation's input.

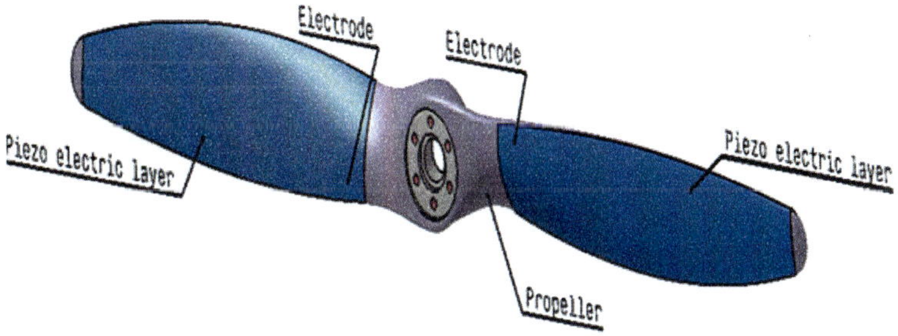

Fig. 4 Propeller with piezoelectric layer and electrode

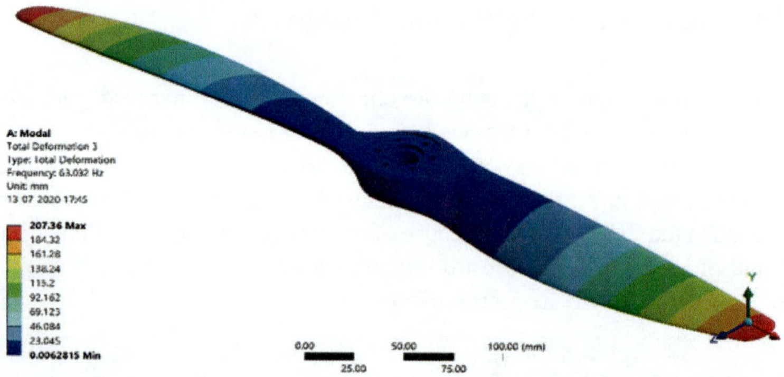

Fig. 5 The displacement variations of the GFRP coated propeller

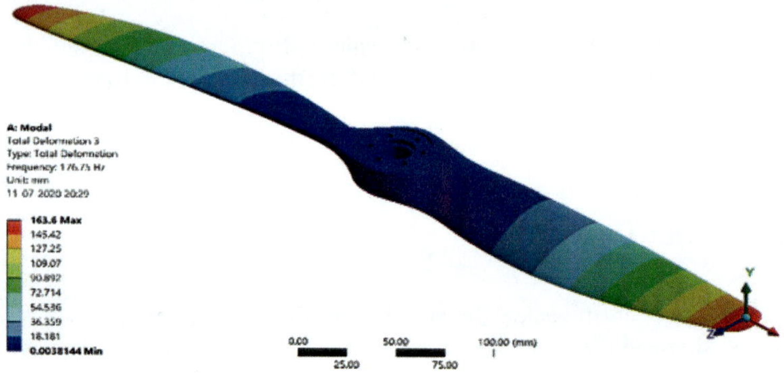

Fig. 6 The displacement variations of the Al Alloy coated propeller

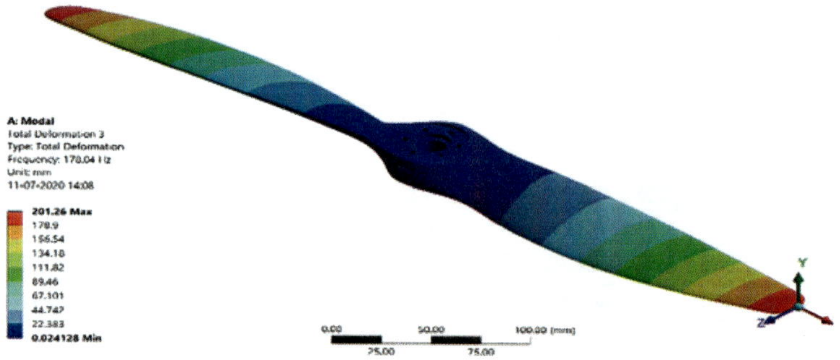

Fig. 7 The displacement variations of the CFRP coated propeller

4.1 Natural Frequencies Results

Figures 8, 9, 10, 11, 12, 13, 14, 15, 16 and 17 are revealed the natural frequencies' details with respect to their mode shapes. First ten mode shapes are obtained for this comparative work, in which first three odes are only capable to occur at the real-time applications. In this work, third mode shape is used for the calculation of energy, which is quite high in very strongest materials and quite low in very lowest strength materials.

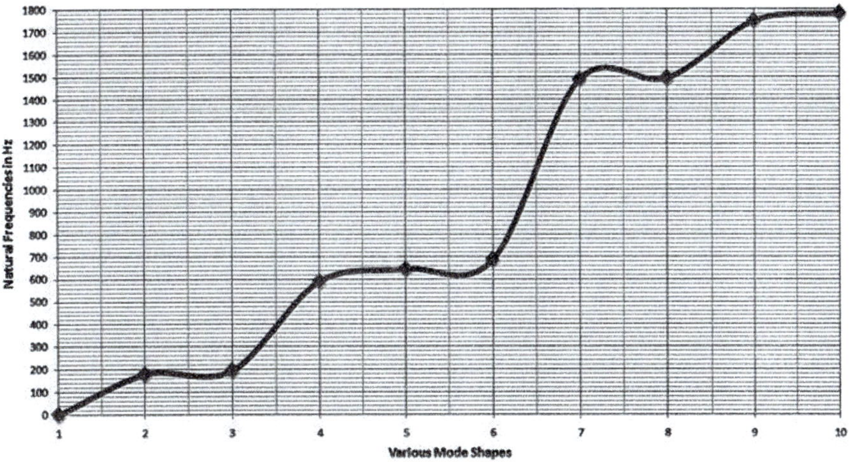

Fig. 8 Aluminium alloy

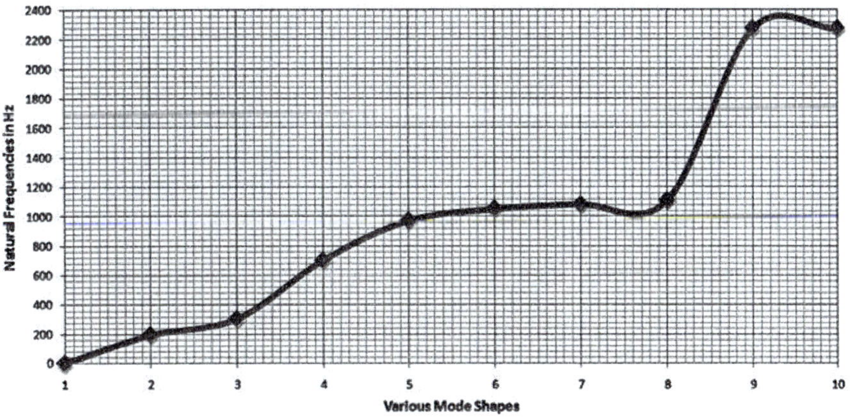

Fig. 9 Epoxy-carbon-UD-230-GPa-wet

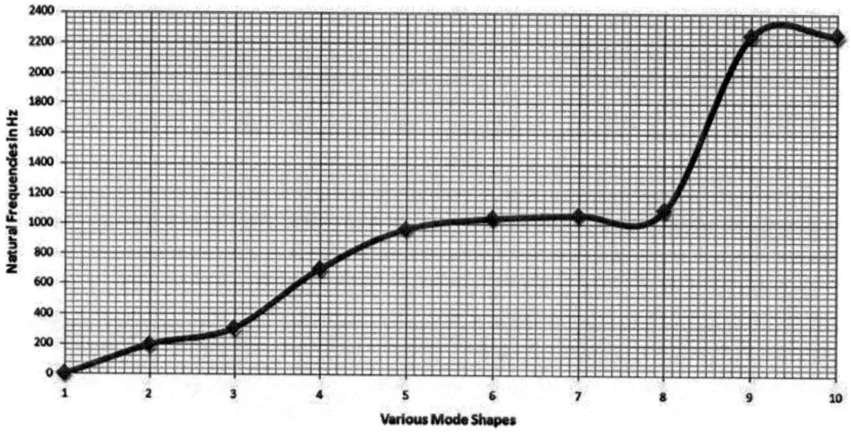

Fig. 10 Epoxy-carbon-UD-230-GPa-prepreg

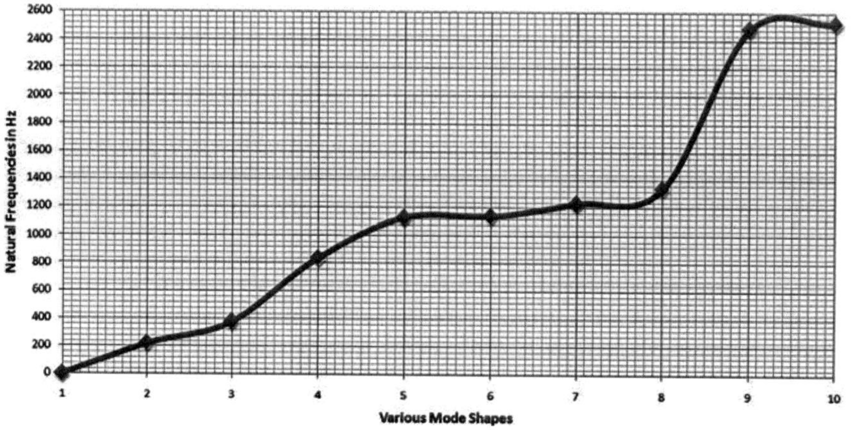

Fig. 11 Epoxy-carbon-UD-395-GPa-prepreg

4.2 CFD Results

Figures 18 and 19 are revealed the velocity variations over propeller and pressure distributions on the propeller respectively.

4.3 Energy Calculation

The alternate energy has been extracted through verified analytical method [1] with the support of advanced computational simulations' inputs. Comparatively, the

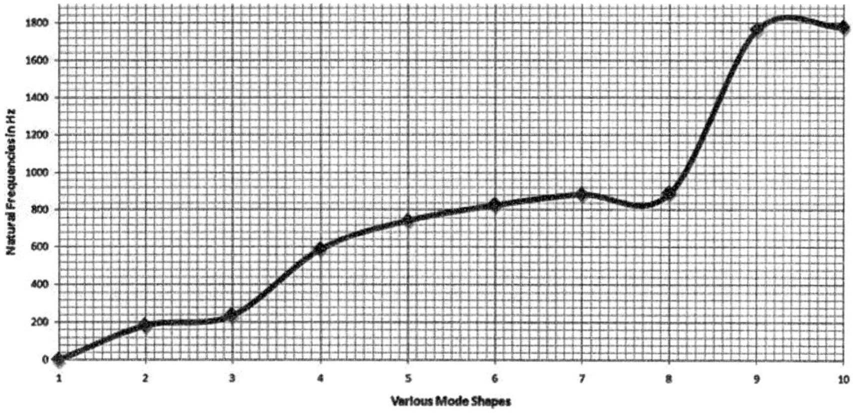

Fig. 12 Epoxy-carbon-woven-230-GPa-prepreg

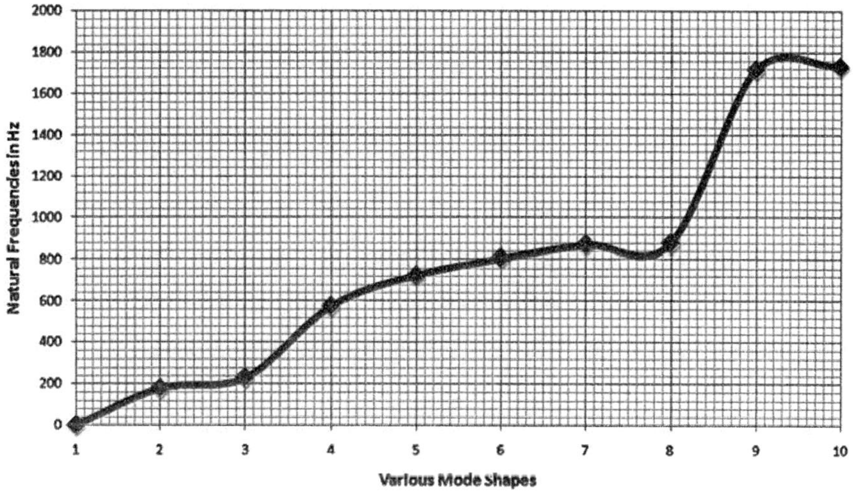

Fig. 13 Epoxy-carbon-woven-230-GPa-wet

proposed PVEH based energy extraction is perfectly matched with the rotodynamic behaviour of the UAV's propeller.

$$B = d_{31}^2 F^2 \omega^2 * \frac{18H}{\left[WL^2(h+H)^4\right]} * \frac{\rho}{[1+\omega\varepsilon\rho]} \tag{1}$$

$$P = B\left(\frac{x^5}{36} - \frac{Lx^4}{6} + \frac{5L^2x^3}{12} - \frac{L^3x^2}{2} + \frac{L^4x}{4}\right) \tag{2}$$

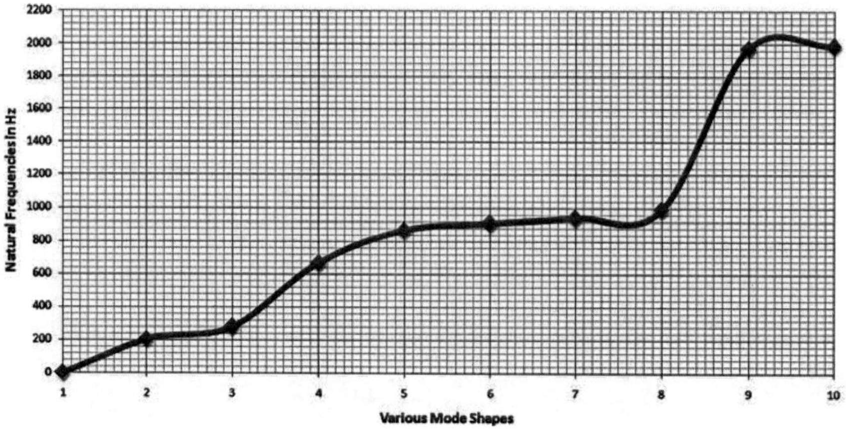

Fig. 14 Epoxy-carbon-woven-395-GPa-prepreg

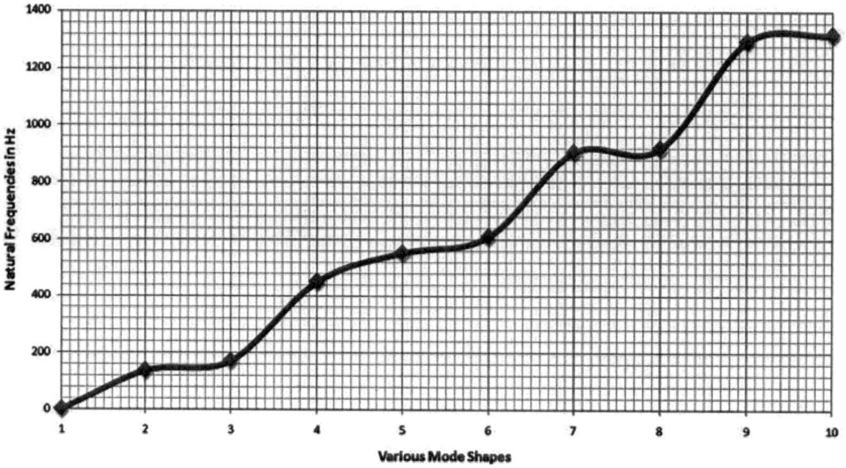

Fig. 15 Epoxy-E-glass-UD-20

where, d_{31} is Piezoelectric material constant (0.12), A is area of the UAV's Propeller (0.0184942 m^2), P is aerodynamic pressure (489.7 N/m^2), F is Force (9.22305754 N), ω is Natural frequency (Hz), W is Width of the wing (0.048409 m), L is Length of the piezoelectric layer (0.472011 m), h is Thickness of the wing (0.011538 m), H is Thickness of the piezoelectric layer (0.000384 m), ρ is Density of the material, ε is permittivity of the materials. The vibrational energy due to aerodynamic load and the material response are calculated through aforesaid coupled engineering approaches and thus the comparative data are listed in Table 2.

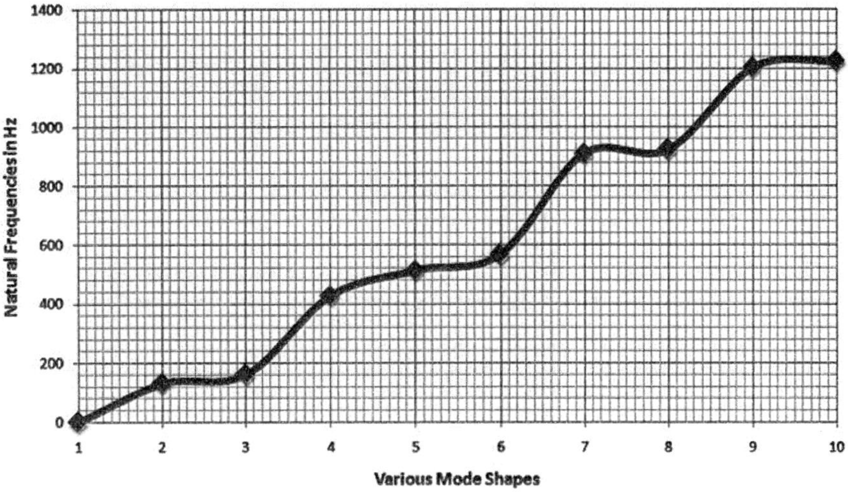

Fig. 16 Epoxy-E-glass-Wet-20

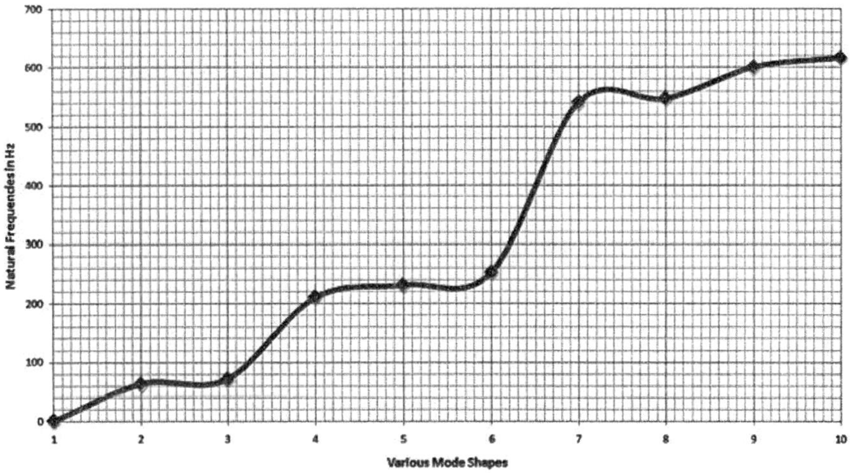

Fig. 17 Epoxy-glass-woven-20

5 Conclusions

The ultimate outcome of this work is determination of alternate energy through PVEH set-up due to the UAV's vibrations imposed by the aerodynamic load. Comparatively, the high payload based UAVs are lack in the endurance than other kinds of UAVs. Thus this work targeted to enhance the endurance of the high payload UAVs by means of increasing propulsive energy with the support of alternative energy source. The platform of this work is propulsive component of high payload UAVs, which is

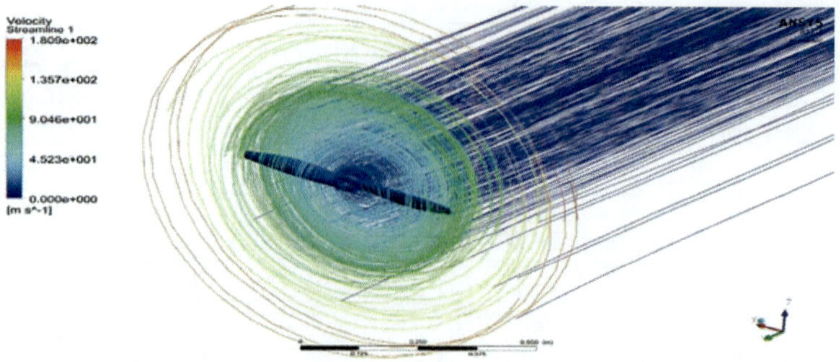

Fig. 18 Velocity variations over the UAV's propeller

Fig. 19 Aerodynamic pressure distribution on the UAV's propeller

Table 2 Comprehensive power estimations on various piezoelectric layers

Material Name	Power (W) through 100%	Power (W) through 75%	Power (W) through 50%	Power (W) through 25%
CFRP-UD-Prepreg	7.08	1.7115886	0.2211094	0.0069097
CFRP-UD-Wet	7.10	1.7175422	0.2218786	0.0069337
CFRP-UD-Prepreg-395	8.5	2.0543474	0.2653883	0.0082934
CFRP-Woven-Prepreg	7.5	1.8173073	0.2347666	0.0073365
CFRP-Woven-Wet	5.82	1.4074423	0.1818187	0.0056818
CFRP-Woven-Prepreg-395	6.74	1.6290962	0.2104528	0.0065766
GFRP-E-UD	5.94	1.4377652	0.1857359	0.0058042
GFRP-E-Wet	5.51	1.3316131	0.1720228	0.0053757
GFRP-E-Woven	2.42	0.5943669	0.0767825	0.0023995
Al Alloy	17.34	4.2011829	0.5427246	0.0169601

20 inch in diameter and 4.5 inch in pitch. The conceptual design of this proposed UAV's propeller is modeled through CATIA with all needful design parameters. Natural frequency and aerodynamic loads are primary factors in the determination of PVEH based energy, so the vibrational analyses are executed for the estimation of first ten natural frequencies of all the advanced composite materials, and coupled MRF-CFD computational approach is employed for the purpose of aerodynamic load's estimation. Both of these computational analyses are computed with the help of reliable working boundary conditions. Finally, the verified analytical formula is involved in the calculation of vibrational energy for all the materials. From these engineering coupled approaches based investigations, it has been understood that the huge amount of power can be extracted from PVEH loaded on UAV's propeller under normal working conditions.

References

1. Sijun Du, Yu Jia and Ashwin Seshia, Maximizing Output Power in a Cantilevered Piezoelectric Vibration Energy Harvester by Electrode Design, Journal of Physics: Conference Series 660 (2015) 012114, pp. 1–5, doi:https://doi.org/10.1088/1742-6596/660/1/012114
2. Sushree S Sahoo, Vijay K Singh and Subrata K Panda, Experimental and simulation study of flexural behaviour of woven Glass/Epoxy laminated composite plate, IOP Conf. Series: Materials Science and Engineering 75 (2015) 012017, pp. 1–10. https://doi.org/10.1088/1757-899X/75/1/012017
3. Nneka Osuchukwu, Leonid Shpanin, Feasibility Study of the Quadcopter Propeller Vibrations for the Energy Product, International Journal of Mechanical and Mechtronics Engineering Vol:11, No:2, 2017, pp. 1–7, doi:https://doi.org/10.5281/zenodo.1340064
4. Saber Mohammadi, Kaveh Cheraghi and Akram Khodayari, Piezoelectric vibration energy harvesting using strain energy method, Eng. Res. Express 1 (2019) 015033, pp. 1–16, doi:https://doi.org/10.1088/2631-8695/ab3f0c
5. Hassan Elahi, Marco Eugeni and Paolo Gaudenzi, A Review on Mechanisms for Piezoelectric-Based Energy Harvesters, Energies 2018, 11(7), 1850, pp. 1–35, doi:https://doi.org/10.3390/en11071850
6. Steven R. Anton and Daniel J. Inma, Performance modeling of unmanned aerial vehicles with on-board energy harvesting, Proceedings of the SPIE Vol. 7977(2011) 79771H, pp. 1–15, doi: https://doi.org/10.1117/12.880473
7. Carlos De Marqui, Wander G. R. Vieira and Daniel J. Inman, Modeling and Analysis of Piezo-electric Energy Harvesting from Aeroelastic Vibrations Using the Doublet-Lattice Method, Journal of Vibration and Acoustics: 2011, Vol. 133 / 011003, pp. 1–9, doi: https://doi.org/10.1115/1.4002785
8. Steven R. Anton, Alper Erturk, Daniel J. Inman, Investigation of a Multifunctional energy harvesting and energy-storage wing spar for unmanned aerial vehicles, Journal of Aircraft, Vol. 49, No. 1, 2012, pp. 292–301, doi: https://doi.org/10.2514/1.C031542
9. Steven Dobbs, Zhen Yn, Kevin R. Anderson, Jonathan A. Franco, Alexander E. Deravanersian, Albert Lin, Andrew Ahn, Design of an In-flight Power Generation And Storage System for UAVs, IEEE Conference on Technologies for Sustainability (2018), Conference Series 978-1-5386-7791-9/18, pp. 1–8, doi: https://doi.org/10.1109/Sus Tech.2018.8671363.
10. Naveen Kumar K, Vijayanandh R, Bruce Ralphin Rose J, Swathi V, Narmatha R, Venkatesan. K, Research on Structural behavior of Composite Materials on different Cantilever Structures using FSI, International Journal of Engineering and Advanced Technology, 8, 6S3, 2019, pp: 1075–1086. https://doi.org/10.35940/ijeat.F1178.0986S319

11. Rajagurunathan. M, Raj Kumar. G, Vijayanandh. R, Vishnu. V, Rakesh Kumar. C & Mohamed Bak. K, The Design Optimization of the Circular Piezoelectric Bimorph Actuators Using FEA, International Journal of Mechanical and Production Engineering Research and Development, 8, 7, 2018, 410–422.

Beneficiation Study of Barsuan Iron Ore Slime to Produce Pellet Grade Concentrate

Nirlipta Nayak

Abstract Indian iron ore is generally friable in nature that results in generation of significant quantity of fines (around 35%) during mining and processing in the country. In India, the ratio of lumps to fines produced is almost 2:3. During washing and sizing of the ore, particles less than 0.21 mm size generated and discarded into the tailing pond. It is found, around 10 million tons of slimes generated in every year during the processing of hematite ore and lost as tailings containing around 48–62% of Fe. The slime sample collected from Barsuan assaying 54% Fe, 8.3% SiO_2 and 11.08% Al_2O_3 analysed for pre-concentration of iron values using hydrocylone for desliming followed by magnetic separation. The results confirmed that Magnetic separation is a better pre-concentration technique as compared to hydrocyclone.

Keywords Iron ore · Magnetic separation · Slime · Pre-concentration · Tailing

1 Introduction

Iron ore deposits are closely associated with schist belts, gneisses and granulite terrain of Archaean age. In India, Iron ore deposits are mostly Banded Hematite Quartzite (BHQ), Banded Hematite Jasper (BHJ) and Banded Magnetite Quartzite (BMQ) in parts of Jharkhand, Karnataka, Chhattisgarh, Maharashtra, Orissa and Tamil Nadu states [1]. India is going to be the second largest steel producer in world with annual production of 200 million tons by the year 2020. India's Iron ore deposits of 28.5 BT of which, 17.88 BT is Hematite and 10.64 billion tons of Magnetite respectively. However, we have Magnetite reserve but we are yet to be explored in detail due to its presence in eco-sensitive areas.

For steel production, India is amongst the highest consumers of CLO (calibrated lump ore). With the projected steel production, requirement of iron ore expected to be around 350–400 million tons per year. As predicted, if the rate of consumption

N. Nayak (✉)
Department of Petroleum Engineering and Earth Sciences, UPES, Dehradun, India
e-mail: npnayak@ddn.upes.ac.in

N. A. Siddiqui et al. (eds.), *Advances in Environment Engineering and Management*, Springer Proceedings in Earth and Environmental Sciences,
https://doi.org/10.1007/978-3-030-79065-3_18

increases in years to come, then good quality reserve will not sustain long. Looking at this crunch of CLO availability, alternative measures need to think of.

The friable nature of Indian iron ore causes lots of fine generation during mining and subsequent stages. In conventional processing of 1 ton of Iron ore, 10–15% slime generated. These fines/slimes however cannot be used for iron making (blast furnace/direct reduced iron (DRI) units. In order to produce 1 ton of lump ore, about 1.5 tons fines are [2] generated of which only 0.5 tons utilized. The rest either dumped at the mine or permanently lost due to lack of proper beneficiation facilities. These fines could be used in the domestic iron and steel industry after suitable beneficiation followed by agglomeration [3]. Slime is ideal for making pellets and can considered for preparation of sinter feed.

Recent studies revealed that, more than 600 million tons of Iron ore fines discarded and remain unutilized and this is due to the predominance of conventional Iron ores. Apart from detrimental effects on environment, they occupy large space. It's unfortunate that slime has more than 50% of Fe values which is in the higher side of IBM guidelines. It is essential [4] to use and recover the additional iron values from slimes to augment mineral wealth.

A detailed Beneficiation scheme followed by pelletization appears to be the best-suited way for utilizing these otherwise waste but valuable mineral. The key reasons for visible growth in iron ore pellet market is due to the factors like adoption of iron-based chemicals, rising awareness pertaining to reduction of carbon emission from steel plants etc. [5]. This can further verified by considering the jump in production, of 69 million tons in the year 2019–20 against 59 MTs in the financial year 2018–19. Pelletization is set to achieve multi-fold growth as India seeks to achieve a target of 300 MT of Steel by year 2025 irrespective of restrictions imposed by Apex Court, high logistic cost. However, the popularity of these routes is yet to gear up/pick up the momentum in Indian iron and steel industry.

2 Methodology

The slime samples dried followed by Sieving using the Vibratory Laboratory Sieve Shaker [6]. For the separation of 200-micron particles, micro-precision sieves are used. From size analysis, it is found that most of the particles in slime are of size less than 50 μm.

Liberation analysis (Table 1) shows that in coarser fractions hematite is highly interlocked with clay. In lower fractions, the gangues are highly liberated whereas for iron it is vice versa. It indicates beneficiation will be effective at lower size range.

Optical microscopic studies revealed that in Barsuan iron ore slime iron principally present in two phases i.e. oxy-hydroxyl and oxide phase. The principal oxide minerals are hematite; martite. Hydroxide minerals such as goethite and limonite are the most commonly found [7]. Goethite having white and light grey features present in very low quantity. The dull appearance of Quartz and clay are the main gangue phases

Table 1 Liberation analysis of iron ore slime

Iron ore slime (size in microns)						
	−500 + 297	−297 + 211	−211 + 150	−150 + 100	−100 + 75	−75 + 65
Free goethite	11.5	11.6	9.3	11.0	12.4	16.7
Free hematite	4.1	5.1	11.6	26.2	35.6	46.0
Interlocked (iron + goethite)	57.1	57.0	52.3	39.3	33.5	23.3
Free Quartz	4.3	3.4	4.2	2.1	4.7	4.0
Percentage of iron mineral liberated	76.9	77.1	77.3	78.6	86.2	90.1

and can be easily distinguished from ore minerals. Iron found interlocked with clay whereas quartz mostly occurs as free liberated grains [8].

In the rim like structure found in goethite outer rim is of ochreous goethite and inter-locked with clay. The inner rim is vitreous goethite, which is hard and crystalline in nature (Fig. 1c). Some of the samples are martitized with relict magnetite (Fig. 1a) grains present in the ore samples indicating oxygen incorporated into magnetite lattice from infiltration water during martite formation. Hematite re-crystallized from magnetite through martite. Martites preserve the skeletal remains of magnetite with internal voids, known as Kenomartite (Fig. 1b).

Liberation analysis along with optical photomicrographs highlighted the concentration of iron and liberation of gangue in finer range [8]. From the set of characterization data, it is imperative that desliming operation removing the ultrafine fraction

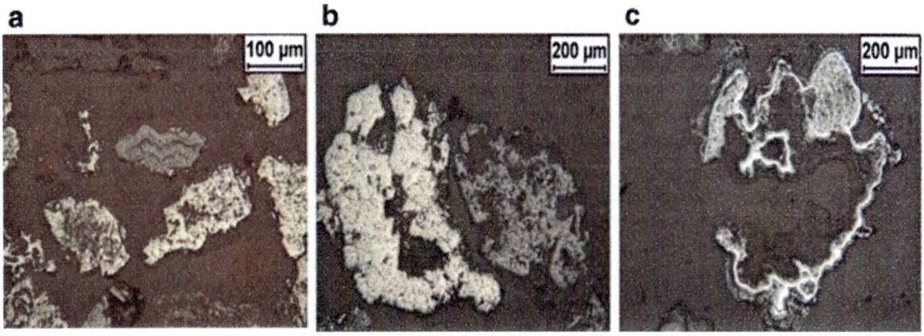

Fig. 1 **a** Rhythmic precipitation of goethite with porous martite, **b** skeletal nature of euhedral martite after magnetite (known as kenomartite) and **c** colloform structure of vitreous goethite with fracture filling, Recrystallized hematite grains in magnetite base

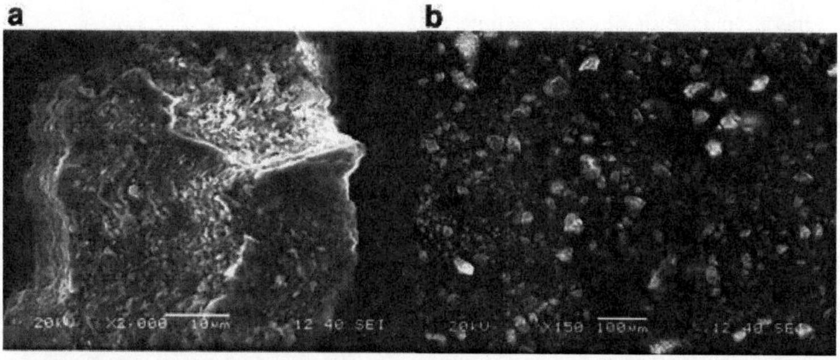

Fig. 2 **a** SEM photomicrographs of massive iron particles, **b** SEM photomicrographs of non-porous iron bearing particles of slime

would improve the grade. Hence, a beneficiation scheme chosen involving classification followed by tabling and WHIMS. The results of the unit operations carried out in these slimes discussed in the following section.

SEM of liberated iron particles containing low silica and alumina are shown in Fig. 2b. Most of the clays (kaolinite and gibbsite) are ferruginous and occur as limonitic kaolinite [9]. Kaolinite in iron ore slime mainly contributes towards the high Al_2O_3. The iron particles (Fig. 2a) are non-porous and relatively compact having smooth surfaces containing very low percentages of impurities.

2.1 *Beneficiation of Slime*

2.1.1 Treatment of slime in Hydrocyclone

The iron ore slime first treated in a 2″-hydrocyclone to remove the ultra-fines [10]. Multiple tests are conducted with different spigot and vortex finder diameters, pulp density, inlet pressure etc. [11]. After each test, both the underflow and overflow collected and analyzed for grade and yield. Two best results shown in Table 2. One

Table 2 Hydrocyclone test results of iron ore slime

Product	Yield	Fe (%)	SiO_2 (%)	Al_2O_3 (%)
Test-1 (15 mm spigot, 14.3 mm vortex and 0.68 MPa feed pressure)				
U/F	69.5	57.5	5.70	3.9
O/F	30	43.5	7.12	10.30
Test-2 (15 mm spigot, 14.3 mm vortex and 0.68 MPa feed pressure)				
U/F	79.6	55	9.32	7.47
O/F	20	37	23.7	23.7
Feed	100	54.70	8.32	11.08

Table 3 WHIMS test with hydrocyclone U/F product of test 1[a]

Product	Fe (%)	SiO_2 (%)	Al_2O_3 (%)	Yield (%)
Magnetic	66.31	3.4	0.7	74
Non-magnetic	22.3	59.5	4.0	26
Feed	57.5	5.7	3.9	100

[a]1 A current, Wash water: 20 L/m, Solid: 10%

of the tests aimed at obtaining high grade of the underflow with a low yield (Test 1), whereas the other aimed at obtaining higher yield of the underflow albeit with a lower grade.

The hydrocyclone underflow of Test 1 treated in Wet High Intensity Magnetic Separator (WHIMS). A number of tests conducted with variable pulp density, current and wash water flow rate. Under best condition of Current: 1Amp, wash water: 20 l/min. and pulp density of 10% solids, the best results obtained (Table 3).

2.1.2 Gravity Separation of Slime Samples Using Wilfley Table

Hematite, clay and silica ae the three principal phases present in slime [11]. The density values of the gangue materials (silica and alumino-silicates) are in the range of 2.5–2.8 g/cc while that of hematite is 5.0–5.5 g/cc. Due to sharp density differences, concentration by gravity separation may be adequate to produce a concentrate grade that is acceptable for iron extraction from slime.

Hydrocyclone underflow of Test-II of Barsuan subjected to concentration in Wilfley Table to check it's efficacy. The concentration grade is about 62.48% that could be much better looking at the raw material quality required for pellet making [11]. Therefore, concentrate product is not a pellet grade material requiring further enrichment (Table 4).

Table 4 Wilfley Table test results of hydrocyclone U/F of Test 2[a]

Product	Fe (%)	SiO_2 (%)	Al_2O_3 (%)	Yield (%)
Concentrate	62.48	2.1	1.5	65.53
Tailing	48.34	48.34	2.3	34.47
Feed	55.00	9.32	7.47	100

[a]10% solids, 0.25 in. inclination, 280 rpm speed and 3 L/m wash water

3 Result and Discussion

A flow sheet involving classification, gravity separation and magnetic separation developed with a view to achieve the grade at reasonably high yield. The hydrocyclone underflow product of test 2 (lower grade but higher yield) is treated by gravity separation technique using Wilfley Table to exploit the differences in specific gravity. It is observed that, quality of the slime could be improved significantly. However, the concentration grade is about 62.48% indicating the requirement of further concentration process. The Tabling results indicate possibility of getting better grade product. The grade improved substantially from 54.93 to 62%. However, to make the concentrate grade for pellet making, further processing is required. In case of U/F product (Test-1) the desired product is suitable for pellet making directly. No further processing is required. From the above discussion it is concluded that, a relatively simple flow sheet may be quite effective in producing pellet grade concentrate from such low- grade iron ore slimes, with a reasonable yield using desliming followed by magnetic separation.

References

1. Acharya, S.: (1986) Textural evolution and mineral paragenesis in Precambrian BIF from sedimentation to metamorphism. Mineral Paragenesis. In Theop. Publ Athens. (1986) 443–469.
2. Das B., Mohapatra B. K., Reddy P. S. R., Das, S.: Characterisation and Beneficiation of iron ore slimes for Further Processing.In: Powder Handling and Process. Volume 7. (1995) 41–44.
3. Chakravarthy S, Bhatacharya P, Chatterjee S. S, Singh B. N. Utilization of Iron ore fines in alternative iron making processes – An Indian Perspective. International Symposium Processing of Fines; (2002) P. 442.
4. Chakravorty, R.: Waste Management by Pelletization of Iron Ores Fines In Indian Context. In: International Journal of Mechanical & Production Engineering. Volume 3. (2015) 120–122.
5. www.pmai.co.in.
6. Iron & Steel-Vision 2020.
7. Nayak, N. P.: Characterization driven processing of sub-marginal grade of Iron ore for value addition. In Ph.D Thesis. (2015).
8. Radhakrishna, B.P., Devaraju T.C., Mahabaleswar B.: Banded Iron Formation of India. In: Journal of Geological Society of India. (1986) 71–91.
9. Raghukumar et al.: Beneficiation of Indian High Alumina Iron Ore Fines – a Case Study, International Journal of Mining Engineering and Mineral Processing. Volume 1. (2012) 94–100.
10. Roy, S., Das, A., Mohanty, M. K.: (2007) Feasibility of Producing Pellet Grade Concentrate by Beneficiation of Iron Ore Slimes in India. In: Separation Science & Technology. Volume 42. (2007) 3271–3287.
11. Upadhyay, R.K.: Iron Ore Resources: Their Conservation, Value addition and Impact on Iron and Steel Industry. In: Tata Search. Volume 1. (2008) 197–205.

Effect of Different Temperature on Host Susceptibility to SMV-Mild and SMV-Severe Isolates of Soybean (Glycine Max)

Nirmala Koranga and Kanchan Deoli Bahukhandi

Abstract Soybean (Glycine max) is a species of legume widely grown for its edible bean which has numerous uses. It got its origin in China from where it migrated to Korea and then to Europe and America. The present study highlights the variation of susceptibility of Soybean plants to mild and severe SMV strains when treated at various temperatures before and after inoculation. Soybean crop can be infected by two isolates of soybean mosaic virus which portray mild and severe symptoms and are hence named as SMV mild and SMV severe. The study of the characteristics of soybean and the mosaic virus strains gave detailed insight into treatment of infection. Experiments conducted in the present study show that post and pre inoculation treatment reduces infection in both isolates of soybean, i.e. mildly and severely affected isolates of soybean. It also indicates that infection gradually decreased as temperature was raised to 30–35 °C.

Keywords Inoculation · Strains

1 Introduction

Soybean originated in China and got introduced in India almost the same time as in the United States. Soybean at present is grown in Hilly and Terai-bhabar areas of Uttar Pradesh, Punjab, Himachal Pradesh, Madhya Pradesh and Maharashtra. It is a Kharif crop, i.e. it is grown as the rainy season is onset in June–July and harvested in October–November. Soybean crop is considered a staple therefore the study of infections is a requisite. It can be seen from previous studies and data showed that Soybean is affected by several bacterial, fungal and viral diseases. Seven viruses, some of which have two or more strains, have been reported naturally occurring in Soybean in various parts of the world. In India Nariani and Pingaly [1] reported soybean mosaic virus in New Delhi. Three virus diseases viz, yellow mosaic, common mosaic and ringspot have been reported from Pantnagar [2]. Soybean has also been

N. Koranga · K. D. Bahukhandi (✉)
Dehradun, India
e-mail: kanchan@ddn.upes.ac.in

N. A. Siddiqui et al. (eds.), *Advances in Environment Engineering and Management*, Springer Proceedings in Earth and Environmental Sciences,
https://doi.org/10.1007/978-3-030-79065-3_19

reported to be affected by yellow mosaic virus possibly caused by mung-bean yellow mosaic virus [3].

After collecting various isolates of mosaic diseases in a survey, two distinct isolates showing mild and severe characteristics were selected for detailed study where physical factors such as temperature were varied for prolonged periods of time to study reduction of infection. The seedlings were exposed to these factors before and after inoculation to find out the effect on the reaction of plants against virus infection. Ram et al. [4] reported new breeding lines of soybean having a gene for resistance to yellow mosaic virus from Glycine soja linn. Singh and Mullick [5] also studied resistance to yellow mosaic in the Soybean. Tripathi and Khare [6] recognised molecular approaches for genetic diversity among Rhizoctonia roots which are rot resistant Soybean genotypes.

Several viral diseases affect soybean crops in different parts of the world. The virus which have been identified to cause diseases in Soybean include Soybean Mosaic virus, Yellow Mosaic virus, Tomato Ringspot virus, Tobacco Ringspot virus, Bean Pod Mottle virus, Cowpea Mosaic virus, Bean Chlorotic Ringspot virus, Bean Yellow Necrosis virus, Bean Local Chlorosis virus, Cowpea Chlorotic Mottle virus, Rugose virus, Vein Necrosis virus, Pea Mosaic virus, Groundnut Mottle virus, Soybean Stunt virus, Lucerne Mosaic virus, Alfalfa Mosaic virus and Top Necrosis virus.

Le D'eau [7] described a virus disease in soybean causing systemic chlorotic steppling which rapidly became necrotic systemic chlorotic steppling and later necrotic which was followed by premature abscission of leaves and flowers. Two viruses, Soja virus-1 (Soybean Mosaic Virus) and Phaseolus virus-2 (Bean Yellow Mosaic Virus) causing mosaic disease on soybean were studied by Canover [8].

Muravera [9] recorded Soybean Mosaic, Bean Yellow Mosaic, Tobacco Ringspot, Lucerne Mosaic, Pea Mosaic, Cowpea Mosaic and Bean Pod Mottle virus on soybean. Lu et al. [10] studied the relationship between soybean mottling seed and strains of Soybean Mosaic Virus (SMV). He reported that analysis of seeds from naturally and artificially infected plants showed that mottled seeds only occurred in plants infected by SMV-Y or SMV-T. The virus was transmitted by both mottled as well as symptom-less seeds from diseased plants. The amount of transmission depends on temperature, cultivars and pathogen strain.

Soybean plants infected with SMV have been observed to contain less moisture content, higher total nitrogen content, increased amino acid content, decreased catalase activity and increased peroxide activity compared to healthy plants as stated by Suteri [11]. Iwai et al. [12] observed the distribution patterns of soybean mosaic virus strains B and D in soybean seeds at different growth stages.

Silodia et al. [13] stated the status and evaluation of soybean varieties against Mungbean Yellow Mosaic (MYMV) disease under changing climatic conditions of Kaymore plateau zone in Madhya Pradesh (India) in Indian Journal of Agricultural Research. Adams et al. [14] studies the molecular criteria for genus and species discrimination within the family polyvirdae. Ross [15] researched about the effect of time and sequences of inoculation of soybean with Soybean Mosaic and Bean Pod Mottle viruses on yield and seed characters as given in Phytopathology 59: 1404–1408.

2 Material and Methods

Experiments were conducted to find out the variation in susceptibility of soybean plants when it was treated at various temperatures before and after inoculation. Six lots, each containing twenty soybean seedlings were taken. The plants of lots 1–5 were kept at 25 °C, 30 °C, 35 °C, 40 °C and 45 °C respectively for 24 h. The 6th lot was taken as control and kept under normal temperatures. After 24 h the treated as well as control plans were inoculated with mild and severe isolates.

3 Discussion

Inoculated plants were kept separately inside insect proof glass chambers. The plants were observed up to 30 days after the date of inoculation. In order to observe the effect of post inoculation temperature treatment on susceptibility of soybean plants, six lots each comprising 20 seedlings were taken. 10 seedlings from each slot were inoculated with mild isolate and rest with severe isolate. Seedlings of the first five lots were then treated at 25 °C, 30 °C, 35 °C, 40°C and 45 °C for 24 h respectively. Plants of the sixth lot were considered as controlled and kept under normal temperatures. Observations were taken upon 30 days of inoculation (Figs. 1 and 2).

In the present study, it was found that under pre inoculation treatment, infection was reduced at 25 °C and it increased gradually between 30° and 45° followed by which the infection decreased again.

In post inoculation temperature treatment the plants inoculated with mild and severe isolate showed a gradual decrease in their susceptibility. Maximum infection was achieved at 25 °C and sharp decrease was noticed after 35 °C. The data obtained are significant at 5% level.

Exposure of seedlings to various temperatures before inoculation showed variable results. Susceptibility of pre-inoculated seedlings to both the strains was reduced at 25 °C and increased between 30 and 35 °C, followed by decrease thereafter. On the

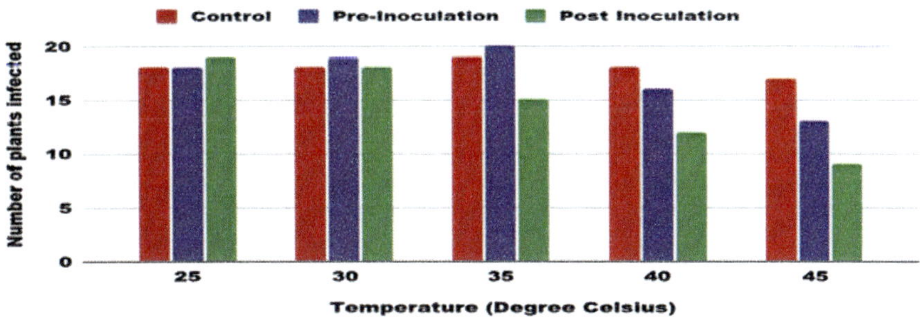

Fig. 1 Effect of temperature on Soyabean susceptibility to SMV-mild isolate

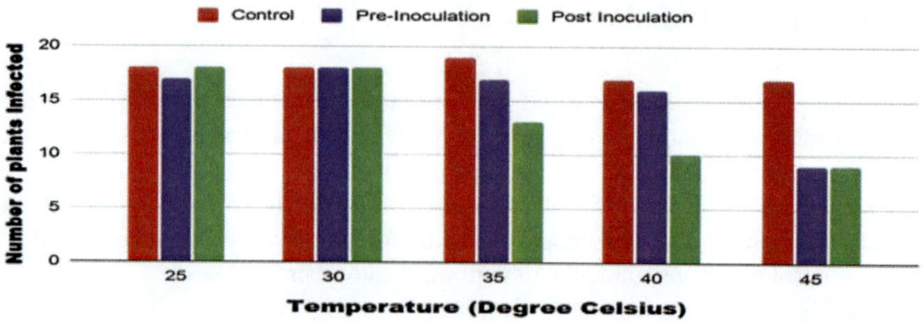

Fig. 2 Effect of temperature on Soyabean susceptibility to SMV severe isolate

other hand gradual decrease in susceptibility of soybean seedlings was recorded with post inoculation temperature treatment.

4 Conclusion

Soybean crop is a kharif crop which is highly consumed by various industries such as food industry, soap industry and various others. Hence, its demand is very high. A majority portion of this crop is lost due to viral infections, which makes their treatment to be a requisite. Various physical factors can be altered that may make the crop immune to different diseases and viruses. In our study we have found that temperature can be used as an effective method for treatment of soybean seedling pre and post inoculation against mild and severe soybean mosaic virus. The infection was found to be reduced at 25 °C temperature in mild Soybean Mosaic Virus and increased between 30 and 35 °C. With gradual increase in temperature the infection decreased again. Seedlings infected by severe soybean mosaic virus were found to be highly susceptible at 25 °C. The susceptibility decreased as the temperature was raised between 30 and 35 °C. After which the susceptibility increased as temperature was increased. Hence, the optimum temperature for decreased susceptibility of soybean crop is dependent on the strain of soybean mosaic virus, i.e. severe soybean mosaic virus infection should be treated at 30–35°C and the mild isolate must be treated at 25 °C.

References

1. Nariani, T.K. and Pingaley, K.U. (1960). A mosaic disease of soybean. Indian Phytopathology, 13(2): 130–136.
2. Anonymous, (1971). Soybean at Pantnagar 1969 and 1970. Research Bulletin No. P.I. February. 1971.
3. Suteri, B.D. (1974). Occurrence of soybean yellow mosaic in Uttar Pradesh. Current Science, 43: 689–690.
4. Ram, H.H., Pushpendra, K., Singh, K., Verma, V.D. (1984). New breeding lines of soybean lines of soybean having a gene for resistance to yellow mosaic virus from Glycine soja Linn. Sieb and Zucc. Indian Journal of Agriculture Sciences. 54:1027–1029.
5. Singh, B.B., Mullick, A.S. (1978). Inheritance of resistance to yellow mosaic in soybean. Indian Journal of Genetics. 34:400–404.
6. Tripathi, N., Khare, D. (2016) Molecular approaching for genetic improvement of seed quality and characterization of genetic diversity in soybean.
7. Le D'eau, F.J. (1947). A virus induced top necrosis in bean (abstract). Phytopathology. 37: 434
8. Canover, R.A.(1948) studies of two viruses causing mosaic disease of soybean. Phytopathology. 38: 724–735.
9. Muravera, M.F. (1968). Virusne bolezni soi. Ser. Khimbiol. 13:156–162.
10. Lu, W., Zhang, W. and Zhong, Z. (1982). Relationship between soybean mottling seed coat and strains of soybean mosaic virus (SMV). Acta Phytopathologica Sinica (1981), 11(2): 31–36.
11. Suteri, B.D.(1985). Metabolic changes in soybean plant infected with two strains of soybean mosaic virus. Indian Pythopath, 38: 714–720.
12. Iwai, H., Ito, T., Sato, K., Wakimoto, S. (1985). Distribution patterns of soybean mosaic virus strains B and D in soybean seeds at different growth stages. Annals Phytopath Soc Japan 51(4): 475–481.
13. Silodia, K., Bhale, U., Bhale, M.S. (2018). Status and evaluation of Soybean varieties against mung Bean yellow mosaic (MYMV) disease under changing climatic conditions of Kaymore Plateau, Madhya Pradesh, India. Indian Journal of Agricultural Research. 52: 686–690.
14. Adams, M.J., Antonio, J.F., Favquat, C.(2005). Molecular criteria for genus and species discrimination within the family polyvirdae. Arch Virol 150: 459–479.
15. Ross, J.P. (1969). Effect of time and sequence of inoculation of soybean with soybean mosaic virus and bean pod mottle viruses on yield and seed characters. Pythopathology. 59(10): 1404–1408.

Development of a Conceptual Safety Culture Framework to Enhance Workplace Safety

Rishi Diwan, Yusuf Faizan, and Sandesh Mishra

Abstract Safety at workplace has always been neglected by firms for greater productivity. From last decades, safety concerns and attention to safety has rapidly increased due to a chain of terrible events occurring across many different industries (e.g., Chernobyl, Upper, Big-Branch Mine, Davis-Besse etc.) which caused havoc in the society. Many organizations have started to focus on promoting a positive safety culture to make people understand and prevent disasters from happening. There is a widespread academic literature devoted to safety culture, and the Department of Energy has also published a significant number of documents stating safety culture. The purpose of the paper was to develop a framework that would build the development and sustenance of sound safety culture in the workplace. After developing a framework for better safety culture, a model is proposed which shows that job satisfaction is the main aspect of safety culture for any individual that ultimately decides their internal psychological factors such as values, beliefs, attitudes and perception towards safety which can be measured and observed through their behavior and moreover the method of assessment and validation of the proposed model with the help of questionnaires has also being highlighted in this report.

Keywords Safety culture · Values · Belief · Attitude · Psychological factor · Framework · Job satisfaction

1 Introduction

Safety at workplace has been neglected from a long period of time for production and profitability. Past industrial disasters like Chernobyl, Big-Branch mine has reflected the outcomes of a poor safety culture. After years of research it has been proven that organizations which are not following safety protocol are developing a negative safety culture. A poor safety culture includes increased accidents, lost time injuries, fatalities and faded image of organization. Researchers have shown that a positive

R. Diwan (✉) · Y. Faizan · S. Mishra
University of Petroleum and Energy Studies, Dehradun, India

N. A. Siddiqui et al. (eds.), *Advances in Environment Engineering and Management*, Springer Proceedings in Earth and Environmental Sciences,
https://doi.org/10.1007/978-3-030-79065-3_20

safety culture sees safety as an investment rather than an expense. Organizations have shown an increased interest in investing in safety to improve training and safety in organization as smaller investments like training and certifications could lead to an improved mindset. Disasters like Chernobyl accounts to a total of $235 billion and all was a result of poor safety culture. One way to understand safety culture is to understand it within context of organizational culture. The effect of organizational culture has been a topic of debate for years. It is proved that organizations with high level safety performance share culture as a common value. As a result many of them achieved thousands of safe working hours. Behavioral-based safety were seen as a solution to problem of safety culture but data always showed that underlying problem is much deeper than it seems. The relationship between organizational culture and safety culture can't be understood without investigating it. Followers [1] of classical theories stated that safety culture from a theoretical perspective of social psychology. Focusing on small groups, management roles and top managers, they looked at how front line managers saw their own working environment. Questionnaires were seen as important part of safety culture as they aim to question on present safety aspects and human-organization relationship. As a result, a well-established safety culture comprises of traits like democratic and unbiased leadership, responsibility and roles to be understood clearly, respect and dedication for the process and procedures, absence of fault finding and blame culture. Safety culture can be classified based on two approaches i.e. Qualitative and Quantitative method. Qualitative methods comprises of observations, discussions and previous information. Qualitative method serves a deep purpose of understanding safety culture by having interactions between the organizations and researchers to give an intensive and in-depth analysis of safety culture. Whereas quantitative approach uses questionnaires and surveys which are highly standardized and regulate. Based on the two approaches qualitative gives in-depth information rather than scores on safety culture. Both of them share a common value to enhancing safety culture at workplace. It is still ambiguous that how safety culture shapes and influences safety behaviors and safety program effectiveness, it is important to define and understand what is meant by safety culture. Each of us has different ideas regarding safety culture. Many followers of classical theories have their own definition and they all share a common theme i.e. to make work place safer and bind everyone with a common theme and belief, attitude and competencies to create a positive safety culture through a value system. Impact of safety culture is still hard to be determined. Research is still going on to develop a conceptual model of safety culture by relating attitudes, belief, behaviors and commitment of organization and the same approach we have tried to shed some light on the topic.

2 Literature Review

2.1 *The Concept Safety Culture*

The term safety culture first appeared in 1987 by International Nuclear Safety Advisory Group on the 1986 Chernobyl disaster. The major conflict raised by them is that the errors and violation of the operating procedures which contribute the incident were due to an ineffective and a poor safety culture of the plant. Then another disaster happened i.e. the Piper Alpha disaster in 1990 where also the root cause was lack of positive safety culture and then the public inquiry in the UK sector of the north sea of this case highlighted the importance of creating a positive safety culture to prevent any future disaster. Although this was the first time when the safety culture came into role and was identified as important aspect of effective safety management system and in all other previous inquiries only addresses the importance of cultural issues without using the term safety culture.

2.2 *Definition of Safety Culture*

Deal Kennedy [2] defines safety culture as "a way in which we act here". Ostrom [3] shows the concept of safety culture based on organizational beliefs and attitudes, shown actions, policy and procedures and thought that these influence the level of safety. Geller [4] then developed an ABC model of safety culture and shows that everyone feels responsible for safety in everyday work. This is a typical behavioral approach.

The second approach which we can identify is based on research of foundations, values for classical (analytical) approach of safety culture. According to this approach of safety culture some researchers has proposed the model such as Cox and Cox [5] for which safety culture is defines as "approaches beliefs, perceptions and values which are shared by employees in relation to safety".

The third approach is based on deeper areas such as values and attitudes and also called as pragmatic approach. One of the example of complex attitude in presenting the definition of safety culture is Pidgeon [6] for which safety culture is the collection of beliefs, values, norms, attitudes, roles, social and technical practices which aimed towards minimizing the exposure of employees, customers and other members of society to danger and threat.

2.3 *Organization Characteristics of Good Safety Culture*

Cooper [7] Researchers make effort to identify some specific organizational characteristics that were able to distinguish low accident companies to high accident companies. At the time of 1960s to the end of 1970 some researchers conducted their research over a wide variety of industries and marked out certain features such as:

- Leadership, senior management commitment and involvement in the safety.
- Better communication, closer contact between all organizational levels.
- Greater hazard control practices and a good housekeeping.
- A mature and stable workforce.
- Job placement, good personnel selection and promotion procedures.
- Good induction and follow-up safety training.
- Reinforcing the importance safety with the help of scheme including near miss reporting.

Recent study also shows almost same signs that were identified at early 1990's which indicate a good and a positive safety culture and to maintain that, these features should be implemented and should always be at first priority:

- Leadership commitment at all level of organization.
- All employees should aware about working knowledge of health and safety.
- Organization must ensure financial investment in health and safety.
- There is regular, facility-wide communication on health and safety topics.
- Managers should spend an adequate amount of time out on the shop floor, where the people are.
- Participation rates should be at all time high which indirectly shows that employees are highly motivate.
- Employees should actively engaged in health and safety initiatives, producing tangible results for the company.
- Employees should feel comfortable in reporting safety issues to their supervisors.
- Regular, detailed audits of the company's health and safety program are conducted by an external auditor.

2.4 *Incipient Issues and Weaknesses in Safety Culture*

Carnino [8] while developing different models of safety culture the four major areas that needs attention includes organization, employee, regulatory and technology. The major issues arise at operating organizations while maintaining the safety culture are as follows:

Organizational Issue

- External environmental pressure
- Inadequate problem resolutions
- Organizational insularity
- Openness.

Regulatory Issues

- Corrective actions
- Patterns of problem
- Procedural inadequacies
- Reality mismatch
- Violations.

Employee Issues

- Duration of work
- Negligence of training
- Efficient hiring of people
- Job roles and responsibilities
- Employment of contractors.

Technology Issues
Plant condition and technological advancement is a major indicator of a sound safety culture. Housekeeping, preventing maintenance and equipment safety are integral part of inherent safety. Technological aspects of plant safety should be properly maintained and understood.

2.5 *Safety Culture Models*

Developing a safety culture mainly depends on continuous manipulation various organizational characteristics that directly affect safety i.e. conducting risk assessment and these manipulations should be such that it is goal directed and if the goal theory concepts are accepted then creation of safety culture can be done by dividing the task into a series of sub-goals. The goals or outcome should be like reduction in accident or injuries, ensuring that safety issues receives proper attention (IAEA), ensuring that organizational members share the same ideas and beliefs about risk, accident and ill health etc. Based on those certain goals and outcome there are some models that have been developed during the period of 1986–2000 and have been adopted as an organizational safety culture. These models are adapted from [9]:

- Cox and Cox [5] model:
  Goal:
  - To further develop the safety culture as part of one means of improving on the company's safety culture.
- Ostrom [3] model:
  Goals:
  - Discussion of concept of safety culture.

- Presenting a survey instrument developed to asses safety culture of organization.
- Discussing how the result of survey instrument used to improve safety culture.

- Geller [4] model:
  Goal:
  - To achieve a Total Safety Culture.
- Berends [10] model:
  Goals:
  - Develop an alternative measure for safety performance.
  - Identify strength and weaknesses.
- Lee [11] model:
  Goals:
  - To identify those aspects of safety culture that is in need of improvement (Table 1).

Table 1 Review of some of the major models of safety culture

Models	Elements considered	Elements lacking/weaknesses
Schein's model of organizational culture	**Basic assumptions**: These are formed over a long period and shape the way that an employee interprets the environment around them **Espoused belief and values**: Ideals, goals, values, aspirations and ideologies **Artefacts**: Visible manifestation such as dress code, signs, structures, written doc, rewards etc	Schein model focus on building general cultural framework rather than defining cultures into specific types Schein model seems to be time-consuming for these outside analysts Due to lack of phyletic classification, straightforward comparison among organizations cannot be proceed
Guldenmund's safety culture framework [9]	Identified three layers of safety culture: **The core**: basic fundamental assumptions which are unconscious and unspecified **The middle layer**: represents safety climate and includes espoused values regarding hardware, software, people and behavior **The outermost layer**: represents the artifacts which are evident appearance of safety culture such as inspection	Major criticism—Safety climate may be assessed through attitudes other researchers identifies perception rather than attitude as psychological feature of safety climate Other criticism—Simple and unidirectional relationship between assumption attitudes and between attitudes and behaviors

(continued)

Table 1 (continued)

Models	Elements considered	Elements lacking/weaknesses
Cooper's reciprocal safety culture model [7]	Represents three factors of safety culture **Person** (safety climate): believes and perception about values of safety culture **Job**: visible safety related behavior **Situation**: objective situation factors such as organizational policies	The amount of resources required to observe safety related behavior might be substantial and come at the cost of other areas such as process safety The observer of unsafe behavior may hesitate to report those seen violating safety rules results in underreporting
Mohamed's model of construction safety culture	**Person**: Safety climate **Behavior**: Safe behavior **Environment/situation**: Safety management system	Not provide a rationale why a separate safety culture is required for construction industry

3 Methodology

3.1 Elements of the Framework

Considering the early studies (late 80s and 90s) and the recent studies on development of conceptual model of safety culture has showed that elements that were common in almost each proposed models are **person, behavior and situation/environment**. For example Heinrich et al [12] identified the interactive relationship between behavior situation and person factor at operator level. After that Bandura [13] reciprocal model appears to offer a perfect framework to analyze safety culture which also consists of **psychological, behavioral and situational** element. Then Reason's pathogen model also recognize **that person, situational and behavioral factors** are the immediate means of unsafe act, strength of each may differ and it may take time for one elements to exert its effect on another. Finally Cooper proposed a theory based on Bandura's reciprocal model [13] elaborating about reciprocal relationship between **person, job and situation**. By observing these models and associated framework this proposed conceptual model also consist certain common attributes and their sub-attributes such as person, job, behavior, organization and accordingly the relationship between them are defined (Fig. 1).

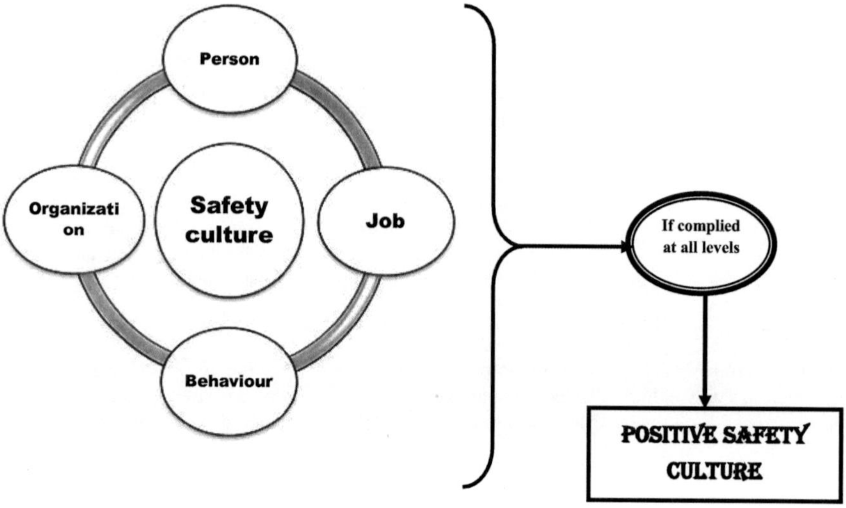

Fig. 1 Cycle of closed loop of the elements of safety culture

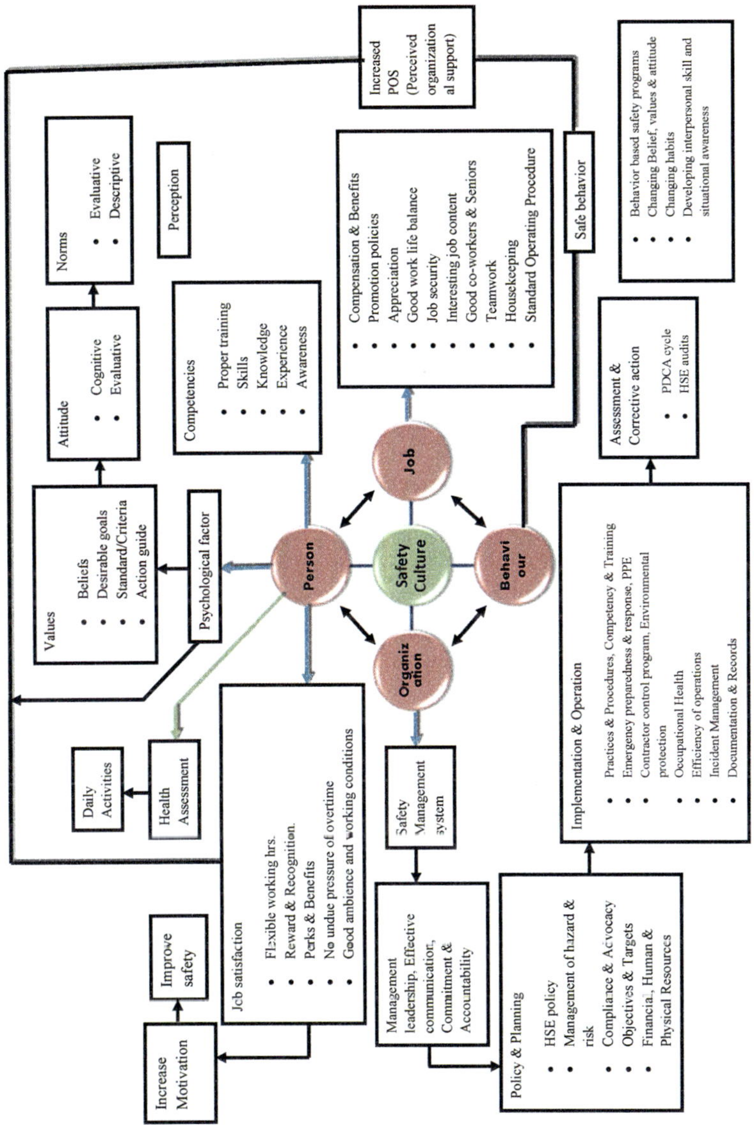
Safety Culture
Person
Job
Behaviour
Organization
Values
Beliefs
Desirable goals
Standard/Criteria
Action guide
Attitude
Cognitive
Evaluative
Norms
Evaluative
Descriptive
Perception
Psychological factor
Competencies
Proper training
Skills
Knowledge
Experience
Awareness
Compensation & Benefits
Promotion policies
Appreciation
Good work life balance
Job security
Interesting job content
Good co-workers & Seniors
Teamwork
Housekeeping
Standard Operating Procedure
Increased POS (Perceived organizational support)
Safe behavior
Behavior based safety programs
Changing Belief, values & attitude
Changing habits
Developing interpersonal skill and situational awareness
Assessment & Corrective action
PDCA cycle
HSE audits
Implementation & Operation
Practices & Procedures, Competency & Training
Emergency preparedness & response, PPE
Contractor control program, Environmental protection
Occupational Health
Efficiency of operations
Incident Management
Documentation & Records
Safety Management system
Management leadership, Effective communication, Commitment & Accountability
Policy & Planning
HSE policy
Management of hazard & risk
Compliance & Advocacy
Objectives & Targets
Financial, Human & Physical Resources
Job satisfaction
Flexible working hrs.
Reward & Recognition
Perks & Benefits
No undue pressure of overtime
Good ambience and working conditions
Daily Activities
Health Assessment
Increase Motivation
Improve safety

3.2 Description of the Framework

The proposed conceptual framework relates four parameters i.e. Person, Job, behavior and Organization. The main focus is to relate a person's perception about job satisfaction to improve behavior related to safety and at last improving organizational behavior. To improve organizational behavior, job satisfaction is seen as a key factor in changing work psychology, job design and workplace safety. It is the degree of satisfaction or perception for one's job.

Job satisfaction shows the level of commitment of employee with the organization and reactive reaction and interaction with working environment. It affects the employee's performance level, employee readiness to participate in problem-solving activities, and the amount of work they do going one step ahead of work assigned. Researchers have found that when a person enjoys his or her work then roles and responsibilities are more clear that would certainly improve safe behavior. As discussed in the model, a person perception regarding job safety could possibly end up changing his behavior. Factors like work environment could be an indicator in determining job satisfaction. Recent research work has proved that there is a deep relationship between behavior, job satisfaction and organizational culture. Details regarding every parameter have been discussed below.

A Norm is "any pattern of behavior or performance that is typical or representative of a group or a society". Break down into 3 types.

- Subjective norms (i.e. the pressure that people perceive from important others to perform, or not to perform, a behavior);
- Descriptive norms (i.e. perceptions of significant others' own attitudes and behaviors in the domain)—exerts the biggest influence on behavior
- Behavioral norms (the consistency of a behavior being enacted among a group of people—i.e. the way people do things around here).

Perceptions refer to a way of regarding, understanding, or interpreting how safety management is being operationalsed in the workplace, at a particular moment in time [14]. Assumption that perceptions serve as a frame of reference for guiding behaviors, but the evidence is not clear.

These psychological factors of a person decides the overall behavior of a person i.e. whether a person inherent the safe, unsafe, or at-risk behavior. Model considers these aspects an important part of behavioral changes from root. The change can be introduced in a person by basic developing sense of satisfaction in them by winning their trust and loyalty which would also increase the productivity.

The role of organization in proposed model is inevitable. An effective organizational culture could easily tackle employee's problem and resolve the conflicts that can certainly help individual to feel a sense of satisfaction in them. Effective leadership can drive company towards an effective culture. People like Managers and CEO's if purpose driven can easily connect with the lower level and make them understand them about their cooperation. By changing policies and providing growth opportunities to employees' organizational culture could be improved. Appreciation and rewarding system for eligible employees can instill a sense of achieving better. The model proposed is directly links an effective organizational culture with employee's behavior and following which can easily achieve a positive safety culture.

3.3 Method of Assessment

The proposed conceptual model depends upon four parameters i.e. person, job, behavior and organization to improve the safety culture. In order to define the relationship between four of them data needs to be collected. This quantitative approach would help to calculate employee satisfaction by proving various questionnaires to employees regarding the organizational culture, working conditions (environment, work stress, co-workers), technology and process, safety aspects and SOP's, perks and benefits (salaries, bonus and growth opportunities, Management issues (leadership issues, commitment etc.) and few more. There is no way to calculate employee satisfaction but ESI (Employee satisfaction index) with seven categories would be the most accurate one. ESI value can be obtained by collecting data from a sample space through online and offline surveys.

After collecting the data, ESI scores can be assigned. Furthermore we can calculate the mean and variance for each question and identify the main issues can be identified. The one with high mean value represents that maximum number of workers are satisfied and if the mean value is low, it needs to be taken under consideration. Similarly high value of variance for a given question shows there is difference in opinion of many employees though average can be higher for the given question. Issues can be easily identified using this method. After performing the ESI and getting the loopholes, next we come to behavior analysis.

Behavior analysis can be easily identified by going through documents and data from previous year records. Records of BBS programs and training can be useful to set a baseline data to compare before and after scenario. After assessing ESI results, loophole needs to fill as soon as possible. Issues should be addressed and any conflict needs to be mitigated.

This can be divided into five phases:

- **Phase 1**: Calculation of ESI and finding the persisting issues
- **Phase 2**: Addressing the issues by finding problem statement
- **Phase 3:** Implementing corrective actions
- **Phase 4**: Comparing the base line data (behavior based or accident data) with new one (after implementation)
- **Phase 5**: Addressing the change and improving organizational behavior (Fig. 2).

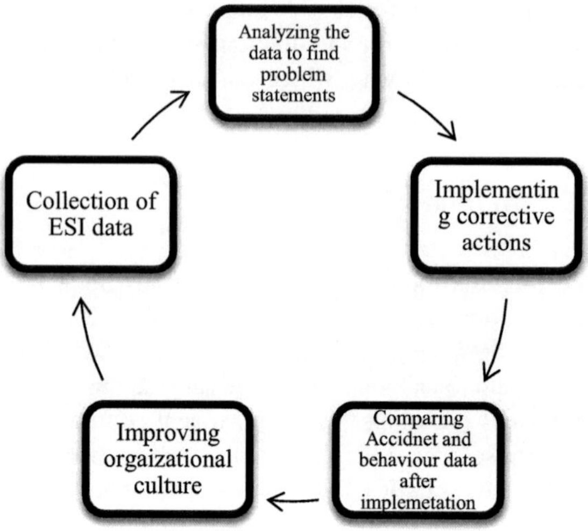

Fig. 2 Steps of the assessment

3.4 Employee Satisfaction Index Survey: (Questionnaires Shown in Appendix)

This approach is quick and easy. It comprises of few questionnaires that would give an overview of present employee satisfaction with the workplace. This approach would not give an in depth analysis of workers satisfaction. This approach can be carried out using three simple questions:

- How satisfied are you with your current workplace and job?
- How well does your workplace meet your expectations?
- How close is your current workplace to your ideal workplace?

Simple formula to calculate ESI (Negitate) is:

$$\text{ESI} = (\text{Mean value of Question}/3) - 1)/9 \times 100$$

The scores gives an ESI value which is out of 100 which shows section of employees with major job satisfaction issues but doesn't give an in depth analysis of workplace. Method can be used for a quick survey just to know if there is any major employee issue persisting in organization.

4 Conclusion

This paper discusses and reviews the literature about the development of different safety culture models in terms of conceptual understanding, measurement methodology and specific case studies for better and clear understanding about the validation of the model and also it has been observe that there is no one correct conceptualization of safety culture, or even one that is generally accepted. Although there are many different safety culture models, there is a significant amount of overlap between models, making it is possible to identify common themes. Given this complexity it is easy for misinterpretations to occur, therefore it is important for regulators to clearly define the term and adopt a conceptual model or framework. Based on the some of the major conceptual models of safety culture and by observing their attributes, elements, sub-elements as well as their weaknesses, a conceptual model has been formed in this paper. This model is anchored in three fundamental conceptual categories that include person, job, behavior and organization; change in any one element could lead to change in other elements also. These elements of the model could be validated by conducting a safety culture survey based on some questionnaires and by analyzing those answers an ESI (Employee satisfaction Index) should be calculated. Surveys like ESI and others can help to address issues which are underlying and untouched. Proper Leadership and care for employees and commitment towards safety can help organizations to take safety culture to a new level. The proposed model focuses on entire benefits of organization consisting of employees, policies, management etc. to improve in such a way that no further difference in opinions exist and workplace safety could be seen as a major source of motivation among employees and goal of "zero harm" and "safety first" could be achieved.

ESI is an effective technique to improve employee satisfaction and organizational culture. In the proposed model employee satisfaction is seen as major cause for distrust and behavioral changes among employees. Improving interactions between employees and work place could show magical changes within a short span of time. Proposed model shows that entire theme of safety culture could be improved though digging deep into employee satisfaction. Human resource needs to be addressed as a major part of organization. To instill safe behavior among employees psychology should be seen and addressed. The kind of loyalty and attachment an employee feels for the organization can be contribute and inevitable contribution in improving overall safety culture. By proving proper training and BBS programs, it can be easier to understand the safe and unsafe behavior among employees and what makes them do that. A proper system of constructive feedback could lead to have a deep insight into worker's psychology and beliefs.

References

1. Kertai-Kiss I. Conceptual Frameworks for Safety Culture and its Manifestation in Organisations. Management, Enterprise and Benchmarking in the 21st Century. 2015:123.

2. Deal, TERRENCE E., and Allen A. Kennedy. "Organization cultures: The rites and rituals of organization life." Reading, UK: Addison-Wesley (1982).
3. Ostrom L, Wilhelmsen C, Kaplan B. Assessing safety culture. Nuclear safety. 1993 Apr 1;34(2):163-72.
4. Geller ES. Ten principles for achieving a total safety culture. Professional Safety. 1994 Sep 1;39(9):18.
5. Cox S, Cox T. The structure of employee attitudes to safety: A European example. Work & stress. 1991 Apr 1;5(2):93-106.
6. Pidgeon N. Safety culture: key theoretical issues. Work & stress. 1998 Jul 1;12(3):202-16.
7. Cooper. "Towards a model of safety culture." Safety Science (2000): 111–136.
8. Carnino, Annick. "Management of safety, safety culture and self assessment." (2000).
9. Guldenmund, F. " The nature of safety culture: a review of theory and research." Safety Sience (2000): 215–257.
10. Berends JJ, van der Schaaf (TW). On the measurement of safety culture. Technische Universiteit Eindhoven; 1996.
11. Lee TR. Perceptions, attitudes and behaviour: the vital elements of a safety culture. Health and Safety. 1996;10:1-5.
12. Heinrich HW, Peterson D, Roos W. Industrial Accidents Prevention McGraw-Hill. New York. 1980.
13. Bandura, A. "The self system in reciprocal determinism." American Psychologist (1978): 344–358.
14. Byrom N, Corbridge J. A tool to assess aspects of an organisations health & safety climate. In Proceedings of International Conference on Safety Culture in the Energy Industries 1997 Sep. University of Aberdeen.

Implementation of Total Quality Management in Construction Industry

Rishi Dewan, Ayush S. Patel, and Md. Obaid Kuraishy

Abstract Quality has become a rising significant means of competition in the market of this world and has developed into a tactical weapon in the tussle for market shares and improved profitability. Implementation of quality management in construction industry provide several benefits such as higher productivity, reduced rework, improve budget performance, improved employee job satisfaction. Since the industry of Construction is dissimilar from other industries in many aspects such as kind of work that they are doing, lack of management support, lack of skilled workers, lack of leadership qualities in employee and also lack of support from higher level, lack of team work and that's why it becomes difficult to implement Total Quality Management in this industry and above this more construction companies found quality as an expensive accessory due to the fact that construction companies are completely unknown to the detail that unimplemented quality is way more elevated than running a fully functional quality program. This paper describes the issues in signifying quality in the majestic industry of construction, a highly dangerous industry and then study the advantages of quality management (QM) implementation in the vast industry of construction and barriers to quality management during its implementation. Data is collected through surveying the contractors. Various Data analysis tools was used for analysis purpose. Result indicate that the contractors know the benefits of quality management but that change is slow and some efforts are needed for effective implementation of quality management in the much required vast industry of construction.

Keywords Quality management · Survey · Industry of construction

R. Dewan (✉) · A. S. Patel · Md. O. Kuraishy
University of Petroleum and Energy Studies, Dehradun, India

N. A. Siddiqui et al. (eds.), *Advances in Environment Engineering and Management*, Springer Proceedings in Earth and Environmental Sciences,
https://doi.org/10.1007/978-3-030-79065-3_21

1 Introduction

India will expected to be the world's third-largest Construction Market by 2025 therefore construction and Infrastructure Sector plays very important role in Indian economy and growth. Construction sector in India contributes nearly 9% of total GDP. This sector involve large number of various category of people hence consider as the sector of large employability. As far as the status of Indian infrastructure sector is consider, it is inadequate to meet the demand of the existing urban population. There is an urgent need of re-generation of urban areas in existing cities and the creation of new, inclusive smart cities for handling the challenge of increasing population and migration from rural to urban areas.

Various unorganized player is involve in this sector which work on subcontracting bases which is responsible for variable profitability across different segment of construction project.

Indian construction market divided into three broad segment (1) Real Estate, This segment involve residential as well as commercial construction of various level (2) Infrastructure, this segment include development of roads, railways, transport projects, power project (3) Industrial construction which involve various oil and gas project, pipeline, refinery etc.

Effective implementation of innovative technology is responsible for reduce the cost and also responsible for speedy completion of project, which is turn increase profitability in system. The current demand of Indian construction industry is "Quality Construction", which involve quality in system, quality in process, procedure, methods etc. which is turn responsible for increase profitability in system.. Reducing the gap between national and international standard bring Indian construction system in pace to beat the challenges of international market.

TQM, brought a new concept which tells not only about quality of product but quality of all matter which are present in the organization. It comprises of all aspects of business, make an objective of continual improvement at all level to increase customer satisfaction [1].

A purely low cost model would not be a sustainable business proposition for Indian construction units in the wake of rising competition. This is the need that current construction industry must focus on developing a strategy, which create a competitive advantage, and ensure its effective implementation. Strategy must percolate from top to each and every employees. Quality actually contribute to the bottom line, in term of reduction in scrap/waste, inventory turnover,productivity,and lead time to execute an order.

TQM aims to establish quality at organizational level to improve effectiveness through eight principles: (1) top level managerial staff, (2) managing customer needs, (3) people needs and wants management, (4) suppliers overseen, (5) extensive quality monitoring, (6) managing of process flows, (7) learning and education, and (8) continuous improvement by training [2]. The process of doing quality work starts from top management by defining quality, its goals, values and by providing resources to operate quality management system. Total Quality Management can be defined as a journey and not as a destination [3] and require a complete turn about overhaul in corporate culture and management approach [4].

A Satisfaction of Customer is the core factor of success in the market. All products, materials, goods must fulfil the needs of customer in order to make them happy and it can only be done through proper communication i.e. by taking feedback or through survey. Resources play a vital role in manufacturing of quality products. Therefore, developing relations between contractor, subcontractor, supplier's increases ability to create valuable product. Various independent processes are carry out in company and these individual process play critical role in the production of final product. So, ultimate motive of TQM is to achieve continual improvement in the organization at technical and managerial level in order to meet customer expectations [5] The implementation of TQM in construction industry has brought reliability, increased productivity decreased product cost [1, 6]. It is affirm that implementation of TQM in construction industry provide benefits such as reduced rework, increasing budget, employee job satisfaction, improve bidding in market, increase market share. However, Lack of Standardization is one of the factors which is creating a barrier during implementation of TQM in construction industry.

This paper aims to identify firstly the problem in defining the quality in construction and then barriers which are found during implementation of TQM in construction industry and after that benefit of implementing TQM in construction industry [7].

According to [8] the implementation of performance of construction related organisations where cost-cutting is of dire need there is a much need of quality cost measurement [9]. But this is a problem even in developed countries as Australia were very few firms or companies have in their arsenal a system that can on ground give a comprehensive and in-depth quality cost feedback of the work they do [10].

It can be hence stated with union of many others that quality management was and is unmistakably a very important component in the proper and successful managing of construction industry related firms [11–13].

2 Literature Review

Building projects are getting more complex and larger; clients are expecting high standard of outcome [14]. In service industries TQM is the successful equipment for management and it hold same for manufacturing sectors. Increase in productivity and quality in construction industry is achieved by TQM. Two case studies conducted for two different construction industry shows successful implementation of TQM [14]. This experiment benefits better job satisfaction for employees, reduction in cost of quality and recognition by clients. TQM execution measures were additionally reflected through best administration duty, client inclusion and fulfilment, representative contribution and strengthening, customer–provider connections, and process enhancement and administration. At long last, a system for executing TQM in development is prescribed.

TQM justifies and mandates the coordination and construction industry existing framework required between the different levels of an industry like organisation, suppliers and customers [15], And here such a facility works on sole needs of customer satisfaction more than what it believes in [16].

TQM can be the answer of problems relating to cost, production, health and safety that the industry is facing [17, 18]. Better construction can be achieved by TQM [19]. Many companies are heart-broken due to the fact of their efforts to improve TQM failed as they focused much on the financial aspect of quality control than to manage quality itself by TQM for better quality [20]. Construction industry slowly and steadily inclines towards TQM, Quality control and assurance believing that these can only be guaranteed by the proper and well use of TQM [21]. Also, management system must be flexible and responsive to effective communication repeatedly improving, which can be an experiential TQM objective that is vibrant in nature and keeps on updating demands as per customer satisfaction [22].

In their survey of writing about quality and security administration in the development business, inspected the meaning of value creators utilized [23]. Specialists in development quality foretold the accompanying definitions for quality execution: 'meeting needs for the client'.

Evolution of quality management is directly proportional to the advert of manufacturing industry [24].

"Development Sourcing" ought to end up perceived by substantial and little customers, including both new and customary clients of the ventures, as middle for solid free counsel on the best way to contract for development, including building, and common, electrical, mechanical and process building [25].

TQM techniques have been initially have said to be made and developed for manufacturing process industries to be used utmost and beneficially for process control and defect prevention pre-happening ultimately helping of cost saving front [24]. Because of the cut-throat bidding and nature of industry in construction, there lies an unsaid but present emphasis on fast work methods and almost complete absence of long-term practicality and quality [23]. Hence TQM methodologies that are used to upgrade quality are hence totally applicable to construction industry in a broad way as well [26].

The examination involves poll improvement and the investigation of its reactions from seventeen driving development associations working in Pakistan amid year 2012 [25]. The investigation likewise fuses information from different auxiliary sources. The examination was completed utilizing fundamental factual instruments, for example, mean, middle and standard deviation and bar graph was utilized to speak to the information. The discoveries uncovered that the idea and importance of value isn't comprehended and deciphered in its actual sense. The conformance to standard details (SOPs) is viewed as proportionate to execution of Quality Management which is a long way from truth. It was likewise confirmed that general absence of mindfulness and untrained workers that are for the most part uneducated work is one of the real causes towards this bleak. It is trusted that the results of this investigation may give a rule to the proprietors of the development business and help them comprehend the idea of Quality Management in its genuine term. It will likewise help the observing and directing organizations to devise strategies in like manner taking into account the need of the advanced period that will help Pakistani development industry to contend locally and all inclusive.

An extensive variety has been seen in connection to particular quality related results [27]. With the end goal to accomplish predictable results, a best practice usage show for administration of value in development should be created by distinguishing the causality of every result. In any case, the improvement of such a model requires the utilization of approved builds particular to the order, something that has not been found in the current writing. An arrangement of builds is proposed for this reason.

Success of TQM can be seen by numerous examples already set in the construction industry which include examples of electric and electronic industry [32], Industries from the UK (Cabe 1996) and HDR Inc., Omaha, Nebraska [19].

However, there is a resistance seen to TQM in construction industries due to: -

- Product Diversity—All construction projects are unique and quality is in reference to personal customer needs [33].
- Organisational Stability—High number of collapses of firms and companies prominently during falling economy [33].
- Misconception of cost—The cost of quality can be seen as cost associated with cost related to non-conformity [28].

People seem to think TQM as extra cost but do not realize that actually the sources the lead to non-achievement of goals, rework and re-do requirements cost, error fixing budget inadequacies due to bad planning and missing crucial deadlines are the ones that cost more [19]. So, the cost in question due to implementation can be high for TQM in big projects but the cost incurred due to under quality can be as high as 12% [22, 29].

3 Methodology

The research design used in this study is descriptive in nature. The objective is to explore all issues related to quality in highly dangerous construction sector. The aim is to study the advantages of Total quality management (TQM) implementation in construction industry. This study is also beneficial to understand the barriers in quality management implementation process.

Data was collected through surveying the contractors at the construction site. A questionnaire-based survey was designed and analysed. The questionnaire was framed after careful perusal of literature. Nine interviews were conducted with contractors of construction industry [30]. The items were measured on a 7-point Likert scale from strongly disagree (1) to strongly agree (7). The questionnaire was prepared to know about quality management in the industry of construction. In first part of questionnaire, questions related to quality were asked in the field of safety in construction and related industries [31]. Second part of questionnaire contains questions linked to protection and excellence in construction industry. Approximately three hundred (300) contractors have participated in this survey.

Data analysis was performed for identifying barriers to quality, gauging quality execution, quality execution enhancement and quality improvement. Data were analysed using SPSS 20.0 and MS-Excel 2010. The extent of relationship between different parameter is studied using correlation and regression analysis. The descriptive analysis was also performed to identify the positive and negative response rate of the respondents.

4 Results

The three type of construction sites was used for data collection and analysis purpose:

1. New construction—Involving the construction of new building from the scratch.
2. Retro-fit construction—Involving fixtures to previously made structure as per the need of customer
3. Upgrade—Involving minor or major changes that can be done to already made structure as per required needs and economical constraints [22].

4.1 Interview

Over three forth of respondents of interview stated that primary measure in construction industry for quality measurement was reputation of contractor in that area. Another important feature without which thing are not going well is customer satisfaction. One of the interviewees is to have said to quote that at ground level we actually come to know that which of the contractors are good and which of them are not. Some quote that getting continuous business of customer related projects to do every now and then can be a proper measure of good quality as people come back for where the work is good enough. But in reality just to measure up how the construction industry can flourish using TQM apart from manufacturing industry is a whole lot dependent on the academic way point of the employees as well as training of quality control that includes initial training, refresher courses and time to time checks and guidance from person skilled enough to train others along with the pre-qualification of workers joining the force of the industry at a worker level as well as managerial level along with the use of state of art and newer techniques of designing and building.

Some people also stated that a major cause of low quality might be due to the bidding structure of construction industry. Most of the time the lower bids is responsible for cost cutting and quality degradation.

Quality can be described as customer satisfaction but many a times customer cannot be the judge of quality as the customer will only be interested in its need fulfilment and then the quality management fall of the shoulders of firm itself.

4.2 Questionnaire Survey

There were 43% respondents who said that they were aware of total quality programmes in the construction industry; 57% said they were not aware of such programmes. 38% of the respondents said that their company has a quality programme; while 62% said that this was not the case. (Table 1; Fig. 1).

51.13% of respondents (contractors) believe that 'satisfaction of customer' is the best measure for quality management. Similarly 36.83% believe that 'commitment to quality by management' is the best measure for quality management. The details shown in Table 2 including all other parameter that are best measure for quality management (Fig. 2).

Table 1 Awarness about TQM programmes

TQM programs	Percentage	
	Yes	No
Are you aware of total quality programs in construction industry?	43	57
Does your construction company has quality programs?	38	62

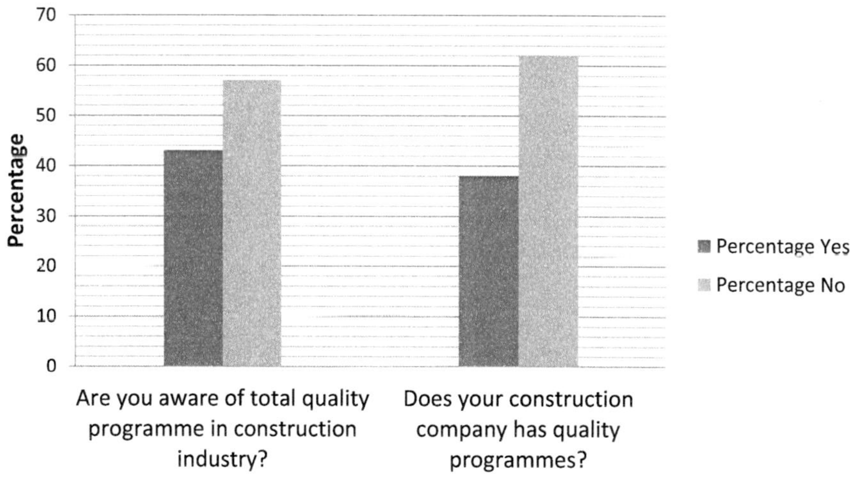

Fig. 1 Awarness about TQM programmes

Table 2 Beat measure for quality management

Best measures for quality management	Percentage (%)
1. Satisfaction of customer	51.13
2. Commitment to quality by management	36.83
3. Returns on business	0
4. Gifted work constrain	0
5. Education training	11.90
6. The measure of call backs/revise	0.02
7. Customary assessments	0.02
8. The length of warranty (in years) companies can give on their work	0.07
9. General construction standards, such as BOCA, CABO, ICC	0.02
10. Quality awards	0.02

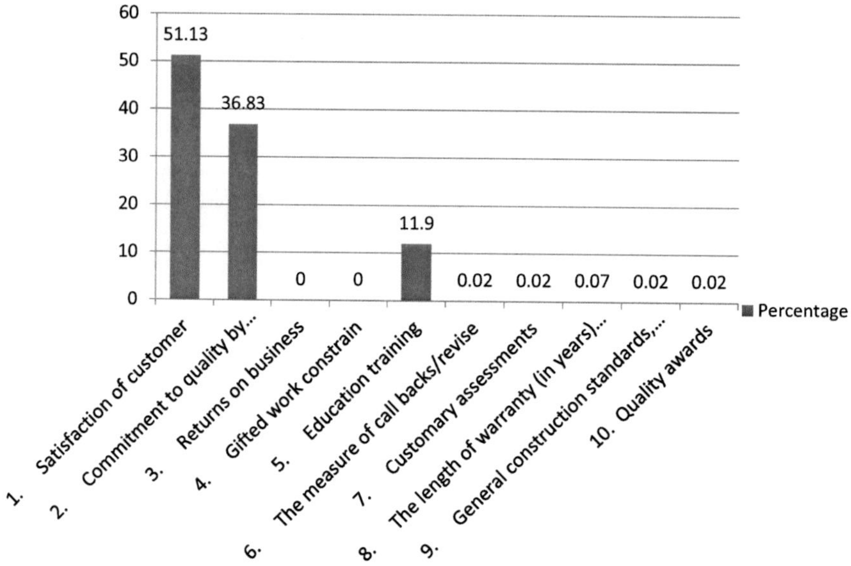

Fig. 2 Best measure for quality management

Table 3 represent the factors for good company characteristics, 38% of respondents believe that 'culture of organization' is a good company characteristic. Similarly, 11.92% believe that 'employee participation' and 'good workforce' are good company characteristics (Fig. 3).

Figure 4 shows that 88% of the respondents responded that their representatives are engaged with endeavours to enhance quality. 76.19% responds positively that a representative quality preparing program does exist at the organization. 78.57% do

Table 3 % Data for good company characteristics (category wise)

Good company characteristics	Percentage
1. Employee participation	11.92
2. Commitment by management	7.14
3. Good workforce	11.92
4. Talk between manager and employee	4.73
5. Training and education	7.14
6. Subcontractors involvement	0
7. Culture of organization	38
8. All around characterized jobs and obligations	4.73
9. Very much characterized jobs and obligations	0
10. Review/analysis used to improve performance	7.14
11. Regular inspections and audits	7.14

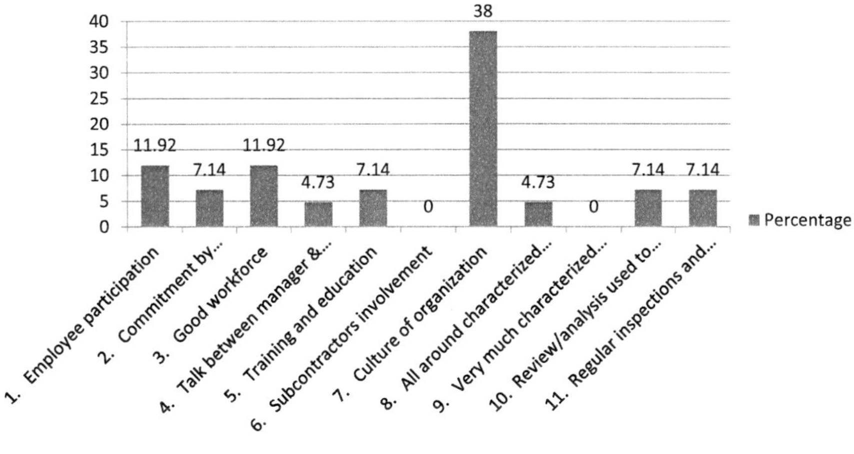

Fig. 3 Good company characteristics

require sellers and subcontractors to utilize a quality program. 73.80% do promote partnering on their projects (Tables 4 and 5).

Figure 5 tells about the few quality improvement deeds done by firm in some kind or another and 78.57% respondents do change the customer's order if there arises an issue with the order. 83.33% respondents work again to correct wrongly placed customer's orders. 90.47% respondents find customer satisfaction as an important thing in their company. 85.71% respondents rectify or try to rectify supplier complains. 80.95% respondents acknowledge complains laid by sub-contractors.

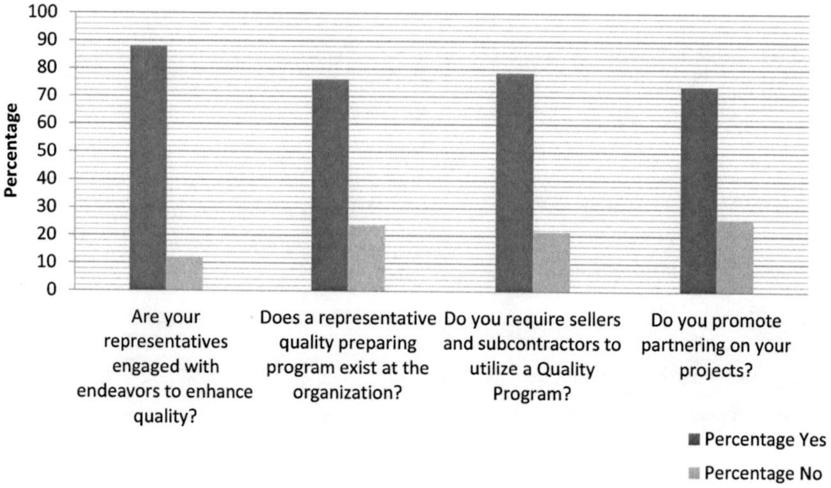

Fig. 4 Company's involvement in quality improvement

Table 4 % wise data related to company's involvement in quality improvement

Company's involvement in quality improvement	Percentage	
	Yes	No
Are your representatives engaged with endeavors to enhance quality?	88	11.9
Does a representative quality preparing program exist at the organization?	76.19	23.8
Do you require sellers and subcontractors to utilize a Quality Program?	78.57	21.42
Do you promote partnering on your projects?	73.8	26.19

Table 5 % Data related to "company's established quality improvement programme"

Company's established quality improvement programme	Percentage	
	Yes	No
Do you change your customer's order if there is an issue in previous order?	78.57	21.42
If the order was wrongly placed then you work again to correct your customer's order?	83.33	16.66
Is customer satisfaction important for your company?	90.47	9.52
Do you rectify supplier complaints?	85.71	14.28
Do you acknowledge complaints of sub-contractors?	80.95	19.04

4.3 *t-statistics Analysis*

t-statistics is a mathematical turned computer tool to standardize value of sample data mainly where there is the requirement of hypothesis testing. That is where we

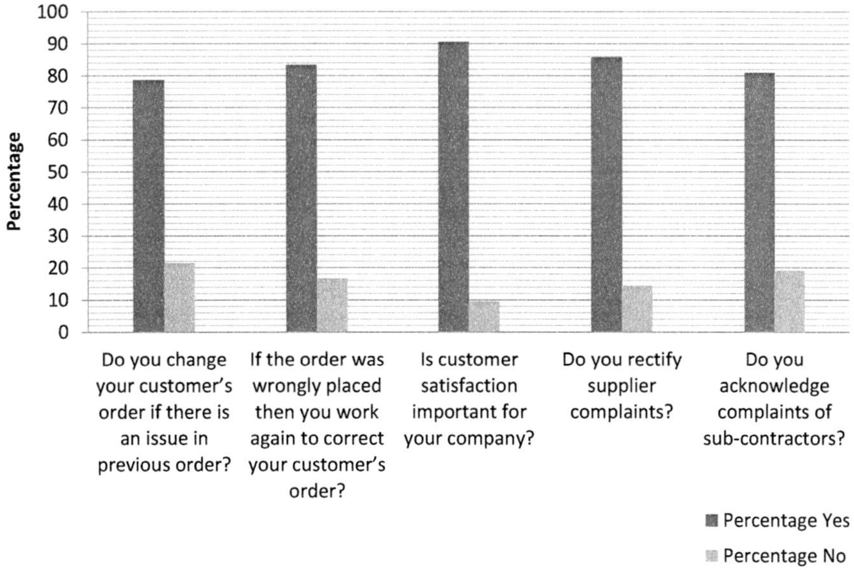

Fig. 5 Company's established quality improvement programme

need to see how our idea is supposed to work when compared to actual hypothesis that we will get in real world application. The various values known as the t-value defines how well the hypothesis matches the tested sample data. 0 indicates sample results exactly matching the sample data and as the value differentiates the t-value changes along as well; generally increases.

In Table 6 the 't' statistics value of 'reputation', 'customer satisfaction' and 'quality programme records' are more than 1.96 (sig. < 0.05). Therefore these are the variables to 'gauge quality execution'. The adjusted R square of the variable 'records of ongoing business and new works' has found to be less than previous variables of

Table 6 't' statistics value of 'reputation', 'customer satisfaction' and 'quality programme records'

Coefficients[a]

Model		Unstandardized coefficients		Standardized coefficients	t	Sig
		B	Std. error	Beta		
1	(Constant)	0.091	0.146		0.625	0.533
	Reputation	0.300	0.015	0.507	20.363	0.000
	Customer_satisfaction	0.316	0.020	0.402	15.557	0.000
	Quality_programme_records	0.357	0.018	0.518	19.984	0.000

[a]Dependent Variable: QualityExecution_scale

Table 7 't' statistics value of 'education plus training', 'work size', 'better craftsmanship' and 'enhanced board and labour relation'

Coefficients[a]						
Model		Unstandardized coefficients		Standardized coefficients	t	Sig
		B	Std. error	Beta		
1	(Constant)	0.418	0.081		5.139	0.000
	Education_plus_training	0.218	0.014	0.310	15.691	0.000
	Work_size	0.244	0.013	0.345	18.374	0.000
	Better_craftsmanship	0.218	0.014	0.289	15.254	0.000
	Enhanced_board_labour_relations	0.235	0.016	0.307	14.830	0.000

[a]Dependent Variable: Quality Execution Enhancement scale

gauging quality execution and hence the independent variable was removed from the model.

As per the findings; the reputation of the company is the highest variable to gauge quality execution followed by quality programme records and customer satisfaction.

In Table 7 The 't' statistics value of 'education plus training', 'work size', 'better craftsmanship' and 'enhanced board and labour relation' are more than 1.96 (sig. < 0.05). Therefore these are the variables which 'enhance quality execution'.

The adjusted R square of the variable 'utilization of configuration, assemble contract and quality requirement' has found to be less than previous variables of 'quality execution enhancement' and hence these independent variable was removed.

As per the findings; the work size for the company is the strongest variable to enhance quality execution followed by education plus training, better craftsmanship and finally enhanced board and labour relations.

Table 8 shows the 't' statistics values of 'product problems', 'turnover in company', 'lack of proper equipment', 'lack of supervision', 'working with new people', 'unreal deadlines' is found to be more than 1.96 (sig. < 0.05). Therefore these variables are a 'barrier to quality'. Bad seed effect and lack of skilled workers had 't' statistics value less than 1.96 hence they were not a barrier to quality.

As per the findings; lack of supervision is the highest barrier to quality followed by unreal deadlines and working with new people.

In Table 9 the 't' statistics values of 'training administration plus workers' and 'consumer loyalty' are more than 1.96 (sig. < 0.05). Therefore these are the variables that affect 'quality improvement'.

The adjusted R square of the variables such as 'means and methods for ensuring continuous improvement', 'obviously characterized objectives identifying with quality work execution', 'systems for collecting and tracking data for ensuring quality objectives' and 'review/analysis process for identifying errors in the system' has

Table 8 A 't' statistics value of 'product problems', 'turnover in company', 'lack of proper equipment', 'lack of supervision', 'working with new people'

Coefficients[a]

Model		Unstandardized coefficients		Standardized coefficients	t	Sig
		B	Std. error	Beta		
1	(Constant)	0.813	0.071		11.419	0.000
	Bad_seed	− 0.029	0.013	− 0.052	− 2.316	0.022
	Lack_SW	− 0.058	0.014	− 0.118	− 4.228	0.000
	Product_problems	0.112	0.012	0.225	9.760	0.000
	Turnover	0.094	0.010	0.176	9.252	0.000
	Lack_ProperEquip	0.113	0.016	0.220	7.160	0.000
	Lack_supervision	0.170	0.015	0.372	10.981	0.000
	Working_newpeople	0.124	0.009	0.290	13.516	0.000
	Unreal_deadlines	0.178	0.012	0.344	14.992	0.000

[a]Dependent Variable: Barriers Quality_Scale

Table 9 't' statistics value of 'training administration plus workers' and 'consumer loyalty'

Coefficients[a]

Model		Unstandardized coefficients		Standardized coefficients	t	Sig
		B	Std. error	Beta		
1	(Constant)	1.379	0.154		8.968	0.000
	Training, administration plus worker	0.447	0.025	0.634	18.170	0.000
	Consumer loyalty	0.320	0.021	0.541	15.499	0.000

[a]Dependent Variable: Quality Improvement Scale

found to be less than previous variables of 'quality improvement' and hence the independent variable was removed from the model.

As per the findings; training administration plus its workers of the company is the strongest variable that affect quality improvement followed by consumer loyalty.

5 Conclusion

From the survey conducted, it was found that most of the respondents (contractors) believe that satisfaction of customer and commitment to quality by management is the best measure for quality management. Many respondents believe that culture of organization, employee participation and good workforce is a good company characteristic. Most of the respondents responded that their representatives are engaged with endeavours to enhance quality. Respondents positively replied to the statement that a representative quality-preparing program does exist at the organization. Many do require sellers and subcontractors to utilize a quality program and most of them do promote partnering on their projects. Many respondents do change the customer's order if there arises an issue with the order and work again to correct wrongly placed customers' orders. Respondents believe that customer satisfaction is an important thing in their company and rectify or try to rectify supplier complains. A high number of respondents acknowledge complains laid by sub-contractors.

Reputation, customer satisfaction and quality programme records are the variables to gauge quality execution. Education plus training, work size, better craftsmanship and enhanced board and labour relation are the variables which enhance quality execution. Product problems, turnover in company, lack of proper equipment, lack of supervision, working with new people, unreal deadlines are found to be the variables that are a barrier to quality. Training administration plus workers and consumer loyalty are the variables that affect quality improvement.

6 Suggestions

In construction industry productivity and quality improvement implementation is new area of interest. In numerous nations dependent on ISO-9000 and ISO-14000 models TQM systems are executing. Anyway, for better usage of TQM, investigations are to be made but to increase effectiveness the following steps can be brought forward like (1) extra effort on education and training of people that might affect the quality of the works of the industry, (2) clear knowledge of customer requirements and planning to satisfy it properly, (3) taking TQM as a work for all and not as responsibility of one, (4) Framework management and development for performance check of TQM in the firm.

References

1. Arditi, D. and H. M. Gunaydin (1997). "Total quality management in the construction process." International Journal of Project Management15(4): 235–243.
2. Koh, T. Y. and S. P. Low (2010). "Empiricist framework for TQM implementation in construction companies." Journal of management in engineering26(3): 133–143.

3. Burati Jr, J. L. and T. H. Oswald (1993). "Implementing TQM in engineering and construction." Journal of management in engineering9(4): 456–470.
4. Quazi, H.A. and S. R. Padibjo (1997)."A journey towards total quality management through ISO 9000 certification- a Singapore Experience." The TQM Magzine.
5. Koskela, L. (1997). "Lean production in construction." Lean construction: 1–9.
6. Besterfield, D. (2003). "H., Besterfield-Michna, C., Besterfield, G., H., Besterfield-Sacre, M." Total quality management 3.
7. Novana, M. and S. Ogunlana (2006). Organizational culture profile of construction companies in Thailand. International Conference on Construction, Culture, Innovation and Management (CCIM), Dubai, UAE.
8. Davis, K., et al. (1989). "Measuring design and construction quality costs." Journal of construction Engineering and Management115(3): 385–400.
9. Abdul-Rahman, H. (1993). "Capturing the cost of quality failures in civil engineering." International Journal of Quality & Reliability Management.
10. Bhuta, C. and S. Karkhanis (1995). Quality performance: How does the Australian construction industry measure it? Proc., Annual Conf. of Australian Institute of Project Managers.
11. Hellard, R. B. (1995). Project partnering: principle and practice, Thomas Telford.
12. Abdul-Rahman, H. (1997). "Some observations on the issues of quality cost in construction." International Journal of Quality & Reliability Management.
13. Love, P. E., et al. (1999). "Rework: a symptom of a dysfunctional supply-chain." European Journal of Purchasing & Supply Management5(1): 1–11.
14. Low, S. P. and H. K. Yeo (1998). "A construction quality costs quantifying system for the building industry." International Journal of Quality & Reliability Management15(3): 329–349.
15. Hunt, V. D. (1993). Managing for quality: Integrating quality and business strategy, Irwin Professional Publishing.
16. Brown, T. J., et al. (1993). "Research note: improving the measurement of service quality." Journal of retailing69(1): 127.
17. Burati Jr, J. L., et al. (1992). "Quality management organizations and techniques." Journal of construction Engineering and Management118(1): 112–128.
18. Burati Jr, J. L., et al. (1992). "Causes of quality deviations in design and construction." Journal of construction Engineering and Management118(1): 34–49.
19. Culp, G., et al. (1993). "Implementing TQM in consulting engineering firm." Journal of management in engineering9(4): 340–356.
20. Torbica, Z. M. and R. C. Stroh (1999). "Impact of total quality management on home-buyer satisfaction." Journal of construction Engineering and Management125(3): 198–203.
21. Jaafari, A. (2001). "Management of risks, uncertainties and opportunities on projects: time for a fundamental shift." International Journal of Project Management19(2): 89–101.
22. Biggar, J. (1990). "Total quality management in construction." AACE International Transactions: Q. 1.1.
23. McCollough, M. and M. Benson (1993). "Five barriers to TQM in construction." Aberdeen's Concrete Construction38: 297–299.
24. Lahndt, L. (1999). "TQM tools for the construction industry." Engineering Mgt. Journal 11(2): 23–27.
25. Latham, S. M. (1994). "Constructing the team."
26. Feigenbaum, A. V. (1989). "How to implement total quality control." Executive Excellence6(11): 15–16.
27. Karim, K., et al. (2005). "Organizational effectiveness model for quality management systems in the Australian construction industry." Total quality management & business excellence16(6): 793–806.
28. Hellard, R. B. (1993). Total quality in construction projects: achieving profitability with customer satisfaction, Thomas Telford.

29. McKim, R. and H. Kiani (1995). "Applying total quality management to the North American cons." Cost Engineering37(3): 24.
30. Loushine, T. W., et al. (2006). "Quality and safety management in construction." Total Quality Management and Business Excellence17(9): 1171–1212.
31. Hoonakker, P., et al. (2010). "Barriers and benefits of quality management in the construction industry: An empirical study." Total quality management21(9): 953–969.
32. Idris, M. A., et al. (1996). "The adoption of ISO 9000 and total quality management in Malaysia." The TQM Magazine 8(5):65–68.
33. Sommerville, J. and H. W. Robertson (2000). "A scorecard approach to benchmarking for total quality construction." International Journal of Quality & Reliability Management.

Study on Effective Management of Total Suspended Particulate Matter Generated at a High-Rise Building Construction Site

S. Sanjay, P. A. Arun, and N. A. Siddiqui

Abstract Particulate matter is a concern and a consequence of the rapid urbanization. These particulate matters of size less than 15 μm are a major deteriorator of public health and wellbeing. Particles size less than 10 microns reach respiratory track and particles size less than 2.5 μm can reach the bronchioles of lungs. Lack of dust management at construction sites may lead to wide spread chronic health diseases linked to non-fatal heart attacks, irregular heartbeat, aggravated asthma, silicosis, lung cancer etc. Average life expectancy of people continuously exposed to Total Suspended Particulates (TSP) reduces over a period of time. A construction site where various activities carried out produces a significant amount of dust particles hence effective management of those particulate matters are of great importance. This paper discusses on the Dust management aspects for the control of total suspended particulate matter generated at a High-rise building construction site. Sampling was carried out using particulate matter measuring equipment and real time monitors. Results were compared with national ambient air quality standards and suggestive measures are proposed for effective dust management in construction sites.

Keywords Total suspended particles · Dust management · Public health

1 Introduction

Airborne Contaminants are a concern due to the diseases associated with it. These airborne contaminants are commonly termed as Dust particles or Particulate matters. As per ISO 4225:1994, these particulate matter sizes are less than 75 μm. The heavier particles settle near the source whereas the lighter size particles are suspended in air for period of time. These particulate matters may have different contaminants and

S. Sanjay · P. A. Arun (✉) · N. A. Siddiqui
School of Engineering, UPES, Dehradun, India
e-mail: arun.pa@ddn.upes.ac.in

N. A. Siddiqui
e-mail: nihal@ddn.upes.ac.in

N. A. Siddiqui et al. (eds.), *Advances in Environment Engineering and Management*, Springer Proceedings in Earth and Environmental Sciences,
https://doi.org/10.1007/978-3-030-79065-3_22

corresponding these are classified as Mineral dust, metallic dust, chemical dust, organic and vegetable dust, biohazards [1]. These particulate matter may be generated by various activities performed at construction sites such as to excavation, bulk material transportation, loading and unloading, open-air material storage, drilling, concrete and mortar making, cut and fill operations and equipment movement. Especially the demolition of concrete or masonry hazardously render workers to respirable crystalline silica [2]. Dust Generated from vehicle traffic on an unpaved road is due to the shearing forces caused at the area between the surface and vehicle tires generates dust from an unpaved road. The generated dust becomes airborne due to turbulence generated by the moving vehicles [3]. Many of the activities at a typical construction site involve cutting, grinding, drilling, excavation, loading and unloading of debris, excavation etc. is the major contributor for generation of dust. Loose silica particles settled on the ground are often disturbed by the wind blowing and carry to the surrounding area.

The Main concern of this particulate matter is the impact on the respiratory system. The fine particles of size equal to and less than 10 μm cause a detrimental effect on the respiratory system. It is estimated globally that approx 3% of cardiopulmonary and 5% of lung cancer deaths are due to particulate matters. Some of the common effects include Silicosis, lung cancer, shortness of breath etc. Since the diseases are chronic and the latency period is high, the risks of dust exposure are often ignored at many of the construction sites. Exposure to particulate matters of 2.5 μm reduces the life expectancy of people of European region has reduced by about 8.6 months on an average. It is also said that Health improvements of those affected due to particulate matter can improve their health almost immediately after reducing the air pollution [4].

Some of the significant sources of particulate matter identified during this course study are from the activities listed in the following,

Drilling and Breaking

In process of the drilling, the drill bit with high rotating speed shatters the concrete and rock forming holes of required sizes and resulting in dust generation due to friction between concrete and drill bit.

Breaking/Chipping

It involves removing the surface layer of the concrete or stone area using machines which linearly vibrates.

Excavation

Excavation involves removing earth materials from the ground surface and moving earth, rock or other materials with tools, equipment which results in fine soil particles flying in air.

Loading and tipping

Loading and tipping activities consists of transporting the excavated materials from one place to another designated place.

Transporting

Transporting construction materials using heavy construction equipment and utility vehicles for transferring materials from activities such as demolishing, abrading, or fracturing silica-containing materials such as brick, block, and concrete can generate respirable crystalline silica dust.

Stockpiling

In stockpiling, the excavated Earth is dumped at a designated place during which the fine particles of Earth materials/ soil are usually eroded by the wind flowing creating particulate matter travel in air. During loading and unloading operations also the fine particles of soil are scattered in air arising dust.

Demolition

This Operation involves removing unwanted structure which are previously existing or built temporarily or which are no longer suitable due to changing conditions. Demolition is carried out manually and use of portable hand tool machine for smaller area. Whereas heavy equipment are being utilized for larger area.

Dry sweeping

Cleaning of Areas covered with dry, loose and fine particles of soil and concrete materials makes the particles fly in air. Cleaning at construction often involves manual sweeping, with the help of compressed air or Portable Air blowers.

Scrabbling or grinding concrete: Grinding the concrete in order to create a smooth surface on the concrete where uneven particles are accumulated. This removal process which involve use of mobile grinding machine. The friction between rotating grinding wheels and the concrete scatters fine dust particles.

Cutting and sanding wood

Wood pieces and Plywood sheets are normally cut with the portable cutting machine. The fine wood particles are generated as the cutting process continues.

Various research work carried out across the globe provides an insight on the particulate matter management and its sustainable solution practiced for an effectiveness. Evelyn et al. [5], states that samples quartz exposure measurement shows that 35% of 728 samples exceeded occupational exposure limit. The authors suggest ways to reduce the occupational exposure of particulate matter and mention that by use of local exhaust ventilation, dust suppression by use cooling water, use of personal protective equipment, awareness and training. Through a short term sampling they found that around 90% of repairable dust can be controlled through Local exhaust ventilation and wet techniques.

As per Johncy et al. [6], explains that when crystalline silica is inhaled, they are settled in and irritate the lungs causing inflammation in small air tubes and sacs of the lungs. When the inflammation heals, scar on the tissue leading to fibrosis.

Nash et al. [7], elaborates that evaluation carried out during a truckpointing operation on a grinder machine with a hose attached to a collection bag found that respirable silica exposure nearly reduced by 93%. John et al. [8] carried out similar evaluation and it was found that when engineering control was used during tuck-pointing reduced the exposure of respirable silica by 98% when the grinding machine was attached a shroud and vacuum system.

Ellen et al. [9], says that Quartz exposure was prominent in some scenarios of most construction activities, out of which five activities having geographical mean exceeding the Threshold Limit Value (surface grinding, tuck-point grinding, concrete cutting, demolition and floor sanding). Risks were highest for surface grinding, which had a geographical mean twelve times the Threshold Limit Value. The amount of enclosure for the work area, equipment utilized and cross draft were prominent factors affecting the exposure.

Zuo et al. [2], conducted a series of surveys and interviews of managers involved in construction projects and When questioned about the opinion of the managers on the attitude and attitudes about dust contamination and their regulation among the employees they oversee, the majority of responses suggested relative ignorance and indifference among the employees. Based on the evaluation by the interviewees of the perception of dust emissions by building workers and its effects, it could be concluded that they are not fully conscious of the issue and sometimes neglect it. Statements varied from "lack of information", "oblivious", "poor perception" and "lack of skill".

Mieke et al. [10], concludes from his research that a prominent factor in the susceptibility to respirable crystalline silica is the nature of building material and its quartz composition. This is most crucial for the drilling operators. Construction material is well established in this occupational category, especially compared with demolition workers, who almost all operate in a mixed dust atmosphere. Drilling operators typically know the form of building material they were drilling, as their preference of grinding wheel dependent on the strength of the material.

2 Methodology

For any particulate matter management, the measurement or the quantification of the amount of dust specifically the total suspended particulate matter which poses a substantial danger to health and wellbeing of human beings. Measurement of the Total Particulate matter of size 10 and 2.5 μm are easily measured with the sampling instrument with the help of the filter paper of 10 and 2.5 μm pore size respectively using gravimetric method. The amount of total suspended particulate matter can be found using the differential weight of the filter paper. The methodology adopted is portrayed in Fig. 1. The sampling equipment used for this study is given in Fig. 2. The sampling was carried out through an agency complied under Legal Regulations.

The equipment was placed at 4 feet above ground and the sampling was done for a period of 24 h. Sampling was done at two points where the building construction work is carried out. At the end of the sampling, the filter paper was removed and sent

Fig. 1 Methodology

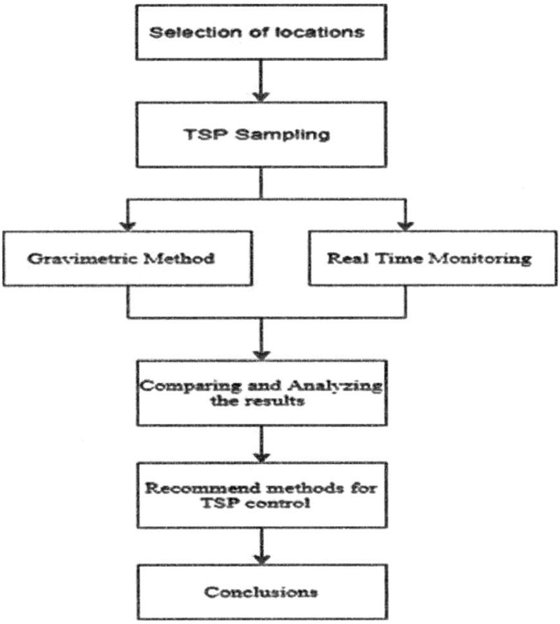

Fig. 2 Particulate Matter Sampling Equipment

Table 3 Real time dust monitoring—post water sprinkling

Sl no	Construction Site	Location	PM 10 ($\mu g/m^3$)	PM 2.5 ($\mu g/m^3$)	PM 1 ($\mu g/m^3$)	AQI
1	Cluster 2	Security Check Post	91	76	50	150
2		Non Tower Area	87	72	49	123
3		Tower 1 Entrance	74	58	39	93
4		Tower Crane	73	58	40	95
5		Steel Yard 1	80	63	44	110
6	Office	Office Entrance	90	74	51	140
7	First Aid Center	Entrance	84	66	46	120
8	Stores	Entrance	82	62	40	92
9	Canteen	Entrance	76	58	39	96
10	Workmen Colony	Middle	80	62	40	110
11	Cluster 4	Steel Yard 2 (in btw Tower 1 & 2)	67	54	36	90
12		Tower Crane 1 (tower 1)	80	60	37	100
13		Entry Point (near Drinking Water)	72	50	36	83
14		Steel Yard 3,4 (behind tower 3)	74	58	38	95
15		Site office (behind Tower 4)	75	57	39	96
16		In front of Tower 3,4	86	73	44	143
17		In front of Tower 3	80	61	43	106
18		In front of Tower 2 (Below Tower Crane 2)	91	72	45	140
19		In front of Tower 1 (near NTA)	90	67	42	123

Table 4 Real-time particulates from activities

S. No.	Activity	PM 10 (μg/m^3)	PM 2.5 (μg/m^3)	PM 1 (μg/m^3)	AQI
1	Dry Sweeping	76	68	42	103
2	Chipping	90	66	42	120
3	Soil Excavation	85	63	40	103
4	Concrete Grinding	183	112	56	270

Table 5 NAAQ standard of particulate matter

Pollutant	Sampling period (h)	Concentration in ambient air	
		Industrial, residential, rural and other areas (μg/m^3)	Ecologically sensitive area (notified by central government) (μg/m^3)
PM 10	24	100	100
PM 2.5	24	60	60

During recording values of Table 3, water was sprinkled on the ground, walkway, pathway etc.

We now have the data of particulate matter at various areas before and after the application of water as a dust suppression agent. There are few activities which majorly generate particulate matter. These activities were specially monitored for the generation of particulate matter. During sampling, a minimum of 4-m distance was maintained to record the values of particulate matter. The recorded data specific to those activities are recorded in Table 4.

Indian Legislation namely National Ambient Air Quality Standard 2009 under the Environmental Protection Act 1986 regulates the Particulate matter generation and prescribed standard values to comply. Table 5 represents the standard set by NAASQ.

Graphical comparison of Sampling results of Particulate matter PM 10 and PM 2.5 levels with prescribed standard under NAAQS. These sampling results were obtained under the normal work conditions where no dust suppressant or dust management was in place (Fig. 4).

The following provides a Graphical comparison of Sampling results of Particulate matter PM 10 and PM 2.5 levels with prescribed standard under NAAQS post water sprinkling on the unpaved roads. It shows that water sprinkler is effective in controlling PM 10 levels but not so effective in controlling PM 2.5 levels (Fig. 5).

Current Dust Management Scenario

The Dust Management is gaining traction since the construction site surrounds a residential neighborhood. Through dust management, the impact on workmen in the site as well as the people in the surrounding areas could be minimized.

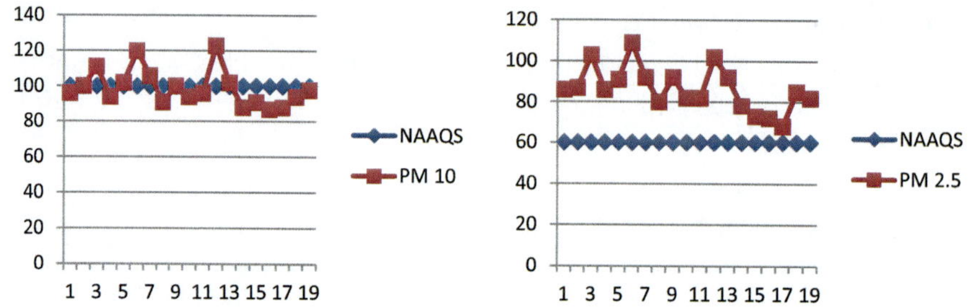

Fig. 4 PM 10 and PM 2.5 levels before water sprinkling

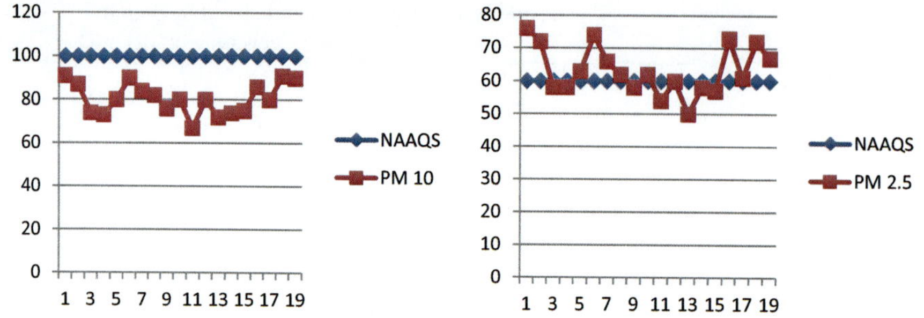

Fig. 5 PM 10 and PM 2.5 levels post water sprinkling

Reduction of Speed

Regulating the vehicle speed to between 10 to 15 km/hr can reduce the fugitive dust from arising in the air. When the vehicle speed increases the there is a shearing force created on the surface of road by vehicle tyre which creates a float of the soil particles. When the vehicle speed is reduced the shearing force acting on the road is also reduced in turn reducing the particulate matter float.The site has restricted the vehicular speed to 10 km/hr through the site and is applicable for all construction vehicles, passenger vehicles and equipment's.

Water Sprinkling

The HSE team at the site is addressing this need for the dust suppression with the application of Water. The site has an adequate supply of Water through groundwater resources. The dust management is carried out on the vehicle pathway in the site. Initially before the hauling of equipment and vehicles starts, water is sprinkled manually on the unpaved pathway. This activity of water sprinkling is being done thrice a day. Water is perhaps the oldest of all dust palliatives; it is readily accessible and is usually spread over the surface of the lane. The capacity to remove the dust is very transient due to evaporation. Water damps the layers with dust. It does need to be used properly, however. That ensures adequate water supplied for the whole period the work is being performed at the correct stage. Wetting the content in advance just doesn't work [11–13].

Water Mist Cannon

Dust suppression is carried out by the fine water mist which moist the dust particles causing them to settle down at a faster rate. The tiny nozzles create a water mist of varying sizes from 10 μm till 100 μm. The water mist guns can spray water molecules for a distance range of 70–100 m. It can be installed on a fixed platform or on a movable wheel base or it can also be mounted on trucks to spray along a required area [14].

Local Exhaust Ventilation

Local Exhaust Ventilation comes under the Engineering control where major Particulate matters generated during the machine operation can be captured. Adapting and using the machine with new technologies can help capture the generated dust along with the required precautions; the dust management can be more effective and sustainable. Portable machines such as grinding machine, Saw machines, drilling machine etc. can be connected to portable vacuum machine to ease the dust collection process. These vacuum machine aids mobility and the waste collected can be easily disposed [16].

Mulch

Mulch is a layer of material applied on the soil surface to aids retention of soil moisture, reduces soil erosion and enhances the visual appearance. Different materials can be used as mulch such as grass, crushed stone etc. depending on the feasibility. Open Areas are prone to soil erosion whenever wind blows causing the top layer of the soil to travel along with the wind and spread around the area. This problem can be managed with growing vegetation or grass can be used as mulch on the open ground where no activities or no vehicle movements are carried out. For pathway where vehicle movement occurs, crushed stones can be used as mulch to prevent dust arising when the vehicle moves.

Dust Nets

Dust nets covering the site, vehicle pathway or the around the activities performed helps to contain the dust. Whenever the dust arises from operation or due to vehicle movement, the material of the net and the design of the net helps to retain the dust for the maximum extent. This also serves the purpose of blocking the winds which disperses the particulate matters into the surrounding places.

Water Dampening System

As per Evelyn et al., based on a brief sampling, it was found that combination of wet techniques and Local exhaust ventilation (LEV) and can minimize exposure to silica and total suspended particulates by > 90%. During operations such as grinding, concrete cutting, chipping etc. application of water reduces dust arising out of the operation since the surface contain moisture, the dust particles are absorbed by the moisture and it also reduces the friction between grinding wheels and the surface. But precautions and supervision to be ensured that water is not accumulated on the working surface to avoid ingress of water into the equipment and instances of electrocutions [15].

Vacuum Cleaning

Conventional and cheaper method followed for cleaning at the site is sweeping. Sweeping generates large amount of dust since activity disturbs the fine particles settled on the surface. Sprinkling of water before sweeping is one way to keep the fire particles settle on the surface, but it is not the optimum technique. Some utilize compressed air to clean the surface which works similar to sweeping by blowing high pressure air on surface. Industrial vacuum cleaners have eased the job of cleaning since the settled particulate matters are extracted without disturbing into the surrounding.

4 Conclusions

Construction sector is the backbone of the infrastructure boost in the county. Particulate matter generation during some construction activities may be inevitable but its exposure to workmen can be mitigated. Firstly adequate awareness is essential on the importance of dust control and its consequence. This motivates people to protect themselves and workers who are at the line of fire. Nowadays dust management is commonly adopted as a part of the project plan and adequate resources are allocated to implement the plans.

In the current scenario, the site HSE team is providing an interim solution of manually watering the unpaved roads thrice a day. The conditions can still be improved by adopting the recommendations provided in this report. These recommendations are examined for their feasibility at the site and can prove to be an effective means of dust mitigation.

References

1. WHO, "Hazard Prevention and Control in the Work Environment: Airborne Dust, Chapter 1 - Dust : Definitions and Concepts," World Health Organization (WHO), 1996, pp. 1–96.
2. J. Zuo, R. Rameezdeen, M. Hagger, Z. Zhou, and Z. Ding, "Dust pollution control on construction sites : Awareness and self-responsibility of managers," *J. Clean. Prod.*, vol. 166, pp. 312–320, 2017.
3. D. L. Barnes, "Alaska Department of Transportation & Public Facilities Alaska University Transportation Center Managing Dust on Unpaved Roads and Airports David Barnes , Ph . D ., P . E . and Billy Connor , P . E .," no. November, 2014.
4. WHO, "Health effects of particulate matter Policy implications for countries in eastern Europe, Caucasus and central Asia." The Regional Office for Europe of the World Health Organization (WHO), 2013.

5. E. T. Nij, S. Hilhorst, T. O. N. Spee, F. Steffens, M. Lumens, and D. Heederik, "Dust Control Measures in the Construction Industry," vol. 47, no. 3, pp. 211–218, 2003.
6. S. S. Johncy, K. T. Ajay, G. Dhanyakumar, N. P. Raj, and T. V. Samuel, "Dust Exposure and Lung Function Impairment in Construction Workers," pp. 9–13.
7. Nancy T. Nash & Donald R. Williams, "Occupational Exposure to Crystalline Silica During Tuckpointing and the Use of Engineering Controls," vol. 15, no. 1, pp. 8–10, 2000.
8. J. D. Meeker, "Engineering Control Technologies to Reduce Occupational Silica Exposures in Masonry Cutting and Tuckpointing," vol. 124, pp. 101–111, 2009.
9. P. Taylor *et al.*, "AIHA Journal," no. August 2013, pp. 37–41.
10. M. E. G. L. Lumensf and T. O. N. Spee, "Determinants of Exposure to Respirable Quartz Dust in the Construction Industry," vol. 45, no. 7, pp. 585–595, 2001.
11. Addo, J. Q., Sanders, T. G., & Ph, D. (2004). Road Dust Suppression : Effect on Maintenance Stability , Safety and the Environment Phases 1–3, (May).
12. Federal, C., & Highway, L. (2013). UNPAVED ROAD DUST MANAGEMENT A Successful Practitioner ' s Handbook, (January).
13. Gas, S. D., Company, E., Diego, S., & Diego, S. (2016). Dust Control Management Plan, (November).
14. KProstański, D. (2013). USE OF AIR-AND-WATER SPRAYING SYSTEMS FOR IMPROVING. *Journal of Sustainable Mining*, *12*(2), 29–34. https://doi.org/10.7424/jsm130204.
15. Wu, Z., Zhang, X., & Wu, M. (2015). Mitigating construction dust pollution : state of the art and the way forward. *Journal of Cleaner Production*. https://doi.org/10.1016/j.jclepro.2015.01.015.
16. Heitbrink, W.A. and Watkins, D.S. 2001. The effect of exhaust flow rate upon the respirable dust emissions for tuckpointing operations. In: In-Depth Study Report: Control Technology for Crystalline Silica Exposures in Construction. U.S. Department of Health and Human Services, Centers for Disease Control and Prevention, National Institute for Occupational Safety and Health. ECTB No. 247–18.

Women Empowerment and Sustainable Development in Uttarakhand: Challenges and Strategies—A Case Study of Selected Villages of Dehradun

Shalini Vohra and Kanchan Deoli Bahukhandi

Abstract India is considered as the country of villages as bulk of its population lives in far flung area/rural area. Survey conducted at village Bhauwala, Saheshpur Block of Dehradun shows that rural women suffer both from economic and social invisibility. Economic invisibility stems from the perception that women are not relevant to the wage and self-employment although they have been playing very important and significant role in the society. In addition, in patriarchal society, the rural women hardly have any decision-making role in the family. In the survey area it has been observed that the women are deprived of development with toughest living condition, the main reason being lack of education, awareness about health and hygiene and lack of self-reliance. Empowerment of women comprises of multidimensional focus and envisages greater access to knowledge, social and economic resources and greater autonomy in the economic decision making process, social equality, personal rights, change in the sexual division of labor, equal access to food, health care education, ownership of assets, employment etc. The present paper thrusts on women empowerment i.e. the social as well as economic wellbeing of women resulting into the sustainable development of society.

Keywords Rural women · Empowerment · Autonomy · Patriarchy · Sustainable development

1 Introduction

Uttarakhand emerged as a 27th state of India on 0th November 2000 leading to the fulfillment of the long cherished dream of the people who are the residents of this hilly region with an area of with an area of 53,483 km^2 and with a population of more than one crore (Raj B.Mathur, Director Environmental Science). Dehradun is the capital of Uttarakhand and the major languages spoken in Dehradun are Garhwali, Kumauni and Hindi. Rural areas of Uttarakhand are deprived of development of

S. Vohra · K. D. Bahukhandi (✉)
UPES, Dehradun, India
e-mail: kanchan@ddn.upes.ac.in

N. A. Siddiqui et al. (eds.), *Advances in Environment Engineering and Management*, Springer Proceedings in Earth and Environmental Sciences,
https://doi.org/10.1007/978-3-030-79065-3_23

development and infrastructural facilities and also provide toughest living condition for the inhabitants of these areas (shodhganga). Women, constituting the weaker section of the society, are suffering from various problems in every field. The Indian society is more prejudiced towards the male child in respect of education, nutrition and other opportunities.

On the other hand women are deprived of development with toughest living condition. Although there is no denying the fact that women in Uattrakhand state have made progress in different fields since independence of our country, but they still have to struggle against many social evils in the patriarchal society, the main reason being lack of education, ignorance about health and hygiene and absence of self-reliance [1]. Swami Vivekanda, One of the greatest son of India quoted on women that" there is no chance for the welfare of the world unless the condition of the women is improved [2]. Thus "Women Empowerment" has become one of the prime goals for each and every nation to enhance the socio-economic status of the women. Across the globe, women empowerment has been burning topic over the last few decades and various efforts are being made to address this issue and to bring social rights, economic independence, political rights and judicial rights to the women such that there is no discrimination between men and women (Dr. Kanikicherla Rani 2019). It is a multidimensional focus which foresees better access to knowledge, social and economic resources and greater individuality in the economic decision making process, social equality, personal rights, change in the sexual division of labor, equal access to food, health care education, ownership of assets, employment.

A number of studies carried out from time to time to understand the position of women in rural areas and different strategies have been worked on to improve their living status [3]. Many programs held for decreasing gender inequalities by improving the status of the woman, Sidhu and Kaur [4] revealed that entrepreneurship is the only solution for the growing employment among rural youth. Bisht [5] argued that the state has to sustain forestry and agriculture for the survival of the people who largely live in rural and hilly areas. A study by Kar (2007) highlights the importance of equitable growth focusing on GDP growth and distribution of growth to all sections of the population and geographical regions of the country [4]. Rana and Masood [6] mentioned that during the last two decades, Indian women have entered the field of entrepreneurship in increasing numbers [1]. Kandari and Bahuguna [7], made a study which showed the impact of patriarchal nature of society in hill rural regions in the form of lack of participation of females in decision making in family, gender biasness in education delivery and their low level of living in the family and in the society [8]. A woman in Uttarakhand is earning Rs. 18.13 per day. This is more than the per capita income of India. But the potential of Uttarakhand women is still not being utilized fully. Thus, there is a need to raise the status and standard of women in Uttarakhand. The study area, Bhauwala is located at a distance of about 19 km from Dehradun. The population of Bhauwala is about 6500 with male population of 4500 and female population of 2500. The area of the village is about 15 km^2. When the survey was carried out, it was found that only about 3% of female is only the workforce.

2 Methodology

Following strategy was adopted to meet the above objectives:

I. **Surveying the villages**: A basic survey conducted in the identified villages for women and adolescent girls' population Dehradun. The survey comprised of the information related to identify their interest area related to English and Computer education.

II. **Identification and Formation of the groups**: Based on survey, we got an idea about the interested women in the field of education. We form a group of 10 women and thrice in a week, we offered classes at least for two hours. Initially we will started with six months—basic course, thereafter exam conducted and women should further motivated for advance course.

III. **Locating the site for camp organization**: A central suitable place is identified for few villages in Bhauwala, Dehradun where the classes can be organized.

IV. **Report compilation**: After basic course, a report is generated compiling the observations, common problems and outcome of the course thereafter women will further motivated and encouraged to join complete course.

According to 2013, UNDP report on Human Development Indicators, all south Asian Countries except Afghanistan, ranked better for women than India (Human Development Report 2013). It also highlighted that: an Indian girl child aged 1–5 years is 75% more likely to die than the boy child is. National Crime Records Bureau (NCRB) has an even more horrifying statistic to share—every 20 min, a woman is raped somewhere in India (NCRB 2018). Not only that: crimes against women have increased by 7.1% nationwide since 2010, and child rape cases have increased by 336% in the last 10 years.

A woman raped once in every 20 min and 10% of all crimes are reported. Women form 48% of India's Population, only 29% of the National workforce, workforce; and only 26% of women have an access to formal credit [9]. In the Families, which lives in interior of Uttarakhand T the male members of the family usually migrate to towns to earn a living for their families. Thus, women in villages become the heads of the family. Even where the men continue to reside in the villages, the women and girls look after the agriculture and cattle [10]. Typically, the leftover food is meagre, considering the families are poor and have little to begin with. This creates a major problem with malnutrition, especially for pregnant or nursing women. Very few women seek medical care while during pregnancy because it is thought of as a temporary condition. This is one main reason why India is maternal and infant mortality rates are so high. Starting from birth, girls do not receive as much care and commitment from their parents and society as a boy would receive. For example, a new baby girl would only be breast fed for a short period, barely supplying her with the nutrients she needs. This is so that the mother can get pregnant as soon as possible, in hopes of a son, the next time.

3 The Condition of Hilly Women

Women plays very important role. They are cores of the family. Though the role of women is very important and crucial in a hill economy, yet they are the invisible workers and lead a tough life. Women are actively engaged in all the activities right from the house to the fields. Interaction among women took place during their work in the field and /or forest, kitchen, during washing clothes etc.

The women face many other problems due to the rigid conduct and behavior pattern:

Health Problems

Health issues are preponderated because of absence of Knowledge about Dietary Pattern: Nutrition requirement depends on the age and sex of an individual. Women have a lack of knowledge about the dietary pattern best suited to different age groups. They do not know how much food they need to consume during the time of pregnancy and about the lactation period for women. Lack of knowledge thus also become one of the cause is a cause of high maternal mortality rate among the women. They suffer from various health problems such as anemia, weakness and vomiting.

Domination Issues: Gender has been the most statistically important determinant of malnutrition among young children, and malnutrition is a frequent direct or underlying cause of death among girls below the age of five. It was also found that women use to eat whatever was left after feeding it to the male members of the family. Women used to consuming the leftover meals of the day for dinner. Girls were breast-fed less frequently and for shorter durations during infancy and childhood and during adulthood, while males were fed first and better.

Communicable Diseases: Communicable diseases occurs due to improper disposal of waste and working barefoot leads to high instances of hookworm infection in rural areas. Hookworm infection is directly responsible for a high percentage of anemia among women.

Lack of Nutritional Education: It was generally found that women have low level of Nutritional education. Low nutrition education explains results in the poor intake of vitamin rich food; especially green vegetables among women, even though they are available in plenty in the Pantnagar nearby area.

Lack of Knowledge of Human Rights.

(a) Child Labor—A considerable amount of girl children are employed in strenuous activities. As They are lack unaware awareness about the Child labor Act and its protectionary measures to prevent it.
(b) Women's Rights—Women, constituting the weaker section of the society, are suffering from various problems in every field. Today, the government has initiated various programs related to Anti-dowry and maternal benefits, but women lack awareness about these laws and rules. The two women who were working in the Pantnagar field had no men to support them in household activities, and many women were on the receiving end of various forms of violence from their male counterparts.

(c) Agriculture Policies—Agriculture, being the backbone of most of the rural population needs particular attention from both, our planners and purveyors of knowledge. Government has initiated various policies and laws in Agriculture.

Rural Women and Entrepreneurship the emergence of rural women entrepreneurship and their contribution to the national economy is quit noticeable in India. The numbers of women entrepreneurs have grown over a period, especially in late twentieth century. Women entrepreneurs need to be lauded for their increased utilization of modern technology, increased investment, finding a niche in the global market, creating an extensive employment for others and setting the trend for other women entrepreneurs in the organized sector. While women entrepreneurs have demonstrated their potential, the fact remains that they are capable of contributing much more than what they already are. Therefore, women entrepreneurs are needed to be studied separately for two more reasons. First, women entrepreneurs have been recognized during the last decade as an important unexploited source of economic growth as they creating new jobs for themselves and others and by being different. They also provide the society with different solutions to management, organization and business problems as well as to the exploitation of entrepreneurial opportunities. Secondly, women entrepreneurs have been neglected largely both in society in general and in the social sciences as well. Not only have women lower participation rates in entrepreneurship than men but they also generally choose to start and manage firms in different industries than men tend to do. Hence, women can effectively undertake both the production and processing oriented enterprises. Entrepreneurship development among rural women helps to enhance their personal capabilities and increase decision-making status in the family and society as a whole. Keeping this perspective we have initiated some projects in Dehradun villages and applied some strategies to implement the programs so that we can move it forward on sustainable basis.

Strategies applied during Projects implementation

Four different projects implemented during 12–16th October 2015 to re-emphasizes commitments towards society. Projects aligned with the policy of women empowerment with skill development for sustainable growth. Village Bhauwala was chosen with the aim to developing it to a model village. Approximately 100 students Volunteers participated in the launch of new initiatives over five days. These new initiatives would enable women folk of village to supplement earnings of their families. The four five project implemented were "Wet Organic Solid waste management", The Technology Enabled Communication Skill Development "Beauty Parlor Training Project", "Tailoring skill development project", "mentoring young children of primary school".

The project "Organic solid waste management" was aligned with the prime minister "Swach Bharat Abhiyan". Training imparted to the women folk. The Project aimed at cleaning the surrounding environment of the village Bhauwala by utilizing the organic wastes (rotten vegetables from the weekly market plus waste foods from different sources viz., kitchen, shops and small restaurants located in the village) and to provide sustainable source of livelihood for villagers by generating organic compost. Two villagers were trained to run the project.

"The Technology Enabled Communication Skill Development" project was launched on 13 October 2015 in the village. This project were linked to Digital India, an initiative of Government of India, to connect rural area with the world of technology. Fifteen women of different age groups from Bhauwala village were enthusiastically participated in the project. This was six months integrated course, which was designed to inculcate the application of information technology to enhance English communication skill, which is vital to develop market linkage for sustainable business and to enhance their personality. The training imparted by corporate professional. The programme is being executed in the village itself and 10 computers have been donated by UPES for the said training programme. The project aimed to empower young women with English Communication using basic computer skill by providing Basics of Grammar, Translation, Vocabulary Development, group discussion, answering ability and to work on MS Office. It was done with the aim that after the course completion, those women could start with teaching their own children and can open up coaching centers in the nearby villages, which is the basic requirement of the developing areas. This project enhanced their overall personality and opened avenue for them to work in different organization.

"Beauty Parlor Training Project" was launched with the fact that every woman, irrespective of her social upbringing, wants to look beautiful gave us enough reason to start the project. In the rural set-ups, the concept of beauty parlors is not yet come to the surface. Through this project, the women trained as beauticians and gained livelihood opportunities and it will be an ice breaker for them. This project was launched on 14 Oct 2015. Through survey, few women from Bhauwala village identified to be professionally trained to take it as their career. After completion of the six months, four women opened their parlor at their home and earning a good income for themselves.

Tailoring skill development project was launched with a concern about Rural Women in our Project Area who suffer from being both economically and socially invisible. Economic invisibility stems from the perception that women are not relevant to the wage and self-employment but they have been playing very important and significant role in the society. There is a need to empower these women by strengthening their economic security. With the ever expansion /growth of primary and secondary schools, the need for school uniform is bound to gear up. Thus, this income generating project of tailoring launched with an aim at equipping the women of Bhauwala village to take up tailoring as an entrepreneurship. Tailoring skill development project was launched on 14 Oct 2015. Four women from Bhauwala village got trained for six months. After completion of income generation training program me, those women encouraged to open their own tailoring shop and became an entrepreneur. Few of them employed at factory working for school and two of them are stitching clothes at home.

Mentoring young children of primary school, the project focused in creative and stimulating methods, which inspired and equipped students to continue their education and to excel. The project was launched on October 17th 2015 and implemented on every Saturdays by UPES students. These projects aimed to provide "Quality education" to the children through interactive session on Mathematics, Science, English and Basics of Computer to develop communication skills of the children, to improve quality of learning by supplemental methods, to support the school system to enable the underprivileged children to cope up with the school work.

We at UPES included various skill-development training programs viz candle, making fruit jam and vegetable preservation, tailoring, soft toy making, Diya decoration, paper bags making etc. from time to time to empower women in the villages by helping them to become economically independent, we are conducting various skill-development training programs.

Education is main concern. Around 45% of adult women are illiterate in India. There is huge gap in higher education system. Attainment of top leadership is still hard nut to crack for women. In rural areas, poverty is main cause that leads to suppression and exploitation. Through paramount concern for the wellbeing of a woman can provide momentous and empower them for leading happy life. Professional inequality and house hold inequality is the major cause of suffering among women fraternity. So through proper concern and by eradicating orthodox pattern also sharing her household work. Our women can lead beautiful and happy life can contribute significantly for the growth of our nation.

References

1. Rita Singh Raghuvanshi (2018) Changing perspective of Home Science Education in India.
2. Reshma Khatun, Nasir Ahmed, Vivekananda's views on women empowerment, International Journal of Advanced Research and Development ISSN: 2455–4030 Impact Factor: RJIF 5.24 www.advancedjournal.com Volume 3; Issue 1; January 2018; Page No. 987–990.
3. Kapur, Radhika. (2019). Status of Women in Rural Areas. Acta Scientific Agriculture. 3. https://doi.org/10.31080/ASAG.2019.03.0558.
4. Kiranjot Sidhu , Sukhjeet Kaur,(2006). Development of entrepreneurship among rural women, J.Soc. Sci., 13(2): 147–149.
5. Bisht, D.S. (2006). Poverty, Planning and Development - A Case Study of Uttaranchal State (submitted to the Planning Commission). Central Himalayan Institute. Dehradun. Trishul Publications, Dehradun.
6. Massod Rana Zehra Masood, 2011, Emergence of women-owned businesses in India-insight. Journal of Arts Science & Commerce. ISSN 2229-4686.
7. Prashant Kandari, Uma Bahuguna, "patriarchy leading to domestic violence and poor social status of females in hill rural areas: a study of district pauri in Uttarakhand", (2015), International journal of social science & interdisciplinary Research; 4 (4); 19-28.
8. JETIR March 2019, Volume 6, Issue 3 www.jetir.org (ISSN-2349–5162) JETIRAP06031 Journal of Emerging Technologies and Innovative Research (JETIR) www.jetir.org 186 Women.
9. Dr. (Smt.) Rajeshwari M. Shettar A Study on Issues and Challenges of Women Empowerment in India IOSR Journal of Business and Management (IOSR-JBM) e-ISSN: 2278–487X, p-ISSN: 2319–7668. Volume 17, Issue 4.Ver. I (Apr. 2015), PP 13–19. www.iosrjournals.org.
10. Women and Development in the Garhwal Himalayas Annpurna NAUTIYAL Keywords Garhwal Himalayas; women's empowerment; panchayats; gram pradhan; women's groups; India, AJWS Vol. 9 No. 4, 2003. pp. 93–113.

11. Napier, Harriet, "Employing Empowerment: Developing the Discourse for Women's Empowerment in Uttarakhand, India" (2011). Independent Study Project (ISP) Collection. Paper 1072. http://digitalcollections.sit.edu/isp_collection/1072.
12. Raj B. Mathur, Director, Environmental Sciences, Tetra Tech, Inc., San Bernardino, California. Coauthor of *A Historical Atlas of South Asia.* https://www.britannica.com/place/Uttarakhand.
13. https://shodhganga.inflibnet.ac.in/bitstream/10603/166288/2/chapter%201.pdf.
14. Tamilselvi, & Vasanthakumar, Jeyarajan. (2008). Entrepreneurship Development among Rural Women. International Journal of Extension Education. 4. 79–84.
15. Surabhi Mittal Gaurav Tripathi Deepti Sethi. (2008). Development Strategy for the Hill Districts of Uttarakhand.
16. Dr. Sahab Singh, Dr. Gaurav Thakur, Dr. P C Gupta - A Case Study on Empowerment of Rural Women through Micro Entrepreneurship Development.
17. http://hdr.undp.org/sites/default/files/reports/14/hdr2013_en_complete.pdf.

Varietal Reaction on the Cultivars of *Cucurbita pepo* Against Two Different Isolates of Pumpkin Mosaic Virus

Shivani Jasrotia and Nirmala Koranga

Abstract *Cucurbita pepo* is important commercial vegetable crop in cucurbitaceae family grown globally with various morphological distinct species and commonly known as field pumpkin. Regardless of this distinction and the crop in request, this significant agriculture crop is prone to invade by pathogens like viruses, lead to devastation with reduction in crop yield in lower altitudes (valleys) sometime up to 95% which is chief component in agriculture perspective and the cause is in adequacy of knowledge about pathogenic infection among farmers and lack of beneficial applications in culturing. The objective of the present study was to figure out distinct cultivars of pumpkin showing effect against mosaic isolates (mild and severe). 19 varieties of pumpkin were taken under consideration including local two varieties to check the susceptibility during year 2019. Out of 19 varieties almost all varieties exhibit some kind of response against isolates of mosaic virus. Extreme infection was observed in cultivars. Cultivars Kashi harit, CM-350, Pusa vikas and Pusa vishwas show resistance in mild isolate and the infection was up to less than 25% while maximum susceptibility in mild isolates was found in pumpkin partap and other two local cultivars of pumpkin and the infection was up to more than 70%. In case of severe isolates none of the cultivar show resistance on the other hand maximum susceptibility was found in cultivars that shows susceptibility in mild isolate include Baramasi. Response of different cultivars differ due to variation in environmental conditions.

Keywords *Cucurbita pepo* · Pumpkin · Cultivars · Varieties · Isolates · Mosaic · Resistance · Susceptibility

S. Jasrotia (✉) · N. Koranga
Department of Botany, P.G College, Dehradun, India

N. A. Siddiqui et al. (eds.), *Advances in Environment Engineering and Management*, Springer Proceedings in Earth and Environmental Sciences,
https://doi.org/10.1007/978-3-030-79065-3_24

1 Introduction

Cucurbitaceae include diversity of important cash crop well recognized by several forenames like pumpkin or gourd family spread all over the world that are natural source of nutrients and contain all the requisites of balance diet. Family cucurbitaceae comprises over 119 genra and 825 species that are distributed predominantly in almost all provinces of the world [1]. Pumpkin (*Cucurbita pepo*) is one of the chief vegetable crops growing in India. Majority of its cultivation in India thrives good in tropical areas with warm and humid environmental conditions. Locally this crop is known as kaddu. The fruit is edible and fulfill the consumers demand and a good source of income to the farmers. Pumpkin is a crop of multiple purpose economically benefits and act as source of food as young tender shoots were used as sag, the edible fruit is used in making desserts and vegetables seeds are used as fresh, uncooked or dry fruits and are loaded with minerals, vitamins, phytosterols, folic acids and other traces that can play a dynamic role in meeting the vegetable shortage and nutritional problem. On the other hand the remaining plant is used as and fodder. The major growing states of pumpkin in India are Uttarakhand, Uttar Pradesh, Andhra Pradesh and Orissa. Pumpkin crop experiences serious damage and its yield is constrained by biotic pathogens like virus, bacteria, fungi and insects as well. Pumpkin and other cucurbits are facing problem globally may result in 100% crop loss [2]. Among viruses CMV is consider as culprit for serious threat to crop due to extensive host range and is a member of Cucumovirus genus of Bromoviridae family [3–5]. Many findings have been done on checking the resistant origin through germplasm monitoring against virus [6]. In addition to pumpkin many other important agricultural crops are also susceptible to this pathogenic mosaic virus. The objective of present work addresses the systematic information to screen out the response on the susceptibility of available commercial pumpkin germplasm on the basis of symptom expression in Dehradun region. The seed samples for study were collected from local farmers and agriculture institutes.

2 Material and Methodology

2.1 Study Area

To figure out the provenance of resistance and transmissibility of two different isolates of virus on pumpkin, a disease monitoring seed bed for trial was settled/formed in experimental field in D.B.S P.G College Department of Botany Karanpur (Dehradun) in 2019. Nineteen seeds of pumpkin cultivars used for inoculation were collected from IARI, G.B Pantnagar university of Agriculture and technology Pantnagar, Nanital and local farmers were sown in plastic pots of 25 cm height containing moist soil under natural environment in field nursery. Distance between each two pots was 20 cm under field condition. The virus isolates were maintained on local two varieties of pumpkin.

To keep the crop in good state, the approved convention agricultural application were applied. With the objective the pumpkin crop were monitor against the two different isolate of virus on different pumpkin varieties.

2.2 *Mechanical Inoculation*

The pumpkin plant were mechanically inoculated at cotyledon stage of seedling or when the plant has two young leaves (seedling germinated after one week). To make inoculation done, the inoculum of two virus isolate mild and severe were taken from natural infected pumpkin plant leaves from the studied field area during survey. Leaves were plucked and crushed in sterilized mortar and sap was expressed by squeezing the pulp through a piece of muslin cloth. The sap was taken in watch glass with silicon carbide powder (600 mesh). Inoculation is done by rubbing the surface of leaves and excess of inoculum is removed by washing with spray of water to prevent the damage to leaf tissues.

2.3 *Screening of Pumpkin Cultivar Against Two Different Isolate of Mosaic Disease*

Nineteen cultivars of pumpkin seedlings were inoculated with mild and severe isolate. the inoculated plant were kept in a insect proof condition and were kept under observation until the appearance of symptoms. On emergence of disease symptoms the infection rate was noted by using given formula.

$$\text{Disease incidence }\% = \text{number of plant infected from disease/total number of plant taken} \times 100$$

The showcase of disease transmissibility of two virus isolate in different cultivars of pumpkin was assessed and classified by following experimental range by Shah et al. [7].

S. No.	Percentage of transmissibility (%)	Severity/Immune response
1	0–10	R
2	11–25	MR
3	26–45	MS
4	46–70	S
5	70–100	HS

R resistant; *MR* moderately resistant; *MS* moderately susceptible; *S* susceptible; *HS* Highly Susceptible

Lines that show complete resistance are all immune to infection that mean no or very less symptoms were seen in those lines while other show susceptibility response against isolate of viral pathogen having variety of symptoms on the surface of leaves and whole plant body. Symptoms were in the form of reduced number of leaves, hypertrophy, stunted growth, leaf deformation (Table 1).

From the above experimental observation it is shown that varieties of pumpkin viz. CM-350, Pusa vikas, Pusa vishwas and kashi harit show high degree of resistance in mild isolate and the infection was up to 10% while maximum susceptibility in mild isolate was found in Narendra upkar, Narendra abooshan, partap, Baramasi Sooraj and two other local varieties (Local I and Local II) and infection was found up to more than 70%. The symptoms shown by plant were reduced height, small foliage and less branches. Cultivar Arka chandan show 70% infection with mild isolate and therefore is susceptible Virus are able to infect susceptible plant only under certain environmental condition. Host plant tend to reach maximum susceptibility at a certain stage during their development and to be resistant to others. On the other hand the plant with severe isolate show symptoms like reduced branches with distorted leaf lamina, hypertrophy and leaf displaying mosaic patches. None of the cultivars was shown to be resistance to severe isolate. Cultivar Kashi harit, Azad pumpkin/ Azad kadoo, Swarna amrit, CM- 350, Pusa vikas and Pusa vishwas were moderately resistance to severe isolate. The infected plant show reduced plant growth. Cultivars Arka chandan, Narendra agrim and Narendra amrit, Sooraj, Pb Samrat were proved to be susceptible to severe isolate. The plant show 70% infection. The inoculated plant developed only a few branches and the leaves show dark green patches with distorted lamina (Fig. 1).

3 Discussion and Conclusion

Cucurbita pepo is known to infected by several viruses. Some of these infectious pathogen have wide host range and can cause havoc outside the family cucurbitaceae while some other have limited host range and can ruin respective cucurbit family. The present experiment were conducted in research study area to screen and monitor the 19 pumpkin varietal seeds including local varieties against two different isolates of cucumber mosaic virus to check the infection. Inoculated plant show response by the appearance of characteristic symptoms against mild as well as severe isolate. Generally plants infected by one strain of a virus are immune or highly resistant to another strain of the same virus. On the basis of these indications screening is done by checking the immune response (Resistance and susceptibility) of plant and its percentage transmissibility. In Dehradun no systematic work has been done to determine the yield loss due to CMV disease on pumpkin crop. Sharma et al. [8] collected 41 assortments of *Cucurbita pepo* including squashes and pumpkin that were screened against pathogenic infection of virus. Post inoculation their variable responses of susceptibility were recorded. Out of 26 pumpkin lines 7 show response of highly resistant against CMV. Virus are able to infect susceptible plants only under

Table 1 Susceptibility response of *Cucurbita pepo* germplasm against two different isolate of Cucumber mosaic virus

Cultivars	Mild isolate				Severe isolate			
	No. of seedlings		% of infection (%)	Category	No. of seedlings		% of infection (%)	Category
	Inoculated	Infected			Inoculated	Infected		
Kashi harit	20	2	10	R	20	5	25	MR
Azad pumpkin -1	20	8	40	MS	20	7	35	MR
Arka chandan	20	14	70	S	20	14	70	S
Arka suryamukhi	20	8	40	MS	20	8	40	MS
Swarna amrit	20	7	35	MS	20	5	25	MR
Narendra agrim	20	4	20	MR	20	14	70	S
Narendra amrit	20	4	20	MR	20	14	70	S
Narendra abooshan	20	15	75	HS	20	7	35	MS
Narendra upkar	20	15	75	HS	20	7	35	MS
Sooraj	20	16	80	HS	20	14	70	S
Pb Nawab	20	8	40	MS	20	18	40	MS
Pb Samrat	20	8	40	MS	20	12	60	S
Pratap	20	17	85	HS	20	8	40	MS
Local-1	20	18	90	HS	20	20	100	HS
Local-2	20	18	90	HS	20	18	90	HS
CM-350	20	2	10	R	20	6	30	MR
Pusa vikas	20	2	10	R	20	6	30	MR
Pusa vishwas	20	2	10	MR	20	6	30	MR
Baramasi	20	15	75	HS	20	18	90	HS

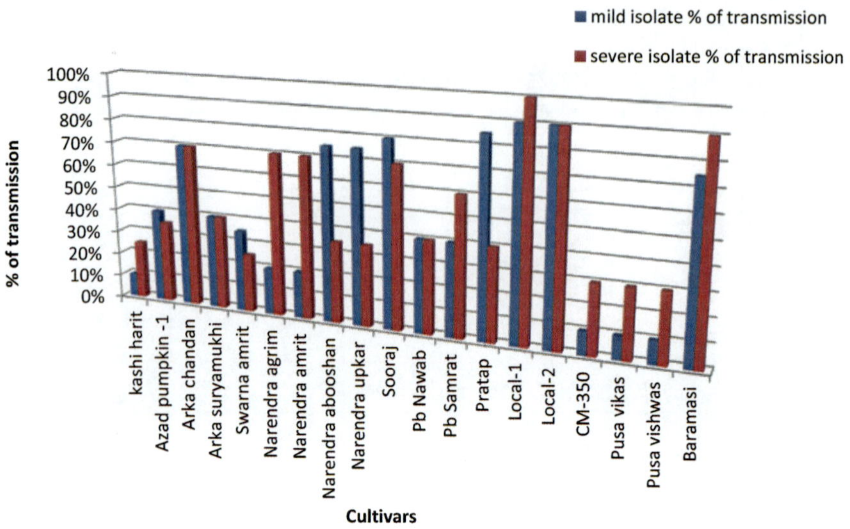

Fig. 1 Relationship between cultivars and their percentage of transmission against two different isolates of virus

certain environmental condition. Host plant tend to reach maximum susceptibility at a certain stage during their development and to be resistant at other. It may therefore possible to manipulate the sowing dates that the crop passes through its susceptible phase at a time when the virus is either absent or unable to reach the crop because of unsuitable environmental condition or absence of vectors.

References

1. Jeffrey, C. Systematics of the Cucurbitaceae: an overview. In: Bates DM, Robinson RW, Jeffrey C (eds.) Biology and Utilization of the Cucurbitaceae. Cornell University Press, Ithaca, NY, USA, (1990) pp 3–9.
2. Ullman, D.E., Cho, J and Germam, L. (1991) Occurrence and distribution of cucurbit viruses in the Hawaii island *plant disease*, 75. (1991) 367–370.
3. Palukaitis, P and Garcia- Arenal, F. Cucumovirus. *Advance virus research,* 62 (2003b):241–323.
4. Garcia- Arenal, F and Palukaitis, P Cucumber mosaic virus. In: Mahy, B.W. J and van Regenmortel, M.H.V(Eds.), Encyclopedia of virology. *Acadmic press of Elsevier.* (2008) Pp.614–619.
5. James, C. K. NG and Keith L .Perry Transmission of plant viruses by aphid vectors. *Molecular plant pathology.*5(5) (2004) pp 505–511.
6. Subodha, J and Polak, J. Preliminary evaluation of squash cultivars for resistance to a Czech isolate of Zucchini yellow mosaic virus. *Eucarpia Meetings on Cucurbit Genetics and Breeding.* (2004) pp 231–5.
7. Shah, H., Yasmin, T., Fahim, M., Hmeed, S., Hague, I.U., Munir, M and K.A. Reaction on exotic and indigenous capsicum genotype against Pakistani isolates of chili venial mottle virus. *Pakistan journal of botany*, Volume 43. (2011) 1707–1711.

8. Sharma, M., Thakur, P.D., Gupta, D and Thakur, A.K Identification of virus and screening of summer squash (*Cucurbita pepo*) germplasm against viral disease under controlled conditions. *Indian journal of agriculture science.* 83(4) 2013 426–430.
9. Andres, T. C. Web site for the plant family Cucurbitaceae & home of The Cucurbit Network. (2004). http://www.cucurbit.org/family.html.

A New Method for Estimation of Staircase Evacuation Time in High Rise Buildings

Suvek Salankar, S. M. Tauseef, and R. K. Sharma

Abstract As cities become more modern and complex, safe building evacuation within the stipulated time during an emergency is one of the major challenges for building designers, building occupants and the government. The most important requirement is to achieve evacuation before tenability is reached, i.e. the RSET (Required Safe Evacuation time) should be less than the ASET (Available Safe Evacuation time). Generally, various evacuation components like staircases, evacuation elevators, sky bridges, fire escape chutes, and controlled lowering devices are available along with evacuation strategies like total evacuation, phased evacuation, stay in placed evacuation and delayed evacuation. However, total building evacuation using the staircase is the most accepted, safe and traditional method of evacuation due to the unexpected behavior of fire and occupants [1–3]. Additionally, to decide on the geometry of the staircase, it is very important to know the required time to evacuate the building, so that the evacuation strategy can be decided. The evacuation time depends on important building parameters like staircase width, the height of the building and the number of occupants. Studies made by Galbreath [4] and Pauls [5] have suggested a relationship between the rates of discharge, the flow of occupants, and occupant density for the calculation of evacuation time. They have suggested two different relationships for two different flows, i.e. free flow and congested flow at the stairs. In this paper, one hundred and twenty building models are studied with different parameters like width of the stairs, occupant load, number of the floors and area of floor with two main objectives. The first objective is to study the impact of various building parameters on the overall building evacuation process, and the second objective is to suggest a single relationship with all parameters considered, which can be applied to calculate evacuation time and to further decide the width of the staircase and the maximum allowed occupants load.

S. Salankar (✉) · S. M. Tauseef
University of Petroleum and Energy Studies, Dehradun, India

R. K. Sharma
India Glycol, Kashipur, India

N. A. Siddiqui et al. (eds.), *Advances in Environment Engineering and Management*, Springer Proceedings in Earth and Environmental Sciences,
https://doi.org/10.1007/978-3-030-79065-3_25

Keywords High-rise building · Evacuation · Evacuation time · RSET · ASET · Means of egress · Staircase width · Effective staircase width · Fire simulation

1 Introduction

With the rapid urbanization in developing countries, there has been a massive influx of people, especially in its metropolitan cities. The scarcity of land demands the construction of high-rise buildings in these cities, especially in the last few decades. It is evident that up to the year 2000, there was not a single building of a height above 150 m in India, but today there are over 64 buildings of height above 150 m. A similar trend is observed globally. Up to year the 2000, there were only 299 buildings above the height of 150 m, but today there are 2643 buildings across the globe. Also, there were only 4 buildings above the height of 300 m, but now there are 51 buildings of height exceeding 300 m globally. This shows a tremendous rise in the number of such tall buildings [6].

The term high-rise building is also referred to as skyscrapers, tall buildings, super tall buildings and mega tall buildings. It is also called tower blocks in Britain and some European countries. But it doesn't have any internationally agreed definition [7] and it varies with the standard and understanding as mentioned below,

- Any structure with a height that can have a serious impact on evacuation [8].
- A high-rise structure is one that extends higher than the maximum reach of available fire-fighting equipment. In absolute numbers, this varies between 75 ft (23 m) and 100 ft (30 m) or about seven to ten stories (depending on the slab-to-slab distance between floors) [9].
- Internal building code, Building Construction and Safety Code and NFPA of the Life Safety Code say '75 ft or higher' measured from the lowest level of access to fire vehicle to the highest floor [10, 11].
- United States General Laws say buildings higher than 70 ft (21 m) [12].
- As per the National Building Code of India, it is a building with a height of 15 m and above.
- In India at a local level, it varies with respective development control rules.
- Apart from "high-rise building", there are a few more similar terminologies. As per Council on Tall Buildings and Urban Habitat (CTBUH) [6], tall buildings are buildings with a height of 50 m or above, super tall buildings are buildings with a height of 300 m or above and Mega tall buildings are buildings with a height of 600 m or above.

Although such high-rise buildings have a smaller footprint than the equivalent low-rise buildings, there are a few challenges in such buildings, from a fire safety perspective, such as long evacuation time, fire department inaccessibility and rapid spread of fire due to the stack effect may be far away from the origin of fire

[13]. As high-rise buildings facilitate high occupancy per unit land area, evacuation becomes one of the main concerns for the building designers, occupants as well as the authorities during an emergency.

Various studies are available on the evacuation characteristics of a high-rise building [14]. It is observed that two factors affect the evacuation characteristics [15], i.e. flow capacity and human behavior. Flow capacity relates to building components such as doorway and stairs, which may form a bottleneck depending on its capacity and occupant load. This is because the occupant flow rates at such locations are influenced by the building geometry characteristics [16] and accordingly the evacuation varies. It has been observed that the relationship between the doorway width and the corresponding achievable flow capacity is not linear but influenced also by its design [17, 18]. Human behaviors include mainly the response and motion characteristics of the evacuees. It also plays an important role in the evacuation process as different people react differently during the evacuation [19]. When the situation threatens their lives, people move swiftly to escape while ignoring polite social norms. However, when the situation is not so bad, the motion characteristics during the evacuation process are very similar to the normal situation [20].

For deciding safe evacuation, there is a well-known and widely used concept of Available Safe Evacuation Time (ASET) and Required Safe Evacuation Time (RSET) which is also used to design the means of egress. The concept of ASET was derived in 1882 by Cooper and Stroup [21]. It provides a good estimate of the time when conditions become untenable due to a fire and occupants must be able to escape before ASET is achieved i.e. RSET < ASET. Cooper did not provide any assistance in the determination of RSET. This was subsequently explained by Sime [22] by adding a coping phase, which is now widely referred to as pre-movement time i.e. time required before actual evacuation starts. So, RSET is the time that needs to evacuate people to a place of safety, including detection time, preparation time and evacuation movement time. RSET is affected by various factors [23] such as the means of escape, the effective width of the staircase, occupant load, speed of occupants, etc. These elements are discussed in detail in this paper and an attempt is made to find out a relationship amongst these elements to estimate the building evacuation time.

2 Evacuation

The basic principle of evacuation is to evacuate all the affected people from any part of the occupancy to an area of safety in a stipulated time through the safe evacuation route. This route ideally should be the same as the path of flow on the daily route. It should be always ready for evacuation, continuous with one or more exits, resistant to fire and smoke, with sufficient capacity, without any obstructions and suitable for all types of occupants.

Available evacuation time depends on tenability factors such as smoke, heat and radiation. The evacuation should be completed before the tenability limit is reached.

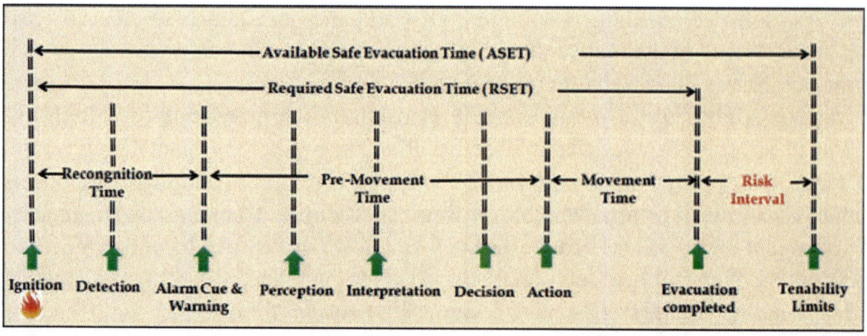

Fig. 1 A time line comparison of the available safe time (ASET) and required Safe egress time (RSET)

So, if we consider smoke as a tenability criteria, the evacuation needs to be completed before the smoke descends to a height at which it obstructs the movement of the people.

Fire evacuations are typically studied in terms of two main components as mentioned by Guylène and Proulx [24], the time delay before moving towards an exit commences which is referred to as pre-movement time and the actual time taken to travel to a place of safety which is referred to as movement time. The universal criteria of a safe fire evacuation timeline are demonstrated in detail in Fig. 1.

RSET is the time required to evacuate to a safe area after the start of a fire. It includes [25]:

(a) Recognition time: The time from fire ignition to detection and from detection to a notification to occupants about a fire emergency,
(b) Pre-movement time: The time from notification until occupants decide to take an action and
(c) The time from decision to take an action until evacuation commences and finally
(d) Movement time: The time from the start of evacuation to its completion.
(e) **Recognition time**: A warning signal is generated after activation of a fire detector due to the release of smoke or thermal radiation during fire. The detection requires few seconds to a few hours, depending on the type of fire, location and availability of detection devices. In some cases, people themselves realize the occurrence of fire. This period is called recognition time, which depends on the type and layout of the detection system, the scale of the fire and its growth rate. Detection may take place either by devices or by human beings. In some cases, there may be a time delay between detection and alarm activation. For example, in the case of manual activation of manual call points, the fire alarm may not be activated due to some reason or is not available, where occupants eventually perceive some cues from the fire or receive a warning by others.
(f) **Pre-movement time**: It is also referred to as a delay time. Occupants take some time to initiate their evacuation movement once they perceive some cues of the

fire. It comprises three subcomponents: perception, interpretation and action. Perception is the time taken by the occupants to perceive the fire alarm signal once they hear it. It could also be a fire cue or warning received from other occupants. Interpretation of this information may vary and people will take decisions accordingly for evacuation. Once the decision to evacuate is made, occupants will engage in other actions before leaving, such as informing others, getting properly dressed, gathering children, finding valuables, etc. The pre-movement time in an evacuation is closely related to their psychological state, age, behavior characteristics and level of familiarity with the buildings and sensitivity to react. All these factors are very dynamic and hence are very difficult to estimate accurately [26]. Observations in real emergencies and fire drills have shown that individual behaviour in the pre-movement phase is responsible for substantial delays in the evacuation times [27]. It was observed that, in unannounced fire drills for the large retail stores, the pre-movement time contributed 30–50% of the total evacuation time [28]. There are several causes of pre-movement delays in evacuations which include difficulties in hearing fire alarms, attempts to get more information and other preparatory actions like collecting personal belongings, putting on warm clothes or waiting for others [29]. It is observed that pre-evacuation time is 2 min 49 s for good audible alarm and may increase up to 5 min 19 s in winter season as it requires some additional preparation like putting on warm clothing. For the buildings, where the alarm is not loud enough, occupants took as long as 25 min to start as mentioned by Proulx [30]. Brennan [31] studied this factor through reports of fire victims. She used the case of a high-rise office building fire that started in a stack of polyurethane chairs stored on the third floor of a 14-story building. From the interviews, it was estimated that the meantime to start the evacuation was approximately 2 min 30 s.

(g) **Movement time**: The movement time is the time required for all the evacuees to evacuate to an area of safety from their position. This period depends on the evacuation parameters such as people density, evacuation distance, the effective width of the staircase and specific flow [26]. Movement time is calculated from the flow time through various elements of the egress system and the travel time to move along an egress route. The evacuation time is calculated from the time of alarm signal activation until the last occupant to evacuate has reached the area of safety. The tenability limit is the time when the fire products like narcotic gases, irritant gases, heat, smoke, toxic gases, etc. affect the health of occupants. So, the evacuation time should be less than the tenability limit and ideally, it should be half [24].

3 Means of Egress

A means of egress is an important parameter for safe evacuation with the basic purpose of getting all the occupants quickly out of a building in a safe manner during a fire emergency. It is a continuous and unobstructed path of vertical and horizontal egress travel from any portion of a building to an area of safety. It consists of three parts: the exit access, exit, and exit discharge [32]. Exit access is the path from any location within a building to the start of an exit e.g. aisle access ways, aisles, exit access doorways, corridors & exterior egress balconies. The exit is an unobstructed part between exit access and exit discharge. In a high-rise building, a stairway along with vertical exit enclosures, exit passageways, horizontal exits and exterior exit stairways/ramps are types of exit.

The exit may be horizontal or vertical. A horizontal exit is an exit that allows occupants to egress from one side of a building to another side through a fire-rated assembly such as a fire wall or fire barrier. Here, once occupants pass through a horizontal exit, they must still have a pathway to another exit that leads to an exit discharge or public way [10]. Exit discharge is the path from the end of the exit to the public way or assembly area. This area is a place that is permanent and dedicated for occupants' use and considered to be safe [32].

There are few egress components like evacuation elevators, platform rescue devices, escape chutes, controlled lowering devices, etc., but all these components have their limitations and disadvantages [2]. So, in this paper, we have considered only the staircase as an exit being the safest and conventional method of evacuation [3].

The staircase requirement is considered mainly on two parameters. One is the width of the staircase and the other is the number of staircases. It varies as per the type of occupancy, its area and height of the building. Following are some of the studies/codes available for the requirement of the staircase,

1. As per the National Building code of India-2016, it is based on the occupant load, width required per person and travel distance with a minimum of two staircases for high rise and special buildings [32].
2. NFPA 5000 / NFPA 101 considers the minimum width of the staircase and the numbers based on total occupants load [10].
3. British Building regulations (Document B) consider occupant load and travel distance for the number of staircases and number of people and number of stories served for the width of the staircase [33].
4. Her Majesty's Stationery Office, fire grading of building, appendix II consider the exit time from the floor, the time required to fill staircase, rate of flow and width of the staircase in a Unit for the total population [34].
5. Galbreath [4] suggested the total time for evacuation of the entire building depends on the number of people above 1st floor, persons per floor, or the number of people who can stand on stairs/area of stairs divided by 0.3 m^2/person (whichever is less), rate of discharge and number of unit widths.
6. Jake Pauls [5] suggested two evacuation times, one for population less than 800 and another for more than 800 based on actual evacuation population per meter of effective stair width.
7. Melinek and Booth suggested total evacuation time for two cases i.e. during congestions and a free walk based on number of floors, number of people per

floor, occupant flow on stairs (person m^{-1} s^{-1}), Effective width of stairs (m) and walking time between adjacent floors [5].

4 Walking Speed of Occupants

An unimpeded walking speed, i.e. the speed at which all the occupants travel across open floor space without causing any obstruction, is an important factor in the evacuation process especially for the travel from any part of the building to exit access. It is directly related to travel distance during an emergency.

A wide range of travel speeds has been reported [35], depending on associated factors such as individual differences, density in the area, sex and environmental conditions. Occupants can move quickly at a normal pace if there is a great space between them. But the closer they are, the more constrained is the movement. There are various studies conducted to estimate the walking speed of occupants.

Nelson and Frederick [24] suggested a walking speed of a mean of 1.19 m/s with a standard deviation of 0.3 m/s. The speed also depends on the population density i.e. number of persons per unit area. It was suggested that the average unrestricted walking speed varies from 1.2 to 1.25 m/s if the density is less than 0.54 people/m^2. But the speed decreases drastically with the increase in density and reaches a standstill when the density reaches 4 or 5 people/m^2, equivalent to a crowded elevator [5, 36]. Wahyu [37], in his experiment, observed that, in horizontal evacuation, the normal walking speed is 1.1 m/s, for a fast walk, the speed is 1.5 m/s and for half running, the speed is 2.1 m/s. Same for vertical evacuation is 0.7 m/s, 0.8 m/s and 2.2 m/s, respectively. In this study, a pathfinder evacuation model is used. It's default setting of occupant's speed is 1.19 m/s [38].

When people start moving towards the exit and when they noticed the blocked exit area by the other people, they reduce their speed to keep a comfortable distance amongst themselves [39, 40] and finally, they stop at one place due to lack of space to move further [35]. They wait in the queue until there is room for them to move forward. Their speed and waiting time depends on the density of the people and the width of the staircase. The person can move freely in lanes having the width of about 22 inches. measured at shoulder height referred to as unit exit width [36].

The speed does not depend only on an engineering approach, but it also depends on behavioural and physiological factors [41]. NIST [42] has estimated that 6% of occupants evacuated during World Trade Centre towers on 11th September 2001 were having mobility impairment that hindered the evacuation process.

5 Unit Exit Width

For evacuation, the exit width is not measured always in meters/feet, but it is measured in the unit of the width occupied by one person. The traditional unit exit width is

560 mm (22 inches), a lane width [36]. This concept originated in the early part of the twentieth century in the US. In the US army, 22 inches were taken as the width of a file of people in motion. A stairway width of 44 inches permits two files of people to move freely down the stairs simultaneously.

The concept of unit exit width is also important from cost perspective as widening staircase width in meters/feet may involve additional expense without getting the expected safety. For example, 44 inches staircase comfortably accommodates two files of people and if we add 4 inches to make a 4 ft staircase, it will not increase its capacity but will only add expenses. However, it has been seen that adding 12 inches to a 44 inches staircase does increase the flow of people by permitting an intermediate staggered file [43]. The default setting of the occupant's shoulder width is 45.58 cm. In the pathfinder egress model [38] which is based on the average of measurements taken of 68 male and female persons from nine countries. The same is refereed by G. Keith Still in his Ph.D. thesis [44].

6 Effective Width of Staircase

The effective width of the staircase is the usable width of a staircase of a building. Often, a certain portion is subtracted from each side of the staircase width to accommodate body sway during evacuation movement or small intrusion into the pathway such as handrails. It does not include the relatively unused area of a stair where people keep some distance like walls and central handrails. The reason is that the occupants try to take their bodies away from these places. The mean flow is directly proportional to the effective stair width. Paul [45] suggested that 150 mm should be deducted from a wall (more for rough surfaces) and 90 mm from the centerline of a handrail.

Pauls [5] suggested that a stair's effective width is 300 mm (12 inches) less than its nominal width. For example, nominal width of 1,120 mm (44 inches) equates to an effective width of 812 mm (32 inches). This means we should reduce 300 mm from the designed width of a staircase to get the respective effective width.

7 Egress Models

To address the challenging issue of fire safety of high rise buildings, the fire safety codes have been changed from prescriptive-based towards performance-based over the last few decades [46]. This is facilitated by various computer models. These models are, normally, categorized into fire models, detector response models, fire endurance models, egress models and miscellaneous models [47] and all these models have been playing an important role to address best possible fire safety issues. Egress models have been used for estimating the RSET, and other egress parameters like the speed of occupants, their behaviors etc.

In this study, we are using the pathfinder simulation model, an agent-based egress simulator that is designed to meet the practical need to study complex buildings. It takes advantage of advances in agent-based approaches to movement modeling that make it possible to capture the more complex behavior and interactions between the occupants. In addition, it also provides tools that make it easy for modelers to create simulation input from existing data and view results using high quality visualization techniques [48].

We are using the default setting of occupant's shoulder width of 45.58 cm. and walking speed of 1.19 m/s is as it is in line with other studies as discussed earlier. Eight steps are considered between landing and mid landing with 17.78 cm of riser and 27.94 cm of tread, which is well within all the National and International standards.

8 Assumptions Made

In the case of a high-rise building, fire is a very complicated scenario with the release of hazardous combustion products like carbon monoxide, carbon dioxide, smoke and heat radiation. This depends on the properties of fuel and available ventilation. Apart from the general complexity of the building, some behavioral factors affect the evacuation process. These factors are very dynamic and vary from person to person depending on the individual's capability, experience, physical health and culture. So, we are considering the following assumptions in this study.

1. Only travel time required for building evacuation is considered from exit access to the last part of the exit.
2. All occupants start their evacuation at the same time once they hear the alarm.
3. All occupants are physically fit and there are no mobility-impaired occupants in the building.
4. There may be more than one exit, but each exit will be used by the equal number of occupants available on the floor. We have kept only one exit for the study purpose.
5. Emergency service people will not use the same staircase during evacuation to avoid the reverse flow.
6. Evacuation time of the last evacuated person is considered as building evacuation time.
7. The width of the staircase landing and mid-landing is the same as that of the width of the staircase.
8. National Building code of India suggests a ceiling height from 2.75 m to 3.6 m as per the type of occupancies. For business occupancy, it is normally 3.6 m. So, the same height has been considered.
9. Although the refuge area is available on the upper floors and the occupant load is slightly less on these floors compared to other floors, we have considered an equal number of occupants on every floor considering visitors' movement.

10. Various codes are available for deciding the width of the exit door. The National Building Code of India mentions 100 cm, NFPA 101 mentions 81 cm, few local codes [49] mentions 100 cm. In this study, we have considered the width of the staircase door to be 100 cm.
11. Basements are not considered in the study as it is not meant for occupancy and it will not affect the overall evacuation process.

9 Simulation Details

The objective of this study is to develop a new and simple mathematical model to determine evacuation time during building evacuation by considering basic parameters like the width of the staircase, occupant load and the number of floors.

We have considered one hundred and twenty model building simulations with different combinations of building parameters like floor area ranging from 400 to 600 m^2, floors from 6 to 51 floors, including the ground floor, a staircase width of 1.5 and 2.0 m with the same landing and mid landing width, maximum occupant load of 10 m^2/person. Only buildings above 5 stories are considered being high-rise buildings as per the National Building Code of India. The width of the staircase and occupant load is as per national and international codes [10, 11, 32].

We are also considering less occupant load than the maximum allowed load for study purposes. The ceiling height is considered as 3.6 m required for the business occupancy.

Details of model buildings are mentioned in Table 1.

The Pathfinder evacuation model is applied to each of the building models for evacuation time and the results are mentioned in Table 2.

Screen shot of pathfinder model is shown in Fig. 2,

Following are the important conclusions from table 2,

1. Evacuation time of the building increases with an increase in the number of floors and number of occupants.
2. Evacuation time is more for a staircase with a width of 1.5 m as compared to the staircase with a width of 2 m.
3. Evacuation time for 51 storied building with 40 occupants per floor of 400 sq. meters from 1.5 m width of the staircase is 1577.55 s (Sr. No.51), similar evacuation for 500 m^2 is 1579.73 s (Sr. No. 41) and for 600 m^2 is 1583.38 s (Sr. No. 21). Similarly, reading for 2 m of staircase width, it is 1223.78 s, 1229.25 s & 1226.93 s respectively (Sr. no. 111, 101 and 81). With these results, it can be concluded that with the change in floor area, there is very little variation in evacuation time and even in a few instances it is slightly less when floor area is more. The reason may be, in the case of a bigger floor area, occupants are taking more time to reach the merging area of the staircase by when the congestion may get clear to some extent.

Table 1 Details of the evacuation model buildings

Sr. No.	Floor area (m^2)	No. of floors	Occupant load per exit	Sr. No.	Floor area (m^2)	No. of floors	Occupant load per exit
Staircase width = 1.5 m; Effective width = 1.2 m				**Staircase width = 1.5 m; Effective width = 1.2 m**			
1	600	51	60	31	500	51	50
2	600	46	60	32	500	46	50
3	600	41	60	33	500	41	50
4	600	36	60	34	500	36	50
5	600	31	60	35	500	31	50
6	600	26	60	36	500	26	50
7	600	21	60	37	500	21	50
8	600	16	60	38	500	16	50
9	600	11	60	39	500	11	50
10	600	6	60	40	500	6	50
11	600	51	50	41	500	51	40
12	600	46	50	42	500	46	40
13	600	41	50	43	500	41	40
14	600	36	50	44	500	36	40
15	600	31	50	45	500	31	40
16	600	26	50	46	500	26	40
17	600	21	50	47	500	21	40
18	600	16	50	48	500	16	40
19	600	11	50	49	500	11	40
20	600	6	50	50	500	6	40
21	600	51	40	51	400	51	40
22	600	46	40	52	400	46	40
23	600	41	40	53	400	41	40
24	600	36	40	54	400	36	40
25	600	31	40	55	400	31	40
26	600	26	40	56	400	26	40
27	600	21	40	57	400	21	40
28	600	16	40	58	400	16	40
29	600	11	40	59	400	11	40
30	600	6	40	60	400	6	40
Staircase width = 2.0 m; Effective width = 1.7 m				**Staircase width = 2.0 m; Effective width = 1.7 m**			
61	600	51	60	91	500	51	50

(continued)

Table 1 (continued)

Sr. No.	Floor area (m^2)	No. of floors	Occupant load per exit	Sr. No.	Floor area (m^2)	No. of floors	Occupant load per exit
62	600	46	60	92	500	46	50
63	600	41	60	93	500	41	50
64	600	36	60	94	500	36	50
65	600	31	60	95	500	31	50
66	600	26	60	96	500	26	50
67	600	21	60	97	500	21	50
68	600	16	60	98	500	16	50
69	600	11	60	99	500	11	50
70	600	6	60	100	500	6	50
71	600	51	50	101	500	51	40
72	600	46	50	102	500	46	40
73	600	41	50	103	500	41	40
74	600	36	50	104	500	36	40
75	600	31	50	105	500	31	40
76	600	26	50	106	500	26	40
77	600	21	50	107	500	21	40
78	600	16	50	108	500	16	40
79	600	11	50	109	500	11	40
80	600	6	50	110	500	6	40
81	600	51	40	111	400	51	40
82	600	46	40	112	400	46	40
83	600	41	40	113	400	41	40
84	600	36	40	114	400	36	40
85	600	31	40	115	400	31	40
86	600	26	40	116	400	26	40
87	600	21	40	117	400	21	40
88	600	16	40	118	400	16	40
89	600	11	40	119	400	11	40
90	600	6	40	120	400	6	40

Apart from the above conclusions, all the results are well studied for their inter-relation between evacuation time, the number of floors, the effective width of the staircase, travel distance in the staircase and number of occupants.

As shown in Fig. 3, the travel distance in the staircase is twice the sum of the total length of the steps from landing to mid landing i.e. B which is 3.08 m and travel distance on the landing and mid landing i.e. C which will be half of the circumference,

Table 2 Evacuation time from each building model

Sr. No.	Floor area (m^2)	No. of floors	Occupant load per exit	Effective width (m)	Evacuation time (s)	Sr. No.	Floor area (m^2)	No. of floors	Occupant load per exit	Effective width (m)	Evacuation Time (Seconds)
	Staircase width = 1.5 m						**Staircase width = 2.0 m**				
1	600	51	60	1.2	2361.98	61	600	51	60	1.7	1822.55
2	600	46	60	1.2	2137.55	62	600	46	60	1.7	1649.40
3	600	41	60	1.2	1892.30	63	600	41	60	1.7	1468.98
4	600	36	60	1.2	1668.35	64	600	36	60	1.7	1284.63
5	600	31	60	1.2	1433.48	65	600	31	60	1.7	1103.00
6	600	26	60	1.2	1199.50	66	600	26	60	1.7	925.23
7	600	21	60	1.2	967.60	67	600	21	60	1.7	750.13
8	600	16	60	1.2	728.18	68	600	16	60	1.7	565.45
9	600	11	60	1.2	495.78	69	600	11	60	1.7	388.53
10	600	6	60	1.2	263.40	70	600	6	60	1.7	210.43
11	600	51	50	1.2	1982.70	71	600	51	50	1.7	1526.93
12	600	46	50	1.2	1781.18	72	600	46	50	1.7	1374.80
13	600	41	50	1.2	1589.13	73	600	41	50	1.7	1226.40
14	600	36	50	1.2	1392.20	74	600	36	50	1.7	1076.50
15	600	31	50	1.2	1191.85	75	600	31	50	1.7	929.08
16	600	26	50	1.2	1000.53	76	600	26	50	1.7	775.90
17	600	21	50	1.2	802.53	77	600	21	50	1.7	626.20
18	600	16	50	1.2	616.93	78	600	16	50	1.7	477.70
19	600	11	50	1.2	417.90	79	600	11	50	1.7	329.80

(continued)

Table 2 (continued)

Sr. No.	Floor area (m^2)	No. of floors	Occupant load per exit	Effective width (m)	Evacuation time (s)	Sr. No.	Floor area (m^2)	No. of floors	Occupant load per exit	Effective width (m)	Evacuation Time (Seconds)
20	600	6	50	1.2	225.00	80	600	6	50	1.7	179.90
21	600	51	40	1.2	1583.38	81	600	51	40	1.7	1226.93
22	600	46	40	1.2	1428.50	82	600	46	40	1.7	1111.13
23	600	41	40	1.2	1283.20	83	600	41	40	1.7	989.70
24	600	36	40	1.2	1118.60	84	600	36	40	1.7	868.08
25	600	31	40	1.2	967.13	85	600	31	40	1.7	747.40
26	600	26	40	1.2	812.50	86	600	26	40	1.7	629.80
27	600	21	40	1.2	654.03	87	600	21	40	1.7	509.93
28	600	16	40	1.2	493.70	88	600	16	40	1.7	386.05
29	600	11	40	1.2	342.30	89	600	11	40	1.7	271.43
30	600	6	40	1.2	184.25	90	600	6	40	1.7	149.40
31	500	51	50	1.2	1991.08	91	500	51	50	1.7	1518.98
32	500	46	50	1.2	1780.00	92	500	46	50	1.7	1365.95
33	500	41	50	1.2	1587.28	93	500	41	50	1.7	1228.65
34	500	36	50	1.2	1392.78	94	500	36	50	1.7	1077.18
35	500	31	50	1.2	1198.85	95	500	31	50	1.7	928.50
36	500	26	50	1.2	1000.50	96	500	26	50	1.7	773.25
37	500	21	50	1.2	811.13	97	500	21	50	1.7	629.80
38	500	16	50	1.2	610.13	98	500	16	50	1.7	479.80
39	500	11	50	1.2	417.40	99	500	11	50	1.7	329.28

(continued)

Table 2 (continued)

Sr. No.	Floor area (m^2)	No. of floors	Occupant load per exit	Effective width (m)	Evacuation time (s)	Sr. No.	Floor area (m^2)	No. of floors	Occupant load per exit	Effective width (m)	Evacuation Time (Seconds)
40	500	6	50	1.2	219.63	100	500	6	50	1.7	177.03
41	500	51	40	1.2	1579.73	101	500	51	40	1.7	1229.25
42	500	46	40	1.2	1432.98	102	500	46	40	1.7	1109.53
43	500	41	40	1.2	1273.00	103	500	41	40	1.7	985.68
44	500	36	40	1.2	1120.55	104	500	36	40	1.7	872.80
45	500	31	40	1.2	956.00	105	500	31	40	1.7	750.33
46	500	26	40	1.2	802.90	106	500	26	40	1.7	629.65
47	500	21	40	1.2	647.30	107	500	21	40	1.7	510.83
48	500	16	40	1.2	493.40	108	500	16	40	1.7	388.40
49	500	11	40	1.2	337.05	109	500	11	40	1.7	272.28
50	500	6	40	1.2	184.73	110	500	6	40	1.7	150.08
51	400	51	40	1.2	1577.55	111	400	51	40	1.7	1223.78
52	400	46	40	1.2	1423.20	112	400	46	40	1.7	1106.63
53	400	41	40	1.2	1270.90	113	400	41	40	1.7	983.65
54	400	36	40	1.2	1118.20	114	400	36	40	1.7	862.88
55	400	31	40	1.2	960.13	115	400	31	40	1.7	747.25
56	400	26	40	1.2	805.13	116	400	26	40	1.7	628.00
57	400	21	40	1.2	651.20	117	400	21	40	1.7	507.78
58	400	16	40	1.2	503.13	118	400	16	40	1.7	388.53
59	400	11	40	1.2	340.90	119	400	11	40	1.7	270.33

(continued)

Table 2 (continued)

Sr. No.	Floor area (m^2)	No. of floors	Occupant load per exit	Effective width (m)	Evacuation time (s)	Sr. No.	Floor area (m^2)	No. of floors	Occupant load per exit	Effective width (m)	Evacuation Time (Seconds)
60	400	6	40	1.2	182.15	120	400	6	40	1.7	149.65

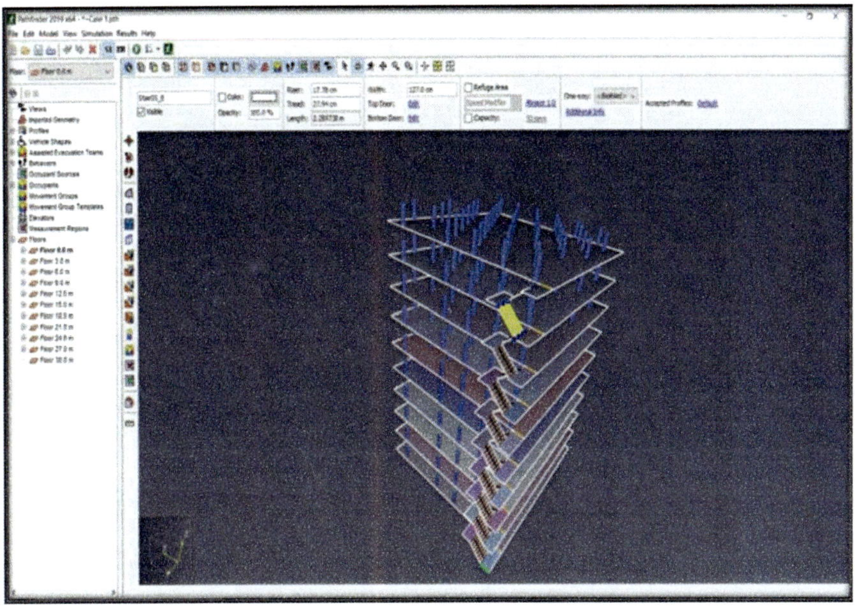

Fig. 2 Pathfinder evacuation model

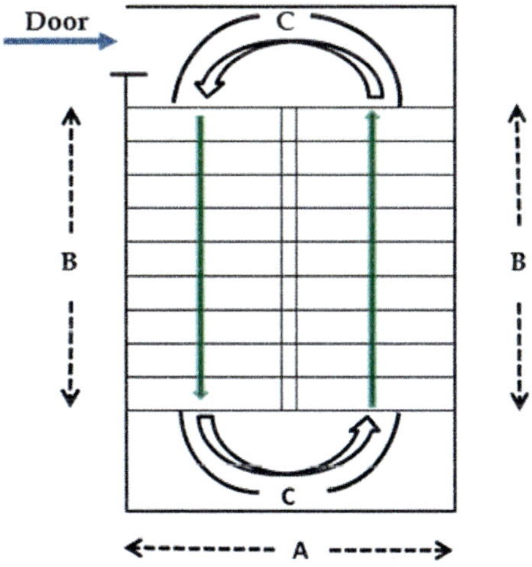

Fig. 3 Staircase details

Here, circumference = πD, D is 1.5 m and 2.0 m.
For 1.5 m staircase width,

$$C = \frac{\pi D}{2} \text{ i.e } \frac{3.14 \times 1.5}{2} = 2.35 \text{ m.}$$

So total travel distance in the staircase will be,

$$2.35 \times 2 + 3.08 \times 2 = 10.86 \text{ m}$$

For 2.0 m staircase width,

$$\text{C} = \frac{\pi \text{D}}{2} \text{ i.e } \frac{3.14 \times 2}{2} = 3.14 \text{ m.}$$

So total travel distance in the staircase will be, $3.14 \times 2 + 3.08 \times 2 = 12.44$ m.

Value of B i.e. 3.08 m can be directly obtained from the pathfinder study and also by using the Pythagoras theorem with the total run of the staircase is 2.5 m and the height difference between landing and mid landing is 1.8 m.

When we study all the above parameters with different combinations, a relationship is obtained with a constant which we can be termed as evacuation factor (α). The relationship is,

$$\alpha = \frac{t \times w}{n \times o \times d} \tag{1}$$

where,

t Maximum time of evacuation from exit access to the end of exit (seconds)
w Effective width of staircase(meters)
n Number of floors in a building
o Number of occupants per floor per exit
d Travel distance in the staircase between two floors (meters).

Applying above mentioned mathematical model, we get an evacuation factor as mentioned in Table 3.

The mean (α) is 0.0836 for the sample of 120 models with standard deviation of 0.001834.

We can rearrange equation no. 1 to get a mathematical model for evacuation time,

$$t = \frac{\alpha \times o \times n \times d}{w}$$

i.e.

$$t = \frac{0.0836 \times o \times n \times d}{w} \tag{2}$$

Equation 2 can be used as a mathematical model to find out evacuation time. This excludes the pre evacuation time required for high-rise buildings.

Table 3 Results after applying above proposed relationship

Sr. No.	Floor area (m^2)	No. of floors	Occupant load per exit	Effective width (m)	Evacuation time (s)	Evacuation factor
	Staircase width = 1.5 m					
1	600	51	60	1.2	2361.98	0.0853
2	600	46	60	1.2	2137.55	0.0856
3	600	41	60	1.2	1892.30	0.0850
4	600	36	60	1.2	1668.35	0.0853
5	600	31	60	1.2	1433.48	0.0852
6	600	26	60	1.2	1199.50	0.0850
7	600	21	60	1.2	967.60	0.0849
8	600	16	60	1.2	728.18	0.0838
9	600	11	60	1.2	495.78	0.0830
10	600	6	60	1.2	263.40	0.0808
11	600	51	50	1.2	1982.70	0.0859
12	600	46	50	1.2	1781.18	0.0856
13	600	41	50	1.2	1589.13	0.0857
14	600	36	50	1.2	1392.20	0.0855
15	600	31	50	1.2	1191.85	0.0850
16	600	26	50	1.2	1000.53	0.0850
17	600	21	50	1.2	802.53	0.0845
18	600	16	50	1.2	616.93	0.0852
19	600	11	50	1.2	417.90	0.0840
20	600	6	50	1.2	225.00	0.0829
21	600	51	40	1.2	1583.38	0.0858
22	600	46	40	1.2	1428.50	0.0858
23	600	41	40	1.2	1283.20	0.0865
24	600	36	40	1.2	1118.60	0.0858
25	600	31	40	1.2	967.13	0.0862
26	600	26	40	1.2	812.50	0.0863
27	600	21	40	1.2	654.03	0.0860
28	600	16	40	1.2	493.70	0.0852
29	600	11	40	1.2	342.30	0.0860
30	600	6	40	1.2	184.25	0.0848
31	500	51	50	1.2	1991.08	0.0863
32	500	46	50	1.2	1780.00	0.0855
33	500	41	50	1.2	1587.28	0.0856
34	500	36	50	1.2	1392.78	0.0855

(continued)

Table 3 (continued)

Sr. No.	Floor area (m^2)	No. of floors	Occupant load per exit	Effective width (m)	Evacuation time (s)	Evacuation factor
35	500	31	50	1.2	1198.85	0.0855
36	500	26	50	1.2	1000.50	0.0850
37	500	21	50	1.2	811.13	0.0854
38	500	16	50	1.2	610.13	0.0843
39	500	11	50	1.2	417.40	0.0839
40	500	6	50	1.2	219.63	0.0809
41	500	51	40	1.2	1579.73	0.0856
42	500	46	40	1.2	1432.98	0.0861
43	500	41	40	1.2	1273.00	0.0858
44	500	36	40	1.2	1120.55	0.0860
45	500	31	40	1.2	956.00	0.0852
46	500	26	40	1.2	802.90	0.0853
47	500	21	40	1.2	647.30	0.0851
48	500	16	40	1.2	493.40	0.0852
49	500	11	40	1.2	337.05	0.0846
50	500	6	40	1.2	184.73	0.0850
51	400	51	40	1.2	1577.55	0.0854
52	400	46	40	1.2	1423.20	0.0855
53	400	41	40	1.2	1270.90	0.0856
54	400	36	40	1.2	1118.20	0.0858
55	400	31	40	1.2	960.13	0.0856
56	400	26	40	1.2	805.13	0.0855
57	400	21	40	1.2	651.20	0.0857
58	400	16	40	1.2	503.13	0.0869
59	400	11	40	1.2	340.90	0.0856
60	400	6	40	1.2	182.15	0.0839
	Staircase width = 2.0 m					
61	600	51	60	1.7	1822.55	0.0814
62	600	46	60	1.7	1649.40	0.0817
63	600	41	60	1.7	1468.98	0.0816
64	600	36	60	1.7	1284.63	0.0813
65	600	31	60	1.7	1103.00	0.0810
66	600	26	60	1.7	925.23	0.0810
67	600	21	60	1.7	750.13	0.0814
68	600	16	60	1.7	565.45	0.0805

(continued)

Table 3 (continued)

Sr. No.	Floor area (m^2)	No. of floors	Occupant load per exit	Effective width (m)	Evacuation time (s)	Evacuation factor
69	600	11	60	1.7	388.53	0.0804
70	600	6	60	1.7	210.43	0.0799
71	600	51	50	1.7	1526.93	0.0818
72	600	46	50	1.7	1374.80	0.0817
73	600	41	50	1.7	1226.40	0.0818
74	600	36	50	1.7	1076.50	0.0817
75	600	31	50	1.7	929.08	0.0819
76	600	26	50	1.7	775.90	0.0816
77	600	21	50	1.7	626.20	0.0815
78	600	16	50	1.7	477.70	0.0816
79	600	11	50	1.7	329.80	0.0819
80	600	6	50	1.7	179.90	0.0819
81	600	51	40	1.7	1226.93	0.0822
82	600	46	40	1.7	1111.13	0.0825
83	600	41	40	1.7	989.70	0.0825
84	600	36	40	1.7	868.08	0.0824
85	600	31	40	1.7	747.40	0.0824
86	600	26	40	1.7	629.80	0.0828
87	600	21	40	1.7	509.93	0.0830
88	600	16	40	1.7	386.05	0.0824
89	600	11	40	1.7	271.43	0.0843
90	600	6	40	1.7	149.40	0.0851
91	500	51	50	1.7	1518.98	0.0814
92	500	46	50	1.7	1365.95	0.0812
93	500	41	50	1.7	1228.65	0.0819
94	500	36	50	1.7	1077.18	0.0818
95	500	31	50	1.7	928.50	0.0819
96	500	26	50	1.7	773.25	0.0813
97	500	21	50	1.7	629.80	0.0820
98	500	16	50	1.7	479.80	0.0820
99	500	11	50	1.7	329.28	0.0818
100	500	6	50	1.7	177.03	0.0806
101	500	51	40	1.7	1229.25	0.0823
102	500	46	40	1.7	1109.53	0.0824
103	500	41	40	1.7	985.68	0.0821

(continued)

Table 3 (continued)

Sr. No.	Floor area (m^2)	No. of floors	Occupant load per exit	Effective width (m)	Evacuation time (s)	Evacuation factor
104	500	36	40	1.7	872.80	0.0828
105	500	31	40	1.7	750.33	0.0827
106	500	26	40	1.7	629.65	0.0827
107	500	21	40	1.7	510.83	0.0831
108	500	16	40	1.7	388.40	0.0829
109	500	11	40	1.7	272.28	0.0846
110	500	6	40	1.7	150.08	0.0855
111	400	51	40	1.7	1223.78	0.0820
112	400	46	40	1.7	1106.63	0.0822
113	400	41	40	1.7	983.65	0.0820
114	400	36	40	1.7	862.88	0.0819
115	400	31	40	1.7	747.25	0.0824
116	400	26	40	1.7	628.00	0.0825
117	400	21	40	1.7	507.78	0.0826
118	400	16	40	1.7	388.53	0.0830
119	400	11	40	1.7	270.33	0.0840
120	400	6	40	1.7	149.65	0.0852

10 Validation and Verification of the Proposed Mathematical Model

To affirm the accuracy of this mathematical model, we need to validate this with evacuation time from actual fire cases or fire drills. Various models are available to simulate the evacuation of occupants from a given building geometry [50], but very limited contemporary data is available to support the model inputs or assumptions. There is very little information available to validate the models for actual emergencies [41]. Few studies are available but with a smaller number of floors. Following are two studies available for validation of the proposed mathematical model.

Validation I:

In the study made by Galbreath [51] on "A Survey of exit facilities in high office buildings", a survey was made of 10 high office buildings to study the relationship between the time of evacuation in practice drills, the number of occupants, and area of stairs. It was also mentioned by him in his paper "Time of evacuation by stairs in high buildings" [4]. Out of ten buildings, we are considering only building numbers 6, 7 and 8 as their staircase width is very close to 150 cm. as compared to other buildings. Details of the Galbreath study and time getting from the proposed model are mentioned in Table 4.

Table 4 Result validation with drill conducted by Galbreath

Sr. No.	Building number	Number of floors	Staircase width (cm)	Person per floor	Number of stairs	Person per floor per stairs	Time of evacuation during drill (s)	Evacuation time by using proposed model (s)
1	6	11	132	110	2	55	390	445
2	7	9	132	111	2	56	330	370
3	8	9	132	133	4	34	270	225

The results are well validated. The slight difference is there, which may be due to difference in ceiling height, which is not mentioned by Galbreath.

Validation II:

In 2002, Eric Rivers [52] had conducted an evacuation drill at 85 Broad Street, New York, a 30-story office building. The evacuation consisted of 1385 people and concluded in approximately 18 min.

As proposed in this paper, after using Eq. 2 with the above mentioned parameters, following result is obtained

$$t = \frac{0.0836 \times 1385 \times 10.86}{1.2}$$

t = 1047.86 s i.e. 17.46 min.

The result is very close to the drill timings.

11 Conclusion and Future Research

In this study, a new method for the estimation of evacuation time is suggested based on important parameters like the number of floors, number of occupants, effective width staircase and travel distance. It is also concluded that evacuation time increases with the decrease in staircase width and an increase in occupant load and the number of floors. But there is very little variation on overall evacuation time with the change in floor area in the case of a high-rise building. As a generic model, it can be used for all types of buildings. The major advantage of this mathematical model is its short computation time and it can be calculated very easily. It will be useful for deciding the Required Safe Evacuation Time, which is helpful to the concerned designer to design the building safety. Also, the acceptable occupant load can be decided once we estimate the required safe egress time. This mathematical model will be more useful for business occupancy buildings where other behavioural aspects are comparatively less.

For estimating the evacuation time, certain assumptions are made in this study like the presence of only physically fit occupants, use of available staircases by the equal number of occupants, no reverse flow by emergency services from the staircase during the evacuation, the start of an actual evacuation once raised the alarm without considering the pre-evacuation time. All these parameters are difficult to judge being very subjective, difficult to identify but they exist. So, future work may explore such situations and analyze the dynamic behavior of the evacuation, which will certainly result into a greater accuracy in estimating the evacuation time.

References

1. Ding N., Zhang H., Chen, T., & Luh, P. B. (2015). Stair evacuation simulations based on cellular automata considering evacuees' walk preferences. Chinese Physics B, 24(6), 1–8.
2. Salankar, S., Tauseef, S. M., & Sharma, R. K. (2018). Need for better high rise building evacuation practices.
3. Ding N., Chen T., Zhang H., & Luh P. B. (2015). Traffic and Granular Flow '13. Traffic and Granular Flow '13, October 2017.
4. Galbreath, M. (1969). Time of evacuation by stairs in high buildings.
5. Pauls J. (1987). Calculating evacuation times for tall buildings. Fire Safety Journal.
6. https://www.statista.com
7. Challinger D. (2008). From the Ground up: Security for Tall Buildings.
8. Shultz J. M., Marcelin L. H., Espinel Z., Madanes S. B., Allen A., & Neria Y. (2013). Encyclopedia of Natural Hazards (Vol. 2).
9. Craighead G. (2003). High-rise Security and Fire Life Safety.
10. NFPA 101 ® Life Safety Code®. (2006).
11. International Code Council. (2018). 2018 International Building Code. In Journal of Chemical Information and Modeling.
12. Kavilkar R., & Patil S. (2014). Study of High Rise Residential Buildings in Indian Cities (A Case Study –Pune City). International Journal of Engineering and Technology, 6(1), 86–90.
13. Kealy M. J., Ceng, H., & Mcibse, F. (2008). Fire Engineering Super tall: A New Approach to Escape, Schirmer Engineering Corporation, Conference proceeding paper.
14. PelechanoN., & Malkawi A. (2008). Evacuation simulation models: Challenges in modeling high rise building evacuation with cellular automata approaches. Automation in Construction, 17(4)
15. Ma, J., Song, W. G., Tian, W., Lo, S. M., & Liao, G. X. (2012). Experimental study on an ultra high-rise building evacuation in China. Safety Science, 50(8), 1665–1674. https://doi.org/10.1016/j.ssci.2011.12.018
16. Cepolina E. M. (2009). Phased evacuation: An optimisation model which takes into account the capacity drop phenomenon in pedestrian flows. Fire Safety Journal, 44(4), 532–544.
17. Gwynne S. M. V., Kuligowsk E. D., Kratchma J., & Milk, J. A. (2009). Questioning the linear relationship between doorway width and achievable flow rate. Fire Safety Journal, 44(1), 80–87.
18. DaamenW.,& Hoogendoorn S. P. (2011). Pedestrian and Evacuation Dynamics. Pedestrian and Evacuation Dynamics.
19. Zhang J., Song, W., & Xu, X. (2008). Experiment and multi-grid modeling of evacuation from a classroom. Physica A: Statistical Mechanics and Its Applications, 387(23), 5901–5909.
20. Frey, B. S., Savage, D. A., &Torgler, B. (2011). Erratum: Interaction of natural survival instincts and internalized social norms exploring the Titanic and Lusitania disasters (Proceedings of the National Academy of Sciences of the United States of America (2010) 107, 11, (4862–4865)
21. Leonard cooper, David Stroup, (1982), Calculating available safe egress time (ASET) – A computer program and user's guide, US department of commerce.
22. Sime J. D. (1986). Perceived Time Available: the Margin of Safety in Fires. 561–570.

23. Lin C. S., & Wu, M. E. (2018). A study of evaluating an evacuation time. Advances in Mechanical Engineering, 10(4), 1–11.
24. Kodu, V. K. R., &Harmathy, T. Z. (2016). Properties of building materials. In SFPE Handbook of Fire Protection Engineering, Fifth Edition.
25. Zhang G., Huang, D., Zhu, G., & Yuan, G. (2017). Probabilistic model for safe evacuation under the effect of uncertain factors in fire. Safety Science, 93, 222–229.
26. Gwynne S., Galea, E. R., Lawrence, P. J., & Filippidis L. (2001). Modelling occupant interaction with fire conditions using the buildingEXODUS evacuation model. Fire Safety Journal, 36(4), 327–357.
27. Zhao C. M., Lo, S. M., Zhang, S. P., & Liu, M. (2009). A post-fire survey on the pre-evacuation human behavior. Fire Technology, 45(1), 71–95.
28. Shields, T. J., & Boyce, K. E. (2000). Study of evacuation from large retail stores. Fire Safety Journal, 35(1), 25–49.
29. Liu, M., & Lo, S. M. (2011). The quantitative investigation on people's pre-evacuation behavior under fire. Automation in Construction, 20(5), 620–628.
30. Proulx, G. (1995). Evacuation time and movement in apartment buildings. Fire Safety Journal, 24(3), 229–246.
31. Brennan, P. (1997). Timing Human Response In Real Fires. Fire Safety Science, 5, 807–818.
32. National Building Code of India -2016 VOL 1.
33. The Building Regulations 2010 B1 Means of warning and escape B2 Internal fire spread (linings) B3 Internal fire spread (structure) B4 External fire spread B5 Access and facilities for the fire service. (2007).
34. Post-war building studies fire grading of buildings part II firefighting equipment personal safety chimneys and flues part III,part IV by a joint committee of the building research board of the department of scientific & industrial research and of the fire offices' committee London: I952 published for the ministry of works by Her Majesty's Stationery office
35. Thompson, P. A., & Marchant, E. W. (1995). A computer model for the evacuation of large building populations. Fire Safety Journal, 24(2), 131–148.
36. Pauls, J. L., Fruin, J. J., &Zupan, J. M. Minimum Stair Width for Evacuation, Overtaking Movement and Counter flow Technical Bases and Suggestions for the Past, Present and Future Consulting Services in Building Use & Safety, USA 2 PED Associates, USA 3 Regional Plan Association, USA.
37. Sujatmiko, W., Dipojono, H. K., & Soelami, F. X. N. (2014). Performance-based Fire Safety Evacuation in High-rise Building Flats in Indonesia – A Case Study in Bandung. Procedia Environmental Sciences, 20, 116–125.
38. Thunderhead Engineering. (2011). Pathfinder User Manual. Springer Reference.
39. Watson, R. (2001). Case Studies : Case Studies : Case Studies : Three-Dimensional Molded Interconnect Devices (3D-MID): Materials, Manufacturing, Assembly, and Applications for Injection Molded Circuit Carriers, 80(2), 1–17.
40. E., G., C., M., & P., B. (1999). Complex evacuation; effects of motivation level and slope of stairs on emergency egress time in a sports stadium. Safety Science, 31(2), 127–141.
41. Averill, J. D., &Kuligowski, E. D. (2014). NIST Technical Note 1624 Stairwell Evacuation from Buildings : What We Know We Don ' t Know. December 2010.
42. Averill, J. D., Mileti, D. S., Peacock, R. D., Kuligowski, E. D., Groner, N., Proulx, G., Reneke, P. A., & Nelson, H. E. (2005). Federal building and fire safety investigation of the World Trade Center Disaster: Occupant behavior, egress, and emergency communications. NistNcstar 1-7, 1–298.
43. Pauls, J. L., Fruin, J. J., & Zupan, J. M. (2007). Minimum Stair Width for Evacuation, Overtaking Movement and Counter flow — Technical Bases and Suggestions for the Past, Present and Future. Pedestrian and Evacuation Dynamics 2005, 57–69.
44. Still, G. K. (2000). 2000_Still Thesis Crowd Dynamic.
45. Pauls, J. (1984). The movement of people in buildings and design solutions for means of egress. Fire Technology.

46. Tavares, R. M., Tavares, J. M. L., & Parry-Jones, S. L. (2008). The use of a mathematical multi criteria decision-making model for selecting the fire origin room. Building and Environment, 43(12), 2090–2100.
47. Name, M., Description, V. S., & References, T. (1995). Computer Models For Fire and Smoke. In Journal of Fire Protection Engineering (Vol. 3, Issue July, pp. 10–12).
48. Thornton, C., Konski, R. O., Hardeman, B., Swenson, D., Ave, P., &Ste, B. (2011). Pedestrian and Evacuation Dynamics. Pedestrian and Evacuation Dynamics, 3–6.
49. MCGM. (1991). Development Control Regulations 1991. 1991(5), 1–110.
50. Kuligowski, E. D., & Peacock, R. D. (2005). A Review of Building Evacuation Models. National Institute of Standards and Technology, 156.
51. Galbreath M. A Survey of Exit facilities in high office buildings (1968). Building research note
52. Rivers, E., Jaynes, C., Kimball, A., & Morrow, E. (2014). Using case study data to validate 3d agent-based pedestrian simulation tool for building egress modeling. Transportation Research Procedia, 2, 123–131.

Study of Carbon Emission During Taxi-Out Operation at International Airport, India

Totakura Bangar Raju, Sheetal Khanka, Arun Kumar Jagadeesan, and Ashwani Khanna

Abstract Global warming is one of the most significant problems faced by all living creatures on planet earth. Greenhouse gases like Carbon dioxide, Nitro oxide, and methane traps the heatwave, which is the cause of the increase in temperature of the earth. This greenhouse gas depletes the ozone layer. Global warming contributed to the problem of Climate Change. The leading cause of global warming is the emission of Carbon dioxide (CO_2). In the year 2016, nearly 174 countries joint their hands together against global warming and signed the Paris Agreement. Under the Paris agreement, each country should regularly report on the contribution that it undertakes to mitigate global warming. CO_2 emission was not a severe problem with Aviation, however, there is a drastic increase in air transportation and thus Aviation sectors' contribution to CO_2 emission increased and the sector has come under scrutiny. Paris Agreement stated that CO_2 emissions are a serious issue and a leading cause of the impact on the life quality of the people living around the Airports. The activities of the aircraft from landing to takeoff known as the Landing takeoff (LTO) cycle which contributes to a sufficient deterioration of air pollution near the airport and the nearby neighborhood. The primary source of CO_2 emission in the airport vicinity is due to the taxiing operation of the aircraft. The movement of the aircrafts on the ground under ints own power is known as taxiing. During the taxi-out operation, the aircraft will move at a speed of 16–19 Kts. Both engines of the aircraft kept operated by the pilot during taxi-out operation may lead to doubling of CO_2 emission. One more reason for increase of CO_2 emission at the airport vicinity is due to aircraft which stand in the queue for take-off operation. This paper discusses CO_2 emissions due

T. B. Raju (✉) · A. K. Jagadeesan
University of Petroleum and Energy Studies, Dehradun, Uttarakhand 248007, India
e-mail: tbraju@ddn.upes.ac.in

S. Khanka
Department International Affair, University of Petroleum and Energy Studies, Dehradun, Uttarakhand 248007, India
e-mail: sheetal@ddn.upes.ac.in

A. Khanna
AeroTech Support Services, Aerocity, New Delhi 110037, India
e-mail: ashwani@aerotechss.com

N. A. Siddiqui et al. (eds.), *Advances in Environment Engineering and Management*, Springer Proceedings in Earth and Environmental Sciences,
https://doi.org/10.1007/978-3-030-79065-3_26

to taxi out operations at Indira Gandhia International Airport, New delhi. This paper also suggests other ideas to improve the taxi-out operation in the airport and gives other ideas to reduce the CO_2 emissions from the aircraft engine.

Keywords Co_2 emission · Global warming · Widebody · Narrow-body · Taxing-out · LTO cycle

1 Introduction

One of the most significant challenges for the human being in this twenty-first century is to stabilize the global climate. The temperature of the earth has been increasing gradually for 38 consecutive years. Everyone around the world has felt the impact of climatic change. The scientists have information about anthropogenic activities that were carried out across the world is the primary source for the increase of GHG emissions in the atmosphere and the primary reason for climate change. The various activities of humans, including usage of fossil fuel in power plants and GHG emission from various industries, transportation and energy supply, etc. are the maojor sources of emissions. Natural factors such as volcanic eruption have an impact on climate change. However, changes in land-use patterns and ozone depletion are also the most important factors contributing to climatic changes.

This climate change plays an essential role in environmental, social, political, and economic repercussions. Therefore, international organization needs an integrated approach to handle this problem across the world. This climate change will affect individuals and business which one have never seen before if climatic changes are not taking seriously and incorporate strategic planning to handle this problem internationally. Every business unit should take climate change seriously, change the business sustainability initiatives, and have objectives to keep GHG at a reduced level. United Nations Climate Change Conference, (COP25), held on December 2 to December 13, 2019, in Madrid, Spain was the continuation Paris 2015 Agreement, to mitigate the emission in member countries. Developing member countries were unhappy because developed countries failed to keep up promise for climate action. However,r they planned to report and monitor the progress in regard to Paris agreement till the year 2021.

1.1 Transportation Sector GHG Emission Globally:

In the whole transportation sector, civil aviation is one that has the fastest growth. From 2001–2005 the growth of scheduled aviation traffic is 3.8%, and now it is growing currently at a rate of 5.9% (Indiaghgp 2015). The below figure shows the contribution of GHG emission by different sectors as a report by the Intergovernmental Panel on Climate Change (IPCC 2014). The figure shows that the

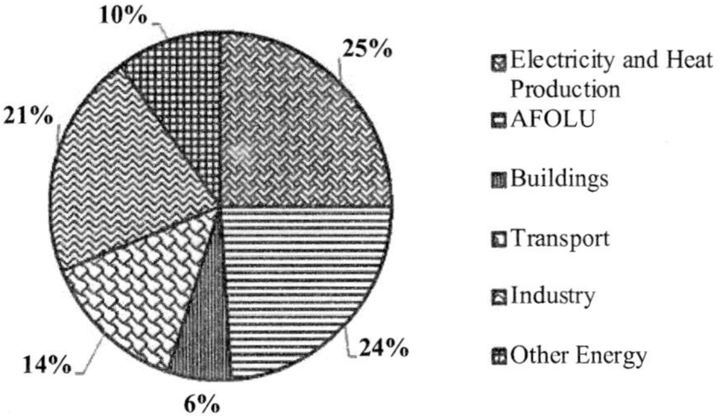

Fig. 1 GHG emission by sectors, 2010

transportation sector is contributing 14% of the total anthropogenic emissions (Fig. 1).

1.2 Aviation GHG Emission Globally:

The GHG emission from civil aviation is majorly from the aircraft operation, which is mainly responsible for greenhouse gas such as Water Vapor (H_2O) and Carbon dioxide (CO_2). There are other products of GHG emissions, such as Nitrogen dioxide (NO_2) and Nitrous oxide (NO), soot and sulfur oxides (SO_x). The movement of the aircraft is a complex process, and the particles are deposited at higher altitudes, which affects climate change. Currently, global aviation contributes nearly 2% of the total global emission, as per IPCC, 2007. ICAO also reported that international aviation contributes 60% of the total aviation emission. IPCC also estimated that by 2050 the emission contributed by the aviation sector would increase to 3% (Fig. 2).

1.3 Indian Aviation

India is one of the largest populated countries in the world. Indian people travel frequently for their business, tourism, festivals, education, and a lot more. The transportation sector is one of the reasons for the growth of the economy. Indians use a different mode of transport like Train, metro, Aircraft, Ships, Cars, buses, etc.. The Indian aviation sector is the fourth largest domestic market in the world and it is the world fastest growing domestic market. Airport transport in India is growing at a very high pace. For the past 4 years, the Indian aviation domestic market has shown double-digit growth. For the year 2018, the Indian domestic aviation sector has grown 18.6%, followed by china 11.7%. According to DGCA, India, the domestic growth in India was 13.9–18.6% in 2018, 11.7% in 2017, 17.3% in 2016, 23.2% in 2015, 20.3% in 2016, which is due to the robust economic expansion in India.

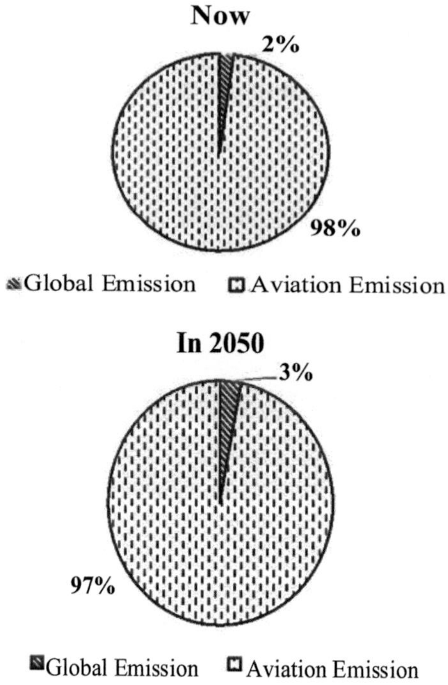

Fig. 2 Global emission versus aviation emission (IPCC 2007)

After the introduction of the LCC model in India, ordinary people in India who used to travel by railways and roadways,ses, they started to travel in aircraft. Indigo is the market leader in India. Apart from these, the growth of airports in India is also rapid. Many Greenfield airports are planned to be build at various states in India. Indian government's decision to develop the Airports under PPP mode has been a major engine driving the passenger growth in the country. The major private airports in India are Indira Gandhi International Airport, Delhi, Chhatrapati Shivaji International Airport, Mumbai, Rajiv Gandhi International Airport, Hyderabad, Kempegowda International airport, Bangalore, Cochin International Airport, etc., Recently Adani group has been awarded six airports namel Ahmedabad, Jaipur, Mangaluru, Trivandrum (Thiruvananthapuram), Lucknow, and Guwahati under PPP mode.

In India, all airports now want to bring the carbon footprint down because of the environmental concern. The government of India is also helping the entire transportation sector to bring down the carbon emission in India.

Figure 3 shows that the contribution of CO_2 by various sectors in India. It shows that in India, the transportation sector contributes nearly 41% of the total CO_2 emission in 2018. Figure 4 shows the contribution of CO_2 emission by various aircraft operations and classes in 2018. The Indian environmental portal reports this information.

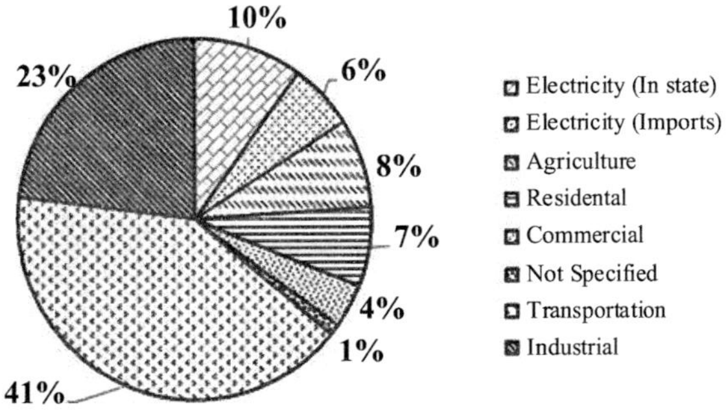

Fig. 3 CO_2 emission by various economic sectors

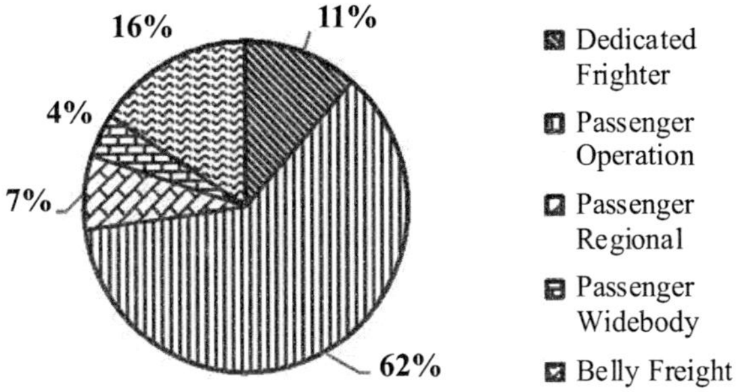

Fig. 4 CO_2 emission in 2018 by operation and aircraft class

2 Problem Statement

The emission of CO_2 gas constitutes the most massive threat to the world, which is affecting the climatic change and increase in the earth's temperature. The environmental scientist has advised maintaining the earth's temperature less than 1.5 °C if the temperature increase to 2 °C; it will make the earth inhabitable. According to the Paris agreement 2016, every country has to contribute to maintain the level of CO_2 emission and make an effort to keep the earth's temperature 1.5 °C. CO_2 gas is emitted from various things like household products, transportation, Industries, etc. Usage of these items has been increased day by day, but the limit of CO_2 emission from these products has been reduced day by day.

The transportation sector is one of the major contributor for the emission of CO_2 gas. World wide, flights produced 915 million tonnes of CO_2 in 2019. Globally, humans produced over 43 billion tonnes of CO_2. Aviation sector is accountable for 12% of CO_2 emissions among all transports sources, compared to 74% from road transport in 2019. The problem authors considered for the project represents the case study about the GHG emission during the taxiing operation at the airport. Each aircraft use approximately 6–9% of fuel for the taxiing operation, which emits tonnes of CO_2 gas. The amount of fuel consumed for the taxiing operation varies between short haul and long haul flights. The CO_2 gas emission at the airport is making the surrounding inhabitable. Apart from the taxiing operation, ground operating vehicles and equipments also emits CO_2 gas.

The authors have collected the data from the airport for estimating the time required for each aircraft for taxiing operation. The amount of time takes for an aircraft for taxiing out serves us to estimate the amount of fuel each airport spends for taxiing operation. The author's have attempted to calculate approximate CO_2 gas emitted at the airport during the taxiing operation and discuss an appropriate solution for reducing CO_2 emission.

3 Literature Review

Čokorilo [1] mainly concentrated on the long-distance passenger transport and environmental issues created due to the aircraft operations like noise emissions and GHG emission from the engine. The paper brings forward that the aircraft CO_2 is are approximately 2% of the Global GHG Emissions, which is expected to increase 3–4% per year. From the paper Čokorilo [1] it is seen that EU transport has a plan to be reduced GHG emission by 30% between 2008–2030, and 60% of the GHG emission is to be reduced between 1990 and 2050. Authors bring forward that the reduction of GHG emission can achieved by various factors such as airport infrastructure, reducing the aircraft age for the usage, technical achievement in the engine design, etc. The author also mentioned that the aviation industry needs to concentrate on alternate fuel implementation, which will provide the solution for the increasing aircraft operating cost and reduce GHG emissions.

Martín-Cejas [2] brought forward a new idea in the airport infrastructure for long-term growth in air transportation. The author proposed to change the pricing of the airport infrastructure by implementing an additional landing fee, which includes the environmental cost. According to the article, the airport pricing for the airline is based on weight-related charges, which may lead to the reduced utilization of the resources. So, the author proposed to implement the environmental cost according to the pollution created by them. The author also suggested a formula for calculated the tonnes of CO_2 emission from the aircraft. The author concluded in that it will be challenging to reduce the carbon emission from the aircraft unless we charge from the airlines.

Şöhret [3] investigates the CO_2 emission from the air transport industry in the UK market. The author discussed that 3–4% of the total carbon emission, which is related to energy, is from air transportation. The study discuses mainly about the CO_2 emission from the domestic flights in the UK market. The author also investigated the GHG emission from different perspectives like financial, thermodynamics, and environmental issues. The result from the analysis indicates that GHG emission is increasing every year. Between the year the 1990s to 2005, the energy is wasted up to 12.1 EJ. The author used two approaches to reduce emissions. The first approach is to reduce the energy from waste, and the second approach is to apply the carbon tax for the fuel.

Ashok et al. [4] examine the quality of air that is polluted by the CO_2 gas due to the aircraft operation at US airports. Also examines that the gas emitted from the aircraft engine CO_2 is the exposure to O_3 and $PM2_{2.5,}$ which decreases the fuel burn and increases the pollution. The authors have used various methods like emissions modeling, regional air quality modeling, CO_2 emissions, and population exposure to O_3 and $PM_{2.5}$. From the analysis, its is derived that there is positive relationship between pollution and Fuel, Population exposure, which is indexed by engine type, thrust setting, etc.

Ashok et al. [5] mainly concentrated on the impact of the CO_2 emission due to the taxi and take-off operation, which affected the air quality and climate change. The author brings forward that we can reduce the CO_2 emission by reducing the fuel burn and optimizing the gate hold and by de-rated take-off. The paper also recommends using pushback control, which minimizes the fuel burn and impact of air quality from taxi operation by 38% of the no gate holds. The author also recommended the pilots to take-off at the de-rated thrust, which will reduce the runway length and minimizes the GHG emissions.

Loo et al. [6] brimgs forward that the CO_2 emission is mainly due to the hub practice in air transport. Loo et al. [6] examines and estimated the CO_2 emission by various factors like airport level, airspace level, and difference flights. The authors conducted a study on the efficiencies of CO_2 emissions at Athens International Airport and the Hong Kong International Airport. It can be found that the Hub airports emits more GHG gas than other small airports. The authors advised that all the International Hub airports should take necessary measures to control the GHG emission at the premises of the airport.

Pagoni and Psaraki [7] discusses about the impacts of the emission of GHG gas from the aircraft on local air quality, which affects climate change. Pagoni and Psaraki [7] developed a tool for calculating the amount of CO_2 emission from an aircraft. This tool will be useful for monitoring real-time CO_2 emission. The output will give a detailed emission map, which included the aircraft routes, airports, aircrafts type, etc.; this tool will help to calculate the aircraft emission during the LTO cycle and the cruise cycle. The author tested the tool at the Greek airport.

Postorino et al. [8] discuss the improvisation of taxi-out operation at the city airport, which will reduce the CO_2 emission at the premises of the airport. The paper establishes that the CO_2 emission created at the premises of the airport by the aircraft operation affects the livelihood of the people who are all living at the surrounding of the airport. This paper mainly brings forth about the CO_2 emission at the airport is due to the LTO cycle and various aircraft operations. The author suggests having an alternate taxi out procedure like on-board Aircraft ground propulsion system, powered wheel tug, and e-taxi system. Various methods were to calculate, which will give an improvised result in reducing the taxi-out time and reduce fuel consumption.

Turgut and Rosen [9] demonstrate that the aircraft movements like LTO in the airports are the main reason for the CO_2 emission near the airport. The authors have collected the data of the aircraft movement for the Atlanta airport and London airport to compare the GHG emission for each airport. The study included eight airports which concluded that LTO operation is the main reason for the emission of GHG of the total flight operation. The author also recommended using fuel cells instead of APUs, which reduce carbon emission.

Vaishnav [10] examines the costs and benefits of the reduction of fuel and emission from the taxiing operation of the aircraft. The author uses various scenarios to test and to reduce the usage of fuel for the taxiing operation. The various scenarios are single-engine taxiing, e taxiing, tug scenario. By using these three methods, the author calculated how much time it is taking for taxiing out operation for each scenario and the amount of fuel used for this operation. He recommended that the tug operation is the most successful in the reduction of usage of fuel as well as CO_2 emission.

Becken & Shuker [11] describes how the booming of tourism directly involved in the blooming of the carbon footprint. The paper mainly concentrated on the combination of the airports where people use to move frequently. Becken and Shuker [11] also estimated the increase of the carbon emission using various factors like passenger volumes, geography, route network, and travel behavior by market influence. Becken and Shuker [11] proposes ten indicators for the carbon risk that help destinations assess their absolute and relative risk to the economic, financial, social, and environmental costs of carbon. The author suggested a penalty for the airline, which emits a huge quantity of carbon. The author found the passenger departing from the following airports are the leading cause of the carbon emission, they are FUK to HND, LAX to JFK, JFK to LAX, DXB to LHR, CTS to HND.

Owen and Lee [12] mainly speaks about the increasing trends of traffic in the aviation sector, which directly affects the increase in GHG emissions into the atmosphere. The author also mentions that even an increase in air traffic does not increase the GHG gas that has been estimated before, because nowadays, the aircraft manufactures aircraft with high performance and fuel-efficient. The author emphasis that even though the manufacturers produce great fuel-efficient aircraft, that does not mean it, will not emit GHG gas. Owen and Lee [12] also brong forth about every compound from the GHG emission and its effect. The authors also suggest that it even can reduce aviation emission by modifying the propulsion system in the aircraft.

Lee et al. [13] describes the roles and responsibilities of the city authorities in mitigation of the CO_2. To facilitate the optimal design for mitigation of the CO_2 emission policy, the city authorities need to understand the magnitudes and source of their CO_2 emission. This information will be useful to understand how CO_2 emission will affect climate change. The author has researched for the Taiwan city to understand the geographical distribution of CO_2 emissions. The author's main aim of doing this research is to study the bottom-up approach for estimating urban CO_2 emissions.

Dray et al. [14] investigates the interaction of various factors like economic, technological, and operational measures, which help to reduce aviation CO_2 emission. The author mention that we can reduce emission by retrofitting winglets, expanding maintenance program by airlines operation. We can also charge carbon price from the airline according to the emission of their aircraft. It can also be reduced by improvising air traffic management and the generation of biofuels. The author took three scenarios, each with three policy cases to generating a result effect on the fuel life cycle of CO_2 emission with and without an ETS standard.

Sherry [15] improvised the method of reporting air quality emission inventories for the GHG emission from aircraft such as CO_2 and NO_x upon the regulation. The author also mentions that in the traditional method of emission inventory calculation, the researchers make many assumptions like using of maximum take-off thrust setting. But in practice to reduce the cost, the airline use take-off thrust 25% lower than an assumption, which will lead to the yield over-estimate. Sherry [15] describes new algorithms that combine data from radar surveillance track, weather, and standardized aircraft performance. This study has been done at Chicago O'Hare airport, which exhibits a normal take-off thrust 86% of the maximum take-off thrust.

Zaporozhets and Synylo [16] speaks about how the aircraft engine emission affects the vicinity of the airport and nearby residential areas during the aircraft landing/take-off operation. The author discusses that the aircraft emission inventory is usually calculated based on the manufacture's certified engine emission indices. However, in real circumstances, the emission from the aircraft engines varies due to various factors like life expectancy of an aircraft, weather condition, and the type of engine installed on the aircraft. The paper also explains that the emission data for aircraft engines should take into account the influence of the real operational and ambient temperature, which will give the more precise result of the emission inventory of the aircraft.

Dray et al. [17] analyses the costs of the CO_2 emission mitigation measures. The authors discuss the relationship between mitigation potential and scenario characteristics. The authors find that the global life cycle of CO_2 emission per passenger-km could be reduced to 1.9 to 3.0%, but the price of the fuel increase from \$75 to \$180 per barrel by 2050. However, we can reduce the price of the oil by 0.1% per year and the carbon prices of \$50 to \$150/tCO_2. However, these assumptions depend on biofuel, which could reduce the emission half by 2050.

Herndon et al. [18] tests the emission from the commercial aircraft engines at ambient conditions. This test is done at Hartsfield-Jackson Atlanta International Airport. The authors have tested different engines with an aircraft to collect sampling 1–2 min downwind from the operational taxiway and runway. The authors found that NO_x from the engine is 18% lower than the certified data with the ambient conditions. But the CO emission from the engine is 100% greater than the certified data for the 7% thrust condition. The authosr observed these significant differences in black carbon. The authors also observed an anticorrelation between the particle mass loading and particle number concentration.

Herndon et al. [19] speaks about the emission of hydrocarbons from commercial aircraft at major airports in the United States. The author used proton-transfer-reaction mass spectrometry (PTR-MS) and turnable infrared differential absorption spectroscopy (TILDAS) to measure the emission of diluted exhaust plumes downwind. The author found that hydrocarbon emission is higher during the taxiing operation and take-off operation. The author found total unburned hydrocarbon from the range in-use idle plumes analyzed in this work is vast.

Hudda et al. [20] measured the pattern of particle number concentration downwind from the Los Angeles International Airport. The author found Lax emission is adversely affected the air quality much further than reported in the previous airport studies. Hudda et al., [20] found that a twofold increase in particle number and the concentration is more during the most hours of the day in the area of about 60 Km^2, which extended to 16 km more. The author brings forth that the airport is the main reason for the impact of air quality and the primary source of PN in the Los Angeles region.

Koudis et al. [21] did an experiment on the reduced thrust take off to know whether it will reduce the fuel emission or not? Because of the experiment, the author found that reduce the take-off of thrust will lead to reduced emission because of the reduced fuel consumption. The authors did this experiment at London Heathrow International Airport. The authors found that the reduced take-off thrust will reduce the nitrogen oxides and black carbon by 1.0–23.2%, 10.7–47.7%, and 49.0–71.7%, respectively, which depends upon the combination of an aircraft engine. The author also suggests that the TOW will affect take-off thrust variability. So, by setting the take-off thrust with the right specification, we can able to achieve the reduced take-off thrust. This will lead us to achieve industry environmental targets and helps us to maintain air quality.

Masiol and Harrison [22] defines that the aircraft operation at the ground station and operation related to aviation affect the environment near the airport. Some of the environmental issues, which the author is mentioning in this paper, are noise-related issues and the deterioration of air quality at ground level due to airport operations. Because of the operation, many pollutants like carbon dioxide, nitrogen oxide, and particulate, which is affecting public health. These pollutions are not only from the aircraft but also the operations like maintenance work, heating facilities, fugitive vapors from refueling operations, kitchens and restaurants for passengers, operators, intermodal transportation systems, and road traffic for transporting people and goods in and out to the airport, etc.,

The author's primary goal is to summarise the key characteristics of pollution in the local vicinity of the airport.

Nikoleris et al. [23] presented a detailed estimation of emission that happens during the taxiing operation and the fuel consumption for the taxi operation using the aircraft position data at Dallas/Fort Worth International Airport. The study took the assumption of the fuel flow rate and the thrust level setting and emission from the ICAO databank. The authors mainly analyzed the emission during the stop-and-go operation. The author concluded that 18% of the fuel is consumed during the taxi operation, which is mainly due to the braking or constant speed of the aircraft, which is due to the congestion-taking place at the airport.

Perl et al. [24] estimated the cost of air pollution that has been created by the aviation sector at Lyon-Satolas airport for the years 1987, 1990, 1994, and 2015. The estimation is done using environmental assessment techniques that yield an emission inventory for aircraft operation. The authors highlight the uncertainty that exists that the economic assessment damages from air pollution and the variation in willingness to pay is different. The authors discuss two scenarios like rural vs. urban impact of pollution and minimal vs. potential preferences.

Petersen and Solberg [25] discuss about the use of glulam beams instead of steel beams in the construction of the airport. The author mentions the production of the steel emits five times higher GHG emission when compared to the manufacturing of the glulam beams. The authors also mention that the glulam beam is favorable or unfavorable compared to steel, which is being landfilled or used for energy production. The mention that the production of the steel involved in using energy sources like electricity, which will also emit the Co_2 gas. By producing the glulam beam, we can avoid carbon emission and fixation on forestland. The author emphasises that 0.24–0.32 tons of CO_2 emission can be stopped by producing the glulam beam instead of steel beam.

Postorino and Mantecchini [26] mentions that airports are an essential part of transportation but also local sources for environmental impacts. In this paper, the author tries to find out the unit carbon footprints that are related to the emission macro sources like land vehicles, on-ground aircraft, airport handling, and terminal equipment. UCFs are defined due to the particular transport variables. The computation for the given airport includes both emission of macro-sources and the effectiveness of transport-related actions that helps to reduce the carbon impact. The author did this experiment at the airport of Bologna in Northern Italy.

Schürmann et al. [27] studies about the impact of the emission at an airport on local air quality. This study is conducted at the Zurich airport. The authors mainly studies the impact of VOC, CO, and NO_x emission on the air quality near the airport. Samples were taken to find the mixing ration of the volatile organic compound. The authors found that the CO emission mainly depends on the aircraft ignition and aircraft movements while NO concentration mainly depends on the ground support vehicles. The authors also found the difference in the emission data, which were published in the ICAO.

Simaiakis and Balakrishnan [28] mainly concentrated on the taxi time and the fuel burnt during the taxi time. The author emphsises that the emission at the airport is mainly due to the increased taxi time, and the aircraft stands in the queue to the aircraft. During this process, more amount of fuel is burnt, which leads to the emission in the vicinity at the airport. The authors found three components that affect the taxi process are unimpeded taxi-out time, time spent in the queue for departure, and the congestion delay. The authors propose a new queuing method for the aircraft movement, which will reduce the taxi time as well as the emission at the airport.

Unal et al. [29] mainly concentrated on the emission related to the airport, which affects the local air quality, especially compounds like ozone and $PM_{2.5}$. The experiment is conducted at Hartsfield–Jackson Atlanta International Airport. The author estimates the emission based on the smoke number. $PM_{2.5}$ emissions are estimated once based on the characteristic SN and a second time using the mode-specific SN. The author found that the impact of $PM_{2.5}$ and ozone of ground support equipment is lower when compared to the impact caused by the aircraft.

Zhu et al. [30] tries to find the aircraft emission, which affects the local air quality from the take-off activities. This study has been carried out at Los Angeles International Airport. The authors try to find out the real-time concentration of emission of ultrafine particles like black carbon, $PM_{2.5,}$ and chemical species. The author found the measurement indicates that during take-off, more than 100-fold difference in the concentration than when there is no take-off occurring. The result of the study shows that the aircraft emits large quantities of UFP at the lower end of currently measurable particle size ranges.

4 Research Methodology

The purpose of the chapter is to provide the reader with an overview of how the study has been carried out. This study is based on the observation that has been made at the airport regarding the aircraft movement in the ground. This study uses five steps for calculation purposes. Those steps are.

4.1 Data Preparation

The data has been segregated according to the narrow-body and wide-body aircraft. Each aircraft is identified and segregated according to the code. Code B—Code C aircraft ares considered as narrow-body aircraft. Code D—Code F aircrafts are considered as wide-body aircraft according to the ICAO nomenclature.

4.2 Data Processing

The data after segregated, the authors use excel to take count of each aircraft. The authors use this excel analysis for every month to count the aircraft according to the code and the time taken for the taxi-out operation. After this, the aircraft are segregated according to the time taken and calculated for fuel consumption. The below conditions are taken into consideration for the calculation of fuel consumption.

Narrow-body planes

- Boeing 737, 700, 800, 900 series including max
- Airbus 320, 318, 319, 320 and neo included, 321
- Average fuel consumption: 720 L/h.

Wide Body planes

- 4 engine wide-body planes: Airbus 380, 340 and Boeing 747
- Fuel consumption: 1800–2000 L/h.
- 2 engine wide-body planes: Airbus 330 new one, 350, Boeing: 787 Dreamliner, 777,
- Consumption is 1000 L/h.

When the authors calculated the consumption, they used those results for a further calculation to estimate the CO_2 emission with the below conditions.

Notes on Conversion

1 kg of ATF Burnt = 3.14 kg of CO_2 emission
1.2 l of ATF = 1 kg of ATF
Formula:
Emission of CO_2 = Fuel Consumption rate in Taxi x Time taken for Taxi.

4.3 Data Analysis and Findings

This data analysis is the continuation of the data collection. The data authors have is raw data, which authors have processed according to our research purpose. There is multiple research design such as descriptive, correlational, Experimental, Review, Meta-analytic. The authors have taken a correlational research design though the research is based on an observational study. The essence of this study is to collect as much data authors could for the perfect result. Furthermore, the data collected is confidential and used only for the research purpose.

Comparison of Taxi Time:
In this topic, the authors are going to compare the taxi-out time for each aircraft. The taxi-out time is compared between wide-body and narrow-body aircraft for the year 2014–2016 as per the data collected from the airport.

Table 1 and Graphs 1 and 2) tell about the taxi-out time comparison for the year 2014–2016. From the graph the author conclude that 23% of narrow body aircraft and 9% of wide body of aircraft take < 10 min for taxi-out operation. 77% of narrow body aircraft and 91% of wide body aircraft takes more than 10 min for taxi-out operation which will directly leads to emit huge amount of CO_2 gas.

Comparison of CO_2Emission:
In this topic, the authors are going to compare the CO_2 emission for each aircraft. The CO_2 emission is compared between wide-body and narrow-body aircraft for the year 2014–2016 as per the data collected.

Table 1 Comparison of taxi-out time

Comparison of taxi time						
Taxi time (min)	Wide body			Narrow body		
	2014 (%)	2015 (%)	2016 (%)	2014 (%)	2015 (%)	2016 (%)
< 10	8	10	9	23	22	20
11–15	24	27	21	42	41	39
16–20	35	32	33	22	23	25
21–25	21	20	22	8	9	10
26–30	7	7	9	3	3	4
31 >	4	4	6	2	2	2

Graph 1 Comparison of taxi time (wide-body)

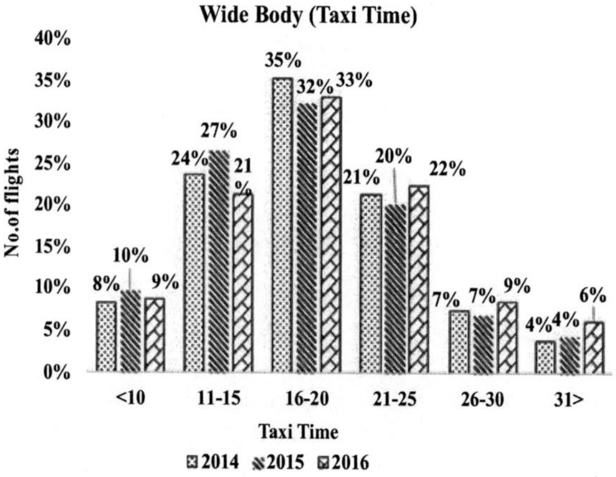

Graph 2 Comparison of taxi time (narrow-body)

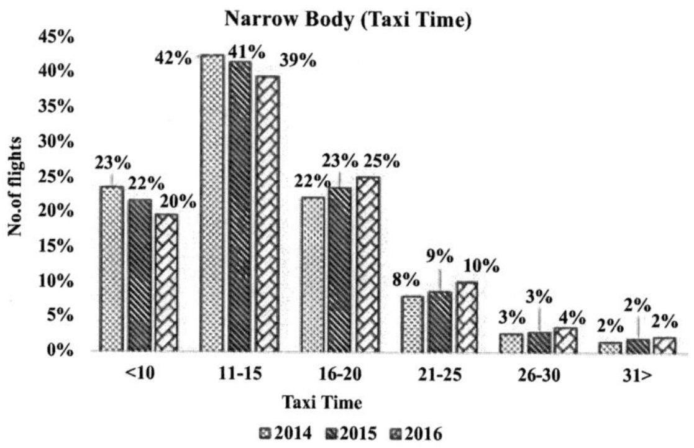

Table 2 and Graphs 3 and 4 tell about the CO_2 emission comparison for the year 2014–2016. From this, authors have found that every year the amount of CO_2 emission is increasing concerning the increase in aircraft movement. The authors have found that every year the CO_2 emission is increasing by 1% when compared to the previous years with irrespective of the aircraft movement.

5 Conclusion and Discussion

Indian aviation has the fastest growing domestic market in the world. The growth of the passenger movement in India is increasing linearly every year, which shows a positive sign of this industry. At the same time, the amount of pollution created by the aviation industry is also increased every year. Like Europe, India can also collect carbon taxes, which help us to create some new technology by using the money from the tax. All the aviation stakeholders to reduce and optimize the GHG

Table 2 Comparison of carbon emission

Comparison of carbon emission						
	Wide body			Narrow body		
Months	2014	2015	2016	2014	2015	2016
Jan	2039.36186	2275.19436	2404.62899	5043.6564	5083.8484	6466.5474
Feb	1769.10401	1784.64004	2138.38148	4741.714	4635.8332	6325.2474
Mar	1935.87302	1974.08864	2197.67697	5196.6686	5157.8896	6513.0194
Apr	1637.19173	1824.78754	2293.44853	4911.8078	5279.6588	6748.0484
May	1677.32555	2253.33044	2027.7303	5240.4716	5928.6968	7169.3736
June	1753.05588	1855.963	1999.10621	5249.295	5686.0062	7094.2962
July	1622.50488	1820.71481	1987.92358	5237.0176	5860.6844	6953.9696
Aug	1880.93834	1952.08073	2031.90037	5249.2636	5831.6708	6759.1954
Sep	1794.5588	1882.73803	2037.15149	5011.4086	5306.1604	6352.7224
Oct	1843.22936	2098.16342	2190.62204	5153.7762	5789.218	6860.8686
Nov	1896.80376	2020.93951	2324.99856	4606.5998	6100.6746	7230.321
Dec	1983.54839	2231.15281	2646.33316	5056.185	6467.5836	7245.7384

Graph 3 Comparison of carbon emission

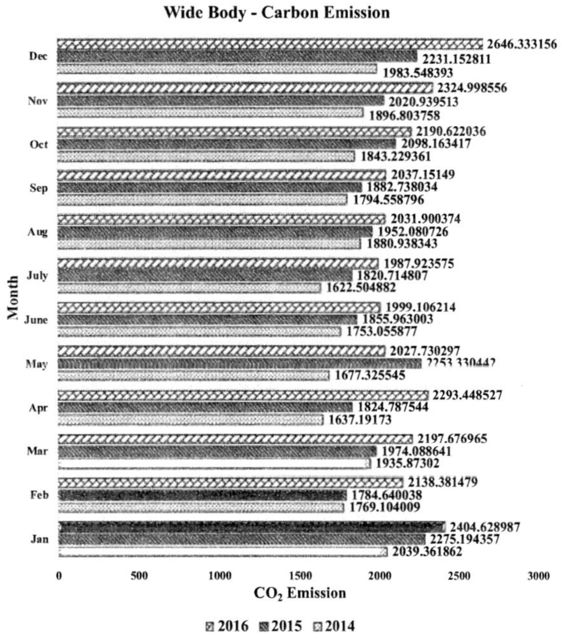

Graph 4 Comparison of carbon emission

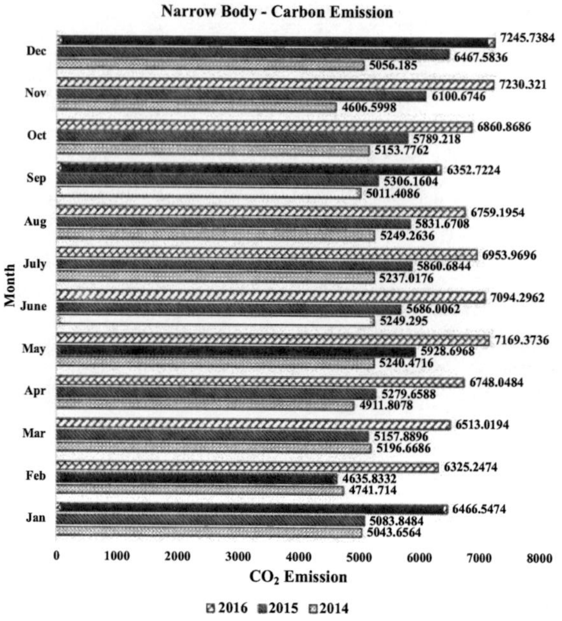

emission from various functions should follow best practices. New technologies can be implemented at the airport to reduce emissions. This report highlights the GHG emission only during the taxi-out operation of the aircraft at the airport.

The study has been completed, which shows that every year the amount of greenhouse gas emitted from the airport is increased. In India, the amount of domestic travel has been increased linearly. In this case, airlines can use some alternative methods to reduce pollution, which is created at the airport.

Alternate Taxiing

Alternate Taxiing is an area which has huge unexplored potential, apart from saving CO2 emmissions, as the engines of the Aircraft are not runningduring most of the Taxiing period the associated Engine Maintanence, Depreciation, Noise, FOD damage, Brake Ware, Apron Congestion are all reduced due to the use of adopting Alternate Taxiing techniques.In addition to the above mentioned savings there is also a saving on the ground time of the Airline therby further enhancing the impact of various savings. All theses savings do also have a financial correlation therby reducing the cost of operations for an Airline.

This also benefits the Airport by making the Apron safter due to the associated Jet Blast, reduced Noise and faster clearence. The CO_2 emmissions per Taxi-Out more than 15 min is in the range of about 550/600 kg per narrow body Aircraft, at busy metro Airports in India on an annual basis more than 400,000 Tons of CO_2 is emitted, this all can be avoided by the use of Alternate Taxiing solutions.

Limitation

In this study, authors have considered only taxi-out time. There are many various factors like ground operation vehicles and equipments taxi-in operation, take-off operation, etc. aren't take into account.

Even in the aircraft type, authors consider only commercial aircraft that are operated from the Indira Gandhi international airport. The authors did not take helicopters and civil aircraft into account because the volume of the operation are minuscule. The authors have calculated only CO_2 emissions for this research.

References

1. Čokorilo, O. (2016). Environmental Issues for Aircraft Operations at Airports. *Transportation Research Procedia, 14*, 3713–3720. https://doi.org/10.1016/j.trpro.2016.05.491.
2. Martín-Cejas, R. R. (2010). Ramsey pricing including CO_2 emission cost: An application to Spanish airports. *Journal of Air Transport Management, 16*(1), 45–47. https://doi.org/10.1016/j.jairtraman.2009.07.001.
3. Şöhret, Y. (2018). Multi-objective evaluation of aviation-induced GHG emissions: UK domestic flight pattern. *Energy and Environment*. https://doi.org/10.1177/0958305X18802778.
4. Ashok, A., Dedoussi, I. C., Yim, S. H. L., Balakrishnan, H., & Barrett, S. R. H. (2014). Quantifying the air quality-CO2 tradeoff potential for airports. *Atmospheric Environment, 99*, 546–555. https://doi.org/10.1016/j.atmosenv.2014.10.024.
5. Ashok, A., Balakrishnan, H., & Barrett, S. R. H. (2017). Reducing the air quality and CO 2 climate impacts of taxi and takeoff operations at airports. *Transportation Research Part D: Transport and Environment, 54*, 287–303. https://doi.org/10.1016/j.trd.2017.05.013.
6. Loo, B. P. Y., Li, L., Psaraki, V., & Pagoni, I. (2014). CO2 emissions associated with hubbing activities in air transport: An international comparison. *Journal of Transport Geography, 34*, 185–193. https://doi.org/10.1016/j.jtrangeo.2013.12.006.
7. Pagoni, I., & Psaraki, V. (2014). A tool for calculating aircraft emissions and its application to Greek airspace. *Transportation Planning and Technology, 37*(2), 138–153. https://doi.org/10.1080/03081060.2013.851510.
8. Postorino, M. N., Mantecchini, L., & Paganelli, F. (2019). Improving taxi-out operations at city airports to reduce CO2 emissions. *Transport Policy, 80* (May), 167–176. https://doi.org/10.1016/j.tranpol.2018.09.002.
9. Turgut, E. T., & Rosen, M. A. (2010). Assessment of emissions at busy airports. *International Journal of Energy Research*. https://doi.org/10.1002/er.1601.
10. Vaishnav, P. (2014). Costs and benefits of reducing fuel burn and emissions from taxiing aircraft. *Transportation Research Record, 2400*, 65–77. https://doi.org/10.3141/2400-08.
11. Becken, S., & Shuker, J. (2019). A framework to help destinations manage carbon risk from aviation emissions. *Tourism Management, 71*(October 2018), 294–304. https://doi.org/10.1016/j.tourman.2018.10.023.
12. Owen, B., & Lee, D. S. (2010). Aviation Emissions. *Encyclopedia of Aerospace Engineering, 1999*, 1–10. https://doi.org/10.1002/9780470686652.eae349.
13. Lee, T. C., Peng, S. K., Yeh, C. T., & Tseng, C. Y. (2018). Bottom-up approach for downscaling CO_2 emissions in Taiwan: robustness analysis and policy implications. *Journal of Environmental Planning and Management, 61*(4), 656–676. https://doi.org/10.1080/09640568.2017.1329714.
14. Dray, L., Evans, A., Reynolds, T., & Schäfer, A. (2010). Mitigation of aviation emissions of carbon dioxide: Analysis for Europe. *Transportation Research Record, 2177*, 17–26. https://doi.org/10.3141/2177-03.

15. Sherry, L. (2015). Improving the accuracy of airport emissions inventories using disparate datasets. *IIE Transactions (Institute of Industrial Engineers)*, *47*(6), 577–585. https://doi.org/10.1080/0740817X.2014.938845.
16. Zaporozhets, O., & Synylo, K. (2017). Improvements on aircraft engine emission and emission inventory asesessment inside the airport area. *Energy*, *140*, 1350–1357. https://doi.org/10.1016/j.energy.2017.07.178.
17. Dray, L. M., Schäfer, A. W., & Zayat, K. Al. (2018). The Global Potential for CO2 Emissions Reduction from Jet Engine Passenger Aircraft. *Transportation Research Record*. https://doi.org/10.1177/0361198118787361.
18. Herndon, S. C., Jayne, J. T., Lobo, P., Onasch, T. B., Fleming, G., Hagen, D. E., Whitefield, P. D., & Miake-Lye, R. C. (2008). Commercial aircraft engine emissions characterization of in-use aircraft at Hartsfield-Jackson Atlanta International Airport. *Environmental Science and Technology*, *42*(6), 1877–1883. https://doi.org/10.1021/es072029.
19. Herndon, S. C., Rogers, T., Dunlea, E. J., Jayne, J. T., Miake-Lye, R., & Knighton, B. (2006). Hydrocarbon emissions from in-use commercial aircraft during airport operations. *Environmental Science and Technology*, *40*(14), 4406–4413. https://doi.org/10.1021/es051209l.
20. Hudda, N., Gould, T., Hartin, K., Larson, T. V., & Fruin, S. A. (2014). Emissions from an international airport increase particle number concentrations 4-fold at 10 km downwind. *Environmental Science and Technology*, *48*(12), 6628–6635. https://doi.org/10.1021/es5001566.
21. Koudis, G. S., Hu, S. J., Majumdar, A., Jones, R., & Stettler, M. E. J. (2017). Airport emissions reductions from reduced thrust takeoff operations. *Transportation Research Part D: Transport and Environment*, *52*, 15–28. https://doi.org/10.1016/j.trd.2017.02.004.
22. Masiol, M., & Harrison, R. M. (2014). Aircraft engine exhaust emissions and other airport-related contributions to ambient air pollution: A review. *Atmospheric Environment*, *95*, 409–455. https://doi.org/10.1016/j.atmosenv.2014.05.070.
23. Nikoleris, T., Gupta, G., & Kistler, M. (2011). Detailed estimation of fuel consumption and emissions during aircraft taxi operations at Dallas/Fort Worth International Airport. *Transportation Research Part D: Transport and Environment*, *16*(4), 302–308. https://doi.org/10.1016/j.trd.2011.01.007.
24. Perl, A., Patterson, J., & Perez, M. (1997). Pricing aircraft emissions at Lyon-Satolas airport. *Transportation Research Part D: Transport and Environment*, *2*(2), 89–105. https://doi.org/10.1016/S1361-9209(97)00005-9.
25. Petersen, A. K., & Solberg, B. (2002). Greenhouse gas emissions, life-cycle inventory and cost-efficiency of using laminated wood instead of steel construction. Case: Beams at Gardermoen airport. *Environmental Science and Policy*, *5*(2), 169–182. https://doi.org/10.1016/S1462-9011(01)00044-2.
26. Postorino, M. N., & Mantecchini, L. (2014). A transport carbon footprint methodology to assess airport carbon emissions. *Journal of Air Transport Management*, *37*, 76–86. https://doi.org/10.1016/j.jairtraman.2014.03.001.
27. Schürmann, G., Schäfer, K., Jahn, C., Hoffmann, H., Bauerfeind, M., Fleuti, E., & Rappenglück, B. (2007). The impact of NOx, CO and VOC emissions on the air quality of Zurich airport. *Atmospheric Environment*, *41*(1), 103–118. https://doi.org/10.1016/j.atmosenv.2006.07.030.
28. Simaiakis, I., & Balakrishnan, H. (2009). Queuing models of airport departure processes for emissions reduction. *AIAA Guidance, Navigation, and Control Conference and Exhibit, August*. https://doi.org/10.2514/6.2009-5650.
29. Unal, A., Hu, Y., Chang, M. E., Odman, M. T., & Russell, A. G. (2005). Airport related emissions and impacts on air quality: Application to the Atlanta International Airport. *Atmospheric Environment*, *39*(32), 5787–5798. https://doi.org/10.1016/j.atmosenv.2005.05.051.
30. Zhu, Y., Fanning, E., Yu, R. C., Zhang, Q., & Froines, J. R. (2011). Aircraft emissions and local air quality impacts from takeoff activities at a large International Airport. *Atmospheric Environment*, *45*(36), 6526–6533. https://doi.org/10.1016/j.atmosenv.2011.08.062.

Healthcare Sustainability Through Technological Innovations

Ruchi Mehrotra and Kishor Joshi

Abstract Technological innovations can help in early diagnosis and management of diseases thereby preventing long term complications of the disease which contribute to morbidity and mortality. These can result in immense savings to the individual and also the healthcare system as a whole by reducing the burden of disease in the population. Cardiovascular diseases such as heart disease and stroke are the leading cause of death and disability in the world. About 20 million people die from Cardiovascular disease every year development of which is increased by diabetes mellitus (DM) and hypertension. Over 422 million of the global population suffer from DM which causes 1.6 million annual deaths and 7.5 million deaths annually from hypertension. In the backdrop of above statistics, the effective control of blood sugar and blood pressure with technological innovation with wearable devices which continuously monitor blood sugar and blood pressure can result in early preventive actions like timely treatment, regular monitoring to control them and can reduce the complications and death rate from diabetes and hypertension and cardiac disease. The research covers key findings indicating technological innovations like electronic medical records, mobile health, telemedicine, self-service kiosks for hospital registrations, payments checking identification, signing paperwork, remote monitoring tools and home monitoring systems and wearable technology and sensors which can greatly help in expediting healthcare delivery and improve quality of health of patients and the population. With digital technology, the healthcare systems which in India will be 372 billion USD by 2022 has enormous potential of usage of technological innovations to prevent early mortality is cases of chronic diseases. The paper concludes with importance of such digital advancement indicating that in less than 25 years, the global average life expectancy at birth has soared from 64 years in 1990 to 71 years in 2013 due to many others factors including early diagnosis of diseases and indicate that technological innovations can lead to better healthcare sustainability.

R. Mehrotra (✉)
University of Petroleum and Energy Studies, Dehradun, India

K. Joshi
Cardio Thoracic Vascular Surgeon, Director, Mahavir Heart Hospital, Patna, India

N. A. Siddiqui et al. (eds.), *Advances in Environment Engineering and Management*, Springer Proceedings in Earth and Environmental Sciences,
https://doi.org/10.1007/978-3-030-79065-3_27

1 Introduction

Technological Innovations Technological innovations can help in early diagnosis and management of diseases thereby preventing long term complications of the disease which contribute to morbidity and mortality. These can result in immense savings to the individual and also the healthcare system as a whole by reducing the burden of disease in the population. Historically too, technological intervention and its evolution saw a dramatic change but with the arrival of internet data, it offered an opportunity to fundamentally reinvent and redefine the link between medicine, healthcare delivery and technology. The 'e-health' era is a digital transformation of the medical practice with a coverage of the commercial side of health industry. Almost two decades back it was said that healthcare is only arriving in the 'Information Economy' [1]. But the transition has been steady and sturdy to handle the burden of growing population as well as the numerous diseases including various epidemics and even pandemics. When we define technological innovation systems we also integrate it with the multi-level frameworks. There exists numerous commonalities between technological innovation and multi-level frameworks and an integration for a radical process and socio-technical transformation is community as a whole [2].

Real time health monitoring and tracking systems and their effective implementation in diagnosis and treatment helps us realize the benefits of bringing medicine and technology in a blended form. These systems being real-time have been able to effectively track, monitor, trace, diagnose, and facilitate treatment and sustain healthcare of fast growing populations [3].

Healthcare Sustainability Healthcare systems face challenges as it is primarily considered to be the responsibility of the state towards the poor and insurance companies are meant to provide for the rich. Hence sustainability becomes a crucial challenge for the future. Likewise to define 'sustainable healthcare system' and identify its strengths and weakness it is important to believe that every business like healthcare business needs financial viability and a regular demand in the markets to sustain itself. But in case with applying technology to healthcare, affordability and adaptability equally becomes imperative [4]. The distinct features also include effectiveness and other interacting factors that connect the other facets of healthcare like diagnostics, surgical care, pharmaceuticals, service delivery including nursing, paramedics, and doctors.

Quality of healthcare holds high importance with aspects relating to integrity, confidentiality and hygiene which has been substantiated by usage of digital devices and standardization of medical devices as well as quality certifications by various national and international agencies. Healthcare professionals themselves attain newer heights to redefine quality in medical field. It is imperative to encompass mental wellness with cultural, social, environmental issues that enhance the likelihood of a sustainable system from a long-term perspective [5].

There are four different aspects of sustainability practices in healthcare, innovations in technology, the forces that drive and sustain these innovations within the industry, implementation of strategies and finally acceptance & adaptability of these innovations. There are various models in the driving force for sustainability in healthcare from a wider perspective and a holistic approach ensures that healthcare services excel in the business and the medical technology self sustains and remains viable in long-term [6].

Burden of Disease Morbidity Monitoring—Reducing Mortality Cardiovascular diseases such as heart disease and stroke are the leading cause of death and disability in the world. As per World Health Organization about 20 million people die from Cardiovascular disease every year development of which is increased by diabetes mellitus (DM) and hypertension. Over 422 million of the global population suffer from DM which causes 1.6 million annual deaths and 7.5 million deaths annually from hypertension [7]. In the backdrop of the above statistics, effective control of blood sugar and blood pressure with wearable devices which continuously monitor blood these parameters can result in preventive actions like timely treatment, regular monitoring to control them and can reduce the complications and death rate from these diseases.

Digital Health Scenario and Markets For the digital health companies looking to expanding their product base, it is an extremely important market with a huge potential, since their underlying base are the developed economies while the emerging markets have immense potential. The emerging market economies have a growing income base and the rapid rise in internet penetration with affordable data tariffs has made digital medical facilities accessible to all. The digital health market currently has fitness trackers and smart watches which form the largest device segment in the wearable category and accounted for 80% of the entire shipments in 2016. By 2022 it is expected to decrease and body sensors be third largest segment in wearable devices. The companies which go beyond the fitness bands and activity trackers and which aim to develop devices for the prevention of chronic diseases like diabetes, blood pressure and cardiac or kidney diseases will have a better business sustenance in the long term [8].

With digital technology, the healthcare system in India will be worth 372 billion USD by 2022 and has further enormous potential. In India, the healthcare sector is quite diversified and has enormous opportunities for consumers, service providers, companies providing technology. The hospital industry in India is expected to increase from Rs 4 trillion (US $ 61.79 billion) in FY17 to Rs 8.6 trillion (US $ 132.84 billion) by FY 22, which accounts to an impressive CAGR of 16–17% [9].

There are many companies across the globe which are market leaders like Apple, Sony, Google and Samsung and many others, which have ventured into the wearable devices market which is growing at a rapid pace. The clothing embedded wearable technology and other accessories have been referred as 'quantified self' and have been useful to educate, aware and motivate people towards better habits and better health. This has propagated the idea of better lifestyle and better habits [10].

Markets have evolved due to the rapid penetration of smart phones and internet connectivity with cost effective data and has made the Indian masses aware of lifestyle changes. Health activity monitoring devices are therefore part of many households now. This helps in monitoring of basic body parameters like temperature, heartbeat, and blood pressure can help in risk mitigation. The ability to maintain historical patients record and monitor them regularly and easily retrieve this data has become a distinctive feature of these devices [3].

Massive Technological Interventions in Healthcare The technological innovations have been pervasive irrespective of the geography and specialty medical facilities are available where there were none before. Technology has creeped into every segments of medical field from diagnosis to treatment and monitoring in follow up. The ease of use of electronic medical records, mobile health, telemedicine and telehealth, and self-service kiosks to expedite hospital registrations reduces queues and waiting time period. These also help in pandemics in reducing contact with the health care providers and maintain social-distancing and have been of immense help to the patients. Digital payments gateways, remote monitoring tools and home monitoring systems and have greatly helped in expediting healthcare delivery. Further development of wearable monitors and sensors can greatly improve the quality of healthcare services.

Innovations in the Medical Technology The authors present in Table 1, the gradual evolution of the medical technology in the last century which has revolutionized healthcare. This table indicates shows the technology its usefulness in diagnosing organ functions.

One of the most important invention in the history of medicine was that of the stethoscope by the French physician Rene Laennec in 1816 to listen to the chest of patients [11]. Dr Crawford Long was the first to perform various surgeries with anaesthesia in 1842 and continued to practice while the first report of use of an anaesthetic for painless surgery was published by WTG Morton, a dentist in Boston [12]. Modern surgery would not be possible without the availability of anaesthesia. Wilhelm Conrad Rontgen, physicist at the Universtity of Strasbourg in Germany discovered X rays in 1895 and was awarded the first nobel prize in physics in 1901[13] Dr Willem Einthoven recorded the first electrocardiograph in 1903 and was awarded the 1924 Nobel prize in Physiology in 1924 [14]. These are only a few of the thousands of progessive innovations and advancements made in the last millennium especially the last 200 years by scientists that have made possible the current healthcare technologies like the Ultrasound, CT Scan, MRI that are so well known today.

Wearables Devices and Healthcare Sustainability Another paradigm shift in medical technological innovations in the last decade has been in the wearable health devices (WHDs) which has made both monitoring as well as historical health database management much easier and user-friendly. WHDs enable the ambulatory monitoring of human vital signs in day to day life, from anywhere including at work and at home or sports field or a clinical environment. This minimizes the discomfort and being wearable these devices do not hinder in normal human activities [15].

The wearable technology is defined as the devices that are worn or mated with the human skin in order to regularly and diligently monitor activities, without any disruption or hindrance in the routine activities of the individual [16].

Table 1 Historical evolution of medical technology

Year	Device/innovation	Description
1815	Stethoscope	Auscultation of heart, lungs, bowel sounds and vascular bruits
1841	Anaesthesia	Reducing the pain and stress of surgery
1895	X ray	Viewing the skeleton internal organs
1903	ECG	Functioning of the Heart
1910	Laparoscopy	Visualisation of the inside of the Abdomen
1924	EEG	Graph of Brain activity as seen by waves
1943	Dialysis	A treatment for Kidney Failure
1947	Cardiac Defibrillation	Normalizing abnormal Heart rhythm
1950	Intraocular lens	Implantable lens after cataract surgery
1953	Heart Lung Machine	Helps in Open Heart Surgery
1954	Kidney Transplant	For Chronic Kidney failure
1963	Artificial Heart	Support the failing Heart
	Liver Transplant	For Liver failure
1965	Ultrasound	Used to visualize Multiple internal organs
1967	Heart Transplant	For Heart failure
1971	CT Scanner	Visualizing Multiple Internal organs
1973	Insulin Pump	For Insulin injection in Insulin dependent Diabetes
1978	MRI	Visualizing Multiple Internal organs without radiation
2000	Human Genome mapping	Mapping of the complete Human genetic Code

Source Compiled by Authors

Wearable technology refers to electronic technologies covering products including smart or advanced personal digital assistants (PDAs) used in watches, clothing, and protective gear equipment as well as quick monitoring systems. The predominant uses for wearable devices are for mainly (a) detecting anomalies in various organ functioning (b) identifying the critical events (c) identify chronology of events (d) monitoring to improve treatment outcome and (e) anticipating future events [17]. These devices have been able to sustain healthcare by timely analysis and interventions so that even in case of a crisis rehabilitation methods have been well adopted [18] (Table 2).

Remarkable Advantages of Technological Interventions in Healthcare The Global average life expectancy in 1900 was 31 years [20]. Where as the growth in medical field which included numerous factors but has made the world average in 2017 was 72.2 years [21]. This indicates the importance of such digital advancement indicating that in less than 25 years, the global average life expectancy at birth has soared from 64 years in 1990 to 71 years in 2013 due to many others factors including early diagnosis of diseases and better sustainability and from 31 to 72 in almost 120 years.

Table 2 Wearable health technologies

Device/Innovation	Activity Monitored / Usage
Smart watches/fitness bands	Heart rate, Blood pressure, Blood Oxygen, skin temperature, physical activity, burned calories, distance travelled, sleep monitoring, arrthymia monitoring
Ear device	Heart rate, oxygen saturation
Wearable heart activity monitors Chest straps, T-shirt, embedded electronics (smart textiles)	Monitor heart rate and rhythm, detect abnormal rhythms and diagnose disease

Source Sensors, July 2018, 2414 [19]

This has improved quality of life for senior citizen by making them cautious of their vitals. As such people now a days are not only well educated but well aware of health parameters and stay health conscious. Receiving proper medical care at the right time can enhance the chances of survival in many cases. Elder people need not to be ignored rather they need extra care. Hence WHDs have provided with amicable solution with their user-friendly design. It creates an alert system in case of any contingency for the friends, relatives or doctors to inform them via SMS or voice call to directly mobile phones. There systems measure crucial aspects like oxygen content in blood, body temperature, blood pressure and heart rate which all form extremely important aspects of critical healthcare systems [22].

2 Future Developments

The transition from illegible written prescriptions to typed ones and digitally saved medical history data has transformed the medical business systems. The automated feedback from the wearer of device to clinicians can help in decision-making and mitigating risks. The future development towards higher sensitivity devices and better predictive analysis to forecast events before they arise by usage of Artificial Intelligence will take the way forward. Future developments should indicate detailed demographic information along with the physiological and even psychological behaviors, cognitive patterns and the geographic location to be more specific for regional health care necessities, planning and monitoring. A multi-dimensional approach with the use of algorithms can enlarge the datasets of healthcare system both geographically and organ specific specialty wise. Societal benefits and certain databases can be made publicly available in order to understand the spread of contagious diseases like SARS COVID, Plague, bird flu or any epidemic/pandemic. Currently an app called as 'ArogyaSetu' developed by the Government of India [23] has helped in identification of COVID19 patients in nearby zones and is useful in alerting people to stay back home to stop spread of pandemic. Many more technological innovations are likely to crop up as newer coding and algorithm techniques are incorporated into medical database management.

The future of robotic surgery may see more advancement with smaller robots, cost effectively and impeccably integrating with the prevalent medical technology. This will enable better availability of surgical care procedures across the globe [24]. The development of da Vinci surgical system combines robotics and computer technology in order to facilitate micro-surgery in a contained environment. Manipulator systems have been developed where a surgeon can sit remotely away from patient in an optimal ergonomic environment [25].

Database management and community healthcare going digital can be extremely fruitful in all levels of societal framework. If individuals share their experiences honestly and diligently and in a timely fashion it can help create robust public health systems. All these will enable sharing of human experiences and in analyzing the wearable device data by database management apps and sites like *PateintsLikeMe* [18].

3 Conclusion

The technological advancements over the last century have not only contributed to the increased life expectancy but have also greatly improved the quality of lives of people. The rapid advancements in digital technology in the first two decades of this millennium are promising. These can lead to sustainability of healthcare by making available devices and technology that can in turn lead to primary prevention of diseases thereby greatly decreasing the burden of disease. These devices are still in their infancy mode and will take considerable advancements to finally deliver what they are intended to do. Wearable devices like smart watches and fitness bands are already in the market and are playing a prominent role in creating greater awareness of the role of exercise in the lives of ordinary people. Lack of exercise and obesity is leading to the biggest pandemic of obesity and metabolic syndromes all across the globe. The incidence of diabetes, hypertension, heart disease and stroke, which affect millions of people, can be greatly reduced by regular exercise and fitness bands or smart watches can help people monitor their daily physical activities. This can lead to huge savings in terms of healthcare expenditure for the individual in long run as well as great savings if reported to doctors in real time manner and the economy as a whole. Real time monitoring of data of patients with chronic diseases can reduce morbidity and mortality by early intervention and therefore lead to healthcare sustainability. The crucial aspects newer technologies need to cover from a healthcare perspective is that they must be affordable, adaptable, efficient, environmentally and economically sustainable and effective. In fact this can be a road to sustainability in healthcare through the path of digitalization and technological innovations.

References

1. R. C. J. Coile, "The digital transformation of health care.," *Physician Exec.*, vol. 26, no. 1, pp. 8–15, 2000.
2. J. Markard and B. Truffer, "Technological innovation systems and the multi-level perspective: Towards an integrated framework," *Res. Policy*, vol. 37, no. 4, pp. 596–615, 2008, doi: https://doi.org/https://doi.org/10.1016/j.respol.2008.01.004.
3. Aziz K, Tarapiah S, Ismail SH, Atalla S (2016) Smart real-time healthcare monitoring and tracking system using GSM/GPS technologies. In: 2016 3rd MEC international conference on big data and smart city (ICBDSC), pp 1–7. https://doi.org/10.1109/ICBDSC.2016.7460394
4. M. Fischer, "Fit for the future? A new approach in the debate about what makes healthcare systems really sustainable," *Sustain.*, vol. 7, no. 1, pp. 294–312, 2015, doi: https://doi.org/10.3390/su7010294.
5. T. V. Health (2018) Sustainability in healthcare : efficiency, effectiveness, vol 5(2), p. 2018.
6. M. Marimuthu and H. Paulose, "Emergence of Sustainability Based Approaches in Healthcare: Expanding Research and Practice," *Procedia - Soc. Behav. Sci.*, vol. 224, pp. 554–561, 2016, doi: https://doi.org/https://doi.org/10.1016/j.sbspro.2016.05.437.
7. WHO (2020) Diabetes. https://www.who.int/news-room/fact-sheets/detail/diabetes
8. Tractica (2017) Wearable device market forecasts. [Online]. Available: https://tractica.omdia.com/research/wearable-device-market-forecasts/
9. IBEF (2020) Indian Healthcare Industry Analysis. [Online]. Available: https://www.ibef.org/industry/healthcare-india.aspx
10. M. S. Patel, D. A. Asch, and K. G. Volpp, "Wearable Devices as Facilitators, Not Drivers, of Health Behavior Change," *JAMA*, vol. 313, no. 5, pp. 459–460, Feb. 2015, doi: https://doi.org/10.1001/jama.2014.14781.
11. A. Roguin, "Rene theophile hyacinthe laënnec (1781-1826): The man behind the stethoscope," *Clin. Med. Res.*, vol. 4, no. 3, pp. 230–235, 2006, doi: https://doi.org/10.3121/cmr.4.3.230.
12. R. Anaya-Prado and D. Schadegg-Peña, "Crawford Williamson Long: The True Pioneer of Surgical Anesthesia," *J. Investig. Surg.*, vol. 28, no. 4, pp. 181–187, Jul. 2015, doi: https://doi.org/10.3109/08941939.2015.1061826.
13. P. B. Riesz, "The life of Wilhelm Conrad Roentgen.," *AJR. Am. J. Roentgenol.*, vol. 165, no. 6, pp. 1533–1537, 1995, doi: https://doi.org/10.2214/ajr.165.6.7484601.
14. M. Rivera-Ruiz, C. Cajavilca, and J. Varon, "Einthoven's String Galvanometer: The first electrocardiograph," *Texas Hear. Inst. J.*, vol. 35, no. 2, pp. 174–178, 2008.
15. Di Rienzo M, Rizzo F, Parati G, Brambilla G, Ferratini M, Castiglioni P (2005) MagIC System: a new textile-based wearable device for biological signal monitoring. Applicability in daily life and clinical setting. In: 2005 IEEE engineering in medicine and biology 27th annual conference, pp 7167–7169. https://doi.org/10.1109/IEMBS.2005.1616161
16. M. Haghi, K. Thurow, I. Habil, R. Stoll, and M. Habil, "Wearable Devices in Medical Internet of Things," *Heal. Informatics Res.*, vol. 23, no. 1, pp. 4–15, 2017, doi: https://doi.org/10.4258/hir.2017.23.1.4.
17. H. Banaee, M. U. Ahmed, and A. Loutfi, "Data mining for wearable sensors in health monitoring systems: A review of recent trends and challenges," *Sensors (Switzerland)*, vol. 13, no. 12, pp. 17472–17500, 2013, doi: https://doi.org/10.3390/s131217472.
18. M. M. Rodgers, G. Alon, V. M. Pai, and R. S. Conroy, "Wearable technologies for active living and rehabilitation: Current research challenges and future opportunities," *J. Rehabil. Assist. Technol. Eng.*, vol. 6, p. 205566831983960, 2019, doi: https://doi.org/10.1177/2055668319839607.
19. Dias D, Cunha JPS (2018) Wearable health devices—vital sign monitoring, systems and technologies. Sensors (Switzerland) 18(8). https://doi.org/10.3390/s18082414
20. Prentice T (2008) Health, history, and hard choices: funding dilemmas in a fast-changing world. Nonprofit Volunt. Sect. Q. 37(1_suppl):63S–75S. https://doi.org/10.1177/0899764007310533
21. World Bank (2019) Life expectancy at birth total years. https://data.worldbank.org/indicator/SP.DYN.LE00.IN

22. Megalingam RK, Radhakrishnan V, Jacob DC, Unnikrishnan DKM, Sudhakaran AK (2011) Assistive technology for elders: wireless intelligent healthcare gadget. In: Proceedings of 2011 IEEE global humanitarian technology conference (GHTC 2011), Oct 2011, pp 296–300. https://doi.org/10.1109/GHTC.2011.94
23. Government of India (2020) Aarogya Setu Mobile App. https://www.mygov.in/aarogya-setu-app/.
24. D. B. Camarillo, T. M. Krummel, and J. K. Salisbury, "Robotic technology in surgery: Past, present, and future," *Am. J. Surg.*, vol. 188, no. 4 SUPPL. 1, pp. 2–15, 2004, doi: https://doi.org/10.1016/j.amjsurg.2004.08.025.
25. M. S. Kavic, "Robotics, technology, and the future of surgery.," *JSLS*, vol. 4, no. 4, pp. 277–279, 2000.

Analysis by Benchmarking the Criteria of GRIHA in Performance Assessment of Green Building

Jaideep Saraswat, Mainak Mukherjee, Nikhil Mall, and Surajit Mondal

Abstract Construction is the living evidence of growth in a country. Construction may be of commercial/residential buildings, roads, bridges, etc. The incessant need for construction is justified by the periodic deterioration of all constructed outcomes. The construction of buildings causes a considerable amount of emissions in the atmosphere. 'Green Building' is a measure to reduce emissions from buildings during all the phases by employing environmentally friendly and energy-efficient techniques. GRIHA stands for Green Rating for Integrated Building Assessment. This paper underlines the prerequisites for GRIHA certification and the process to be followed to get one. Moreover, the post-certification period has seen challenges that undermine the very purpose of the certification. The problem lies in the negligence of the building authorities once it is ratified as a green building. Certain measures and precautions are ambiguous which are overlooked afterward. A detailed study of a GRIHA certified building helped to identify these problems and paved the way to come up with a solution to address these issues. Green Building as a concept has gained prominence in the last decade. Both developed and developing nations are trying to reap its benefits. In some countries, the new construction sector is rising in this space while in some the retrofit projects are gaining importance. The current global trends in green building are also exhibited in this paper.

Keywords Green building · Global scenario · GRIHA · Audit instruments · Energy efficiency

J. Saraswat · M. Mukherjee · N. Mall · S. Mondal
Department of Electrical and Electronics Engineering, University of Petroleum and Energy Studies UPES, Bidholi Campus, Dehradun, Uttarakhand 248007, India

S. Mondal (✉)
Department of Research and Development, University of Petroleum and Energy Studies UPES, Bidholi Campus, Dehradun, Uttarakhand 248007, India

N. A. Siddiqui et al. (eds.), *Advances in Environment Engineering and Management*, Springer Proceedings in Earth and Environmental Sciences,
https://doi.org/10.1007/978-3-030-79065-3_28

1 Introduction

According to a study in the United States of America, buildings contribute to about 39% of CO_2 emissions in totality [1]. It encompasses phases like construction, maintenance, renovation, and demolition. This CO_2 emission will ultimately lead to an increase in the greenhouse effect and hence global warming. The temperature of the earth has increased by 0.8 °C since 1880 [2]. It will increment in the future as well if enough precautionary measures are not undertaken at the very earliest.

Green building or sustainable design, is a process leading to a decrease in emissions and increase in the efficiency of the building. This in turn benefits human health and the entire lifecycle of the building. A Green Building has 34% less CO_2 emissions, 25% less energy consumption than the standard commercial building [3]. The growth and development of the building have a sustainable impact on our natural environment. Green Building strengthens the notion of minimum waste disposal by reducing the generation of waste, recycling generated waste to the maximum possible extent and reusing the recycled waste. Though some of the waste generated might not be recycled or reused, its disposal is closely monitored to ensure it doesn't affect our environment. The collaboration of local bodies like municipal corporations, health departments, and building authorities can make this effort go a long way.

GRIHA or Green Rating for Integrated Habitat Assessment is a council for ratification of buildings to be certified as 'Green Buildings'. It is a credit-based system that gives points based on 34 criteria out of which 8 are mandatory. The criteria are based on efficient usage of natural resources available at the site like land, water and ensures a reduced impact on the environment. GRIHA has a unique strategy to award points for outcomes as well as for the implemented strategies to realize those outcomes. A point window is created as shown in Fig. 1 for granting star rating. Eligibility criteria for GRIHA certification are minimum of 50 points and 2500 m^2 area [4].

Fig. 1 GRIHA ratings [4]

2 Methodology Followed for the Process of Getting a Building Ratified

Steps	Process in brief
1. Registration	• Online registration of the project in the GRIHA council • Registration to be done at the commencement of the project
2. Documentation	• The relevant documents related to the attempted criteria are reviewed • Queries on the rating are sent to info@grihaindia.org
3. Evaluation	• This includes a three-tier process; the online submitted documents are evaluated by a team of experts in the GRIHA secretariat • Evaluated scheme is sent to renewed sectors and reviewed independently to award points • Report consists of comment on each individual criterion that was followed and an overall comment on the complete project • The final report is awarded to the Advisory Committee for the approval and award of a rating • Validity of the rating is for 5 years from the commissioning of the project instead of the day of award of a rating

Audits mean to conduct an official inspection of the project based on various tenets depending upon the requirement. The audit is conducted based on energy, finance, accounts, manufacturing process, etc. The main objective of the audit lies in the fact that it is done to improve the efficiency of the project and should not be misconstrued as a fault finding opportunity.

The methodology adopted in this audit is a comparative approach. The method of direct observation has been used to analyze the area of study and evaluative information has been reported. The information gathered is compared with the required baseline data.

Equipment Used in Auditing

Hygrometer—used to analyze humidity in the room.

Lux meter—used to measure lumens in an area.

Voltmeter—used to measure voltage.

Clamp ammeter—used to measure current.

Power factor meter—used to measure power factor.

DB meter—used to measure noise levels.

The observations after the survey of a certified green building are mentioned in Table 1.

3 A Case Study of GRIHA Certified Educational Institution

S. No.	GRIHA criteria	Followed/not followed	Reason	Baseline requirement
1	Site selection	Not followed	Already constructed and deforestation has taken place	1. Less transportation 2. Less deforestation 3. Top soil conservation
2	Preserve and protect landscape during construction	Followed	Topography of the place retained	Minimum damage to landscape
3	Soil conservation	Followed	Garden and non-hard paved area consist of all top soil	Top soil preserved during construction for future use
4	Design to include existing site features	Partially followed	Top soil conservation using efficient farming	Top soil conservation using terrace farming
5	Reduce hard paving on site and/or provide shaded hard paved surface	Followed	45% of area is hard paved	50% or more non hard paved
6	Enhance outdoor lighting system efficiency and use renewable energy	Partially followed	1. Timers for on/off at certain places 2. No renewable based outdoor lighting	1. Solar and wind energy equipment installed on site 2. Automatic on/off
7	Plan utilities efficiently and optimize on-site circulation efficiency	Partially followed	Consolidated corridor not present everywhere	1. Reduce transportation corridor out-side 2. Utility corridor to be collaborated
8	Provide minimum level of sanitation/safety facilities for construction workers	Partially followed	Didn't pay need to safety facility	1. Safety facilities like helmet, shoes, etc. 2. Proper sanitation facility 3. Portable drinking water
9	Reduce air pollution during construction	Not followed	1. Uncovered stockpiles 2. Smoking carried out on-site	1. Fine mesh sieve at dust source 2. Fine water sprays to damper 3. Covering stock piles

(continued)

(continued)

S. No.	GRIHA criteria	Followed/not followed	Reason	Baseline requirement
10	Reduce landscape water requirement	Followed	1. Drip and sprinklers used 2. Sewage treatment plant providing water for irrigation	1. Drip and sprinkler based irrigation 2. Water recycling plant
11	Reduce the water used by the building	Partially followed	Flow rate for all taps not as prescribed	Low flow rate faucets such that
12	Efficient water use during construction	Followed	1. Dry sweep 2. Waste water utilisation	1. Dry sweep 2. Not allowing dust particles to enter sewage line 3. Waste water for cement mixing
13	Optimize building design to reduce conventional energy demand	Not followed	1. Improve ventilation 2. Excessive load on artificial lighting	1. Less use of artificial lighting such that Led, CFL 2. Proper ventilation
14	Optimize energy performance of the building within specified comfort limits	Not followed	1. Classroom-138 lx 2 Library-400 lx 3. Corridor-6.2 lx	1. Classroom-250 lx 2. Library-500 lx 3. Workshops-750 lx
15	Utilisation of fly ash in building structures	Followed	Fly ash bricks used	1. PPC 2. Fly ash bricks 3. Mosaic tiles
16	Reduce volume and weight, and time of construction by efficient technologies	Not followed	1. Modularisation was not considered 2. Prefabrication was not considered	1. Modularisation, prefabrication 2. Offsite fabrication 3. Use of natural materials
17	Use of low energy materials in interiors	Not followed	High embodied energy material used	1. Recycled glass, aluminium etc 2. Reclaimed lumber, cork, bamboo
18	Renewable energy utilisation	Followed	Energy consumption reduction 42.73%	1. Solar photo-voltaic system 2. Solar water heater
19	Renewable energy based hot water system	Followed	95.17% reduction in energy consumption for hot water system, 70,000 l	Solar hot water system

(continued)

(continued)

S. No.	GRIHA criteria	Followed/not followed	Reason	Baseline requirement
20	Waste water treatment	Followed	1. Sewage treatment plant 2. 250 klpd capacity	1. Sewage treatment plant 2. Grey/black water treatment plant
21	Water recycle and reuse (including rain water)	Followed	1. Sewage treatment plant 2. 250 klpd irrigation	1. Recycle and reuse of water from building complexes 2. Rain water harvesting
22	Reduction in waste during construction	Followed	1. Reusing glass from earlier project 2. Collaboration with suppliers	1. Procurement of materials that much is required 2. Collaboration with suppliers for return
23	Efficient waste segregation	Followed	1. Green bin for biodegradable 2. Yellow for non-biodegradable	Different colour bins
24	Storage and disposal of waste	Followed	1. Biodegradable and toxic waste goes to Bharat oil 2. Rest waste to municipal corporation	1. Reduction, reuse and recycling 2. Composting and fermentation 3. Landfills -incineration
25	Resource recovery from waste	Not followed	No recovery of energy	Biodegradable waste in biogas plant
26	Use of low VOC paints/adhesives/sealants	Not followed	VOC containing paints used	1. Milk white paint 2. Benjamin Moore Natural/Aura
27	Minimize ozone depleting substance	Followed	HCFC, CFC free refrigerants and air conditioning	1. Reduction in use of HCFC, CFC for air conditioners 2. Less release halons form fire extinguishers

(continued)

(continued)

S. No.	GRIHA criteria	Followed/not followed	Reason	Baseline requirement
28	Ensure water quality	Followed	1. Portable drinking water for occupants 2. Effectively treated water for bathrooms, kitchens, basins 3. Test performed regularly at NABL laboratory	Properly treated water as per requirement
29	Acceptable outdoor and indoor noise levels	Followed	Outdoor: 64.4 db Indoor: 53 db	Residential: 1. Indoor-45 db 2. Outdoor-55 db Commercial: 1. Indoor-55 db 2. Outdoor-65 db
30	Tobacco smoke control	Followed	1. Smoke free area 2. No designated smoking zone	1. Smoke free area 2. Designated smoking zone
31	Provide at least minimum level of accessibility for persons with disabilities	Partially followed	1. No appropriate toilets 2. No elevator in all buildings 3. No reserve parking spots	1. Slope apart from stairs for passage 2. Elevators to approach all rooms 3. Appropriate toilets and reserve parking
32	Energy audit and validation	Followed	Execution of audit	Validation and execution of audit
33	Operation and maintenance	Followed	1. Horticulture maintenance 2. Day to day, yearly maintenance	1. Day to day and yearly maintenance 2. Horticulture and landscape maintenance 3. Pre and post monsoon maintenance
34	Innovative ideas	Followed	1. 100 KW solar pv system greater than prescribed value 2. Repetitive seminars to educate people	Utilisation of renewable energy

Table 1 Green building activity in different sectors globally [5]

S. No.	Sectors	2008 (%)	2012 (%)
1	New commercial construction	48	63
2	Existing/building retrofits	45	50
3	New institutional construction	N/A	45
4	Communities	34	30
5	New high rise residential	24	29
6	Commercial interiors	32	25
7	New low rise residential	28	20

4 Green Buildings Globally

Countries are required to reduce their GHG emissions to meet the targets set under the Paris Accord. To meet their targets many countries have made their green building regulating and authoritative bodies and have started undertaking green building projects. It is clear from Table 1 that the green building activity has increased in recent times.

South Africa, Singapore, India, Mexico etc. are having green building activities above 24% which is the global average [6]. Green Building projects are likely to grow more in developing countries like India, Brazil than in developed countries like U.S.A., U.K. etc. New commercial construction sector is ranked higher for green building projects in Brazil, Mexico etc., while retrofit projects tops in U.K., South Africa etc.

5 Conclusion

The post-certification period of a green building sees non-compliance on many criteria. This non-compliance can be because of the following reasons:

- Economic aspects—initial investment not attractive for a system
- Flaws in initial planning and design
- Overlooking safety of construction workers
- Not reaping benefits from various campaigns initiated by the government
- Lack of technological advancement

Building authorities must ensure compliance with all criteria even in post-certification period to meet the very need for which it was conducted.

Myriad merits of Green Buildings have made them a widely accepted concept across the globe. Most nations are escalating their efforts towards green buildings in order to achieve their climate targets.

Acknowledgements The authors would like to thank the Department of Electrical Engineering at UPES, for the coursework on 'Green building' and consequent case studies.

References

1. Buildings and Climate Change. Available at: http://www.eesi.org/files/climate.pdf
2. National Aeronautics and Space Administration, Global Temperatures. Available at: http://earthobservatory.nasa.gov/Features/WorldOfChange/decadaltemp.php
3. U.S. Green Building Council, Benefits of Green Building. Available at: http://www.usgbc.org/articles/green-building-facts
4. Project Registration. Available at: http://www.grihaindia.org
5. Cherry Bekaert LLP, Real Estate and Construction. Available at: http://www.cbh.com/study-on-global-growth-trends-in-green-building/
6. World Green Building Trends 2016: More nations, more projects. Available at: http://www.usgbc.org/articles/world-green-building-trends-2016-more-nations-more-projects

Dependence of Physiochemical Features on Marine Chlorophyll Analysis with Learning Techniques

Subhrangshu Adhikary, Sudhir Kumar Chaturvedi, Saikat Banerjee, and Sourav Basu

Abstract Marine chlorophyll which is present within phytoplankton are the basis of photosynthesis and they have a high significance in sustaining ecological balance as they highly contribute toward global primary productivity and comes under the food chain of many marine organisms. Imbalance in the concentrations of phytoplankton can disrupt the ecological balance. The growth of phytoplankton depends upon the optimum concentrations of physiochemical constituents like iron, nitrates, phosphates, pH level, salinity, etc. and deviations from an ideal concentration can affect the growth of phytoplankton which can ultimately disrupt the ecosystem at a large scale. Thus the analysis of such constituents has high significance to estimate the probable growth of marine phytoplankton. The advancements of remote sensing technologies have improved the scope to remotely study the physiochemical constituents on a global scale. The machine learning techniques have made it possible to predict the marine chlorophyll levels based on physiochemical properties and deep learning helped to do the same but in a more advanced manner simulating the working principle of a human brain. In this study, we have used machine learning and deep learning for the Bay of Bengal to establish a regression model of chlorophyll levels based on physiochemical features and discussed its reliability and performance for different regression models. This could help to estimate the amount of chlorophyll present in water bodies based on physiochemical features so we can plan early in

S. Adhikary
Department of Computer Science and Engineering, Dr. B. C. Roy Engineering College, Durgapur, West Bengal 713206, India
e-mail: subhrangshu.adhikary@spiraldevs.com

S. K. Chaturvedi (✉)
Department of Aerospace Engineering, UPES, Dehradun 248007, India
e-mail: sudhir.chaturvedi@ddn.upes.ac.in

S. Banerjee
Department of Mechanical Engineering, Cubicx, Kolkata, West Bengal 700070, India
e-mail: saikatbanerjee@cubicxindia.com

S. Basu
Department of Electrical Engineering, Cubicx, Kolkata, West Bengal 700070, India
e-mail: souravbasu@cubicxindia.com

N. A. Siddiqui et al. (eds.), *Advances in Environment Engineering and Management*, Springer Proceedings in Earth and Environmental Sciences,
https://doi.org/10.1007/978-3-030-79065-3_29

case there arises a possibility of disruption in the ecosystem due to imbalance in marine phytoplankton.

Keywords Chlorophyll · Remote sensing · Machine learning · Deep learning · Classification

1 Introduction

Phytoplanktons are marine organisms coexisting with other marine organisms and together they establish a balance of the ecosystem as they become part of the food chain for many other sea creatures as well as they are responsible for primary productivity and contributes significantly toward global oxygen production. Phytoplanktons are autotrophs which means they can prepare their food by using organic sources like carbon dioxide, water, etc. with the help of sunlight. The photosynthesis undergone by phytoplankton means that they have a pigment in their cell called chlorophyll. A chlorophyll can reflect near-infrared light bands by which satellites can remotely detect them [1]. Therefore remote sensing could be useful to remotely monitor phytoplankton blooming. Remote sensing could also be used to detect many other physiochemical features. EU Copernicus Program has used these remote sensing abilities to create a publicly available reanalysis data which we have used in this experiment. The basic remote sensing data are used in different combination to create reanalysis data which contains features like primary productivity, pH levels, salinity, phosphates, nitrates, etc.

The amount of phytoplankton in the ocean directly or indirectly depends upon the concentration of many physiochemical features and phytoplanktons require a balanced amount of these features for optimum growth. Deviation of these features could deplete the phytoplankton levels. Therefore reanalysis data of remote sensing can be useful to detect volumes of both chlorophyll and other physiochemical features, this leaves an immense opportunity to remotely monitor them. On the other hand, due to the advancement of machine learning and deep learning technologies, we can monitor the chlorophyll levels automatically utilizing remote sensing.

In this experiment, we have tried to establish a relationship between chlorophyll with physiochemical features utilizing different machine learning and deep learning techniques to detect chlorophyll levels based on these features remotely and automatically. We have chosen the Bay of Bengal as our study location as it holds lots of ecological importance because of the Sundarbans Delta, which is home for many rare animals and sea creatures that co-exist. The imbalance of the ecosystem for either of land or sea could affect one another directly or indirectly. The coordinates of the study location are from 5° N to 15° N and 82° E to 92° E (Fig. 1).

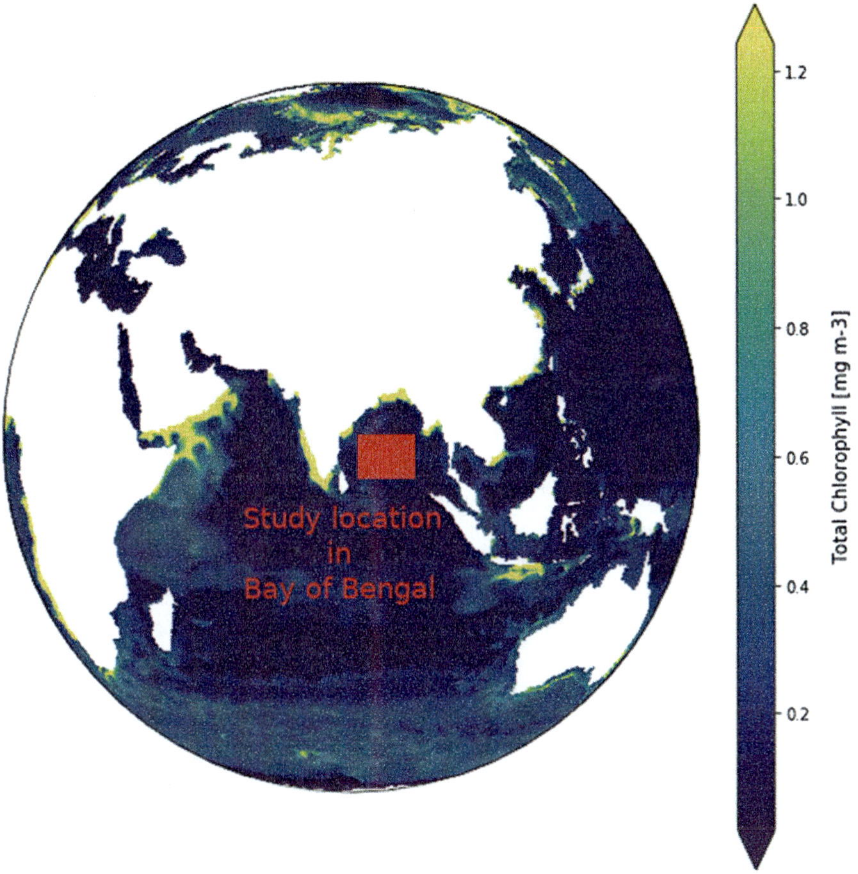

Fig. 1 A map of chlorophyll distribution at depth 0.5 m of sea-level indicating our study location in the Bay of Bengal [2]

2 Recent Advancements in Marine Chlorophyll Study

Here in this section, we will discuss the advancements in the field of chlorophyll study, remote sensing, machine learning and deep learning studies and will discuss why we have chosen this study area.

Chlorophylls are present within phytoplankton which makes it capable of performing photosynthesis. Chlorophyll present within phytoplanktons could be detected by phaeophytin by fluorescence method [2]. Earlier, *N*,*N*-dimethylformamide has been introduced as an upgraded method for phytoplankton chlorophyll detection [3]. The phytoplankton distribution in the southern ocean was studied to understand the primary production capacity of marine phytoplanktons [4]. A model was prepared in [4] for the carbon ratio and the conversion factor between productivity and growth rate of phytoplankton. The thermal structure of the ocean and its consequences of non-uniformity in chlorophyll profile was discussed in [5]. The nutrient requirements and importance of nitrogen requirements as well as its storage in macroalgae was established [6].

Earlier, an early-warning protocol was developed for prediction of chlorophyll-a concentrations with machine learning for the estuarine reservoirs, Korea with a 7-year monitoring data [7]. Geostationary Ocean Color Imager satellite data was earlier used to monitor coastal water quality with help of machine learning technique and obtained low accuracy of 1.74 mg m^{-3} root mean squared error [8]. The application of machine-learning techniques was used to create a consistent and calibrated global chlorophyll concentration baseline dataset by utilizing remote sensing data for ocean colour [9].

Random Forest Classifier was used for classification of remote sensing data obtained from Landsat Enhanced Thematic Mapper Plus (ETM+) and its performance was compared with support vector machine classifier [10]. The same classifier along with Decision Tree Classifier was used for assessment of chlorophyll sufficiency levels of mature palm oil utilizing hyperspectral remote sensing technology [11]. Decision tree was also used to classify transparent plastic-mulched landcover utilizing Landsat-5 TM images [12]. Artificial neural network-based approach with the help of multi-layer perceptron to predict leaf chlorophyll content by analyzing cotton plant images [13]. MLP was also proven to be effective for estimation of chlorophyll concentration index of lettuce and obtained coefficient of determination 0.98 and a mean squared error 0.006 during the validation process. Convolution neural network was used for the estimation of seawater chlorophyll-a by analyzing hyperspectral images [14]. Image-based canopy reflectance model was developed for remote estimation of one-sided leaf area and leaf chlorophyll content by utilizing CNN technologies [15]. Deep neural network with the help of CNN was also used for chlorophyll-a concentration model with an extremely imbalanced dataset and showed that proper pre-processing techniques and oversampling can improve prediction accuracy in these cases [15]. Deep learning was used for investigation of chlorophyll-a and total suspended matter concentration in Taihu Lake, China utilizing Landsat ETM and field spectral measurements [16].

3 Methodology

Here in this section, we have talked about the process of data collection, preprocessing and classification of the data in details.

3.1 Data Availability and Preprocessing

EU Marine Copernicus Program is a widely accepted platform for accessing publicly available remote sensing data. They have used basic satellite remote sensing data

Table 1 The summary of the dataset

Feature	Min	Max	Mean	Standard deviation	Unit
spco2	28.3267	41.4043	35.877050	1.647791	Pa
O_2	194.291	219.503	202.106733	2.703826	mmol m^{-3}
Chl	0.0726752	1.31307	0.120117	0.058221	mg m^{-3}
NO_3	0.00155384	7.94053	10.000000	0.419270	mmol m^{-3}
PO_4	4.00281e−05	0.414416	0.022617	0.031002	mmol m^{-3}
si	0.720682	11.9874	3.283496	1.289600	mmol m^{-3}
pH	8.01069	8.13985	8.059485	0.016567	–
nppv	1.21979	82.9279	4.686437	4.723566	mg m^{-3} day^{-1}
Fe	3.91981e−05	0.00294045	0.001006	0.000363	mmol m^{-3}
mlotst	10.5289	66.2252	20.445362	7.716653	m
thetao	25.4276	31.463	28.749488	0.846414	°C
bottomT	0.654163	20.5935	1.620214	1.627649	°C
vo	−1.5656	1.20792	0.007621	0.165873	ms^{-1}
uo	−0.913724	1.37638	−0.000650	0.218761	ms^{-1}
so	28.0969	35.4183	33.248507	0.702603	10^{-3}
zos	0.146489	0.979339	0.505828	0.088441	m

to create a reanalysis data which indicates the concentrations of different physiochemical items like surface carbon dioxide pressure, dissolved oxygen, chlorophyll, dissolved silicates, pH levels, total primary productivity, etc. at a depth of 0.5 m. We have downloaded these data for the year 1993–2018 in NetCDF format. This data contains the historical data in the form of time series for each location coordinates at 4 km pixel resolution. Now, we have converted this file to a time series statistical table for each location totalling up to over thirty-seven thousand rows to train on. To indicate low levels of chlorophyll, we have marked 1/4th of the dataset having the lowest amount of chlorophyll as "Critical" and marked remaining as "Normal". Later we have randomly split up the table into training and testing set containing 75% and 25% data respectively. The summary of the time series data is mentioned in Table 1 and a portion of the time series is represented by Fig. 2.

3.2 *Classifiers*

Random Forest Classifier (RFC)

Random Forest Classifiers are powerful ensemble-based classifier which creates several trees to make a decision with each and then the output of the classifier becomes the class which have received the majority of these trees' decisions. To decide the number of nodes of the trees to make a decision, the Gini Index or Entropy function is used. We have used the Gini Index method for this experiment to create branches

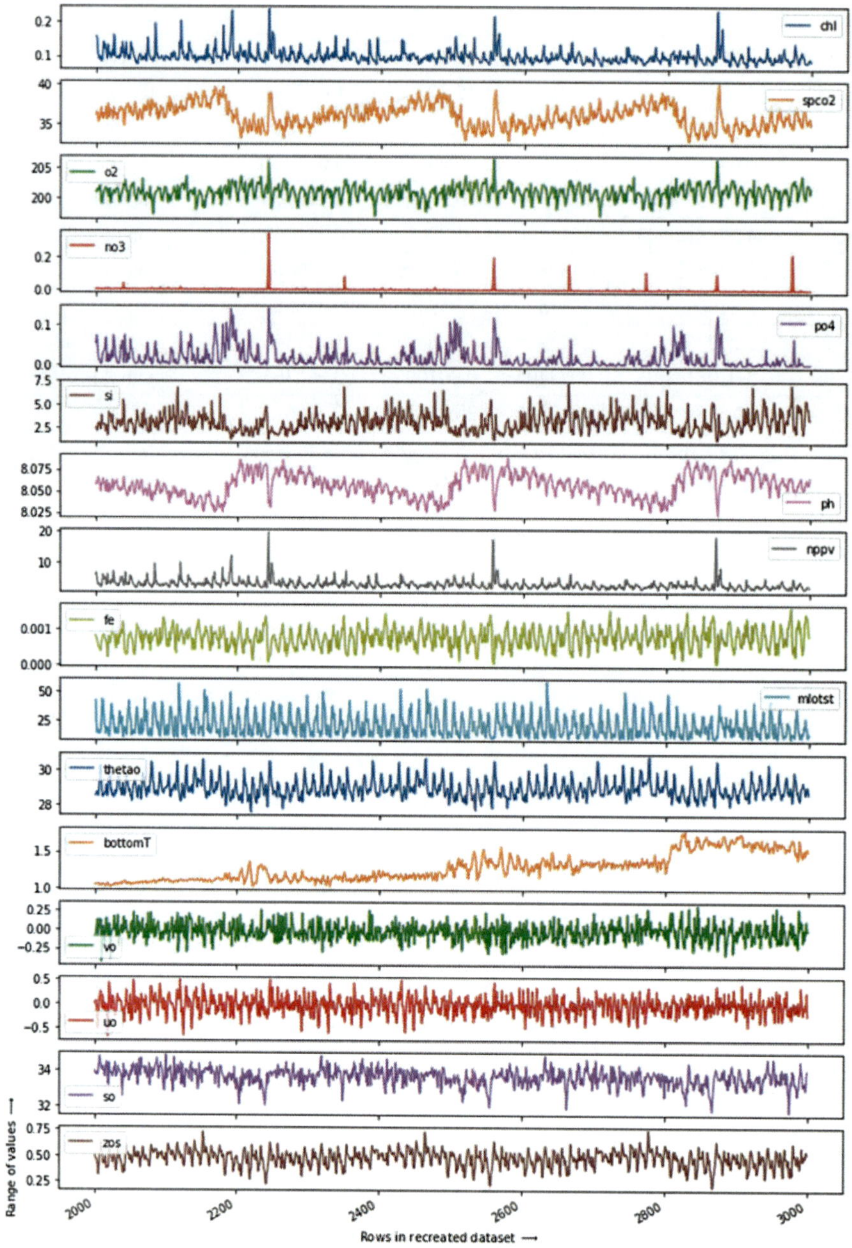

Fig. 2 A graph of all the features from a small portion of the entire dataset as a sample

of the trees. As RFC classifier works by making several trees and reach a decision based on their output, it becomes very slow compared to other classifiers for the purpose but its decisions become very accurate and stable.

Decision Tree Classifier (DTC)
A decision tree is a supervised machine learning algorithm based on ensemble classification method. The features of the dataset are represented as nodes of a DTC and the branches of the tree represent a decision that follows and leaf nodes of the tree represent outcomes. Similar to RFC, DTC also uses the Gini Index and Entropy function method to decide nodes of the tree. A decision tree learns from the pathway of the decision made during the training process, therefore DTC often seems to overfit the model, which means it attempts to fit as many training data as possible and if the values of the testing set are too much different from the training set, DTC might start showing erroneous results. Therefore for an ideal DTC, the training data should have a close match for a testing data and in such situation, there needs to have enough amount of training data so a model could properly recognize close matches while predicting outputs.

Multi Layer Perceptron (MLP)
A perceptron is an artificial neuron and works similar to how a natural neuron works. They get signal from one end, it is modified and reach the other end from where it is transmitted to other perceptrons. An artificial neuron gets an input, multiply with some weight, add some bias and pass the result to next artificial neuron. When this process is repeated over a range of neurons forming stacks of different layers to create a network, it is called a Multi Layer Perceptron and it is the simplest form of neural network. The weights and biases of each node are first initialized with random values and then the cost of the network is monitored to tune the weights and biases to minimize loss. The goal is to create a path in the network passing through which inputs will mimic the ideal output. Dropout functions are often used to introduce randomness in the network by randomly activating and deactivating different nodes in the network to make the model learn and not memorize. It is a very powerful technique which mimics the decision making process of a human brain.

Convolution Neural Network (CNN)
Convolution neural network is an extension of MLP which primarily specializes to deal with image data however they are very well suited for statistical classification as well. A CNN first extracts several features from the dataset like grey level cooccurrence matrix, different edge detection techniques, *k*-means clustering and several more. However all extracted data are not necessarily useful therefore a max-pooling concept is used which extracts the most promising features from the feature extraction matrix. This discards unnecessary features and only keeps important ones so the model does not consume too many resources in the process. After a convolution layer, MLP is added to the base of it to make it more suitable for classification, this includes adding several hidden dense layers, dropouts, and activation functions which modifies the output of a node or a layer based on given rules. Finally, the output of the CNN network is used to make the decision.

4 Results and Discussion

Here in this section, we will discuss the performance of our models in two stages. First, we will use all available features to train our model and second we will use only the top 4 features which influence classification the most and we will discuss them in details.

4.1 Classification with All Available Features

When used all the features available in our dataset, we have observed that RFC obtained the highest classification accuracy of 93.92% followed by CNN, DTC and MLP obtaining 92.62%, 91.52% and 90.78% classification accuracy respectively. Now, accuracy is a metric which is very useful when using a balanced dataset for training means both the classes in the classifier should have an equal number of items. It's like if the dataset has 90 items of one class and 10 items of other class and our model labels all of them as the first class, it will still show 90% accuracy while it cannot identify any of the items from other class. Our dataset is unbalanced as it's 1/4th data are labelled as "Critical" and other 3/4th data are labelled "Normal", therefore we need more metrics to decide how our model performed. Two such important metrics are precision and recall. Precision is the ability of the classifier not to detect a false-positive item and recall is the ability of the classifier not to detect a false-negative item in the dataset. Closer the score is to 1, better the model performs.

We can see that RFC has obtained a precision score of 0.923 which is quite impressive, and CNN received 0.916. However, both DTC and MLP have a lower precision score which indicates that they have labelled many items as false-positive. This indicates that RFC and CNN have better prediction than DTC and MLP in terms of precision. Now when we consider recall, we can see that RFC has the highest recall but DTC, MLP and CNN have a lower recall score indicating that they have labelled many items as a false negative. As far as accuracy, precision and recall are concerned, RFC stands out to be the best classifier for the situation.

Now when it comes about resource optimization, we can see that DTC was the fastest to train by training the model in 400.63 ms, followed by RFC, MLP and CNN with 7878.39 ms, 11,351.09 ms and 18,057.22 ms respectively. For generating output, we can see that DTC again becomes the fastest to predict results in 1.668 ms, followed by MLP, RFC and CNN which took 5.22 ms, 139.494 ms and 207.026 ms respectively. From this, we can see that DTC by far is the fastest model for both training and testing. However, given its low precision and recall score, we choose RFC as an alternative when we need fast training. But RFC fails to generate results fast, therefore looking at the testing time, MLP ranked second. However, accuracy, precision and recall of MLP are lower than DTC and therefore it cannot be fully reliable either. Therefore, going by the third in the ranking, although it performs slower than the other classifiers, we conclude RFC to be the best classifier based

Table 2 Classification performance while using all features

Classifier	Training time (ms)	Testing time (ms)	Accuracy (%)	Precision	Recall
Random forest	7878.39	139.494	93.92	0.923	0.917
Decision tree	400.63	1.668	91.52	0.884	0.89
Multi layer perceptron	11,351.09	5.22	90.78	0.882	0.875
Convolution neural network	18,057.22	207.026	92.62	0.916	0.894

on overall performance, and when speed is the ultimate goal, DTC could also be alternatively used by compromising around 2.4% accuracy. The performances are recorded in Table 2.

4.2 *Using Only the Best 4 Features to Increase Model Speed*

Using each of features available in our dataset to train the model, although increases overall accuracy, reduces the model building time and prediction time. Therefore to reduce this time, we can use fewer but more relevant features which although reduces training and testing time, it trades off some of the accuracies in the process. Our goal is to minimize this accuracy loss and maximize resource optimization. For this purpose, we have used the pair plot technique to find the most relevant features for our training. By pair plotting all the features against one another, we can observe that more separate the clusters of classes are, better it works as a classification feature. Figure 3 shows the pair plot for 4 features whose clustered classes were most distinguishable among the rest of the models. Therefore by using them, we can expect a reasonable classification performance.

The performances of the best four features for classification which are phosphate concentration, the concentration of dissolved silicate, total primary productivity per day and ocean temperature are recorded in Table 3. From this, we can see that using these four features, RFC has obtained the best accuracy of 91.51%, followed by CNN, MLP and DTC which obtained 90.12, 89.59 and 88.86% classification accuracy. We can see that accuracies of all the classifiers have slightly reduced. Looking at the precision scores, both RFC and CNN have equal precision scores of 0.89 followed by MLP and DTC which obtained 0.868 and 0.852 respectively. We can see a drop in precision for all the classifiers but DTC dropped the most which indicated that losing features, it now generating lots of false-positive errors. For recall, we can see that RFC has the highest recall score of 0.884 followed by CNN, MLP and DTC which have recall scores of 0.859, 0.857 and 0.849. We can see that RFC still have decent recall score but all remaining classifiers started to obtain many false-negative errors. Based on accuracy, precision and recall, even after using only 4 features, RFC still proves to be a good classifier for the given problem.

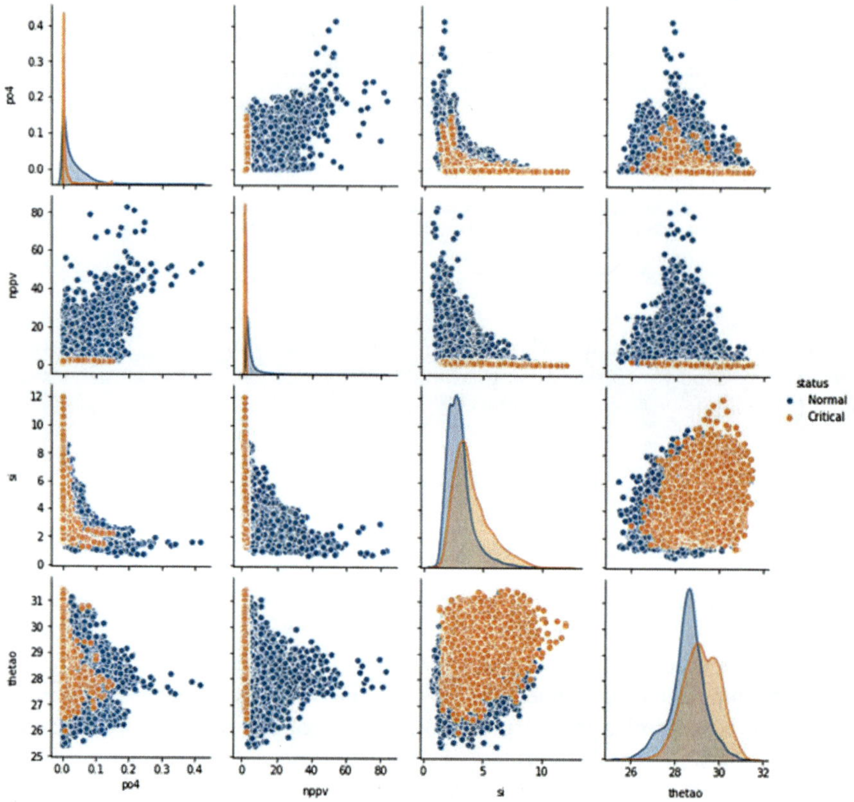

Fig. 3 A pair plot for top 4 features which have the highest influence in the classification

Table 3 Classification performance while using all features

	Training time (ms)	Testing time (ms)	Accuracy (%)	Precision	Recall
Random forest	5927.74	138.989	91.51	0.89	0.884
Decision tree	206.08	2.346	88.86	0.852	0.849
Multi layer perceptron	8514.07	5.186	89.59	0.868	0.857
Convolution neural network	16,262.27	200.318	90.12	0.89	0.859

Now for training time, there was a reduction in training time for all the classifiers. However, for testing time, we can see that although RFC, MLP and CNN have retained almost similar testing time, testing time for DTC has increased. For this classification, RFC performs the best among the remaining classifiers.

From these two observations, we can conclude that RFC proves to be the best classifier based on overall performance. But when used only 4 classifiers, RFC's accuracy was approximately equal to DTC built with all features. But in this situation, using DTC with all feature trains around 14 times faster than RFC with 4 features. Therefore, when accuracy is the ultimate goal, RFC could be safely used with all features to obtain an optimum classification accuracy but when resource consumption is concerned, DTC with all features is a much better choice for classification rather than using RFC with lesser features.

5 Conclusion

Phytoplanktons are an essential part of the ecosystem and their conservation has a high priority for the sake of maintaining an ecological balance as they are primary producer significantly contributing toward global oxygen production and also as they come under food chain of many marine creatures. They contain chlorophyll which conducts photosynthesis and chlorophylls are visible under near-infrared light bands. For this property, they can be monitored with remote sensing technologies. Remote sensing could also be reanalyzed to estimate many physiochemical features which affect phytoplankton growth.

In this experiment, we have used remote sensing data to establish a classifiable relationship between different physiochemical features to estimate chlorophyll levels based on these features. For our experiment we have constructed time-series data from EU Marine Copernicus Program data and used different classifiers like Random Forest, Decision Tree, etc. and analysed their performances.

From our experiment, we can observe that RFC performs the best for overall performance obtaining classification accuracy of 93.92%, however, its speed is slow. To overcome this, instead of reducing training features, using DTC could provide very fast and reliable performance. DTC obtained 2.4% lower accuracy but trains 19.66 times faster and produces results 83.62 times faster than RFC.

The model could be deployed as an inexpensive global monitoring solution of worldwide chlorophyll levels. This model could be upgraded by training with data from more locations and other features. The extension of the work could be to monitor chlorophyll levels at a much greater depth along with their relation with zooplanktons could be established with the help of remote sensing and learning techniques. Also, their relation with pollutants could be established. All of these together will help toward sustaining an ideal ecosystem.

6 Code Availability

The code of the project have been made available on GitHub Repository under MIT License of open source distribution of programs and the codes can be freely reused for commercial and non-commercial purpose [17].

Table 4 Abbreviations

Key	Full-form	Key	Full-form
DTC	Decision Tree Classifier	nppv	Total Primary Production of Phytoplankton
CNN	Convolution Neural Network	Fe	Dissolved Iron
RFC	Random Forest Classifier	mlotst	Density Ocean Mixed Layer Thickness
MLP	Multi Layer Perceptron	thetao	Temperature
spco2	Surface CO_2 Pressure	bottomT	Sea Floor Potential Temperature
O_2	Dissolved Oxygen	vo	Northward Velocity
Chl	Total Chlorophyll	uo	Eastward velocity
NO_3	Dissolved Nitrate	so	Salinity
PO_4	Dissolved Phosphate	zos	Sea Surface Height
si	Dissolved Silicate	pH	pH

7 Abbreviations

We have used the following abbreviations throughout the paper (Table 4)

References

1. Datt, Bisun. "A new reflectance index for remote sensing of chlorophyll content in higher plants: tests using Eucalyptus leaves." Journal of Plant Physiology 154, no. 1 (1999): 30–36.
2. Yentsch, Charles S., and David W. Menzel. "A method for the determination of phytoplankton chlorophyll and phaeophytin by fluorescence." In *Deep Sea Research and Oceanographic Abstracts*, vol. 10, no. 3, pp. 221–231. Elsevier, 1963.
3. Suzuki, Reiko, and Takashi Ishimaru. "An improved method for the determination of phytoplankton chlorophyll using N, N-dimethylformamide." *Journal of the Oceanographical Society of Japan* 46, no. 4 (1990): 190–194.
4. Moore, J. Keith, and Mark R. Abbott. "Phytoplankton chlorophyll distributions and primary production in the Southern Ocean." *Journal of Geophysical Research: Oceans* 105, no. C12 (2000): 28709–28722.
5. Lewis, Marlon R., John J. Cullen, and Trevor Platt. "Phytoplankton and thermal structure in the upper ocean: consequences of nonuniformity in chlorophyll profile." *Journal of Geophysical Research: Oceans* 88, no. C4 (1983): 2565–2570.
6. Pedersen, Morten Foldager, and Jens Borum. "Nutrient control of algal growth in estuarine waters. Nutrient limitation and the importance of nitrogen requirements and nitrogen storage among phytoplankton and species of macroalgae." *Marine Ecology progress series* 142 (1996): 261–272.
7. Park, Yongeun, Kyung Hwa Cho, Jihwan Park, Sung Min Cha, and Joon Ha Kim. "Development of early-warning protocol for predicting chlorophyll-a concentration using machine learning models in freshwater and estuarine reservoirs, Korea." *Science of the Total Environment* 502 (2015): 31–41.

8. Kim, Yong Hoon, Jungho Im, Ho Kyung Ha, Jong-Kuk Choi, and Sunghyun Ha. "Machine learning approaches to coastal water quality monitoring using GOCI satellite data." *GIScience & Remote Sensing* 51, no. 2 (2014): 158–174.
9. Kwiatkowska, Ewa J., and Giulietta S. Fargion. "Application of machine-learning techniques toward the creation of a consistent and calibrated global chlorophyll concentration baseline dataset using remotely sensed ocean color data." *IEEE Transactions on Geoscience and Remote Sensing* 41, no. 12 (2003): 2844–2860.
10. Pal, Mahesh. "Random forest classifier for remote sensing classification." *International journal of remote sensing* 26, no. 1 (2005): 217–222.
11. Amirruddin, Amiratul Diyana, Farrah Melissa Muharam, Mohd Hasmadi Ismail, Mohd Firdaus Ismail, Ngai Paing Tan, and Daljit Singh Karam. "Hyperspectral remote sensing for assessment of chlorophyll sufficiency levels in mature oil palm (Elaeis guineensis) based on frond numbers: Analysis of decision tree and random forest." *Computers and Electronics in Agriculture* 169 (2020): 105221.
12. Lu, Lizhen, Liping Di, and Yanmei Ye. "A decision-tree classifier for extracting transparent plastic-mulched landcover from Landsat-5 TM images." *IEEE Journal of Selected Topics in Applied Earth Observations and Remote Sensing* 7, no. 11 (2014): 4548–4558.
13. Suo, Xing-mei, Ying-Tao Jiang, Y. A. N. G. Mei, Shao-kun Li, Ke-rum WANG, and Chong-tao Wang. "Artificial neural network to predict leaf population chlorophyll content from cotton plant images." *Agricultural Sciences in China* 9, no. 1 (2010): 38–45.
14. Awad, Mohamad. "Sea water chlorophyll-a estimation using hyperspectral images and supervised artificial neural network." *Ecological informatics* 24 (2014): 60–68.
15. Houborg, Rasmus, Martha Anderson, and Craig Daughtry. "Utility of an image-based canopy reflectance modeling tool for remote estimation of LAI and leaf chlorophyll content at the field scale." *Remote Sensing of Environment* 113, no. 1 (2009): 259–274.
16. Ma, Ronghua, and Jinfang Dai. "Investigation of chlorophyll-a and total suspended matter concentrations using Landsat ETM and field spectral measurement in Taihu Lake, China." *International Journal of Remote Sensing* 26, no. 13 (2005): 2779–2795.
17. S. Adhikary, S. Banerjee, S. K. Banerjee, S. Basu. "Chlorophyll and Physiochemical Features Relationship with Machine Learning, Deep Learning and Remote Sensing", *GitHub*, https://doi.org/10.5281/zenodo.3951632.

COVID-19 Spreading Prediction and Impact Analysis by Using Artificial Intelligence for Sustainable Global Health Assessment

Subhrangshu Adhikary, Sonam Chaturvedi, Sudhir Kumar Chaturvedi, and Saikat Banerjee

Abstract The COVID-19 pandemic is considered as the most alarming global health calamity of this century. COVID-19 has been confirmed to be mutated from coronavirus family. As stated by the records of The World Health Organization (WHO at April 18 2020), the present epidemic of COVID-19, has influenced more than 2,164,111 persons and killed more than 146,198 folks in over 200 countries across the globe and billions had confronted impacts in lifestyle because of this virus outbreak. The ongoing overall outbreak of the COVID-19 opened up new difficulties to the research sectors. Artificial intelligence (AI) driven strategies can be valuable to predict the parameters, hazards, and impacts of such an epidemic in a cost-efficient manner. The fundamental difficulties of AI in this situation is the limited availability of information and the uncertain nature of the disease. Here in this article, we have tried to integrate AI to predict the infection outbreak and along with this, we have also tried to test whether AI with help deep learning can recognize COVID-19 infected chest X-Rays or not. The global outbreak of the virus posed enormous economic, ecological and societal challenges into the human population and with help of this paper, we have tried to give a message that AI can help us to identify certain features of the disease outbreak that could prove to be essential to protect the humanity from this deadly disease.

S. Adhikary
Department of Computer Science and Engineering, Dr. B. C. Roy Engineering College, Durgapur, West Bengal 713206, India
e-mail: subhrangshu.adhikary@spiraldevs.com

S. Chaturvedi
Department of Health Safety Environment and Civil Engineering, University of Petroleum and Energy Studies (UPES), Bidholi Energy Acres, Dehradun 248007, India

S. K. Chaturvedi (✉)
Department of Aerospace Engineering, UPES, Dehradun 248007, India
e-mail: sudhir.chaturvedi@ddn.upes.ac.in

S. Banerjee
Department of Mechanical Engineering, Cubicx, Kolkata, West Bengal 700070, India
e-mail: saikatbanerjee@cubicxindia.com

N. A. Siddiqui et al. (eds.), *Advances in Environment Engineering and Management*, Springer Proceedings in Earth and Environmental Sciences,
https://doi.org/10.1007/978-3-030-79065-3_30

Keywords COVID-19 · Deep learning · Prediction · Pandemic · Health assessment · Sustainable · Global health assessment

1 Introduction

The widespread outbreak of COVID-19 from the city of Wuhan, China has spread all over the world has affected humanity drastically. Millions of people have been infected by the virus and over five hundred thousands of people have already died because of the virus. The remaining people all over the world is at risk of contamination. COVID-19 is a single-stranded RNA virus from the coronavirus family. It has spiked glycoproteins around its body surface which attach to receptor cells of the host and infects primarily lungs causing auto-immune response around alveoli which in turn swells the surrounding area and lungs filled up with fluids. This makes the gaseous exchange of the lungs difficult and severely infected patients require a life support system to continue a sufficient oxygen supply. There is no specific cure available to fight against the virus and only generic treatments are being used to reduce the damage caused by the infection. Researchers are working hard to discover a vaccine or antiviral treatment to fight against the disease. Until then, prevention of the disease from spreading is the best strategy to minimize the contamination.

Artificial Intelligence (AI) can be a very useful tool to control the pandemic situation. It could be used in many situations like aiding researchers to simulate the vaccination process, aid doctors to monitor patients and alert doctors if a patient requires special care so doctors can attend as many patients as possible, aid shopping malls and markets to monitor the crowd and enforce social distancing measures, many more. Usage of AI in all these segments could help to optimize the pandemic control capacity in an inexpensive and fast deployable approach. As AI is a software program which can be transmitted quickly and can work with a wide variety of infrastructure setup available worldwide, it can be deployed rapidly. However, the uncertain nature of the pandemic and limited availability of quality data to train on is the biggest hurdle for AI. However, implementations of advanced techniques like usage of generative algorithm and transferring of pre-trained weights for deep learning can be very useful to train and predict outcomes with high accuracy even with a very limited dataset. In this paper, we have demonstrated the use of AI in two cases.

First, we have used generative machine learning algorithm with Bayesian Ridge regressor technique which works on a probabilistic approach to reconstruct a wide range of data with given inputs and Gaussian Ridge regressor technique which is an extension of the earlier. By these, we have built a machine learning model which can predict the upcoming future trends of the virus based on currently available data so that early measures could be taken to prevent as much contamination as possible. The model trains on two sets, to estimate the amount of confirmed COVID-19 cases and to estimate the number of death cases going to take place in the near future. The model is described in Sect. 3.2 and its performances are discussed in Sect. 4.1.

Second, we have used deep learning on chest X-Ray images to classify COVID-19 infected patients from normal healthy lungs as well as lungs infected by other lung diseases like Streptococcus pneumonia infection, Acute Respiratory Distress Syndrome (ARDS), etc. COVID-19 causes swelling in the lungs which are spotted in X-Rays. Normal lungs would not have such swelling and should be easily classified by the model. However, other lung infections as stated above also form swelling and fluid build-ups in their unique pattern. The model that we have built, has analysed all these features and successfully classified the two classes. The model is described in Sect. 3.3 and its performances are discussed in Sect. 4.2.

The paper is arranged in the following manner. The recent advancements in studies related to COVID-19 and AI are listed in Sect. 2, Sect. 3 contains the availability of the dataset and its preprocessing, procedure of making our models, Sect. 4 contains the results observed from our experiment and discussed their performances and Sect. 5 consist of the conclusions we have drawn from the experiment.

2 Background Works

Here in this section, we have discussed the origin, evolution and biological background of COVID-19 along with usages of AI for controlling the pandemic, specifically the usage of machine learning and deep learning for the same purpose.

2.1 *Background of COVID-19 Pandemic*

The epidemiology and pathogenesis of COVID-19 have been discussed in [1]. In this, the phylogenetic analysis of the virus revealed that it is potentially a zoonotic virus which means it spreads between animal and people and patients develop intestinal symptoms like diarrhoea which was rarely evident in the case of MERS-CoV and SARS-CoV. A summary report based on 72,314 cases from the Chinese center for disease control and prevention has been made which shows the epidemiological characteristics of the disease [2]. A summarized report on the origin, transmission medium and clinical therapies on COVID-19 have been specified in [3]. The utilization of critical care for the COVID-19 outbreak in Lombardy, Italy for an emergency response to the situation is described in [4]. Migrations have been one of the major reason for the virus outbreak in different places and the effect of travel restriction on the spread of the virus is discussed in [5]. The social distancing method to avoid human to human contact is an effective strategy to control the virus outbreak and the feasibility of this has been discussed in [6]. The demographic science was used to understand the spread and fatality rate caused by the virus [7].

COVID-19 is a single-stranded RNA virus and the structure of RNA-dependent polymerase of the virus is discussed in [8]. The virus has a round body with a spiked glycoprotein structure which attaches itself to the receptors of host cells. The glycan shield and the spikes' interaction with human have been predicted in [9, 10]. Genotype is the genetic characteristic of a cell and phenotype is the corresponding physical characteristic of the cell. Genotypic and phenotypic characterization of the COVID-19 cell and their roles in pathogenesis was discussed in [11]. The virology epidemiology, pathogenesis and the control measures for the pandemic have been discussed in details in [12].

The Food and Drug Administration (FDA) assures the safety and effectiveness of drugs for public use, the virtual screening of FDA approved drugs to give relief to COVID-19 patients were conducted by authors and discussed in [13]. The use of chloroquine phosphate for the treatment of COVID-19 associated pneumonia has shown positive results [14]. The usage of randomized drugs on clinical trials and their outcomes were studied and are recorded in [15]. COVID-19 causes inflammation within the lungs with the build-up of fluids. Therefore the usage of anti-inflammatory drugs for the treatment of COVID-19 patients with the severe condition can give relief for breathing. This clinical immunologists from China have provided their perspective about the matter in [16]. Anti-inflammatory drugs are often steroidal, but to study the effects of non-steroidal anti-inflammatory drugs on COVID-19 patients, experiments were conducted and the outcomes were recorded in [17].

2.2 Usage of AI for Controlling the COVID-19 Pandemic

Since the evolution of better processing capacity of computers, AI has become a very essential tool to assist human in multiple aspects. AI has intervened the processes where human intervention is not practically feasible or is expensive. AI has the decision-making capabilities which can be primarily in the form of classification and regression. AI was earlier used to form a framework for quicker identification of COVID-19 in cities and towns under quarantine with a mobile phone-based survey model [18]. Chest CT scan reports were used to distinguish COVID-19 from community-acquired pneumonia with AI [19]. The authors in [20] have reviewed the use of AI to fight against COVID-19 and given prominent perspective about the matter based on all recent studies. AI was coupled with universal data sharing standards to monitor human health in smart city networks [21].

3 Methodology

Here we have discussed the methods we have implemented to test different artificial intelligence models which can be useful for controlling the spread of the COVID-19 outbreak. This section is divided into two primary sub-sections. One discusses the use of machine learning regression model to predict the trend of the virus outbreak, that is the prediction of future deaths and confirmation of COVID-19 positive cases based on current data. This is based on two generative machine learning regressors known as Bayesian Ridge Regressor and Gaussian Process Regressor. The other section discusses the implementation of deep learning technology for the detection of COVID-19 positive patients with the help of chest X-Rays as an alternative to traditional lab test. This is performed by deep convolution neural network model. Details of the procedures for both are given in following sub-section. The performances of all these techniques are recorded in Sect. 4.

3.1 Data Availability and Pre-processing

The regression model for estimating future confirmed cases and death cases have been made with the help of publicly available records provided by Johns Hopkins University [22]. For the X-Ray, classification has used Kaggle dataset [23, 24] which were cited by many peer-reviewed articles. These datasets combined contained over four thousand image data for chest X-Ray including normal subjects, COVID-19 patients, pneumonia patients, ARDS patients and more. The datasets, however, contained some noise in the form of CT-Scan images which we have manually removed. Later we have split them into two sets for training and testing purpose.

3.2 Virus Outbreak Trend Prediction with Machine Learning

The trend prediction of the virus is to estimate confirmed cases and deaths in future based on present data we have used two popular generative machine learning regressor known as Bayesian Ridge Regressor and Gaussian Process Regressor.

Bayesian Ridge Regressor (BRR)
Bayesian Ridge Regressor is a generative regressor which can reconstruct missing data or poorly distributed data by creating a linear regression model with the help of probability distributors instead of estimating from available points for training. As the number of deaths and confirmed cases will grow with time and will not oscillate within a specific range, a reconstruction of points from available points are necessary and BRR algorithms fit perfectly with this situation. Based on currently available data, it builds a probabilistic model estimating the change in the dependent feature (future trend) with respect to the independent feature (past trend) and this probability measure is used to estimate newer sets of data.

Gaussian Process Regressor (GPR)
GPR is a non-parametric regressor which works on the principle of the probability distribution for all functions that fit the data instead of calculating the probability distribution of parameters of a specific function. GPR can train with a very small dataset and provide good outcomes from them. It works like a multivariate Gaussian

distribution of infinite-dimension and the space of functions could be incorporated by the selection of the mean and covariance function. A GPR has multiple kernels and our regressor is primarily based on two kernels, dot product kernel and white kernel. A dot product kernel can be obtained from linear regression by tuning the coefficients and biases. It is invariant to coordinate rotation about the origin but not translation. A white kernel, on the other hand, is a part of sum kernel which explains the noise of the signal as an independent and identical normal distribution.

3.3 Detection of COVID-19 from Chest X-Rays with Deep Learning

Deep learning has been evolved as a powerful tool for classification and regression problem for both supervised and unsupervised learning. The deep convolution neural network (CNN) is the most promising algorithm to work with image data. A CNN traverses through all images in a dataset and extracts several features from it like the colour cluster concentration, texture, boundary, etc. out of which the most contrasting features are filtered and establish a classifiable relationship between the different classes or categories of the image datasets. We have trained the classifiers to figure out certain contrasting features between two classes of data, that is, one dataset containing chest X-Ray of patients infected with COVID-19 and other dataset contains a combination of normal chest X-Rays and other lung infection apart from COVID-19. The performances are recorded in Sect. 4.2.

4 Results and Discussions

The performance of the outbreak trend predictor model is discussed in Sect. 4.1 and performance analysis of COVID-19 detection with chest X-Rays model is discussed in Sect. 4.2.

4.1 Performance Analysis of Outbreak Trend Predictor Model

The daily recorded values of total confirmed cases and death cases of India were recorded from the first detected confirmed case up to the next 100 days and these data were split into two parts. The records of confirmed cases were split and first 67% of the data were used to train the model and the other 33% of the data were used to validate the accuracy of the model. Due to the uncertain nature of the disease, we needed more data to train the model to predict the future trend of death case and because of this, we have used 75% of the death record data for training and 25% of the data for validation purpose. To estimate the performance of the data, we have used Root Mean Squared Error (RMS Error) metric along with the lower and upper bound of the testing dataset and also the training and testing time for each model. The performances are recorded in Table 1 and the graph of the prediction is plotted in Figs. 1 and 2.

From Table 1, we can observe that to determine the performance of the regressors to predict the estimated confirmed COVID-19 positive cases, we can see that the

Table 1 Performance of the classifiers

Regressor	Training time (ms)	Prediction time (ms)	Prediction root mean squared error (RMSE)	Lower bound of test set	Upper bound of test set
Bayesian ridge (BRR) (confirmed cases)	1.3067	0.1402	2031.08	9152	52,952
Gaussian process (GPR) (confirmed cases)	14.948	0.1478	2089.13	9152	52,952
Bayesian ridge (BRR) (death cases)	1.7235	0.3426	29.50	718	1783
Gaussian ridge (GPR) (death cases)	62.226	0.4079	29.71	718	1783

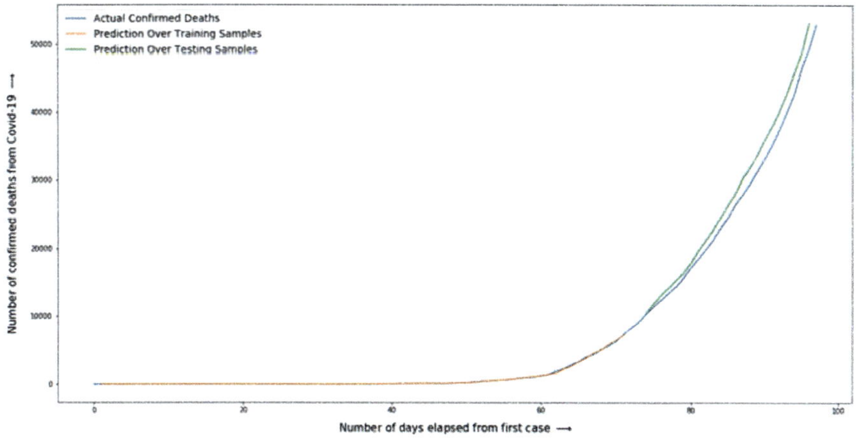

Fig. 1 The predicted trend of COVID-19 confirmed cases from 1st confirmed case till 100 days

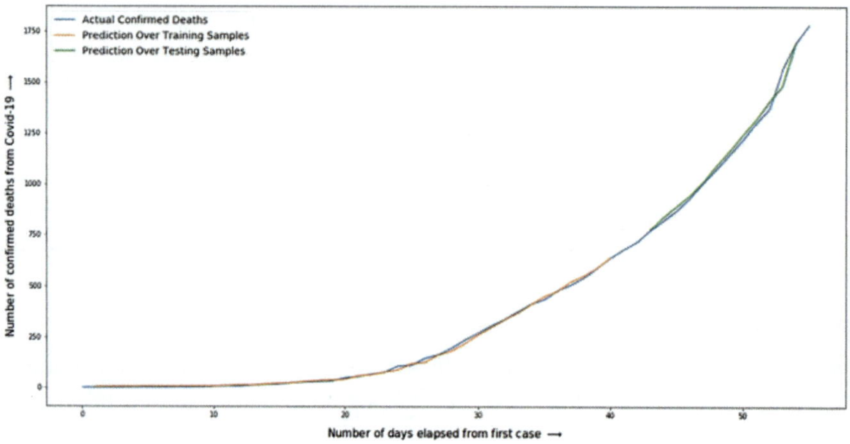

Fig. 2 The predicted trend of COVID-19 confirmed death cases from 1st confirmed case till 100 days

root mean squared error (RMSE) for BRR is 2031.08 and GPR is 2089.13 where the lower and upper bounds of the test sets are 9152 and 52,952 respectively. Now as the RMSE for both the regressors are very close, to test the performance of the classifiers, we need to check which one among them consumes lesser resources. For this, we can see that BRR trains the model in 1.3067 milliseconds (ms) and generates output in 0.1402 ms whereas GPR trains the model in 14.948 ms and generates output in 0.1478 ms. Both of the regressors' prediction times are very close but we can observe that BRR trains approximately 10 times faster than GPR. Based on all the parameters, we can confirm that BRR performs the best to determine the estimated confirmed COVID-19 positive cases based on currently available statistics.

Now for the death cases prediction, we can see that RMSE of BRR and GPR are 29.50 and 29.71 respectively for the test dataset having lower bound of 718 and upper bound of 1783. Now as the RMSE of both the classifiers is close, looking at the training and prediction time we can see that BRR trains 36 times faster and predicts 1.1 times faster than GPR. Therefore we see that in both the cases, BRR outperforms GPR in almost every aspect and we can reliably use BRR algorithm for the prediction of both COVID-19 confirmed cases as well as death cases caused by the virus.

4.2 Performance Analysis of COVID-19 Detection with Chest X-Rays Model

As discussed in Sect. 3.3, a convolution neural network extracts several properties from the image to establish a classifiable relationship between the classes of images.

Figure 3 gives an overlook of the intermediate feature extraction process of CNN algorithm. A COVID-19 infected chest X-Ray was passed through the model, and CNN extracted several features from that image and those are reflected in each frame of Fig. 3, this is how our model sees the image. After extraction of several such features, our model figured out the most important features from the dataset which can be used to classify the data. After training the model for several rounds until a consistent validation loss was obtained, our model was trained to classify COVID-19 from normal as well as lungs infected with other infections apart from COVID-19 by utilizing chest X-Ray images.

From Table 2, we can see that for the classification of the dataset, our model has obtained a classification accuracy of 95.02%. Now as the number of images in the two classes were different, therefore we use other metrics to understand the model performance at more depth. The precision and recall of the model are 0.944

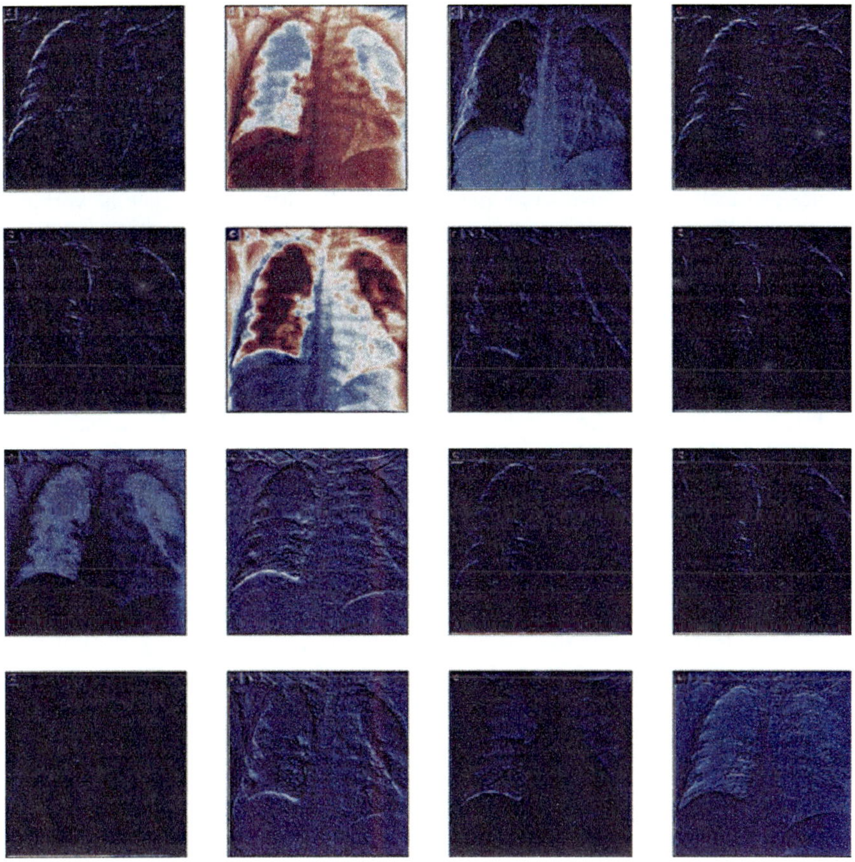

Fig. 3 How our deep learning model sees the COVID-19 infected chest X-Ray image to extract classifiable features

Table 2 The performance of the COVID-19 versus others classifier

	Accuracy (%)	Precision	Recall	*F*1 score
Performance	95.02	0.944	1.000	0.968

and 1.000 respectively which means our model has not labelled any image as false-negative but few images were labelled as false-positive. The closer the score to 1, the better the model should perform. Same goes for F1 score which is the harmonic mean of precision and recall. The high value of accuracy, precision, recall and F1 score indicates that our model performed with high accuracy. Further development and testing with this model could be done to deploy as an inexpensive alternative COVID-19 testing strategy.

5 Conclusion

COVID-19 is the most serious disease outbreak of the twenty-first century and there is no specific cure available yet. To stop the spread of COVID-19, the best strategy is to prevent the community spread of the virus. To control the outbreak of the virus, science and technology have an immense role to play. Medical science is working very hard to discover a specific cure and other branches of science can also contribute to stop the spread of the virus. In this paper, we have discussed the role of Artificial Intelligence in controlling the pandemic. We have developed two strategies which can help in controlling the virus outbreak.

First, we have developed a machine learning algorithm with the help of Bayesian Ridge and Gaussian Process regressor techniques. We have trained it to predict the future trends of the virus outbreak based on currently available statistics. To estimated confirmed COVID-19 positive cases and death cases likely to happen in the near future, Bayesian Ridge Regressor performs the best in almost all aspect. By understanding the upcoming trends, it would be very useful to flatten the curve of reported incidence for a specific place.

After this, we have developed a deep learning model which could be useful to replace traditional tests. As the traditional test takes a long time and the testing centres are sparsely located, alternative testing strategies have high importance. With the help of our deep learning model, we have successfully classified Covid-19 positive patients from a dataset containing normal healthy individual and patients affected by other lung diseases by analyzing chest X-Ray images. For this purpose, our model has obtained a classification accuracy of 95.02%. This could help us to identify COVID-19 infected patients in a widely available and inexpensive manner.

Artificial intelligence (AI) could be a very useful tool to prevent the virus outbreak. Further development in AI could be done to predict the geographical expansion of the disease or aid officials to plan to serve an optimum amount of patients by monitoring the severity of their condition.

6 Code Availability

The code for the chest X-Rays classification model to detect COVID-19 has been made available in GitHub with MIT License for open source distribution and it can be freely reused for commercial or non-commercial purpose [25].

References

1. Rothan, Hussin A., and Siddappa N. Byrareddy. "The epidemiology and pathogenesis of coronavirus disease (COVID-19) outbreak." *Journal of autoimmunity* (2020): 102433.
2. Wu, Zunyou, and Jennifer M. McGoogan. "Characteristics of and important lessons from the coronavirus disease 2019 (COVID-19) outbreak in China: summary of a report of 72 314 cases from the Chinese Center for Disease Control and Prevention." *Jama* 323, no. 13 (2020): 1239–1242.
3. Guo, Yan-Rong, Qing-Dong Cao, Zhong-Si Hong, Yuan-Yang Tan, Shou-Deng Chen, Hong-Jun Jin, Kai-Sen Tan, De-Yun Wang, and Yan Yan. "The origin, transmission and clinical therapies on coronavirus disease 2019 (COVID-19) outbreak–an update on the status." *Military Medical Research* 7, no. 1 (2020): 1–10.
4. Grasselli, Giacomo, Antonio Pesenti, and Maurizio Cecconi. "Critical care utilization for the COVID-19 outbreak in Lombardy, Italy: early experience and forecast during an emergency response." *Jama* 323, no. 16 (2020): 1545–1546.
5. Chinazzi, Matteo, Jessica T. Davis, Marco Ajelli, Corrado Gioannini, Maria Litvinova, Stefano Merler, Ana Pastore y Piontti et al. "The effect of travel restrictions on the spread of the 2019 novel coronavirus (COVID-19) outbreak." *Science* 368, no. 6489 (2020): 395–400.
6. Hellewell, Joel, Sam Abbott, Amy Gimma, Nikos I. Bosse, Christopher I. Jarvis, Timothy W. Russell, James D. Munday et al. "Feasibility of controlling COVID-19 outbreaks by isolation of cases and contacts." *The Lancet Global Health* (2020).
7. Dowd, Jennifer Beam, Liliana Andriano, David M. Brazel, Valentina Rotondi, Per Block, Xuejie Ding, Yan Liu, and Melinda C. Mills. "Demographic science aids in understanding the spread and fatality rates of COVID-19." *Proceedings of the National Academy of Sciences* 117, no. 18 (2020): 9696–9698.
8. Gao, Yan, Liming Yan, Yucen Huang, Fengjiang Liu, Yao Zhao, Lin Cao, Tao Wang et al. "Structure of the RNA-dependent RNA polymerase from COVID-19 virus." *Science* 368, no. 6492 (2020): 779–782.
9. Vankadari, Naveen, and Jacqueline A. Wilce. "Emerging COVID-19 coronavirus: glycan shield and structure prediction of spike glycoprotein and its interaction with human CD26." *Emerging microbes & infections* 9, no. 1 (2020): 601–604.
10. Ibrahim, Ibrahim M., Doaa H. Abdelmalek, Mohammed E. Elshahat, and Abdo A. Elfiky. "COVID-19 spike-host cell receptor GRP78 binding site prediction." *Journal of Infection* (2020).
11. Mousavizadeh, Leila, and Sorayya Ghasemi. "Genotype and phenotype of COVID-19: Their roles in pathogenesis." *Journal of Microbiology, Immunology and Infection* (2020).
12. Jin, Yuefei, Haiyan Yang, Wangquan Ji, Weidong Wu, Shuaiyin Chen, Weiguo Zhang, and Guangcai Duan. "Virology, epidemiology, pathogenesis, and control of COVID-19." *Viruses* 12, no. 4 (2020): 372.
13. Kandeel, Mahmoud, and Mohammed Al-Nazawi. "Virtual screening and repurposing of FDA approved drugs against COVID-19 main protease." *Life sciences* (2020): 117627.
14. Gao, Jianjun, Zhenxue Tian, and Xu Yang. "Breakthrough: Chloroquine phosphate has shown apparent efficacy in treatment of COVID-19 associated pneumonia in clinical studies." *Bioscience trends* (2020).

15. Kalil, Andre C. "Treating COVID-19—off-label drug use, compassionate use, and randomized clinical trials during pandemics." *Jama* 323, no. 19 (2020): 1897–1898.
16. Zhang, Wen, Yan Zhao, Fengchun Zhang, Qian Wang, Taisheng Li, Zhengyin Liu, Jinglan Wang et al. "The use of anti-inflammatory drugs in the treatment of people with severe coronavirus disease 2019 (COVID-19): The experience of clinical immunologists from China." *Clinical Immunology* (2020): 108393.
17. Little, Paul. "Non-steroidal anti-inflammatory drugs and covid-19." (2020).
18. Rao, Arni SR Srinivasa, and Jose A. Vazquez. "Identification of COVID-19 can be quicker through artificial intelligence framework using a mobile phone–based survey when cities and towns are under quarantine." *Infection Control & Hospital Epidemiology* 41, no. 7 (2020): 826–830.
19. Li, Lin, Lixin Qin, Zeguo Xu, Youbing Yin, Xin Wang, Bin Kong, Junjie Bai et al. "Artificial intelligence distinguishes COVID-19 from community acquired pneumonia on chest CT." *Radiology* (2020): 200905.
20. Naudé, Wim. "Artificial Intelligence against COVID-19: An early review." (2020).
21. Allam, Zaheer, and David S. Jones. "On the coronavirus (COVID-19) outbreak and the smart city network: universal data sharing standards coupled with artificial intelligence (AI) to benefit urban health monitoring and management." In *Healthcare*, vol. 8, no. 1, p. 46. Multidisciplinary Digital Publishing Institute, 2020.
22. Dong E, Du H, Gardner L. An interactive web-based dashboard to track COVID-19 in real time. *Lancet Infect Dis*; published online Feb 19. https://doi.org/10.1016/S1473-3099(20)30120-1.
23. C. M. M. Refat, "Chest xray images pneumonia and covid-19," apr 2020.
24. A. Hamada, "Covid-19 chest x-ray dataset," apr 2020.
25. S. Adhikary, S. Banerjee, S. K. Chaturvedi, S. Basu, S. Chaturvedi, "Classification of covid-19 vs normal with chest x-ray and transfer learning," jul 2020, GitHub, https://doi.org/10.5281/zenodo.3941518.

Smart Mucoadhesive Bio-flexy Film Former from Pulp of *Magnifera indica* and Its In-Built Properties for Pharmaceutical Applications

Sugandha Varshney and N. V. Satheesh Madhav

Abstract This research work deals with isolation of biopolymer from *Magnifera indica pulp* as film former for Formulating nanosized Tiagabine loaded bio-flexy films for epilepsy treatment. Biopolymer isolated by simplified economic process and used as bio-excipient. Bio-flexy films containing six ratios of nanosized Tiagabine: Magnifera indica biopolymer (1:0.5, 1:1, 1:3, 1:5, 1:6, 1:10) (FIT1–FIT6) were prepared solvent casting method and compared with that of nanosized Tiagabine: Sodium CMC standard polymer films (1:0.5, 1:1, 1:3, 1:5, 1:6, 1:10) (FET1–FET6). The biopolymer which was isolated from *Magnifera indica* showed percentage yield of 20.04% ± 0.001, yellowish brown in color, sweet odour, water soluble, colour changing point was found to be 225 °C ± 4. It tested positive for proteins and carbohydrates. Biopolymer possessed in-built biodegradability, biocompatibility, non-toxic, non-irritantancy and non-reactiveness on soft palatal surface, filmability, mucoadhesivity. Formulations prepared by solvent casting method and evaluated. Thickness of nanosized Tiagabine loaded bio-flexy films containing Magnifera indica (FIT1–FIT6) 0.046 mm ± 0.003 to 0.078 mm ± 0.002, Folding Endurance: 81–108, Surface pH: 7.01 ± 0.03 to 7.01 ± 0.02, Weight Uniformity: 0.032 ± 0.04 to 0.040 ± 0.03, Drug Content Uniformity: 71.0% ± 0.35 to 80.4% ± 0.30, Swelling Percentage: 61% ± 0.2 to 72% ± 0.1, Percentage Moisture Uptake (PTU): 2.0% ± 0.12 to 2.3% ± 0.10. Mucoretentivity: 150–360 min, Mucoadhesivity: 180–420 min. Drug release pattern for formulations FIT1–FIT6 containing Magnifera indica biopolymer based on the $T_{50\%}$ and $T_{80\%}$ was FIT5 (1:6) > FIT6 (1:10) > FIT2(1:1) > FIT4 (1:5) > FIT3 (1:3) > FIT1 (1:0.5). Formulations showed sustained release behavior. Based on all evaluation parameters, FIT5 (containing Tiagabine: Magnifera indica biopolymer (1:6)) Bio-flexy film having $R^2 = 0.9312$, Higuchi Matrix as best fit model, follows Fickian Diffusion (Higuchi Matrix) release mechanism, $T_{50\%}$: 21.25 h, $T_{80\%}$: 44.71 h using

S. Varshney (✉)
Guru Nanak College of Pharmaceutical Sciences, Jhajra, Banshiwala, India

N. V. Satheesh Madhav
Formulation Research Development, Vital Therapeutics and Formulations, Pvt. Ltd., Hyderabad, Telangana 500003, India

N. A. Siddiqui et al. (eds.), *Advances in Environment Engineering and Management*, Springer Proceedings in Earth and Environmental Sciences,
https://doi.org/10.1007/978-3-030-79065-3_31

BITS Software 1.12. Stability study revealed stable formulations. Prepared formulations of Tiagabine loaded bio-flexy films containing Magnifera indica biopolymer were suitable for Soft Palatal Delivery.

Keywords Magnifera indica biopolymer · Film former · Nanosized tiagabine · Bio-flexy films · Soft palatal delivery

Abbreviations

mm	Millimeter
cm^2	Centimeters square
min	Minutes
mL	Milliliter
g	Grams
mg	Milligram
nm	Nanometer
μg	Microgram
μM	Micrometer
d (nm)	Diameter in nanometer
mV	Millivolt
rpm	Revolutions per minute
°C	Degree centigrade
GABA	Gamma amino butyric acid
KBr	Potassium bromide
GIT	Gastro intestinal tract
API	Active pharmaceutical ingredient
U.V.	Ultraviolet visible spectroscopy
λ_{max}	Maximum absorbance
pKa	Dissociation constant
C_{max}	Maximum concentration
T_{max}	Time to attain peak concentration
$t_{1/2}$	Half life
SEM	Scanning electron microscopy
IR	Infra-red spectroscopy
DSC	Differential scanning calorimetry
NMR	Nuclear magnetic resonance spectroscopy
GIT	Gastro Intestinal Tract
Sodium CMC	Sodium Carboxyl Methyl Cellulose Standard Polymer
FIT1–FIT6	6 Bio-Flexy Films Formulations of nanosized Tiagabine with *Magnifera indica* biopolymer in ratios of (1:0.5–1:10)
FET1–FET6	6 Flexy Films Formulations of nanosized Tiagabine with Sodium Carboxyl Methyl Cellulose Standard polymer in ratios of (1:0.5–1:10)

RH	Relative humidity
CDR	Cumulative drug release
T50%	Time during which 50% drug is released
T80%	Time during which 80% drug is released
ICH	International Conference on Harmonization

1 Introduction

Epilepsy is a chronic neurological disorder affecting 65 million people globally as stated by World Health Organization. It can be Idiopathic i.e., with no identifiable cause or Symptomatic i.e., due to known cause such as brain damage, genetic abnormality, stroke, meningitis, encephalitis, brain tumor etc. There is no cure for epilepsy, but the disorder can be managed with medications and other strategies. Seizures are the main symptom of epilepsy. It occurs due to excessive electrical discharges in brain cells. Two or more unprovoked seizures leads to Epilepsy. There are two types of Neurotransmitters in brain. Excitatory or Glutamatergic Neurotransmitter and Inhibitory or GABAminergic Neurotransmitters. Excess of Excitatory neurotransmitter discharges lead to epilepsy. It is characterized by involuntary movements involving partial or complete brain that might lead to unconsciousness [1].

Many existing pharmaceuticals are rendered ineffective in the treatment of cerebral diseases due to inability to effectively deliver and sustain them within the brain. Drug targeting to brain by circumventing the physiological barriers prerequisite for drugs acting on central nervous system (CNS) [2].

Tiagabine, anticonvulsant drug ($t_{1/2}$:7–9 h; bioavailability: 90–95%; protein binding: 96%; water solubility: 22 mg/L). Enhances activity of Gamma Amino Butyric Acid (GABA), acts as selective GABA reuptake inhibitor. Dosage form of Tiagabine available: 2–16 mg tablets [3].

Existing oral therapeutic treatment of epilepsy requires prolonged increased dose frequency of antiepileptic drugs which leads to unexpected adverse effects in patients. This causes cumulative dose dumping in patient's body.

This approach offers low dosing level up to 25–100 folds which in turn causes minimization or devoid of adverse reactions offered by Tiagabine (like Sudden Unexpected Death, Syncope, Depression) respectively upon oral administration.

In this study, bio-flexy films formulations loaded with nanosized Tiagabine consisting of novel biopolymer isolated from *Magnifera indica* pulp were formulated and evaluated. Biopolymer was incorporated as bio-excipient and mucoadhesive film former that can provide controlled release for prolonged time period. Biopolymer have numerous advantages being edible, biocompatible, biodegradable in nature, nontoxic, and low cost etc. The potentiality of Biopolymer can be explored and can be substituted for current synthetic expensive polymers. Biopolymer depicts promising and tremendous use in drug delivery due to its novelistic characteristics and minimized side effects. It displayed significant mucoadhesivity and mucoretentability with optimal polar group which helps in polymer wetting [4, 5].

Till date, anticonvulsant molecule delivery to brain is a significant challenging task, so, a novelistic attempt has been made to deliver anticonvulsant through Soft Palatal platform. The research work focused to exploit a novelistic route for brain specifically via a trans-soft palatal route. Soft palate is a part of oral mucosa. It is a smart route for drug delivery as it crosses BBB. When an active pharmaceutical ingredient suitably formulated is placed in the soft palate, significant amount of drug can reach to the brain via neural pathway. Soft palatal route also avoids drawbacks of oral route, increases drug's therapeutic effect and minimizes its side effects. It is mucoadhesive site for systemic drug delivery, non-keratinized histology, acts as sustained and controlled drug delivery system. Tongue activity, salivary secretion does not affect drug performance via soft palatal mucosa. Due to lack of taste buds, bitter tasting drugs can be administered by this route. Does not interfere with patient's regular activities like talking, eating, drinking, etc. Drug wastage due to First Pass Metabolism is avoided by this pathway. It is has no bone, non-invasive, non-mobile and provides high mucoretentivity and bioavailability. It is enriched with blood and nerve supply, drug can directly reach into systemic circulation. It comprises of Ascending Palatine branch of Facial Artery, Ascending Pharyngeal Artery, Middle Meningeal Artery, Greater Palatine branch of Maxillary Artery and Accessory Meningeal Artery. It is innervated by Mandibular branch of Trigeminal Nerve, Motor Nerves, Glossopharyngeal Nerve, Nasopalatine Nerve, Lesser Palatine Nerve and Greater Palatine Nerve. Since Trigeminal Nerve directly connects soft palate to brain, nanosized drug can directly reach into brain via inter and intra neural pathway. Soft Palate has Surface area: 200 cm^2, Thickness: 158–224 μm, Blood flow: 0.89 mL/min/cm, pH: 7.34 ± 0.38.

Anticonvulsant molecule Tiagabine can be effectively delivered to brain via Oro Soft Palatal Route by formulating Bio-flexy Films containing nanosized drugs molecule for the management of epilepsy. This approach can also provide complete patient compliance, economic, safer to patients with lesser API burden in the body [5] (Fig. 1).

2 Materials and Methods

Drug: Tiagabine Hydrochloride (Sun Pharmaceuticals Industries Ltd., Gujarat).

Polymers: Procured Mango fruits from local market. Sodium Carboxyl Methyl Cellulose (Central drug House (P) Ltd. New Delhi) all other reagents used were of highest purity and analytical grade. Double distilled water was used throughout the experimental work.

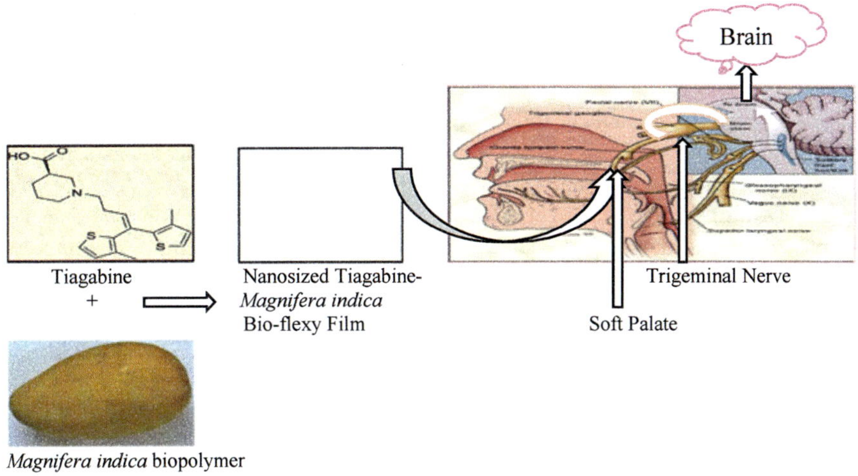

Fig. 1 Tiagabine-*Magnifera indica* bio-flexy film for trans-soft palatal drug delivery

2.1 Isolation of Biomaterial from Magnifera Indica [5]

Procured Mango fruits from local market. 500 g Ripened mango fruits taken. Removed seed. Mashed. Pulp treated with Distilled Water. Magnetic stirring at 8000 rpm for 10 min. Centrifuged at 3500 rpm for 15 min. Supernatant liquid treated with 1000 mL of propan-2-one in ratio of 1:2. Precipitate obtained was naturally dried for 24 h. Biomaterial obtained was powdered, passed through Sieve No. 120. Optimized 6 times, calculated % yield. Stored in well closed container for further use.

2.2 Physicochemical Characterization of Isolated Biomaterial [6, 7]

The isolated bio-material was characterized for physicochemical parameters such as odor, color, melting point, solubility along with chemical tests.

(i) **Texture**, (ii) **Color**, (iii) **Odor** were examined physically.

(iv) **Color Changing Point**: Determined by capillary method by Melting point apparatus. 10 mg of Bio-polymer was placed in a capillary tube. Fitted in a Melting point apparatus. Determined temperature by thermometer.

(v) **Solubility**: Dispersion of 10 mg of *Magnifera indica* biopolymer in distilled water was prepared and added to various solvents like Distilled Water, Methanol, Acetone, Ethanol, Methanol, Chloroform, Carbon Tetrachloride and Isopropanol taken in different test tubes. Reported solubility of all the isolated biopolymers.

(vi) **Test for carbohydrates: Molisch Reagent Test**: 2 mL of *Magnifera indica* biopolymer solution (0.1 g dissolved in 2 mL of distilled water) was taken in a test tube. Added 2 drops of Molisch reagent (Solution of α-naphthol in 95% Ethanol). Concentrated sulphuric acid (2 mL) was taken in a test tube and biopolymer solution was gradually poured over it leading to the formation of two separate layers. Change in color was observed.

(vii) **Test for proteins: Biuret Test**: Determines the presence of peptide bonds in protein content in isolated biomaterial. In a test tube, 2 mL of *Magnifera indica* biomaterial solution (0.1 g *Magnifera indica* biopolymer dissolved in 2 mL of distilled water) was taken. Added 1 mL of sodium hydroxide solution (1%) and 1% of copper (II) sulphate solution to above biomaterial solution drop wise. Allowed the mixture to stand for 5 min and observed the color change.

(viii) **Test for starch**: Added 2 drops of Iodine solution to a test tube containing 2 mL of *Magnifera indica* biopolymer solution (0.1 g *Magnifera indica* biopolymer dissolved in 2 mL of distilled water) observed the change in color.

(ix) **Test for reducing sugar**: Incorporated 1 mL of Fehling's A and 1 mL of Fehling's B solutions to a test tube containing 2 mL of *Magnifera indica* biopolymer solution (0.1 g *Magnifera indica* biopolymer dissolved in 2 mL of distilled water). Heated at 60 °C for few minutes and observed change in color.

2.3 *Spectral Studies of Isolated Biopolymer [6, 7]*

2.3.1 IR Spectroscopy by KBr Disc Method

The IR spectroscopy of isolated *Magnifera indica* biopolymer in solid form was performed by using Potassium Bromide Disc Method. 1 mg of sample was finely admixed with about 100 mg of Potassium Bromide (KBr) in mortar. Pressure of 10 t was applied to mixture using hydraulic pump. Small pellet of 1–2 mm in diameter was formed. The prepared pellet was kept in path of IR radiation and recorded the spectrum within the range of 4000–200 cm^{-1}. IR Spectra of isolated biopolymer was recorded.

2.3.2 DSC (Differential Scanning Calorimetry)

Amount of the heat difference of Sample and Reference was measured against temperature. It was performed for determination of Glass Transition temperature (GTT or Tg). For DSC the Perkin Elmer Instrument, Model-JADE DSC was used, with the Heat flow of 50–250 °C at the rate of 10 °C/min and Nitrogen rate of flow of 20 mL/min was used. DSC Spectra of isolated biopolymer was recorded.

2.3.3 NMR (Nuclear Magnetic Resonance) Spectral Analysis

Exploits the magnetic properties of atomic nuclei, determines the physical and chemical properties of atoms or the molecules in which they are contained. It relies on the phenomenon of Nuclear Magnetic Resonance and can provide detailed information about the structure, dynamics, reaction state, and chemical environment of molecule. Solvent used was DMSO (Dimethyl Sulfoxide). The spectrometer was connected to flow cell of 5 mm diameter. High flow rates were applied to the biopolymer sample, a valve switch was activated to stop the flow for quick measurement. When the valve switches back, the flow cell in the instrument was rinsed again with the reaction mixture. The spectrum was sent to the automation computer where it can be processed and analyzed.

2.3.4 SEM (Scanning Electron Microscopy) Analysis

Morphological examination of the surface and internal structure of the biomaterial was performed by using a scanning electron microscope (SEM) (Model: PHILIPS PSEM 515). A small amount of biomaterial was fixed on aluminum studs and it was coated with gold using a sputter coater under vacuum (pressure: 1 mm Hg). The biomaterial was then analyzed by SEM and reported.

2.4 *In-Vitro Mucoadhesivity of Isolated Biopolymer*

In-Vitro Mucoadhesivity of isolated *Magnifera indica* biopolymer was determined by Modified Shear Stress Apparatus. Different concentrations 1, 2, 4, 6, 8 and 10% of biopolymer solutions were placed between two glass plates of Modified Shear Stress Apparatus. Subjected to shear stress for assessment for In-Vitro Adhesive Strength in terms of weight required for breaking Adhesive bonds between the biomaterial and the glass plate after specified contact time from 0 to 30 min. The results were reported in tabulated form.

2.5 *Standard Graph of Drug [8]*

(a) **Preparation of Standard Curve of Tiagabine in Distilled Water**
10 mg of Tiagabine was dissolved in 30 mL of distilled water in a 100 mL volumetric flask and diluted up to the mark with distilled water (100 μg/mL). Dilutions of Concentrations (0.5,1, 2, 3, 4 and 5 μg/mL) were prepared in 10 mL volumetric flasks. Volume made up to 10 mL with distilled water (λ_{max} = 257 nm). Absorbance was measured against solvent blank.

(b) **Preparation of Standard Graph of Tiagabine in Phosphate Buffer of pH 7.4**
10 mg of Tiagabine was dissolved in 30 mL of Phosphate Buffer (pH 7.4) in a 100 mL volumetric flask and diluted up to the mark with Phosphate Buffer (100 μg/mL). Dilutions of Concentrations (1, 2, 3, 4, 5, 8, 10, 20, 30, 40, 50 μg/mL) were prepared in 10 mL volumetric flasks. Volume was made up to 100 mL with Phosphate Buffer (pH 7.4) ($\lambda_{max} = 396$ nm). Absorbance was measured against solvent blank.

2.6 *Drug Biopolymer Interaction Studies [9, 10]*

Tiagabine: isolated *Magnifera indica* biopolymer in ratios of 1:1, 1:3 and 3:1 were taken. Measured Absorbance and compared with pure Tiagabine.

(a) **Dry method**: Tiagabine: Magnifera indica biomaterial in above mentioned ratios were taken in dry form in three petridishes. Kept for two hours at room temperature. Diluted the mixtures with 2 mL of Methanol. Absorbance was measured, observed shift in λ_{max} in comparison with pure drug and reported.
(b) **Wet method**: Tiagabine: Magnifera indica biomaterial in above mentioned ratios were taken in dry form in three petridishes 1 mL of distilled water was added in each petridish. Dried in oven for 30 min at 50 °C. Diluted with 2 mL of Methanol. Absorbance was measured, observed shift in λ_{max} in comparison with pure drug and reported.
(c) **Colorimetry Method**: Tiagabine: Magnifera indica in ratio of 1:1 were mixed with Potassium Permanganate on glass plate. Observed color change, diluted suitably with distilled water, analyzed by UV. Repeated with Drug: Distilled Water and Drug: Potassium Permanganate.

2.7 *Preparation of Tiagabine from Tiagabine Hydrochloride by Precipitation Method*

Tiagabine is available as Tiagabine Hydrochloride salt form. Thus, in order to enhance absorption and bioavailability of drug, to avoid ill-effects it is converted to its pure form by Precipitation method. To 100 mg of Tiagabine hydrochloride, 20 mL of distilled water was added in a test tube and shaken vigorously. Mixture was subjected to sonication for 1 cycle (each cycle of 3 min) in ultrasonic bath sonicator. 10 mL of 1 N sodium hydroxide solution was incorporated drop wise in above Tiagabine solution. Precipitate was formed at bottom of test tube. Mixture was centrifuged for 15 min at 3500 rpm. Tiagabine was separated, washed with 10 mL distilled water and air dried. Isolated Tiagabine was analyzed using U.V. Spectrophotometer.

2.8 Nanosizing of Drug [9]

1. **Solvent Evaporation Method**: 100 mg Tiagabine was admixed with 5 mg of Fructose, 10 mg of Dextrose and 10 mL of Methanol in mortar pestle. Sonication of mixture was performed for 5 cycles (180 s/cycle) in ultrasonic bath sonicator. The mixture was then diluted with 50 mL distilled water and sonicated up to 15 cycles. Measured % Transmittance, Absorbance, % Blockage (100 − % Transmittance) after every 5 cycles. The residue was then dried and stored for further use.
2. **Sonication method**: 100 mg Tiagabine was admixed with 5 mg of Fructose, 10 mg of Dextrose and 10 mL of Distilled water in mortar pestle. Sonication of mixture was performed for 5 cycles (180 s/cycle) in ultrasonic bath sonicator. The mixture was then diluted with 50 mL distilled water and sonicated up to 15 cycles. Measured % Transmittance, Absorbance, % Blockage (100% Transmittance) after every 5 cycles. The residue was then dried and stored for further use.
   The main purpose of Nanosizing Tiagabine by two different methods was to compare Novel Sonication Method with published Standard Solvent Evaporation Method.
3. **Nano size range determination by Preliminary U.V. spectroscopic method**: It is a novel primarily screening method for nano-size range particles by U.V. Spectroscopy. Transmittance is based on the concept of Tyndall Effect. When light of specified wavelength passes through the media containing particles less than or greater than the specified particle range, the % Blockage represents particles beyond the size range whereas the % Transmittance is considered that the particles lies above the size range at particular range.

2.9 Solvent Casting Method as Formulation Technique of Bio-flexy Films [11]

100 mg of Nanosized Tiagabine (Anticonvulsant) was triturated with 50 mg of biopolymer (Mucoadhesive, film forming cum retarding agent) (in ratio of 1:0.5) for 2 min using pestle mortar. Added 10 m of Distilled Water (Solvent).To this dispersion, incorporated 10 mg of Dextrose (Flexicizer), 5 mg of Fructose (Flexicizer) and 10 μL of Glycerine (1% solution v/v) (Plasticizer) with continuous stirring. 0.6 g of Pectin (Film Initiator) was added. Mixture was further uniformly triturated for 5 min. Volume was made up to 20 mL using Distilled water. Mixture was subjected to magnetic stirring for 15 min, followed by sonication for up to 5 cycles (each cycle 3 min). Clear dispersion obtained was poured into petridish. Kept for drying at room temperature for 24 h. Removed prepared nanosized drug loaded Bio-flexy film from petridish. Bio-flexy film formulation obtained was cut in 1 cm^2 dimension, packed in well closed air tight container for further use. Similarly, six different formulations

Table 1 Formulation of nanosized tiagabine loaded bio-flexy films using *Magnifera indica* biopolymer

Formulation Drug:biopolymer ratios	FIT1 (1:0.5)	FIT2 (1:1)	FIT3 (1:3)	FIT4 (1:5)	FIT5 (1:6)	FIT6 (1:10)
Nanosized tiagabine (mg)	100	100	100	100	100	100
Magnifera indica biopolymer (mg)	50	100	300	500	600	1000
Dextrose (mg)	10	10	10	10	10	10
Fructose (mg)	5	5	5	5	5	5
Glycerine (μl)	10	10	10	10	10	10
Pectin (g)	0.6	0.6	0.6	0.6	0.6	0.6
Distilled water (mL)	20	20	20	20	20	20

Table 2 Formulation of nanosized tiagabine loaded flexy films using sodium carboxyl methyl cellulose standard polymer

Formulation Drug:polymer ratios	FET1 (1:0.5)	FET2 (1:1)	FET3 (1:3)	FET4 (1:5)	FET5 (1:6)	FET6 (1:10)
Nanosized tiagabine (mg)	100	100	100	100	100	100
Sodium carboxyl methyl cellulose standard polymer (SCMC) (mg)	50	100	300	500	600	1000
Dextrose (mg)	10	10	10	10	10	10
Fructose (mg)	5	5	5	5	5	5
Glycerine (μl)	10	10	10	10	10	10
Pectin (g)	0.6	0.6	0.6	0.6	0.6	0.6
Distilled water (mL)	20	20	20	20	20	20

of nanosized Tiagabine with different isolated biopolymers and Standard Sodium Carboxyl Methyl Cellulose Polymer in different ratios of 1:1, 1:3, 1:5, 1:6 and 1:10 were prepared. Total 12 nanosized Tiagabine loaded bio-flexy films formulations were prepared and evaluated (Tables 1 and 2).

2.10 *Evaluation of Formulated Bio-flexy Films [12, 13]*

2.10.1 Thickness

Determined the average thickness of formulations by standard digital micrometer and reported with appropriate standard deviation.

2.10.2 Surface pH Study

Surface pH of formulated films was determined by using Digital pH meter. It should be Neutral or close to soft palatal pH otherwise formulation might cause irritation to the Soft Palatal mucosa. The formulations were kept in contact with 1 mL of Distilled Water at room temperature for 1 h. pH was then measured in triplicate and reported. Compatibility of formulations with soft palatal pH is essential.

2.10.3 Ex-Vivo Mucoadhesion Study of Formulations by Rotating Cylinder Method

The Mucoadhesivity of best selected formulations was further evaluated on Soft Palatal mucosa of *Capra aegagrus* (i.e., Goat). Bio-flexy films of area 1 cm^2 of each formulation were cut down using sharp blade. Tied the goat soft palatal mucosa over the rotating basket of I-Dissolution Apparatus. The Dissolution media was 900 mL of buffer (pH 7.4), maintained at 37 °C, subjected for rotation at 50 rpm. Films were applied over the inner surface of goat soft palatal mucosa until they got dislodged. The dislodgement and detachment of films from mucosal surface was observed at regular intervals and reported (n = 6).

2.10.4 Ex-Vivo Mucoretention Study of Formulations

The Mucoretention study of best selected formulations was further evaluated on Soft Palatal mucosa of *Capra aegagrus* (i.e., Goat). Bio-flexy films of area 1 cm^2 of each formulation were cut down using sharp blade. Tied the *Capra aegagrus* (Goat) soft palatal mucosa over slanting condenser over which buffer was allowed to flow from a burette. It was applied over the inner surface of Goat intestinal mucosa until it got dislodged. The detachment and Dislodgement of film from mucosal substrate was noted at regular intervals and reported (n = 6).

2.10.5 Weight Uniformity of Formulated Nanosized Drugs Loaded Bio-flexy Films

Weight uniformity of formulated films was determined by weighing formulations of 1 cm^2 diameter. Determined average weight and reported (n = 3).

2.10.6 Drug Content Uniformity of Formulated Nanosized Drugs Loaded Bio-flexy Films

Drug Content Uniformity of formulated films was calculated by dissolving the films in Phosphate Buffer (pH 7.4) (100 mL) for 24 h with occasional shaking. Diluted 5 mL of solution with phosphate buffer pH 7.4 up to 20 mL. Filtered through Whatman filter paper of 0.45 mm. The drug content determined by UV analysis at λ_{max} 750 nm for nanosized Tiagabine loaded formulations.

2.10.7 Folding Endurance of Formulated Nanosized Drugs Loaded Bio-flexy Films

Folding endurance of flexy film was determined by repeatedly folding one of the film at the same place till it broke or folded up to 300 times manually, which was considered satisfactory to reveal good properties. The number of times of film could be folded at the same place without breaking will give the value of the folding endurance. This test was done on randomly selected three formulations from each Drug:Biopolymer ratio (n = 3).

2.10.8 Swelling Percentage Study of Formulated Nanosized Drugs Loaded Bio-flexy Films

It was determined as increase in weight and area because of Swelling. $1 \times 1\ cm^2$ sized films were weighed, transferred in petridish and added 10 mL of distilled water. After 1 h, reweighed the films. Absorption of water and swelling of films caused increased in weights of films. The study was performed for 24 h. Calculated % Swelling Index and reported (n = 3).

2.10.9 Percentage Moisture Uptake (PMU) of Formulated Nanosized Drugs Loaded Bio-flexy Films

Percentage Moisture Uptake of Formulations was determined so as to check the physical stability of the prepared bio-flexy films in high moist conditions. Bio-flexy films of 1 cm diameter were kept in saturated solution of Aluminum Chloride in desiccator. The humidity inside the desiccator was maintained at 79.5%. Removed the films after 3 days, weighed, calculated Percentage Moisture Absorption and reported (n = 3).

$$\text{Percentage Moisture Uptake} = \left(\frac{\text{Final weight of Films} - \text{Initial weight of films}}{\text{Initial weight of Films}}\right) \times 100$$

2.10.10 In-Vitro Drug Release Study of Formulated Nanosized Drugs Loaded Bio-flexy Films Using Modified M.S. In-Vitro Diffusion Apparatus [14]

In-Vitro Drug Release Study of Formulations was performed by using Modified M.S. In-Vitro Diffusion Apparatus. Buffer pH 7.4 was filled in 36 vials (receiver compartment). These were kept in thermostatically controlled compartment. Tied Egg membranes to Donor compartment (containing formulations). Donor compartments were inserted into receiver compartments. Temperature was kept constant at 37 °C with orbital shaker incubator. Sampling was done at regular intervals from 10 min to 48 h. Buffer was completely replaced after every sampling. Performed Ultra Violet Spectral analysis of every sample.

2.10.11 Stability Studies of Formulations as Per ICH Guidelines (Q1B)

Stability Studies of prepared Films were conducted as per ICH Guidelines Q1B. Stability testing of pharmaceutical product is done to ensure the Efficacy, Safety and Quality of active drug substance and dosage forms and shelf life or expiration period. The Stability Studies of the formulations were performed at 40 °C ± 2 °C with ± 45 ± 5% RH, at 25 ± 2 °C with 60 ± 5% RH and at 2 ± 5 °C conditions of temperature and relative humidity for 3 months. Observed for change in pH, Folding Endurance, In-Vitro Drug Release of formulations at room temperature and after stability study (25 ± 2 °C with 60 ± 5% RH).

3 Results and Discussion

3.1 Yield of Isolated Biopolymer

Biopolymer was isolated from natural edible source of *Magnifera indica* by simple and economical method. The isolated biopolymer was optimized repeatedly for six times. Calculated and reported % yield. The % yield of *Magnifera indica* biopolymer was found to be 20.04% ± 0.001.

3.2 Physicochemical Properties of Isolated Biomaterial

The biomaterial obtained from the pulp of *Magnifera indica* and showed following characteristics:

(a) **Texture**: Rough; (b) **Color**: Yellowish Brown; (c) **Odor**: Characteristic;

(d) **Solubility**: soluble in methanol and water; (e) **Color Changing Point**: 225 °C ± 4;

(f) **Molisch Reagent test for Carbohydrates**: Purple color appeared at interface of two layers because of formation of 5-hydroxy methyl furfural. This indicated presence of carbohydrates.

(g) **Biuret test for Proteins**: Change in color was observed as Cu(II) ions formed a chelate complex of violet color which absorbed light at 540 nm. This confirmed the presence of Proteins.

(h) **Starch Test**: Intense blue black color did not appeared confirmed the absence of Starch in isolated biomaterial.

(i) **Test for Reducing Sugar**: Appearance of brick red precipitate indicated reducing sugar.

3.3 Spectral Studies of Isolated Biopolymer

3.3.1 IR Spectroscopy

IR Spectroscopy was performed for the isolated biomaterial to determine the presence of Functional Groups in biopolymer. IR Peaks of *Magnifera indica* biopolymer were obtained at 3647 cm^{-1}, 2479 cm^{-1}, 1224 cm^{-1}, 1006 cm^{-1} which indicated functional groups C=C-COOH, RCH_2OH, RNH_2, RCOOH (Fig. 2).

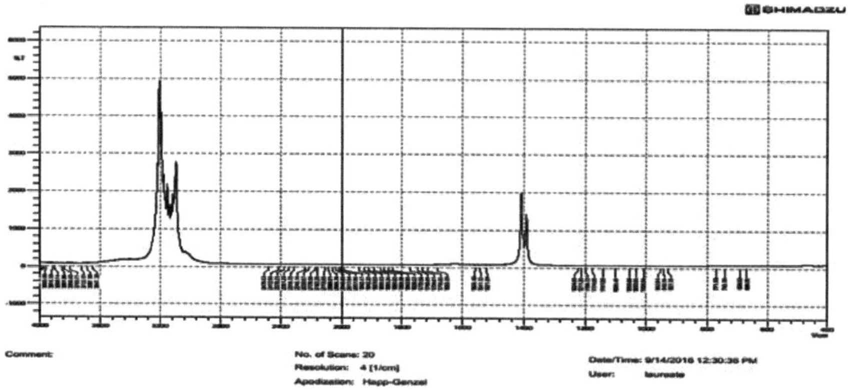

Fig. 2 Infra-red spectroscopy (IR) spectra of *Magnifera indica* biopolymer

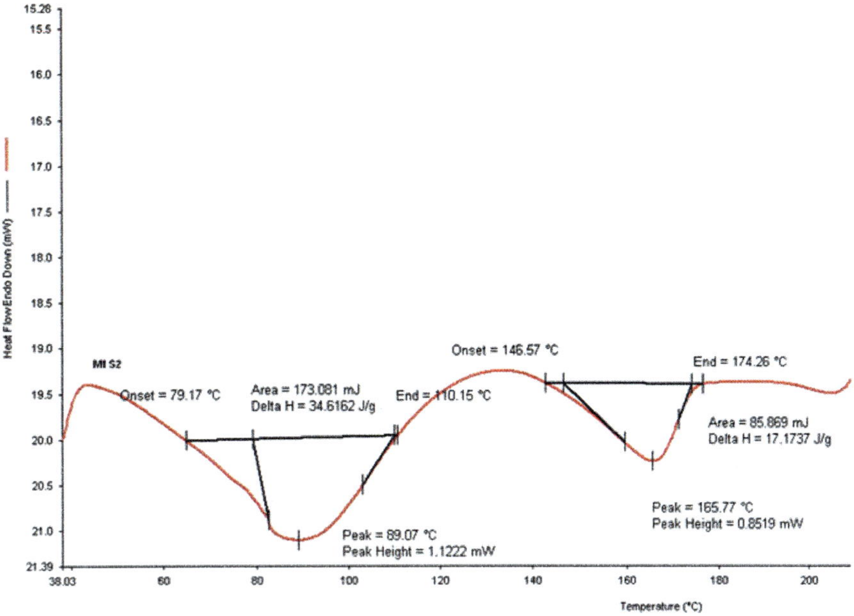

Fig. 3 Differential scanning calorimetry (DSC) spectra of *Magnifera indica* biopolymer

3.3.2 Differential Scanning Calorimetry (DSC)

DSC Peak of *Magnifera indica* biopolymer was obtained at 89.07 °C, peak height is 1.1222 mW, delta H is 34.6162 J/g, Onset depicts boiling point at 79.17 °C and Glass Transition temperature is 110.15 °C (Fig. 3).

3.3.3 Nuclear Magnetic Resonance Spectroscopy (NMR)

^{1}HNMR Spectra of *Magnifera indica* biopolymer confirmed the presence of carbohydrates residue within the biopolymer extracted as shift of carbohydrate protons were 3–6 ppm and the spectra when compared reflected the peak at 3.4383 ppm (Fig. 4).

3.3.4 Scanning Electron Microscopy (SEM) of Isolated Biopolymer

SEM image of *Magnifera indica* biopolymer showed size range of 50 μm, porous and irregular shaped particles with smooth surface morphology (Fig. 5).

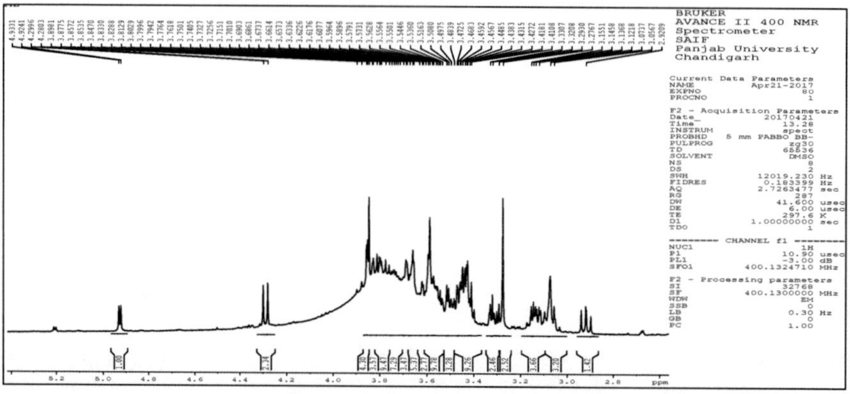

Fig. 4 Nuclear magnetic resonance spectroscopy (NMR) spectra of *Magnifera indica* biopolymer

Fig. 5 Scanning electron microscopy (SEM) of *Magnifera indica* biopolymer

3.4 In-Vitro Mucoadhesivity of Isolated Biopolymers by Shear Stress Method

In-Vitro Mucoadhesion Study data of *Magnifera indica* biopolymer revealed that 1, 2 and 4% biopolymer concentrations showed significant results with p value < 0.05 when compared to 10% biopolymer concentration. 1% biopolymer concentration showed significant results with p value < 0.05 when compared to 1% Sodium Carboxyl Methyl Cellulose Standard polymer. Order of Mucoadhesivity of all concentrations of Magnifera indica biopolymer was 10% Magnifera indica

Table 3 In-vitro mucoadhesivity of *Magnifera indica* biopolymer by shear stress method

S. No.	Time (min)	Concentration of *Magnifera indica* biopolymer solutions (%w/v) (g)						Sodium CMC (g)
		1%	2%	4%	6%	8%	10%	1%
1	0	125.48***,a1	150.94***	162.42 ***	188.06	219.75	238.85	186.85
2	10	155.53***,a1	182.62***	194.65***	220.72	250.22	262.7	222.84
3	20	180.42***,a1	195.54***	212.08***	252.05	284.82	290.02	260.06
4	30	210.98***,a1	250.37***	268.5***	280.02	299.84	310.04	300.04

***$p < 0.05$ as compared to 10% w/v biopolymer
***, a1$p < 0.05$ as compared to 1%w/v sodium carboxyl methyl cellulose standard polymer significance level at 0.05, one way ANOVA using T test calculator

biopolymer > 8% Magnifera indica biopolymer > 6% Magnifera indica biopolymer > 4% Magnifera indica biopolymer > 2% Magnifera indica biopolymer > 1% Magnifera indica biopolymer (Table 3).

3.5 *Standard Graphs of Tiagabine*

(a) **Standard Graph of Tiagabine in Distilled Water**: Calibration Curve of Tiagabine was prepared in Distilled Water. The Standard Graph of Tiagabine was obtained by plotting Concentration versus Absorbance. The Standard Curve of Tiagabine showed linearityat a λ_{max} of 257 nm. R^2 value was found to be 0.9993 (Fig. 6a).

(b) **Standard Graph of Tiagabine in Phosphate Buffer pH 7.4**: Calibration Curve of Tiagabine was prepared in Phosphate Buffer pH 7.4. The Standard Graph of Tiagabine was obtained by plotting Concentration versus Absorbance. The Standard Curve of Tiagabine showed linearity at λ_{max} of 396 nm. R^2 value was found to be 0.9967 (Fig. 6b).

3.6 *Drug–Polymer Interaction Study of the Isolated Biopolymer*

(1) **Wet method**: λ_{max} was observed at 260 nm, and there was no significant difference from that of the pure drug Tiagabine at 257 nm. Therefore, drug-excipient interaction did not occur as there was no shift in λ_{max}.
(2) **Dry method**: λ_{max} was observed at 260 nm, and there was no significant difference from that of the pure drug Tiagabine at 257 nm. Therefore, drug-excipient interaction did not occur as there was no shift in λ_{max}. Drug polymer

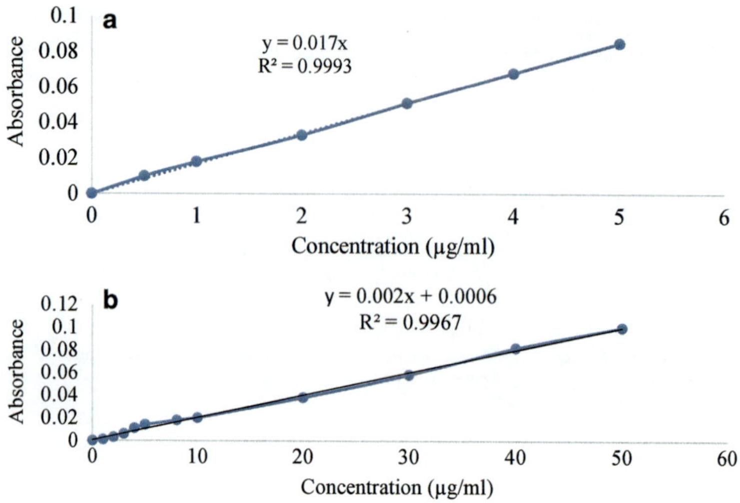

Fig. 6 **a** Standard graph of tiagabine in distilled water [8]. **b** Standard graph of tiagabine in phosphate buffer pH 7.4 [8]

interaction was not observed because no change in wavelength of pure drug and drug to biopolymer ratio

(3) **Colorimetry**: 10 mg of drug taken and mixed with biopolymers in ratio of Drug:Polymer 1:1 on glass plate. Added 10 μL of 1% solution of Potassium Permanganate. Observed color change, diluted suitably with distilled water, subjected for U.V. analysis. Scanned in 200–800 nm range and determined absorbance. Similarly repeated with Drug:Distilled Water and Drug:Potassium Permanganate. Drug showed color change from pink to brown with Potassium Permanganate while polymer showed no color change. No significant difference in shift of λ_{max} than that of pure drug observed.

3.7 *Nanosizing of Tiagabine*

See Fig. 7.

3.8 *Evaluation Parameters of Formulations*

3.8.1 Thickness of Formulated Bio-flexy Films

As polymer concentration was increased, thickness of films increased proportionately. The thickness of nanosized Tiagabine loaded Bio-Flexy films containing

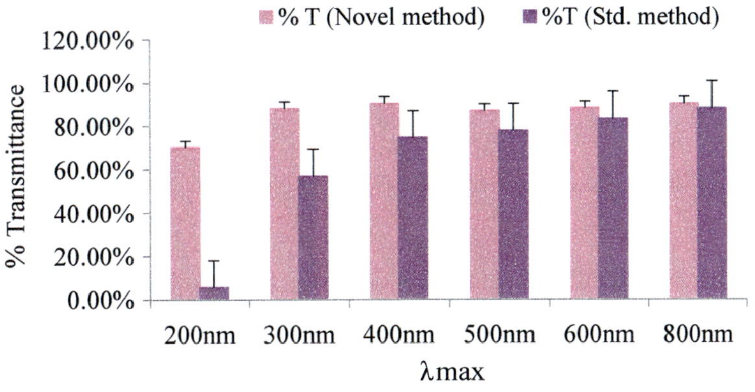

Fig. 7 Comparative graph between %Transmittance and λ_{max} of nanosized tiagabine (by novel and standard solvent evaporation methods) [9].

Magnifera indica biopolymer (FIT1–FIT6) was found to be in range of 0.046 ± 0.003 mm to 0.078 ± 0.002 mm.

3.8.2 Surface pH of Formulated Bio-flexy Films

The Surface pH of nanosized Tiagabine loaded Bio-Flexy films containing Magnifera indica biopolymer (FIT1–FIT6) was found to be in range of 7.01 ± 0.03 to 7.01 ± 0.02.

3.8.3 Ex-Vivo Mucoadhesion Study of Formulated Bio-flexy Films Using Capra Aegagrus (Goat) Soft Palatal Mucosa

Ex-Vivo Mucoadhesion Study by Rotating Cylinder method revealed nanosized Tiagabine loaded Bio-Flexy films containing Magnifera indica biopolymer (FIT1–FIT6) showed mucoadhesivity on Capra aegagrus mucosal surface for time period of 150–360 min.

3.8.4 Ex-Vivo Mucoretention Study of Formulated Bio-Flexy Films Using Capra Aegagrus (Goat) Soft Palatal Mucosa

Ex-Vivo Mucoretention Study revealed nanosized Tiagabine loaded Bio-Flexy films containing Magnifera indica biopolymer (FIT1–FIT6) were mucoretentive on Capra aegagrus mucosal surface for time period of 180–420 min.

3.8.5 Weight Uniformity of Formulated Bio-flexy Films

The Weight Uniformity of all the Formulations was proportionally increased as polymer concentration was increased. The Weight Uniformity of nanosized Tiagabine loaded Bio-Flexy films containing Magnifera indica biopolymer (FIT1–FIT6) was found to be in range of 0.032 $\pm$ 0.04 mg to 0.040 $\pm$ 0.03 mg.

3.8.6 Drug Content Uniformity of Formulated Bio-flexy Films

The Drug Content Uniformity of nanosized Tiagabine loaded Bio-Flexy films containing Magnifera indica biopolymer (FIT1–FIT6) was found to be in range of 71.0% $\pm$ 0.35 to 80.4% $\pm$ 0.30.

3.8.7 Folding Endurance of Formulated Bio-flexy Films

Folding Endurance of all the formulations was measured and it showed that flexibility was proportionately increased significantly as concentration of polymer in formulation was increased. The Bio-flexy films were devoid of brittleness showing significant folding endurance due to presence of dextrose and fructose as excipients in optimized ratio. The Folding Endurance of nanosized Tiagabine loaded Bio-Flexy films containing Magnifera indica biopolymer (FIT1–FIT6) was found to be in range of 81–108.

3.8.8 Swelling Percentage of Formulated Bio-flexy Films

The Swelling Percentage of nanosized Tiagabine loaded Bio-Flexy films containing Magnifera indica biopolymer (FIT1–FIT6) was found to be in range of 61% $\pm$ 0.2 to 72% $\pm$ 0.1.

3.8.9 Percentage Moisture Uptake of Formulated Bio-flexy Films

The Percentage Moisture Uptake of nanosized Tiagabine loaded Bio-Flexy films containing Magnifera indica biopolymer (FIT1–FIT6) was found to be in range of 2.0% $\pm$ 0.12 to 2.3% $\pm$ 0.10.

3.8.10 In-Vitro Release Study of Formulated Bio-flexy Films by Modified M.S. Diffusion Apparatus

The drug release pattern for formulations FIT1–FIT6 containing Magnifera indica biopolymer based on the $T_{50\%}$ and $T_{80\%}$ was found to be FIT5 (1:6) > FIT6 (1:10) >

FIT2(1:1) > FIT4 (1:5) > FIT3 (1:3) > FIT1 (1:0.5). Based on all above mentioned evaluation parameters, FIT5 (containing Tiagabine: Magnifera indica biopolymer (1:6)) Bio-flexy film was selected as the Best formulation as it showed significant values of $T_{50\%}$: 21.25 h; $T_{80\%}$: 44.71 h and having $R^2 = 0.9312$, Higuchi Matrix as best fit model, follows Fickian Diffusion (Higuchi Matrix) release mechanism in comparison to other formulations of same biopolymer (Fig. 8a, Table 4).

The drug release pattern for formulations FET1–FET6 containing Sodium Carboxyl Methyl Cellulose Standard polymer based on the T50% and T80% was found to be FET5 (1:6) > FET1 (1:0.5) > FET2(1:1) > FET3 (1:3) > FET4 (1:5) > FET6 (1:10). Based on all above mentioned evaluation parameters, FET5 (containing Tiagabine: Sodium Carboxyl Methyl Cellulose standard polymer (1:6)) Flexy film was selected as the Best formulation as it showed significant values of $T_{50\%}$: 40.66 h, $T_{80\%}$: 43.79 h and having $R^2 = 0.9301$, Higuchi Matrix as best fit model, follows

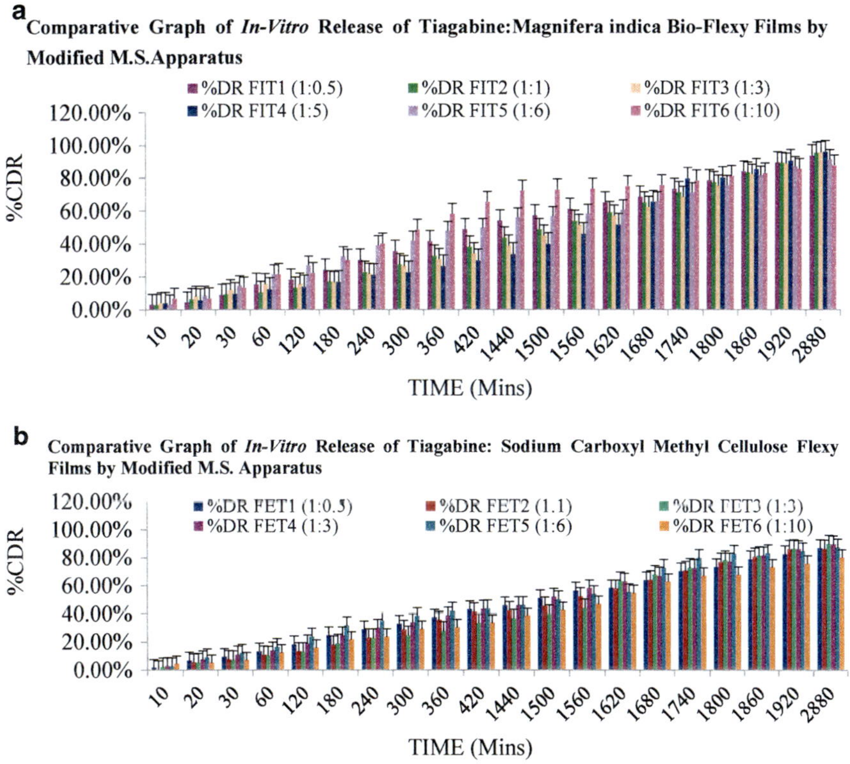

Fig. 8 **a** In-vitro drug release graph of nanosized tiagabine loaded bio-flexy films using *Magnifera indica* biopolymer by modified M.S. diffusion apparatus [11]. **b** In-vitro drug release graph of nanosized tiagabine loaded bio-flexy films using sodium carboxyl methyl cellulose standard polymer by modified M.S. diffusion apparatus [11]

Table 4 Kinetics release of tiagabine-*Magnifera indica* polymer bio-flexy films

Release kinetics analysis dynamic method formulation of tiagabine: *Magnifera indica* bio-flexy films

Formulations	R^2					Best fit model	Mechanism of action
	Zero order	1st order	Higuchi matrix	Peppas	Hixon Crowell		
FIT1 (1:0.5)	0.8931	0.8935	0.9402	0.9572	0.8934	Peppas Korsmeyer	Anomalous transport
FIT2 (1:1)	0.9180	0.9181	0.9390	0.9665	0.9181	Peppas Korsmeyer	Anomalous transport
FIT3 (1:3)	0.9102	0.9102	0.9329	0.9476	0.9102	Peppas Korsmeyer	Fickian diffusion (Higuchi matrix)
FIT4 (1:5)	0.8796	0.8794	0.9365	0.9430	0.8795	Peppas Korsmeyer	Anomalous transport
FIT5 (1:6)	**0.8418**	**0.8424**	**0.9312**	**0.9055**	**0.8422**	**Higuchi matrix**	**Fickian diffusion (Higuchi matrix)**
FIT6 (1:10)	0.7988	0.7994	0.9486	0.9450	0.7992	Higuchi matrix	Fickian diffusion (Higuchi matrix)

Fickian Diffusion (Higuchi Matrix) release mechanism in comparison to other formulations of same standard polymer (Fig. 8b, Table 5).

3.8.11 Stability Studies of Formulated Nanosized Drugs Loaded Bio-flexy Films as per ICH Guidelines Q1B

The stability studies of the formulations revealed stable films (Fig. 9).

4 Conclusion

The objective of the research work is to formulate nanosized bio-flexy films loaded with Tiagabine (anticonvulsant) and using novel biopolymer from *Magnifera indica* (mango) fruit pulp. Isolated biopolymer by simple economic method using optimized concentration non-solvent acetone. Biopolymer being of edible was bio-safe, biodegradable, biocompatible, non-toxic, inert, non-irritant nature, non-reactive on soft palatal surface. It showed excellent film forming properties along with mucoadhesive and mucoretentive properties. *Magnifera indica* biopolymer showed

Table 5 Kinetics release of tiagabine-sodium CMC flexy films

Release kinetics analysis dynamic method formulations of tiagabine: sodium CMC flexy films							
Formulations	R^2					Best fit model	Mechanism of action
	Zero order	1st order	Higuchi matrix	Peppas	Hixon Crowell		
FET1 (1:0.5)	0.8894	0.8897	0.9356	0,9300	0.8896	Higuchi-matrix	Anomalous transport
FET2 (1:1)	0.8852	0.8853	0.9324	0.8424	0.8853	Higuchi-matrix	Anomalous transport
FET3 (1:3)	0.8868	0.8868	0.9377	0.9550	0.8868	Peppas Korsmeyer	Anomalous transport
FET4 (1:5)	0.8906	0.8908	0.9361	0.9514	0.8908	Peppas Korsmeyer	Anomalous transport
FET5 (1:6)	**0.8360**	**0.8363**	**0.9301**	**0.9084**	**0.8362**	**Higuchi-matrix**	**Fickian diffusion (Higuchi matrix)**
FET6 (1:10)	0.8960	0.8962	0.9372	0.9692	0.8961	Peppas Korsmeyer	Anomalous transport

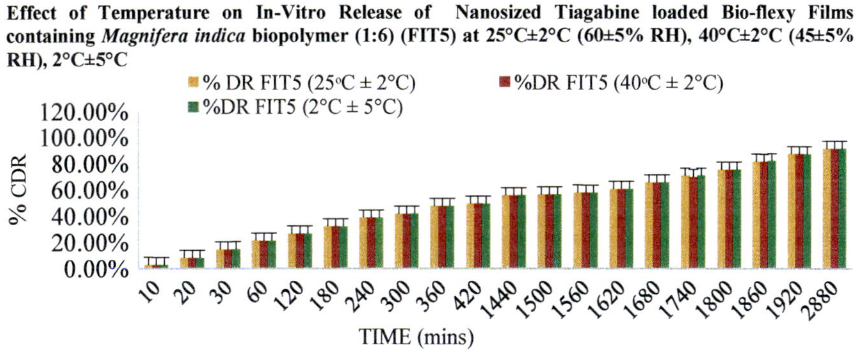

Fig. 9 Stability study graph of best formulations of nanosized tiagabine loaded bio-flexy films containing *Magnifera indica* biopolymer

percentage yield of 20.04% ± 0.001, yellowish brown in color, sweet odour, soluble in water. Its colour changing point was found to be 225 °C ± 4. It was tested positive for proteins and carbohydrates, amino acids were not present. It was characterized by physicochemical methods and by spectral analysis. It revealed their suitability as bio-excipients in films formulations.

Ratios were chosen at six levels for nanosized Tiagabine: *Magnifera indica* Biopolymer (1:0.5 to 1:10) and six levels for nanosized Tiagabine: Sodium Carboxyl Methyl Cellulose (1:0.5 to 1:10) for both drugs and each biopolymer for formulating flexy-films. Tiagabine unidirectional bio-flexy films were formulated by Solvent

Casting Method and evaluated. Bio-flexy Films formulations showed promising evaluation parameters performance comparable to that of standard flexy films. This is due to the fact that biomaterials had high molecular weight exhibited higher adhesion and better mucoadhesion than the synthetic polymer at the same concentration. This may be due to the presence of numerous disulphide bridges and carboxyl and hydroxyl groups, which adopt more favorable macromolecular conformation, and accessibility of its hydrogen- binding groups.

Soft Palatal Delivery can be preferred as an alternative therapy for treatment of convulsions in low dose with sub-minimal side effects, economic to patients and manufacturers directly and to physicians indirectly. 100% patient compliance can be achieved. It is an approach to deliver antiepileptic molecules to brain at lesser dose than oral dose to the patients by completely bypassing oral therapy. Since soft palate is devoid of taste buds, bitter tasting API molecules can be administered through this platform.

Prepared formulations were screened for mucoadhesivity, sustainability, *In-Vitro* performance. Thickness of nanosized Tiagabine loaded bio-flexy films containing *Magnifera indica* (FIT1–FIT6) was ranging from 0.046 mm ± 0.003 to 0.078 mm ± 0.002, Folding Endurance: 81–108, Surface pH: 7.01 ± 0.03 to 7.01 ± 0.02, Weight Uniformity: 0.032 ± 0.04 to 0.040 ± 0.03, Drug Content Uniformity: 71.0% ± 0.35 to 80.4% ± 0.30, Swelling Percentage: 61% ± 0.2 to 72% ± 0.1, Percentage Moisture Uptake (PTU): 2.0% ± 0.12 to 2.3% ± 0.10. Mucoretentive Study by Dynamic method revealed that Nanosized Tiagabine loaded bio-flexy films containing Magnifera indica biopolymer were mucoretentive for time period of 150–360 min Mucoadhesion Study revealed that Nanosized Tiagabine loaded bio-flexy films containing Magnifera indica biopolymer were mucoadhesive for time period of 180–420 min. The drug release pattern for formulations FIT1–FIT6 containing Magnifera indica biopolymer based on the $T_{50\%}$ and $T_{80\%}$ was found to be FIT5 (1:6) > FIT6 (1:10) > FIT2(1:1) > FIT4 (1:5) > FIT3 (1:3) > FIT1 (1:0.5). In-vitro drug release was performed for all the formulations and the data indicate that drug loaded formulations show the sustained release behavior. Graph was plotted between %CDR and time, the R^2 value, $T_{50\%}$ and $T_{80\%}$ were calculated from graph. Based on all above mentioned evaluation parameters, FIT5 (containing Tiagabine: Magnifera indica biopolymer (1:6)) Bio-flexy film having $R^2 = 0.9312$, Higuchi Matrix as best fit model, follows Fickian Diffusion (Higuchi Matrix) release mechanism, $T_{50\%}$: 21.5 h, $T_{80\%}$: 44.71 h using BITS Software 1.12. Stability study revealed stable bio-flexy films with no significant change in physical appearance and stable pH. Prepared formulations of Tiagabine loaded bio-flexy films containing *Magnifera indica* biopolymer were suitable for Soft Palatal Delivery.

Acknowledgements We would like to acknowledge Mr. Anuj Aggarwal (Chairman, DIT University), Prof. K.K. Raina (Vice Chancellor, DIT University) for providing platform for conducting the research work.

Conflict of Interest There is no potential conflict(s) of interest(s).

Note: This research did not receive any specific grant from funding agencies in public, Commercial, or not-for-profit sectors.

References

1. Khambhati AN., et al.: Dynamic Network Drivers of Seizure Generation, Propagation and Termination in Human Neocortical Epilepsy. In: PLOS Computational Biology Volume 11, Issue 12. (2015) 1-19.
2. Mohanachandran, P.S., et al.: Recent Trends in Brain Targeted Drug Delivery Systems: An Overview. In: Pharmacie Globale, International Journal of Comprehensive Pharmacy (IJCP). Volume 3, Issue 10. (2012) 1–9.
3. Vossler, D.G., et al.: Tiagabine in clinical practice: effects on seizure control and behavior. In: Epilepsy and Behavior Journal. Volume 28, Issue 2. (2013) 211-216.
4. Madhav, N.V.S., et al.: Recent Advances in Oral Mucoadhesive Drug Delivery Systems: A Review. In: International Journal of Pharmacy Research and Development. Volume 3, Issue 2. (2010) 1-15.
5. Madhav, N.V.S., et al.: A Smart Soft Palatal Platform for Oral Trans-mucosal Delivery by using Drug loaded *Mangifera indica* bio-plate. In: Proceedings of Indian Pharmaceutical Congress, BHU. (2007).
6. Madhav, N.V.S., et al.: Review on Biopolymers as Novel Bio-Excipients in Drug Delivery System. In: European Journal of Pharmaceutical and Medical Research. Volume 4, Issue 6. (2017)247-250.
7. Singh S, et al.: Targeted Nanomedicines: Effective Treatment Modalities for Cancer, AIDS and Brain Disorders. In: Nanomedicine Journal. Volume 4, Issue 1. (2008) 1748-6963.
8. Patil S., et al.: Method Development and Validation for Quantitative Analysis of Tiagabine Hydrochloride by Ultraviolet Spectrophotometry. In: International Journal of Chemical Sciences. Volume 6, Issue 1. (2008) 413-416.
9. Varley H.: Introductory Collection of Specimens and some General Techniques. In: Practical Clinical Biochemistry, 4th Edition, New Delhi, C.B.S. Publishers and Distributors. (2005) 21-29.
10. Nilani P., et al.: Formulation and Evaluation of Polysaccharide based Biopolymer—An Ecofriendly Alternative for Synthetic Polymer. In: Journal of Pharmaceutical Sciences and Research. Volume 2, Issue 3. (2010) 178-184.
11. Varshney, S., Madhav, N.V.S.: Bio-flexy Films Formulation for Delivery of Tiagabine via Oro Trans-Soft Palatal Route and Its *In-Vitro* Stability Study Approach. In: Asian Journal of Nanoscience and Material. Volume 2, Issue 3. (2019) 327-349.
12. Karki S., et al.: Thin films as an emerging Platform for Drug Delivery. In: Asian Journal of Pharmaceutical Sciences. Volume 11. (2016) 559-574.
13. Madhav, N.V.S., et al.: A Novelistic Trans-Soft Palatal Route for Smart Drug Delivery of Gentamicin by Mucoadhesive Plates. In: Proceedings of 17th Singapore Pharmacy Congress, Singapore. (2005).
14. Varshney, S., Madhav, N.V.S.: Development and Evaluation of Unidirectional Mucoadhesive Bio-Flexy Films loaded with Nanosized Topiramate using a novel biopolymer from Glycine max. In: Indian Journal of Pharmaceutical Education and Research. Volume 54, Issue 6. (2020) 618-629.

Varietal Evaluation in Okra for Yield and Yield Attributing Traits Under Mid-Hill Conditions of Garhwal Himalayas

Udit Joshi, D. K. Rana, T. S. Bisht, and Vivek Singh

Abstract Okra is a traditional and one of the most popular crops among various groups of vegetable crops, generally cultivated extensively in summer and rainy season in India. Although at present, a considerable number of commercial cultivars and hybrids of okra are available in the market, yet farmers are facing difficulties in production as the available cultivars are not well adapted to country's specific agro-climatic conditions. Hence, the present investigation was conducted for performance-based evaluation of twenty-six okra cultivars, including a standard check Arka Anamika (C) thus, selecting elite cultivars from the evaluated ones. A Randomized Block Design (RBD) with three replications was employed for aligning all the cultivars under study. The data for 32 quantitative traits were taken using five plants from each treatment corresponding to every replication. A fair amount of variation was observed in all the cultivars in terms of all the yield and its attributing traits under study. The analysis of variance depicted significant differences among genotype for all the traits under investigation where the yield per hectare ranged from 231.87 q/ha (VL Bhindi-2) to 86.95 q/ha (LC-1) with a mean value of 155.55 q/ha. Out of twenty-six cultivars, seventeen genotypes viz., Agri Bahar, Chanda, Hisar Naveen, Hisar Unnat, Kashi Kranti, Kashi Mohini, Kashi Pragati, Kaveri, LC-3, LC-4, LC-6, Lucky-666, Pusa A-4, Pusa Sawni, VL Bhindi-2, Vandana-241, and Varsha Uphar performed better than check cultivar Arka Anamika in terms of yield and its attributing traits. Therefore, these seventeen genotypes can be recommended for further study, future breeding aspects and cultivation in the mid-hill conditions of Garhwal Himalaya.

Keywords Okra · Cultivars · Evaluation · Mean performance · Quantitative · Yield

U. Joshi (✉) · D. K. Rana · T. S. Bisht · V. Singh
Department of Horticulture, School of Agriculture and Allied Sciences, H.N.B. Garhwal University, Srinagar (Garhwal), Uttarakhand 246174, India

N. A. Siddiqui et al. (eds.), *Advances in Environment Engineering and Management*, Springer Proceedings in Earth and Environmental Sciences,
https://doi.org/10.1007/978-3-030-79065-3_32

1 Introduction

Okra [*Abelmoschus esculentus* (L.) Moench 2n = 2x = 130] is the most crucial member and the vegetable crop of the Malvaceae family [25]. The ancient Egyptians cultivated it by the twelfth century [27]. Okra has its name derived from an African word and is considered a native to Northern Africa, including Ethiopia, Sudan, and areas around the Nile River [26]. Southeast Asia is regarded as the centre of diversity for cultivated and wild species of okra, as this region clearly showed overlapping for both forms [24]. Okra is summer and rainy season crop cultivated from tropics to subtropics widely [14]. Okra needs an average optimum temperature of 20–35 °C [15, 24] with 15 and 42 °C minimum and maximum temperature, respectively, for growth and development [26]. Fruit/pods of okra having high iodine content, are very useful for controlling goitre disease. Stem and fully ripened fruits of okra contain fibres used in the paper industry [5, 8]. Mucilage is obtained from the dry roots and stems soaked in water overnight and is used for clarifying sugarcane juice in preparation for quality jaggery manufacture in India.

Edible fruits/pods of okra carry 89 g water, 2–4 g protein, 0.3 g fat, 7.6 g carbohydrate, 92 mg calcium, 51 mg phosphorus, 0.6 mg iron and 249 mg potassium and vitamins like 88 I.U. vitamin-A, 13 mg vitamin-C and 0.6 mg nicotinic acid per 100-g of its fresh weight. Okra is also rich in iron and vitamin A, B and C [8, 11]. In India, Okra has the largest area and production followed by Nigeria in the world. India produces 72% of the total world okra. In India, okra is commercially grown in Gujrat, Maharashtra, Uttar Pradesh, Andhra Pradesh, West Bengal, Rajasthan, Tamil Nadu, Madhya Pradesh, Assam, Haryana, Punjab, and Karnataka. During 2018 the okra crop had average productivity of 11.97 t/ha having area and annual production of 0.509 million hectares and 6.095 million tonnes respectively all over India [4, 11].

At present, a vast number of commercial cultivars and hybrids of okra are available in the market. However, farmers face difficulties in production as the available cultivars are not well adapted to the country's agro-climatic condition and region, which demands an area and climate-specific recommendation about the suitability of genotypes a significantly lesser extent in the country [11]. Growers and farmers always face problems in selecting region-specific genotypes/cultivars for commercial cultivation. As okra is a highly preferred crop among the vegetable consumers both at the domestic and commercial level, it has great importance, and there is a necessity to improve and develop varieties that are well adapted to complex of agro-ecological conditions and for varieties with unique features. Hence the study was conducted to select elite cultivars from the twenty-six cultivars to make them available to farmers for cultivation in the local growing conditions. Thus the evaluation of twenty-six cultivars taking various yield and its attributing quantitative traits was done to identify promising genotypes suitable for local growing mid-hill conditions of Garhwal Himalaya.

2 Materials and Methods

2.1 Experimental Site

Investigation site for the present study was H.N.B. Garhwal University, Srinagar (Garhwal), Uttarakhand (India), Horticultural Research Centre, Chauras campus. The experimental site is situated in Alaknanda valley, which lies between 78°47′30″ E longitude and 30°13′0″ N latitude, at an elevation 540 m above M.S.L., in the lesser Himalayan region. The site of the experiment has a sub-tropical climate, showing an extreme range of the temperature during both winter and summer seasons. The temperature reaches up to 35–40 °C from May to June. The experimental site exhibits dry summer and rigorous rain in the month from April to July. The soil status of the area was sandy loam in texture and having pH 5.97. Nutrient status of the soil was 3.21% carbon, 238 kg/ha available nitrogen, 16.86 kg/ha available phosphorus, and 98.7 kg/ha available potassium.

2.2 Experimental Design and Plant Material Used

Twenty-six genotypes including one check cultivar (Arka Anamika) viz., Agri Bahar, Chanda, Hisar Naveen, Hisar Unnat, Kashi Kranti, Kashi Mohini, Kashi Pragati, Kashi Vibhuti, Kaveri, King Bhindi, LC-1, LC-2, LC-3, LC-4, LC-5, LC-6, Lucky-666, Parbhani Kranti, Punjab-8, Pusa A-4, Pusa Sawni, Super Anamika, VL Bhindi-2, Vandana-241, and Varsha Uphar were analysed in R.B.D. (Randomized block design) with three replications during the Kharif/Summer season of 2019 from month June to September.

2.3 Management and Cultural Practices

The field was prepared to a fine tilth by giving three ploughings followed by the harrowing. The whole experimental area was divided into seven blocks, each having length and width 23.05 m and 2.4 m respectively, every block was separated at 50 cm from each other. The plots of size 2.4 m × 1.8 m were prepared at 50 cm apart as per the layout plan with a rope, measuring tape, and spade. Before the sowing of seeds, the seeds were soaked overnight to facilitate faster germination. Two to three healthy seeds were sown by dibbling them 2 cm deep at a spacing of 60 cm × 60 cm. Well-decomposed farmyard manure (F.Y.M.) @ 25 tonnes per hectare was incorporated during the preparation of the field. The inorganic fertilizers were applied as per the recommendation, i.e. 150:60:50 kg N.P.K. per hectare in the form of urea, single super phosphate, and muriate of potash. Half a dose of nitrogen and a full amount of phosphorus and potassium were applied in the soil before the sowing of seeds. The

Remaining half dose of N was used in two split doses, i.e. one fourth after 30 days of sowing and one fourth at the time of flowering and fruiting.

2.4 Data Recording and Analysis

During the fieldwork, various Qualitative and Quantitative traits were recorded. Selected five plants from each treatment were randomly tagged for recording the multiple observations. Chlorophyll content was recorded by chlorophyll meter/SPAD meter. The quality analysis Ascorbic Acid content and ash content were estimated as per procedure given by Ranganna [21] and A.O.A.C. [10]. For each character observed in the study, statistical analysis was carried out using MS-Excel spreadsheets and method given by [23].

3 Result and Discussion

3.1 Analysis of Variance

The ANOVA studies of the experiment showed that the mean sum of square attributable to genotypes was highly significant for all characters studied viz., the first emergence of seedling (Days), plant height, number of primary branches per plant, internodal length, stem girth, petiole length, petiole diameter, leaf length, number of epicalyx segments, days to first flowering, number of nodes at first flowering, flower diameter, fruit length, fruit diameter, average fruit weight, number of fruits per plant, number of ridges per fruit, pedicel length, pedicel diameter, days taken to first fruit set, days taken to first fruit harvest, flesh thickness, number of seeds per fruit, seed index, yield per plot, yield per hectare, total soluble solids, ascorbic acid, chlorophyll content, ash content, moisture content and shelf life which confers the presence of a good range of variations in the germplasms under study. Similar results were reported by the earlier workers like [1, 6, 7, 12, 19, 20, 22] while it was in contrast with [13] in case of no. of ridges per fruit which is because there is lesser variability of this trait in most of the genotypes used in that study.

3.2 *Mean Performance*

3.2.1 Mean Performance of Okra Genotypes for Growth and Yield Traits

The data regarding the mean performance of the okra cultivars in furnished in Tables 1, 2, and 3 for growth, yield, and quality parameters. Days to the emergence of seedling decides the maturity time of the crop, which is related to market price fetching by the crop as early available products in the market fetches reasonable price; hence minimum days to emergence is a desirable parameter for crop selection. Minimum days to the first emergence of seedling were recorded in the genotype Parbhani Kranti. In contrast, the genotype Kashi Mohini had taken maximum days to the first emergence of the seedling. Plants with taller stature are desirable as they lead to a higher number of branches and increased yield. The maximum height was recorded in Punjab-8, while the minimum height of the plant was observed in VL Bhindi-2 similar results, but in other genotypes [3, 9, 20].

A higher number of branches per plant leads to a higher number of fruits and ultimately resulting in increased productivity hence becomes a desirable parameter. Highest primary branches in number were recorded in LC-2, and the minimum number of primary branches was observed in Vandana-241. The maximum internodal length was recorded in Pusa Sawani, whereas the minimum internodal length was recorded in VL Bhindi-2. Stem girth is an important parameter because it prevents damage due to severe winds that can cause lodging in plants. Maximum stem circumference was observed in Pusa A-4, while minimum stem girth was recorded in Kaveri. Higher petiole length is a desirable trait as it decides the exposure of leaf to the sunlight, which leads to more accumulation of photosynthates. Maximum petiole length was recorded in VL Bhindi-2, whereas minimum petiole length was observed in LC-1. Higher petiole diameter leads to the stability of the leaf against strong winds and prevents leaves from falling off. The maximum diameter of petiole was recorded in LC-1, whereas the minimum diameter of petiole was observed in Parbhani Kranti [2, 3, 20].

Higher leaf length is a desirable trait as it leads to better exposure of leaves to sunlight, which ultimately leads to an increase in productivity. The maximum length of the leaves was recorded in LC-1, whereas it was minimum in LC-6. The maximum number of epicalyx segments was recorded in LC-5, while the minimum number of epicalyx segments was observed in King Bhindi. Earliness in flowering is one of the most critical factors in deciding earliness in the fruit harvesting stage. Minimum days to first flowering were observed in the genotype Varsha Uphar, whereas LC-2 genotype took maximum days to first flowering, which confers that Varsha Uphar can be taken as a genotype for the exploitation of earliness, similar results were also observed by [2, 16].

The lowest number of nodes at first flowering was recorded in the genotype Kashi Pragati, while the highest was observed in the genotype Arka Anamika. Higher flower diameter leads to increased area exposure to pollinators hence resulting in increased

Table 1 Mean performance of okra genotypes for growth and yield traits

S. No.	Genotypes	Days to the first emergence of seedling	Plant height (cm)	No. of primary branches	Internodal length (cm)	Stem girth (mm)	Petiole length (cm)	Petiole diameter (mm)	Leaf length (cm)	No. of epicalyx segments	Days to first flowering
1	Agri Bahar	5.33	214.87	3.80	7.20	37.91	35.82	6.95	21.17	11.00	45.33
2	Chanda	5.33	239.47	4.80	9.17	37.21	36.22	6.31	21.46	10.67	46.73
3	Hisar Naveen	5.33	218.07	4.53	6.62	37.53	38.36	6.65	20.30	10.87	48.80
4	Hisar Unnat	5.33	215.67	5.33	8.81	36.98	35.67	6.77	20.53	10.40	47.13
5	Kashi Kranti	7.33	194.47	5.13	6.88	35.48	35.11	6.36	21.57	11.80	49.07
6	Kashi Mohini	7.67	236.20	4.53	6.11	36.87	34.34	6.14	21.22	10.87	50.20
7	Kashi Pragati	5.33	231.20	4.73	8.15	36.00	37.09	6.36	22.15	11.73	46.13
8	Kashi Vibhuti	7.67	166.07	3.87	5.97	35.10	37.17	6.13	22.27	11.60	52.53
9	Kaveri	6.33	213.87	4.80	7.94	26.22	34.25	7.01	20.61	11.40	48.33
10	King Bhindi	5.67	247.40	5.87	9.09	40.65	38.37	7.01	21.44	10.07	50.00
11	LC-1	6.33	224.20	10.07	9.53	38.37	31.49	7.13	22.76	11.67	53.40
12	LC-2	5.67	192.40	11.47	7.82	38.17	32.91	6.55	20.43	10.87	53.87
13	LC-3	5.67	234.47	5.20	7.01	36.83	34.10	5.91	21.77	10.53	49.00
14	LC-4	5.67	216.80	4.13	7.37	34.54	36.01	6.19	22.11	11.87	45.20
15	LC-5	5.67	243.93	7.27	8.39	36.92	38.12	6.00	20.96	11.93	48.87
16	LC-6	5.33	211.60	5.87	7.51	35.87	34.96	6.22	19.75	11.73	49.67
17	Lucky-666	6.67	201.23	4.07	6.35	34.63	33.17	6.29	20.88	10.40	45.93
18	Parbhani Kranti	4.67	223.27	7.80	9.37	33.83	34.48	5.57	21.15	10.93	51.93
19	Punjab-8	5.33	277.87	5.87	8.78	41.29	35.82	6.58	20.65	10.80	51.60

(continued)

Table 1 (continued)

S. No.	Genotypes	Days to the first emergence of seedling	Plant height (cm)	No. of primary branches	Internodal length (cm)	Stem girth (mm)	Petiole length (cm)	Petiole diameter (mm)	Leaf length (cm)	No. of epicalyx segments	Days to first flowering
20	Pusa A-4	5.67	241.53	5.93	6.19	42.34	34.41	6.52	21.35	10.53	47.20
21	Pusa Sawni	5.33	245.93	4.93	10.05	34.85	34.67	6.41	22.03	11.33	50.60
22	Super Anamika	5.33	258.60	5.00	9.01	32.94	36.58	6.83	21.96	10.13	48.87
23	VL Bhindi-2	5.33	138.20	4.80	4.71	39.09	40.01	6.79	22.33	10.40	48.87
24	Vandana-241	5.67	203.27	3.80	9.27	38.07	36.32	6.38	20.39	10.93	46.93
25	Varsha Uphar	5.33	205.33	4.13	7.76	35.12	35.42	6.69	20.77	11.40	44.87
26	Arka Anamika (C)	5.00	247.13	7.73	9.32	40.60	34.41	6.36	22.45	10.53	51.33
	Mean	**5.77**	**220.89**	**5.59**	**7.86**	**36.67**	**35.59**	**6.47**	**21.33**	**11.02**	**48.94**
	Range (Min.)	4.67	138.20	3.80	4.71	26.22	31.49	5.57	19.75	10.07	44.87
	(Max.)	7.67	277.87	11.47	10.05	42.34	40.01	7.13	22.76	11.93	53.87
	SE(d)	0.62	2.39	0.30	0.30	0.95	1.28	0.30	0.77	0.17	1.37
	C.D	1.24	4.82	0.61	0.60	1.92	2.58	0.60	1.55	0.35	2.77
	C.V	13.09	1.33	6.65	4.62	3.18	4.40	5.68	4.42	1.91	3.44

Table 2 Mean performance of okra genotypes for growth and yield traits

S. No.	Genotypes	No. of nodes at flowering	Flower diameter (cm)	Fruit length (cm)	Fruit diameter (mm)	Average fruit weight (g)	No. of fruits per plant	No. of ridges per fruit	Pedicel length (cm)	Pedicel diameter (mm)	Days to first fruit set	Days to first fruit harvest
1	Agri Bahar	6.07	8.10	16.32	15.72	19.49	45.67	5.13	4.37	6.88	46.00	52.60
2	Chanda	6.13	8.07	17.77	16.73	18.73	69.27	5.00	5.43	5.95	47.87	53.67
3	Hisar Naveen	6.33	9.13	18.36	15.35	19.49	53.53	5.00	4.46	6.56	50.13	56.00
4	Hisar Unnat	6.33	8.20	16.76	16.16	19.82	62.33	5.00	4.99	6.32	48.67	55.60
5	Kashi Kranti	6.47	8.33	15.65	14.78	19.61	48.73	5.00	4.33	6.56	50.13	56.40
6	Kashi Mohini	6.00	9.50	18.29	16.36	18.91	60.20	5.40	5.27	6.13	51.40	57.20
7	Kashi Pragati	4.80	8.97	19.62	15.80	19.48	51.40	5.00	4.17	6.11	51.07	56.33
8	Kashi Vibhuti	5.67	9.00	18.49	16.26	17.91	56.67	5.67	5.13	6.90	53.40	58.47
9	Kaveri	5.60	7.23	17.39	15.71	19.85	57.00	5.00	3.53	6.47	49.80	55.47
10	King Bhindi	6.67	7.67	16.74	15.52	20.04	51.20	5.53	5.20	6.30	51.13	56.67
11	LC-1	6.67	9.09	14.10	16.38	18.99	42.40	6.00	5.57	6.79	54.60	60.80
12	LC-2	6.93	8.90	19.83	15.89	18.11	48.00	6.00	5.17	6.19	54.87	60.40
13	LC-3	5.80	7.37	17.08	15.50	21.47	47.20	5.20	5.15	6.28	50.27	55.67
14	LC-4	6.07	8.00	17.25	16.21	18.91	59.13	5.00	3.66	7.39	46.60	52.47
15	LC-5	6.53	7.53	17.53	15.90	18.47	50.00	5.20	5.24	6.54	51.67	57.40
16	LC-6	6.20	6.77	18.77	14.84	18.73	57.07	5.00	4.10	6.17	50.87	56.67
17	Lucky-666	5.00	7.70	18.88	15.78	19.15	50.47	5.33	4.31	6.17	48.41	53.73
18	Parbhani Kranti	7.00	7.82	17.44	14.65	19.78	51.60	5.00	4.25	6.04	53.00	59.33

(continued)

Table 2 (continued)

S. No.	Genotypes	No. of nodes at flowering	Flower diameter (cm)	Fruit length (cm)	Fruit diameter (mm)	Average fruit weight (g)	No. of fruits per plant	No. of ridges per fruit	Pedicel length (cm)	Pedicel diameter (mm)	Days to first fruit set	Days to first fruit harvest
19	Punjab-8	6.80	9.50	16.98	15.31	21.35	46.47	5.53	5.17	6.00	52.70	58.53
20	Pusa A-4	6.07	8.53	18.60	15.18	20.21	57.53	5.00	4.24	5.61	50.13	55.80
21	Pusa Sawni	6.47	7.31	16.97	14.37	19.95	52.93	5.00	4.21	6.43	51.73	57.53
22	Super Anamika	6.40	7.30	18.14	15.26	19.10	50.13	5.00	3.67	6.64	50.07	54.67
23	VL Bhindi-2	6.33	9.30	22.32	18.26	23.16	59.73	5.00	4.41	7.43	49.67	56.27
24	Vandana-241	6.60	7.61	17.03	15.43	20.36	70.87	5.00	5.14	6.44	48.53	54.53
25	Varsha Uphar	4.80	8.00	18.17	15.47	19.13	48.73	5.00	4.26	6.44	46.47	53.00
26	Arka Anamika (C)	7.47	7.59	17.87	15.68	18.99	54.40	5.00	5.45	6.34	52.80	58.40
	Mean	**6.20**	**8.17**	**17.78**	**15.71**	**19.58**	**53.95**	**5.19**	**4.65**	**6.43**	**50.46**	**56.29**
	Range (Min.)	4.80	6.77	14.10	14.37	17.91	42.40	5.00	3.53	5.61	46.00	52.47
	(Max.)	7.47	9.50	22.32	18.26	23.16	70.87	6.00	5.57	7.43	54.87	60.80
	SE(d)	0.48	0.13	0.76	0.38	1.03	4.99	0.11	0.16	0.24	0.81	1.12
	C.D	0.98	0.26	1.54	0.77	2.08	10.05	0.23	0.33	0.49	1.64	2.25
	C.V	9.57	1.96	5.26	2.97	6.46	11.33	2.67	4.29	4.61	1.97	2.43

Table 3 Mean performance of okra genotypes for yield, seed and quality traits

S. No.	Genotypes	Flesh thickness (mm)	No. of seeds per fruit	Seed index	Yield per plot (kg)	Yield per hectare (q)	T.S.S (° Brix)	Ascorbic acid (mg/100 g)	Chlorophyll content (SPAD)	Ash content	Moisture content	Shelf life
1	Agri Bahar	1.53	52.60	6.34	8.08	186.95	9.63	17.19	58.33	7.65	90.23	4.67
2	Chanda	1.71	73.20	6.36	8.57	198.44	10.13	17.47	55.64	9.44	90.60	5.33
3	Hisar Naveen	1.56	71.60	4.84	7.78	180.19	10.13	17.68	54.22	10.77	88.60	6.00
4	Hisar Unnat	1.63	70.87	6.60	7.33	169.59	12.57	18.31	59.85	9.75	90.60	7.00
5	Kashi Kranti	1.59	78.13	6.79	6.77	156.71	12.27	18.44	59.61	9.41	90.87	5.33
6	Kashi Mohini	1.60	53.53	7.45	6.74	155.94	12.10	18.19	63.08	11.24	91.80	4.00
7	Kashi Pragati	1.46	65.20	6.76	7.48	173.07	9.53	18.15	59.39	8.98	92.00	5.33
8	Kashi Vibhuti	1.74	53.07	5.99	5.59	129.72	9.00	16.50	59.46	10.35	87.87	5.00
9	Kaveri	1.49	54.07	7.27	6.26	144.90	10.77	17.24	55.49	10.27	90.53	5.33
10	King Bhindi	1.47	55.67	6.46	4.73	109.41	9.90	17.40	59.83	9.78	86.60	4.67
11	LC-1	1.67	86.80	5.71	3.84	88.84	11.80	16.77	59.48	9.58	88.80	4.67
12	LC-2	1.63	80.67	5.16	5.27	121.83	12.53	15.40	54.66	9.63	89.60	4.67
13	LC-3	1.50	61.67	6.34	6.80	157.40	12.50	18.25	61.37	8.50	89.87	4.33
14	LC-4	1.52	61.53	5.63	7.29	168.75	8.63	18.20	54.79	9.64	90.13	5.67
15	LC-5	1.99	62.33	6.46	3.76	86.95	10.40	19.16	59.68	9.66	91.13	3.33
16	LC-6	1.57	48.93	5.16	7.51	173.76	10.33	15.77	54.05	8.61	91.40	4.67
17	Lucky-666	1.67	44.73	6.97	8.74	202.31	10.37	17.16	59.07	9.71	91.90	6.00
18	Parbhani Kranti	1.27	79.13	5.62	5.08	117.59	13.90	17.38	57.95	9.80	90.00	4.67
19	Punjab-8	1.48	83.73	6.91	5.40	124.92	12.63	18.52	55.91	11.54	91.53	4.33

(continued)

Table 3 (continued)

S. No.	Genotypes	Flesh thickness (mm)	No. of seeds per fruit	Seed index	Yield per plot (kg)	Yield per hectare (q)	T.S.S (° Brix)	Ascorbic acid (mg/100 g)	Chlorophyll content (SPAD)	Ash content	Moisture content	Shelf life
20	Pusa A-4	1.54	68.40	6.01	8.74	202.31	9.91	16.89	61.06	11.23	92.40	6.00
21	Pusa Sawni	1.58	68.07	6.67	5.97	138.11	10.57	16.18	57.04	10.11	90.40	6.33
22	Super Anamika	1.56	57.67	6.48	5.12	118.44	8.07	16.46	60.01	7.71	89.00	5.00
23	VL Bhindi-2	1.87	83.27	7.40	10.02	231.87	9.90	16.78	54.67	10.73	91.90	3.33
24	Vandana-241	1.44	50.27	7.62	9.17	212.19	12.23	15.55	55.13	7.41	87.93	3.67
25	Varsha Uphar	1.72	66.93	5.96	6.78	157.02	10.23	15.90	54.05	10.79	92.37	5.67
26	Arka Anamika (C)	1.59	60.60	6.50	5.94	137.22	12.37	18.73	59.71	10.71	91.17	3.33
	Mean	**1.59**	**65.10**	**6.36**	**6.72**	**155.55**	**10.86**	**17.30**	**57.83**	**9.73**	**90.36**	**4.94**
	Range (Min.)	1.27	44.73	4.84	3.76	86.95	8.07	15.40	54.05	7.41	86.60	3.33
	(Max.)	1.99	86.80	7.62	10.02	231.87	13.90	19.16	63.08	11.54	92.40	7.00
	SE(d)	0.10	3.59	0.40	0.31	7.11	0.37	0.47	2.66	0.62	0.90	0.39
	C.D	0.21	7.23	0.81	0.62	14.32	0.74	0.94	5.37	1.25	1.81	0.79
	C.V	7.96	6.76	7.74	5.60	5.60	4.13	3.30	5.64	7.82	1.22	9.68

pollination and improved yield. The maximum flower diameter was recorded in Kashi Mohini, whereas the minimum flower diameter was observed in LC-6 [16]. Length of the fruits becomes an important trait as it contributes directly to the weight of fruit, thereby affects the yield of fruits. Maximum fruit length was recorded in VL Bhindi-2, whereas minimum fruit length was recorded in LC-1. The higher diameter increases the fruit weight and hence increase the fruit/pod yield. The maximum diameter of fruit was recorded in VL Bhindi-2, whereas the minimum diameter of fruit was observed in Pusa Sawani. The present findings were following the results obtained by [2] but disputes with the findings of [9, 16] as they recorded a higher value of fruit diameter.

Fruit weight directly affects fruit/pod yield; hence it becomes a useful parameter for selection. Maximum average fruit weight was observed in the genotype VL Bhindi-2 whereas, minimum average fruit weight was followed in Kashi Vibhuti. The highest number of fruits per plant was observed in the genotype Vandana-241, while the lowest number of fruits per plant was recorded in the genotype LC-1 [3, 13]. Ridges in fruits are a feature that aids in the seed dispersal and facilitates the removal of seeds during thrashing [17]. The highest ridges per fruit were recorded in LC-1 (6.00). In contrast, the lowest ridges per fruit were recorded in Chanda, Hisar Naveen, Hisar Unnat, Kashi Kranti, Kashi Mohini, Kaveri, LC-4, LC-6, Parbhani Kranti, Pusa A-4, Pusa Sawani, Super Anamika, VL Bhindi-2, Vandana-241, Varsha Uphar and Arka Anamika as most of the genotypes have more or less similar no. of ridges in the fruit [13].

The maximum pedicel length was recorded in LC-1, whereas the minimum pedicel length was recorded in Kaveri. A higher pedicel diameter is desirable as it supports the pod/fruit, attached to the stem or branches. The maximum diameter of the pedicel was recorded in VL Bhindi-2, whereas the minimum diameter of fruit was observed in Pusa A-4 [19, 22]. Lesser days to fruit set decides earliness, which is one of the most important factors determining the harvesting stage of fruits and is desirable for fetching better market price. Minimum days to first fruit set was recorded in the genotype Agri Bahar, whereas the genotype LC-2 had taken maximum days to the first fruit set. Minimum days to first fruit harvest was recorded in the genotype LC-4. In contrast, the genotype LC-1 had taken maximum days to first fruit harvest; the results were disputed with the findings of as the earliness in harvesting depends on both the genotypes and environmental conditions [16].

3.2.2 Mean Performance of Okra Genotypes for Yield, Seed, and Quality Traits

Higher flesh thickness is a desirable trait as it directly contributes to the fruit/pod yield. Maximum flesh thickness was recorded in LC-5, while minimum flesh thickness was observed in Parbhani Kranti. The higher no. of seed per fruit is desirable for raising a crop for seed production practices. The highest number of seeds per fruit was observed in the genotype LC-1, whereas the lowest number of seeds per fruit was observed in Lucky-666. The maximum seed index was recorded in Vandana-241, while the minimum seed index was followed in Hisar Naveen [9, 20].

Getting the highest yield per unit area for better returns is the primary objective of crop production. Getting a higher fruit yield is the prime and ultimate goal of any plant breeder; therefore, it requires the highest consideration. The maximum yield per plot was observed in the genotype VL Bhindi-2, while the minimum yield per plot was recorded in the genotype LC-5. Yield per hectare gives a good idea about the yield that can be obtained in a large area. Yield per hectare was recorded minimum in LC-5 and maximum in VL Bhindi-2 [13, 16, 23].

Higher total soluble solid is a desirable trait as it is a critical quality parameter and determines the quality of the produce. Maximum total soluble solids were observed in the genotype Parbhani Kranti, whereas minimum total soluble solids were observed in Super Anamika. Ascorbic acid content is an important quality parameter as it increases the desirability and consumer acceptability of the produce. Maximum ascorbic acid was observed in LC-5 while minimum ascorbic acid was recorded in genotype LC-2. The higher chlorophyll content is desirable as it directly influences the yield due to the increased accumulation of photosynthates. In the present study, chlorophyll content was found to be low as it was recorded when the plant was going to senescence phase. Chlorophyll content (SPAD) was minimum in Varsha Uphar and LC-6 and maximum in Kashi Mohini [9, 19, 22].

Ash content depicts the number of inorganic minerals present in the product; hence higher ash content is desirable for the produce. Maximum ash content was observed in the genotype Punjab-8, whereas minimum ash content was observed in Vandana-241, which was similar to the results obtained by [18]. Higher moisture content adds to the fruit weight but reduces the shelf life. Maximum moisture content was observed in the genotype Pusa A-4, whereas minimum moisture content was observed in King Bhindi. Shelf life has a crucial role in determining keeping quality of the pods/fruits. Fruits with longer shelf life are desirable as they can be transported to markets at longer distances.

In contrast, fruits with lower shelf life are not desirable as they are vulnerable to long-distance transport and disease injury. Maximum shelf life was observed in the genotype Hisar Unnat. In contrast, minimum shelf life was observed in Arka Anamika, LC-5, and VL Bhindi-2; the results were in close conformity with the previous findings [16, 19].

4 Conclusion

From the present investigation, based on mean performance, it can be concluded that seventeen genotypes viz., Agri Bahar, Chanda, Hisar Naveen, Hisar Unnat, Kashi Kranti, Kashi Mohini, Kashi Pragati, Kaveri, LC-3, LC-4, LC-6, Lucky-666, Pusa A-4, Pusa Sawni, VL Bhindi-2, Vandana-241 and Varsha Uphar gave a high yield of fruit and also performed better for other yield attributing traits than check cultivar Arka Anamika. The cultivar VL Bhindi-2 might act as a strong substitute for growing in Himalayan conditions. Further testing of these genotypes will be a great aid before releasing as an alternative for various presently existing okra varieties. Along with this, the cultivars can be involved in the future breeding programs for superior varieties or hybrids development for yield and quality improvement of okra under the subtropical Himalayan regions of Uttarakhand.

References

1. Akotkar, K. P., De, D. K., and Pal, A. K. 2010. Genetic Variability and Diversity in Okra (*Abelmoschus esculentus* L. Moench). El. J. P. Breed. 1(4):393–398.
2. Aminu, D., Bello, O. B., and Gambo, B. A. 2016. Varietal performance and correlation of okra pod yield and yield components. Agri. and Env. 8:112–125.
3. Amiteye, S., Amitaaba, T., and Amoatey, H. M. 2019. Morphological Characterization of Accessions of Okra (*Abelmoschus Spp* L.) Int. J. Pure App. Biosci. 7(1):1–13.
4. Anonymous. 2018. Indian Horticulture Database, National Horticulture Board, Gurgaon www.nhb.gov.in
5. Bell, A. L. 1988. Plant fibers for paper making. Liliaceae, Oregon, pp. 60.
6. Chandramouli, B., Shrihari, D., Rao, A. V. D., and Rao, M. P. 2016. Studies on genetic variability, heritability, and genetic advance in okra [*Abelmoschus esculentus* (L.) Monech] genotypes. Plant Arch. 16(2):679–682.
7. Chaukhande, P., Chaukhande, P. B., and Dod, V. N. 2011. Genetic variability in okra. National Symposium on Vegetable Biodiversity. pp. 30.
8. Choudhary, B. R. 2015. Vegetables, (4th Ed.) *National Book Trust. New Delhi*, pp. 62–64.
9. Deepanshu, and Shamd, A. 2017. Genetic variability, heritability, and correlation coefficient in okra (*Abelmoschus esculentus* (L.) Moench) in Allahabad agroclimatic condition. Plant Arch. 17(2):1597–1602.
10. International AOAC., 1990. Official methods of analysis of A.O.A.C. International. Arlington, VA: A.O.A.C. Intl. Pv (Loose-Leaf).
11. Joshi, U., Rana, D.K., and Singh, V. 2020. Characterization study based on the morphology of various okra [*Abelmoschus esculentus* (L.) Moench.] genotypes. Journal of Emerging Technologies and Innovative Research. 7(5):701–710. https://doi.org/10.1729/Journal.23754
12. Kandasamy, R. 2015. Variability studies in okra (*Abelmoschus esculentus* L.). The Asian Journal of Horticulture. 10(1):60–63.
13. Khajuria, R. K., Sharma, J. P., Samnotra, R. K., Kumar, S., and Ranjit, K. 2015. Variability studies in Okra (*Abelmoschus esculentus* L. Moench). El. J. Pl. Breed. 10: 226–227.
14. Kochar, S. L. 1986. Tropical Crops. A Text Book of Economic Botany. Macmillan Indian Ltd. pp. 263–264.
15. Martin, F. W. 1982. Okra, potential multiple-purpose crop for the temperate zones and tropics. Econ. Bot. 36(3): 340–345.
16. Mishra, A., Mishra, H. N., Senapati, N., and Tripathy, P. 2015. Genetic variability and correlation studies in Okra (*Abelmoschus esculentus* (L.) Moench). El. J. Pl. Breed. 6(3):866–869.
17. Ogwu, M. C., Ohwo, O. U., and Osawaru, E. M. 2018. Morphological Characterization of Okra (*Abelmoschus* [Medik.]) Accessions. Makara J. Sci. 22(2):67–76.
18. Olivera, D. F., Mugridge, A., Chaves, A. R., Mascheroni, R. H., and Viña, S. Z. 2012. Quality Attributes of Okra (*Abelmoschus esculentus* L. Moench) Pods as Affected by Cultivar and Fruit Size. J. Food Res. 1(4):224–235.
19. Pachiyappan, R., and Saravannan, K. 2016. Studies on genetic variability and correlation for fruit yield and fruit quantity characters of okra. The Asian J. Hort. 11(1):101-104.
20. Rajesh, J., Prasad, V. M., and Kerketta, A. 2018. Evaluation of Different Okra [*Abelmoschus esculentus* (L.) Moench] Hybrids for Yield and Yield Attributes under Allahabad Agro-climatic Condition, Int. J. Pure App. Biosci. 6(5):1343–1346.

21. Ranganna, S. 2015. Analysis and quality control for fruits and vegetable products. Tata Mc Grew Hill Publication. New Delhi, India. 2nd Ed. pp. 110–112.
22. Sawant, S. N., Nagre, P. K., Gudadhe, P. S., and Narkhede, G. W. 2014. Correlation coefficient and path analysis studies in okra (*Abelmoschus esculentus* L. Moench). Int. J. Trop. Agri. 32(3/4):341–347.
23. Sheoran, O. P., Tonk, D. S., Kaushik, L. S., Hasija, R. C., and Pannu, R. S. 1998. Statistical Software Package for Agricultural Research Workers. Recent Advances in information theory, Statistics & Computer Applications by D.S. Hooda & R.C. Hasija Department of Mathematics Statistics, CCS HAU, Hisar 139–143.
24. Singh, D. K., and Jain, S. K. 2012. Performance of okra hybrids for quantitative attributes. Pantnagar J. Res. 10(1):123-129.
25. Sorpong. 2012. Okra [*Abelmoschus esculentus* (L.) Moench] as a valuable vegetable of the world. Ratar. Povrt. 49:105–112.
26. Tripathi, K. K., Govila, O. P., Warrier, R. and Ahuja, V. 2011. Biology of *Abelmoschus esculentus* L. (Okra). Series of Crop Specific Biology Documents (35 p.), Department of Biotechnology, Ministry of Science & Technology & Ministry of Environment and Forest, Govt. of India.
27. Voss, R., and Bell, M. 2007. Okra. World Vegetables, 2nd ed. Aspen Publications. pp. 843.

Population Assessment and Phyto-chemical Screening of *Meconopsis aculeata* Royle an Endangered Medicinal Plant of Western Himalaya

Vandana Shukla, Ankit Singh, and A. R. Nautiyal

Abstract Alpine areas are hot spot of biodiversity and harbor high value medicinal plants. However, habitat destruction and anthropogenic disturbance have led many plant species to various levels of threat. *Meconopsis aculeata* Royle belonging to family Papaveraceae is one such alpine plant species with known medicinal properties. It is analgesic and febrifuge and has cooling potency. The plant is used in Tibetan medicine and used for the treatment of bones near the ribes. The flowers are used to treat asthma, body pain and fever. In the alpine areas of Uttarakhand the species has been found to have restricted distribution confined to specific habitats. *M. aculeata* has been reported to be endangered as per the IUCN red list and thus there is a need to make a population assessment of the species in the region. Thus, present study was conducted in areas of its occurrence representing the natural populations in Districts Chamoli and Rudraprayag. The sites were Valley of Flowers, Hemkund, Madhmaheshwar, Rudranath and Tungnath. The study revealed a substantial variation in frequency, density and IVI in the five sites. Highest Frequency was 47.22% in Tungnath and lowest 25.81% in Rudranath whereas maximum IVI 9.55 was recorded in Valley of Flowers and minimum 2.96 in Rudranath. Remaining three sites had IVI ranging from 3.72 to 4.05. Contrary to this, the highest density with 2.75 individuals/m^2 was recorded in Madhmaheshwar followed by 2.00 individuals/m^2 in Tungnath, 1.15 ind/m^2 in Valley of Flowers and minimum 0.71 and 0.58 ind/m^2 in Rudranath and Hemkund respectively. On overall assessment basis the species is critically endangered. Phytochemical constituents like terpenoids, tannins, alkaloid and phenols were determined in leaves from all the populations that also revealed considerable variations. The mature individuals of the species in the populations were in low number that further indicates need for devising conservation strategy for the species.

Keywords *Meconopsis aculeata* · Endangered · Himalaya · Population · Phytochemical

---

V. Shukla (✉) · A. Singh · A. R. Nautiyal
High Altitude Plant Physiology Research Centre (HAPPRC), H.N.B. Garhwal University, Post Box: 14, Srinagar, Garhwal, Uttarakhand 246174, India

N. A. Siddiqui et al. (eds.), *Advances in Environment Engineering and Management*, Springer Proceedings in Earth and Environmental Sciences,
https://doi.org/10.1007/978-3-030-79065-3_33

1 Introduction

The Himalayan region expand around 595,000 km^2 and harbor rich biodiversity with nearly 18,440 plant species, out of these 1800 are medicinal plants, 675 wild edible and about 30% are endemic to the region [1]. For the protection of the biodiversity 5 Biosphere reserves, 28 National parks and 98 Wildlife sanctuaries with an area about 51,899 km^2 have been established in Indian Himalayan region [2].

In the last few decades, much attention has been paid to the conservation of the medicinal plants, due to their increasing demand and economic point of view [3].

Most of the high value medicinal plants of the Himalayan region are facing high threat and listed as endangered and critically endangered in Indian Red data book [4–6]. Over-exploitation, climate change, habitat degradation, pollution, and introduction of alien species are major threats for useful plant species.

High altitude alpine meadows contain rich herbaceous species diversity i.e., *Bistorta macrophylla* (D. Don) Sojak, *Danthonia cachemyriana* Jaub. & Spach, *Potentilla microphylla* D.Don, *Trachydium roylei* Lindl etc. In the Himalayan region several medicinal plants species found which are in use since time immemorial i.e. *Aconitum heterophyllum* Wall ex Royle, *Picrorhiza Kurrooa* Royle ex Benth, *Dactylorhiza hatagirea* (D.Don) Soó, *Meconopsis aculeata* Royle, etc. These plants naturally grows in alpine meadows, slopes and on the rock and boulder.

Meconopsis aculeata Royle belonging to family papaveraceae commonly known as Himalayan poppy or Blue poppy and Kalihari in Garhwal, is an endemic endangered medicinal plant that naturally grow on moist rocks and boulders.

The plant is perennial and monocarpic with a height between 1 and 2 feet, prickly, erect, unbranched stem. It is distributed in the Western Himalayan region from Kashmir to Kumaon and found at an elevation range of 3300–4600 m [7]. In Uttarakhand, the plant grows between 3000 and 4700 m. It has analgesic, febrifuge, cooling potency, used in Tibetan medicine [8] and also to treat the bones near the ribs related problem [9]. Flower of the species used for the treatment of asthma, fever, cough, and commonly used by inhabitants of Himachal Pradesh [10] and Uttarakhand [11].

Several studies reveal that plants are rich source of natural antioxidants [12, 13] and polyphenolics which can be utilised further as antioxidant supplements [14]. Due to the high medicinal demand, it is over-exploited from wild and thus it is banned for wild collection [15].

No ecological work has been conducted in Garhwal Himalaya. Therefore, the present study was carried out to evaluate population status, habitat assessment and categorization of threat for valuable medicinal plant *Meconopsis aculeata* (Fig. 1).

Fig. 1 *Meconopsis aculeata*'s whole plant and flower

Table 1 Study sites of meconopsis aculeata in Uttarakhand, Western Himalaya, India

Study stations	Latitude	Longitude	Altitude-range (m)
Valley of flowers (SA1)—BV*	30.7224	79.59156	3160–3248
Hemkund (SA2)—BH*	30.70049	79.61198	4310–4400
Madhyamaheswer (SA3)—AM*	30.64502	79.22939	3611–3697
Rudranath (SA4)—NP*	30.48222	79.33835	3610–3736
Tungnath (SA5)—BT*	30.48985	79.21671	3425–3600

BV* Between Valley of flowers, BH* Below Hemkund, AM* Above Madhyamaheswer, NP* Near Panar, BT* Between Tungnath

2 Material and Method

2.1 Study Sites

This research work was conducted in an alpine and subalpine region of two districts of Uttarakhand namely Chamoli and Rudraprayag. Five sites were randomly selected for population assessment of the species (Table 1, Figs. 2 and 3).

2.2 Community Analysis

Community analysis was carried out in 2018–19 from June to September when most of the plants attain their flowering growth. Vegetation sampling was done using belt transect method, for this 100 × 100 m area was selected in each site considering the habitat of *M. aculeata*. Further transect was divided into twenty-five sub transects of 20 × 20 m, out of total three sub transects were selected randomly and 10 quadrats

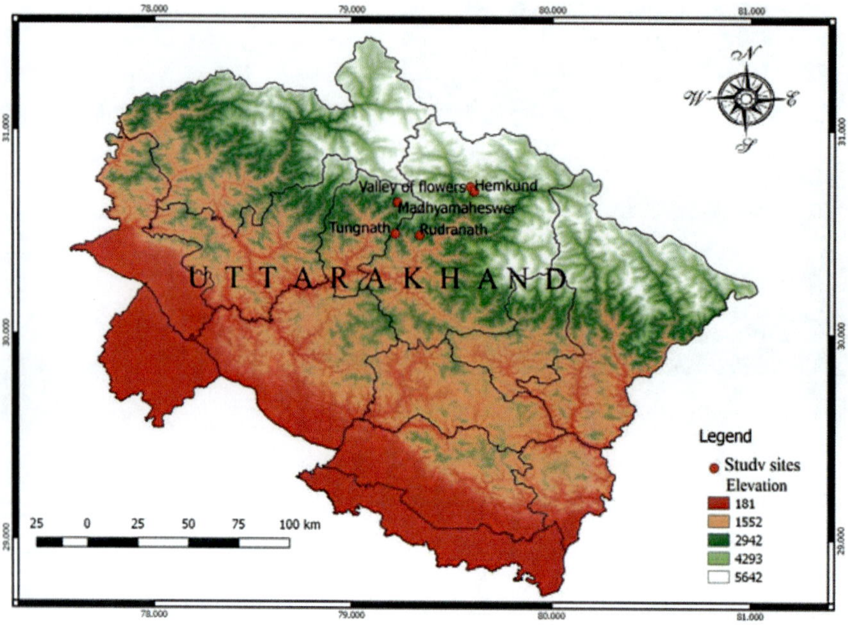

Fig. 2 Study sites of *Meconopsis aculeata* in Uttarakhand, Western Himalaya, India

Fig. 3 **a** Valley of flowers; **b** Hemkund; **c** Madhyamaheswer; **d** Rudranath; **e** Tungnath

of 1 × 1 m size were laid randomly. Basal area cover of each species was determined by measuring the plant diameter from above 5–10 cm of the surface using a vernier calliper. Species were identified with the help of Flora of Garhwal [16], Flora of Chamoli [17] and Plants of Kedarnath Wildlife Sanctuary [8].

2.3 *Quantitative Analysis*

The frequency percentage, density, abundance, relative frequency, relative density, relative dominance and importance value index (IVI) were determined through the formula given by Mishra [18].

Parameters of Phytosociological attributes:

$$\text{Frequency }\% = \frac{\text{Total number of quadrats in which species occurred}}{\text{Total number of quadrats studied}} \times 100$$

$$\text{Density (ind/m}^2) = \frac{\text{Total number of individuals of a species in all quadrats}}{\text{Total number of quadrats studied}}$$

$$\text{Abundance} = \frac{\text{Total number of individuals of a species in all quadrats}}{\text{Total number of quadrats in which the species occurred}}$$

$$\frac{\text{A}}{\text{F}}\text{ ratio} = \frac{\text{Abundance}}{\text{Frequency}}$$

$$\text{Relative Frequency }\% = \frac{\text{Species's Frequency}}{\text{Total frequency of all the species}} \times 100$$

$$\text{Relative density }\% = \frac{\text{Species's Density}}{\text{Total density of all species}} \times 100$$

$$\text{Relative dominance }\% = \frac{\text{Total basal cover of the species}}{\text{Total basal cover of all species}} \times 100$$

$$\text{Total Basal Cover (TBC)} = \text{Mean Basal Cover} \times \text{Density}$$

$$\text{Importance Value Index (I.V.I)} = \text{Relative frequency} + \text{Relative density} + \text{Relative dominance}$$

$$\text{Simpson' Index of Dominance (D)} = 1 - \frac{\left(\sum n(n-1)\right)}{N(N-1)}$$

where n = the total number of individual of a particular species.

N = the total number of individual of all species [19].

$$\text{Shannon-Wiener Index (H)} = -\sum_{i=1}^{i=s} (\text{pi*ln(pi)})$$

where H = Species diversity Index,
N = Number of all individuals,
ni = Number of individuals in species,
pi = The proportional of individuals in species (pi = ni/N) [20].

Species Richness Index (RI) was calculated by using formula Margalef (1958) as:

$$\text{Richness Index (R.I)} = \frac{\text{S} - 1}{\text{Ln(N)}}$$

where S = Total no. of species in community-1.
N = Total no of individuals of all species in a community [21].

Species Evenness Index (EI) was calculated by using formula Pielou (1975) as:

$$\text{Evenness Index(E.I)} = \frac{\text{H}'}{\text{Ln(S)}}$$

where H′ = Shannon diversity index.
S = Total species number of a community [22].

The Index Similarity (S) was calculated to compare the species of two Stands (Sorenson 1948) as:

$$\text{Similarity Index (S)} = \frac{2\text{C}}{\text{A} + \text{B}} \times 100$$

where A and B = total species in community A and B respectively.
C = number of common species in both community [23].

2.4 *Distribution Pattern and Threat Assessment*

Distribution pattern of the species was assumed on the basis of abundance to frequency (A/F) ratio. A/F ratio < 0.025 was considered as regular distribution, between 0.026 and 0.050 as random and > 0.050 contiguous kinds of dissemination [24].

Two criteria were used for threat assessment as per IUCN criteria of red list [25]. (1) Estimation of Population on basis of Density and mature individuals number. (2) On the basis of extent of occurrence (number of population/plots). Flowering and fruiting plants were considered as mature individuals and used for population assessment.

A species found < 250 mature individuals was considered as critically endangered, < 2500 as endangered, and < 10,000 as vulnerable. When a species having only a single population was assessed as critically endangered, < 5 population endangered and < 10 as vulnerable. Furthermore, separate status was also assigned for every natural site as well as for the entire Garhwal region of western Himalaya India.

2.5 *Collection and Extraction of Plant*

For phytochemical investigation leaves were collected from each site and stored in the icebox at 4°C temperature until reach to HAPPRC laboratory, further plant sample washed three-four times with tap water and two times by distilled water and shade dried at room temperature.

After fully drying, the plant leaves samples were crushed into fine powder utilizing Willey processor Grinder Mill Machine (Micro Scientific, India) and stored in an airtight boxes at room temperature in obscurity then the dried powder was filled in the thimble (200 ml of a solvent with 20 g of *Meconopsis aculeata* leaves test) of soxhlet unit and extraction with acetone, methanol and aqueous for 72 h at Boiling temperature 56, 64.7, 100°C. The dissolvable concentrates sifted with Whatman filter paper (grade 1) then concentrated with Rotary Vacuum evaporator Quickvap (PATENT REGD. NO. 229440) pressure and preserved at 4°C in the airtight container till further use.

2.6 *Qualitative Screening*

2.6.1 Alkaloid Procedure

Add 1.5 ml of HCl (1%) with 2 ml of solvent extract, heated in the water bath for a few minutes and add Six drops of Mayor's/Wagner's/Dragendroff's Reagent. Development of an orange precipitate (ppt) confirmed the positive outcome [26].

2.6.2 Tannins

2 ml of an extract with 2 ml of $FeCl_2$ (5%). Formation of yellow–brown ppt indicates the positive result of tannin [27].

2.6.3 Glycoside

A little amount of prepared extract was taken in test tube with 1 ml of distilled water. Two–three drops of aqueous Sodium hydroxide were added. A yellow colour indicates the positive result of glycoside [28].

2.6.4 Cardiac Glycosides

2 ml of extract added 2 ml Glacial Acetic acid. Added a few drops of $FeCl_3$ and 1 ml of conc. H_2SO_4. The appearance of brown rings in the upper surface and violet rings in the lower surface indicates the presence of cardiac glycosides [28].

2.6.5 Saponin

1 ml leaves extract was added in 20 ml of DH_2O and shaken in a graduated cylinder for 15 min. The arrangement of a 1–2 cm layer of froth demonstrates the presence of saponin [28].

2.6.6 Phenolics

To 2 ml extract, 1 ml of Ferric chloride (1%) was added. Blue and green colour appearances confirmed the presence of the phenol [29].

2.6.7 Terpenoids

Take 2 ml chloroform in test tube then add 0.5 ml of extract. After that three ml concentrated H_2SO_4 was added which forms a layer in tube. The reddish-brown colouration shows terpenoid presence [28].

2.6.8 Steroids

One gram of *Meconopsis aculeata*'s leaves sample was dissolved in 10 ml of chloroform and 1 ml of concentrated H_2SO_4 into the test tube from wall sides. The top layer shading turned in red colour and the sulphuric acid layer indicate yellow with green fluorescence confirm the steroid positivity [30].

2.6.9 Flavonoids

1–5 drops of Concentrated Hydrochloric acid (HCl) were included to a small quantity of leaves extract. It gives immediate red colour formation and showed the flavonoid presence [28].

3 Results

Meconopsis aculeata is one of the majestic flowering plants species of Himalaya its grow in boulders rocks, and near water stream due to restricted habitat population and phytosociology assessment was done that shows that maximum IVI value of *Meconopsis aculeata* was found in Valley of flowers (IVI 9.55) followed by Hemkund (IVI 4.05), Tungnath (IVI 3.74), Madhyamaheswer (IVI 3.72), Rudranath (IVI 2.96). The frequency % was maximum in Tungnath (47.22%) and minimum in Rudranath (25.81%). Density was found maximum in Madhyamaheswer (2.75 plants/m^2) and minimum in Hemkund (0.58 plant/m^2). Result revealed that the Distribution pattern is contiguous in all five selected sites (Table 2; Fig. 4).

Diversity index was highest in Tungnath and lowest in Hemkund region and dominance index was highest in Madhyamaheshwar however rest all four sites had 0.99. Species richness was maximum in Tungnath region followed by Rudranath, Madhyamaheswer, Valley of Flowers and Hemkund. Species evenness and similarity index of all five study sites shown in Tables 3 and 4.

Some dominant associated species of *M. aculeata* are *Danthonia cachemyriana* Jaub. & Spach., *Salix denticulata* Andersson, *Carex sentosa* Boott, *Bistorta affinis* (D.Don) Greene, *Bistorta macrophylla* (D.Don) Sojak, *Viola biflora* L. (Table 5). However *M. aculeata* natural surroundings are also connected in study zones with exceptionally high potential therapeutic plant's living habitat like *Picrorhiza kurroa* Royle ex Benth., *Rheum emodi* Wall. Ex Meisn., *Aconitum violaceum* Jacquem. ex Stapf., *Megacarpaea polyandra* Benth., and *Angelica glauca* Edgew.

According to research work in selected sites, it has been observed that the plant is found only in moist, boulders and rocky area somewhere under canopy of *Betula utilis* D.Don, *Abies spectabilis* (D. Don) Spach and *Rhododendron campanulatum* D.Don. From this it can be deduced that such elements which are present in rocks are favourable for plant growth instead of soil elements.

The result of preliminary analysis of leaf extract in various solvents has shown remarkable variations are tabulated in Table 6. Phyto-constituents variability was found in all selected sites. None of the sites from which the plant was obtained did contain all phytochemical i.e., alkaloids, tannins, glycosides, cardiac glycosides, phenols, terpenoids, steroids and flavonoids. The resulting study revealed that in aqueous extract few bioactive constituents were effectively dissolvable, in methanol extract most of the bioactive were strongly dissolvable and present but in acetone extract bioactive moderately present.

Table 2 Threat category assessment and population status of *Meconopsis aculeata* in five selected study sites in Garhwal Himalaya

Study sites	F (%)	R Fre	D (plant/m^2)	R Den	A/F	TBC	R Dom	IVI	DP	MI	P	Status
SA1	43.90	4.53	1.15	3.44	0.06	0.06	1.58	9.55	C	33	3	EN[a] CE[b]
SA2	27.27	3.37	0.58	0.57	0.08	0.06	0.11	4.05	C	13	1	CE[a] CE[b]
SA3	43.75	2.43	2.75	1.28	0.14	3.08	0.01	3.72	C	25	2	EN[a] CE[b]
SA4	25.81	1.74	0.71	0.73	0.11	0.06	0.50	2.96	C	10	1	CE[a] CE[b]
SA5	47.22	1.95	2.00	0.81	0.09	2.00	0.98	3.74	C	70	4	VN[a] CE[b]
Overall status of *M. aculeata* in Garhwal region												EN CE

F%-Frequency percentage, R Fre-Relative Frequency, D-Density, R Den-Relative density, A/F-Abundance/Frequency, TBC-Total basal cover, R Dom-Relative dominance, IVI-Importance value index, DP Distribution pattern, MI Mature individuals, P Number of populations, C Contiguous

[a]On basis of extent of occurrence

[b]On basis of Population estimation, EN-Endangered, CE-Critically Endangered, VN-Vulnerable

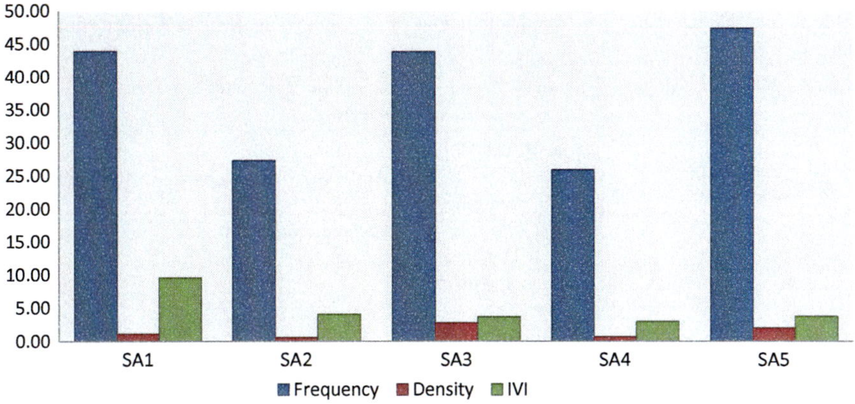

Fig. 4 Frequency, density and IVI values in five sites of studied plant *Meconopsis aculeata*

Table 3 Species diversity and related parameters of *Meconopsis aculeata* in all five selected study sites

	Diversity index (Shannon (H′)	Dominance index (Simpson Cd)	Species richness index	Species evenness index
SA1	3.1	0.99	3.60	0.940
SA2	1.99	0.99	1.97	0.702
SA3	2.98	1	4.29	0.813
SA4	3.37	0.99	5.11	0.902
SA5	3.5	0.99	5.39	0.903

Table 4 Percent similarity index of different sites at high elevations

Sites	SA2	SA3	SA4	SA5
SA1	18	24	20	15
SA2		18	24	33
SA3			34	36
SA4				46

4 Discussion and Conclusion

Present study aims to find out population status and phytosociological attributes, habitat structure, type of *M. aculeata* in different study sites in Garhwal Himalaya, thus, it gives useful information regarding composition, structure, species diversity, resource share of the community.

M. aculeata has now assessed as endangered, however recent research shows the status critically endangered in Kashmir Himalaya as per IUCN criteria [31]. The

Table 5 Site characteristics of selected populations of *Meconopsis aculeata* in Western Himalaya, India

Study sites	Aspect	Slope	Habitat	Dominant associated species	Threats
(SA1)	SE	20–25°	Open Rocky slopes in the subalpine-alpine region with *Betula utilis* D.Don and associated with *Rheum emodi* Wall ex. Meissn endangered spp.	*Salix denticulata* Andersson, *Viola biflora* L., *Danthonia cachemyriana* Jaub. & Spach.,	Human disturbance
(SA2)	NE	40–50°	Open Steep Dry rocky slopes in alpine with *Aconitum violecum* Jacquem ex. Stapf vulnerable spp.	*Danthonia cachemyriana* Jaub. & Spach., *Bistorta affinis* (D. Don) Greene, *Bistorta macrophylla* (D.Don) Sojak, *Euphorbia stracheyii* Boiss	Degradation of habitat, interference by human
(SA3)	NW	30–35°	Moist grassy slopes in open Alpine meadows	*Danthonia cachemyriana* Jaub. & Spach., *Carex setosa* Boott, *Viola biflora* L	Habitat degradation, human interference
(SA4)	NE	25–30°	Gentle grassy moist slopes in open Alpine meadows	*Danthonia cachemyriana* Jaub. & Spach., *Veronica cana* Wall. Ex Benth., *Bistorta vaccinifolia* Wall. Ex Meisn	Grazing and interference by human
(SA5)	NW	20–30°	Moist rocky shady slopes in alpine with *Nardostachys jatamansi* DC. and *Picrorhiza kurroa* Royle ex. Benth both are an Endangered spp.	*Danthonia cachemyriana* Jaub. & Spach., *Carex setosa* Boott., *Synotis kunthiana* (Wall. ex DC.)	Grazing, interference by human, degradation of habitat

information showed that although species occurrence is seldom (frequency was < 50% in each site), the area of occurrence is narrowing down with very few mature individuals present and along these lines, the species are sorted as vulnerable based on occurrence and endangered based on population estimation. According to IUCN classifications, on the basis of occurrence plant categorized in endangered and on the basis of population estimation, it is categorized as critically endangered. The qualitative investigation revealed that the secondary metabolites used for medicinal purpose are found in the plant and these categories are Alkaloid, Tannin, Glycoside, Cardiac

Table 6 Qualitative phytochemical screening of leaves extract of *Meconopsis aculeata*

	Valley of flowers			Hemkund			Madhyamaheswer			Rudranath			Tungnath		
	Aq	Ace	Meth	Aq	Ace	Meth	Aq	Ace	Meth	Aq	Ace	Meth	Aq	Ace	Meth
Alkaloid	+	–	++	–	+	++	–	+	+++	–	+	–	+	–	++
Tannin	–	+	+++	++	–	+	+	+	++	+	+	++	+	++	+++
Glycoside	–	–	+	–	++	++	–	–	++	–	+	–	–	+	+
Cardiac glycoside	++	–	–	–	+	++	+	++	–	–	++	+++	–	++	++
Phenols	–	+	++	–	++	+++	–	+	++	–	–	++	+	–	+++
Terpenoids	–	+	++	–	–	++	–	+	++	+	–	++	+	–	++
Steroids	+	–	++	–	–	++	–	–	+++	–	+	+	–	++	–
Flavonoids	–	+	+	–	++	++	–	+	++	+	+	++	–	+	–
Saponin	–	+	++	+	+	++	–	++	–	–	+	–	+	+	++

–: absent, +: present, ++: moderate present, +++: strongly present glycoside, Phenols, Terpenoids, Steroids, Flavonoids, and Saponin are present On the basis of number of populations which are low according to phytosociology study, the immediate conservation is required for plant conservation due to its therapeutic potential worth. The study revealed that the strong presence of all most phytocompounds was seen mostly in methanolic and acetone leaves extraction of *Meconopsis aculeata*.

Acknowledgements Authors are thankful of Research scholars Miss. Vaishali Chnadola and Mr. Abhishek Jamloki, Department H.A.P.P.R.C., Srinagar Grahwal for their support and guidance in field work.

References

1. Singh, D.K. and Hajra, K.: Floristic diversity. In Gujral GS and Sharma V (eds), Changing Perspectives of Biodiversity Status in the Himalaya. New Delhi: British Council Division; (1996);23–38
2. Mathur, V.B., Kathayat, J.S. and Rath, D.P.: Envis Bulletin: Wildlife and Protected Areas, Dehradun: Wildlife Insititute of India; Vol.3 (1) (2000).
3. Dhar, U., Rawat, R.S. and Uprety, J.: Setting priorities for conservation of medicinal plants - A case study in the Indian Himalaya, Biological Conservation, 95.,(2000) pp. 57-65.
4. Nayar, M.P. and Sastary, A.R.K.: Botanical Survey of India, Calcutta. Red Data Book of Indian Plants, Vol. I, 1987.
5. Nayar, M.P. and Sastary, A.R.K.: Botanical Survey of India, Calcutta. Red Data Book of Indian Plants, Vol. II, 1988.
6. Nayar, M.P. and Sastary, A.R.K.: Botanical Survey of India, Calcutta. Red Data Book of Indian Plants, Vol. III, 1990.
7. Hooker, J.D.: Flora of British India. Reprint by Bishen Singh Mahendra Pal Singh, Dehradun., Vol 1: 118 (1872).
8. Rai, I. D., Singh, G., and Rawat, G. S.: Plants of Kedarnath Wildlife Sanctuary, Western Himalayas: A Field Guide (2017) 77.
9. Tsarong, T.J.:Tibetan medicinal plants (India: Tibetan Medical Publications).(1994) ISBN No. 81–900489–0–2.
10. Singhal, V.K. and Kumar, P.: Impact of cytomixis on meiosis, pollen viability and pollen size in wild populations of Himalayan poppy (Meconopsis aculeata Royle). J Biosci 33: (2008) 371–380.
11. Rawat, B., Sekar, K.C. and Gairola, S.: Ethnomedicinal plants of Sunderdhunga valley, western Himalaya, India-traditional use, current status and future scenario. Ind For 139:(2013) 61–68.
12. Do, Q.D., Angkawijaya, A.E., Tran-Nguyen, P.L., Huynh, L.H., Soetaredjo, F.E., Ismadji, S., Ju YH.: Effect of extraction solvent on total phenol content, total flavonoid content, and antioxidant activity of *Limnophila aromatica*. J Food Drug Anal 22:(2014) 296–302.
13. Rawat, S., Jugran, A.K., Bhatt, I.D., and Rawal, R.S.: Influence of the growth phenophases on the phenolic composition and anti-oxidant properties of *Roscoea procera* Wall. in western Himalaya. J Food Sci Technol 55: (2018) 578–585.
14. Bahukhandi, A., Sekar, K.C., Barola, A., Bisht, M. and Mehta, P.: Total Phenolic Content and Antioxidant Activity of *Meconopsis aculeata* Royle: A High-Value Medicinal Herb of Himalaya. Proc. Natl. Acad. Sci., India, Sect. B Biol. Sci. Springer (2018). https://doi.org/10.1007/s40011-018-1054-1
15. Nautiyal, M.C. and Nautiyal, B.P.: Agrotechniques for High Altitudes Medicinal and Aromatic plants. Bishen Singh Mahendra Pal Singh, Dehradun:(2004) 4, 6.

16. Gaur, R.D.: Flora of the district Garhwal northwest Himalayas (with Ethnobotanical notes). Srinagar, Garhwal: Transmedia (1999).
17. Naithani, B.D.: The Flora of Chamoli. Volumes 1–2. Botanical Survey of India, Howrah (1984–1985).
18. Misra, R.: Ecology Work Book, Oxford and IBH Publishing Co., New Delhi (1968).
19. Simpson, H.E.: Measurement of diversity, Nature, 163,(1949) p. 688.
20. Shannon, C.E. and Wiener, W.: The Mathematical Theory of Communication. University Press, Illinois, USA (1963).
21. Margaleff, D.R.: Information theory in ecology. Year book of the Society for General Systems Research, 3: (1958) 36-71.
22. Pielou, E.C.: Ecological Diversity. John Wiley & Sons, New York (1975).
23. Sorenson, T.: A Method of Establishing Groups of Equal Amplitudes in Plant Sociology Based on Similarity of Species Content and Its Application to Analyses of the Vegetation on Danish Commons. Kongelige Danske Videnskabernes Selskab, Biologiske Skrifter, 5, (1948) 1-34.
24. Cottam, G. and Curtis, J. T.: The use of distance measures in phytosociological sampling, Ecology, 37, (1956) pp. 451-460.
25. IUCN, Draft IUCN Red List Categories, Gland Switzerland, (2000).
26. Oguyemi, A.O.: The Origin of Herbal Cure and its Spread. In: African Medicinal Plants: Proceedings of a Conference, Sofowora, A. (Eds.). University of Ife Press, Ife-Ife,(1979) pp: 20–22.
27. Parekh, J. and Chanda, S.V.: In vitro antimicrobial activity and phytochemical analysis of some Indian medicinal plants. Turk. J. Biol., 31: (2007) 53-58.
28. Trease, G. E. and Evans, W. C.:Trease and Evan's Textbook of Pharmacognosy. 13th Edition. Cambridge University Press, London. (1989) 546.
29. Martinez, A. and Valencia, G.: Manual de practicas de Farmacognosia y Fitoquimia: 1999.1. Universidad de Antiquia, Medellin, (2003) pp: 59–65.
30. Ahmad, M., Kaloo, Z. A., Bashir A., Ganai, H. A., Ganaie and Singh, S.: Phytochemical Screening of *Meconopsis aculeata* Royle an Important Medicinal Plant of Kashmir Himalaya: A Perspective. Research Journal of Phytochemistry.(2016). https://doi.org/10.3923/rjphyto. https://www.researchgate.net/publication/283214041
31. Majid, A., Ahmad, H., Saqib, Z., Ali, H. and Alam, J.: Conservation status assessment of *Meconopsis aculeata* Royle: a threatened endemic of Pakistan and Kashmir. Pak. J. Bot. 47:(2015) 1–5.

Multi-disciplinary Analyses on Hydro Rotor Using Computational Hydrodynamic Simulation (CHS)

R. Vigneshwaran, B. Surya, S. Tharani Tharan, M. Senthil Kumar, M. Ramesh, and R. Vijayanandh

Abstract The energy extraction from hydrodynamic fluid is one of the primary tasks in the hydro energy sector. Because of the unstable availability of water, the efficient tool is compulsory to provide required average energy. Comparatively, all other tools in hydro energy sector, the rotor and its dynamic nature play a top role to generate high energy from hydro fluid. In this work deals an innovative and integrative conceptual design of hydro rotor and its hydrodynamic performance estimations. Fundamentally, this proposed rotor is oriented horizontally added with twisted wing component. Comparative analyses also executed with normal twisted plates. Because of the implementation of twisted wing, the torque production is quite higher than other models. Computational hydrodynamic analysis is the methodology used in this work with the help of CFD tool i.e., ANSYS Fluent 17.2. In which, the conceptual design of all the complicated hydro rotors are modelled in CATIA.

Keywords Conceptual design · CFD · Hydro-energy · Hydro rotors · Optimization

1 Introduction

The requirement of energy is increasing constantly in current days because of its large utilization in all fields. Non-renewable energy sources are incapable to fulfil the current and future needs of the world in the perspective of electricity utilization, transportation, etc. So the research on energy extraction from renewable energy sources such as hydro, wind, solar, etc. are emerged everywhere. The platform of this research work is chosen as hydro environment and its energy converting unit. Generally, two primary groups of components are available in the renewable energy sector, which are Mechanical group and Electrical group. In electrical component group, generator, rectifier are plays a vital role in the electrical conversion and its utilization. But mechanical group, a single and solid component solely plays a top role, which is rotor. Rotor and its design parameters are have the drastic capability

R. Vigneshwaran · B. Surya · S. Tharani Tharan · M. Senthil Kumar · M. Ramesh · R. Vijayanandh (✉)
Aeronautical Engineering, Kumaraguru College of Technology, Coimbatore, Tamil Nadu, India

N. A. Siddiqui et al. (eds.), *Advances in Environment Engineering and Management*, Springer Proceedings in Earth and Environmental Sciences,
https://doi.org/10.1007/978-3-030-79065-3_34

to extract huge amount of energy from hydro environments. Also, the high energy extraction is totally depends upon the efficient way of dealing of hydrodynamic fluid by the support of rotor profile. Naturally, the hydro fluid is high dense one, which may affect the lifetime of a rotor hugely. Therefore, in this work deals an innovative conceptual design of a hydro rotor, which can able to provide high torque induction in the high resisting force hydro environments.

2 Literature Survey

The various hydro rotors and its designs are studied through this literature survey. Generally, it is important to utilize renewable energy in an efficient way for energy management, in which hydro rotors are one of the innovative ways of producing energy which is a renewable source. In the olden days, Francis hydro turbine units were used for turbines. The runners' in these turbines had serious cavitation damage. The major drawback of turbine's recital is fully relayed on the cavitation damage occurrence in the continuous manner, which majorly affect the linear working environments. In this section, several researches are studied and come up with the new idea on the conceptual design of rotors.

The author [1] used computational fluid dynamics for the optimization of the hydro turbine's design for better performance. Computationally, CFD codes have been solved the design based optimization on hydro turbines in an easy manner through its advanced capacity. The airfoil-shaped rotor blade was used and solved through CFD, thus concluded that the all the flow cum design profile angles are having the full effect on the increment of the hydro-turbine's propulsive efficiency.

The authors [2] used the principle of BEM theory for its analysis, which dealt the fundamental design and its procedures of all kind of rotors. The two-dimensional lift-drag data was implemented to assess the Hydrodynamic forces on rotor elements with local flow conditions. The summation of each elemental force can provide the consolidated force of the rotating component. In addition to the BEM theory, the momentum theory also used, which deals the rotor plane's pressure loss due to the fluid's work done executed on the rotor. In an iterative manner, it solved the coupled equations to predict the roto-hydrodynamic forces across the short-listed rotor.

The authors [3] targeted to work on the turbine rotor's optimization and guide vane's performance enhancement because the efficiency of the VLH axial flow turbine was a major impact. The axial turbine was firstly tested with fluid flow simulations, then turbine's efficiency enhanced through the concept of regression. An advanced CFD methodology was implemented in the prediction of this complicated turbine, in which the computational simulations based sensitivity tests i.e., grid refinement, are organized to enhance the trustworthiness of the CFD's outcome.

The author [4] used the airfoil shape in hydro-turbine, which gave the tremendous benefits on the characteristics of flow. Normally, the design parameters of aerofoil such as aerofoil shape, leading edge's radius, maximum thickness's position and maximum camber's location have been played the great role in the fluid movement.

Local high-speed acceleration is possible only when there is a Small leading-edge radius. Computational simulations were executed on the rotor blades with different working environments, in which two rotors were constructed with a varied chord and pitch ratio. For the common rotodynamic initial cum boundary conditions, it needed a various design profiles and produces a different power. Because the interaction of fluid on the turbine blades with different velocities have been produced the different potential energy in a linear manner.

The author [5] dealt the Hydropower production technologies through Hydrokinetic energy, which is one of their major techniques because of low environmental impacts. In this research, the major finding was that the output hydro power has been proportional to the square of the diameter of the rotor. Therefore the turbine rotor diameter was optimized to get better efficiency.

In this article [6] the blade shape was provided by Bezier-curve which is a parameterization method. A rotor profile design was invented to avoid unnecessary calculations. It was based on the manufacturing constraints to make soon the optimization process. The proposed blade multi-elements have higher C_L. As the flap increases, the flow separation decreased and thereby a decreased in C_D was occurred, finally, the C_L/C_D and torque were increased. The multi-element hydrofoils are highly recommended for the blade design instead of standard hydrofoils because the manufacturing process costs are higher for standard hydrofoils when compared to multi-element hydrofoils.

2.1 Summary

From the survey of research papers, this paper coming up with an idea that the rotor design optimization is one of the primary requirements in implementing better performance of hydro rotors. In that, the rotor diameter, thickness, blade angles, are important and the same inputs cum boundary conditions are necessary for computational simulations as well as other engineering approaches to analyse the performance of the hydro-rotors. Comparatively, the CFD based solution is important that for faster computing results rather than theoretical calculations so the same the methodology is approached for this existing work. Commonly, the water is an efficient and reliable fuel available enormously in the earth thus handling the hydrodynamics matter in an efficient way with design parameters will give us a vital energy source. In the phase of CFD simulation, control volume, mathematical modelling, computational modelling, discretization, boundary conditions and solution methods have been important, which were extracted from this standard literature survey. With the help of this survey, the efficient cum advanced hydro rotor will come up with new computational investigations on hydrokinetic turbines, which help in the improvement, development, and implementation of the rotor to the sustainable use of water resources [7–10].

3 Methodology Used—CHS

3.1 Conceptual Design

The good hydrodynamic performance and thereby efficiency in high level are primarily relayed on design profile of the hydro rotor. Generally, conceptual design in a subordinate phase involved in the design stage of hydro rotor, in which the output of conceptual design must be a three dimensional model. Apart from the three dimensional data, the critical facilities such as pitch angle, hydrofoil shape, number of leaves are contributes a lot in the conceptual design phase [11, 12]. This paper proposed an innovative hydro-rotor with comparative study through CFD. Therefore, the conceptual design phase is an unavoidable procedure in this work. Totally, four hydro rotors are constructed with the help of CATIA. All the four models are oriented in the length-wise (Horizontal) shaft as base, added with twisted profiles such as thin plate, wing, etc. An innovative idea involved in the conceptual design is torque enhancement in the hydro rotor with same and normal hydro environments [13–18]. For the comparative purpose, Vertical axis oriented hydro rotors are also modelled with same aforesaid twisted profiles. Thus totally eight hydro rotors are constructed for hydro energy extraction. The draft of all the hydro rotors are revealed in Figs. 1, 2, 3 and 4, in which the length of all the rotors are maintained as 100 mm and the internal diameter maintained as 50 mm for all the hydro rotors and the diameter of the external twisting profile maintained as 27.98 mm for all the models. The

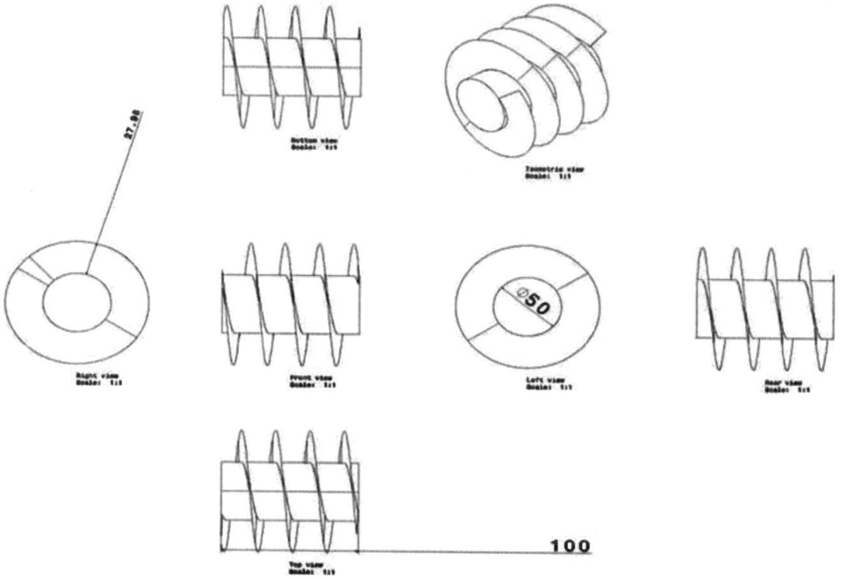

Fig. 1 Conceptual design of hydro rotor—Model-I

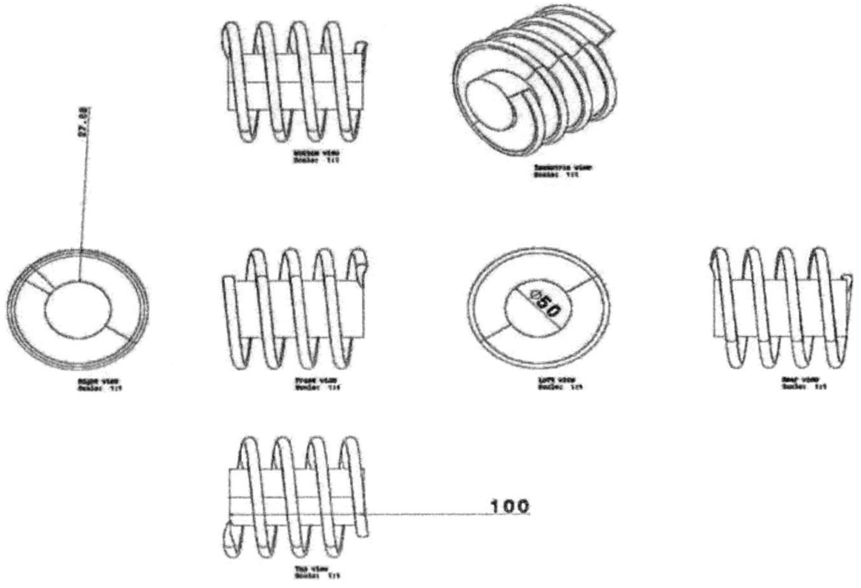

Fig. 2 Conceptual design of hydro rotor—Model-II

ultimate design of this work is model-II, which is shaft with twisted wings. The model-II is equipped hydrodynamic efficient related profile, which may increase the hydro rotors.

3.2 Mesh

The nature of the flow over hydro rotor is comes under external computation so the rectangular based control volume is imposed in this work with the dimensions of 4000 mm in the primary direction and 2000 mm in the secondary directions. The fine un-structural mesh elements are used in this grid generation, which can easily solved through Finite Volume Method based computation. To enhance the quality on the mesh, the refinement facility is applied over the hydro rotor. Finally, the total nodes and elements involved in this mesh construction are 929,847 and 4,876,258 respectively. The mesh quality is attained as 0.99983, which has been greatly accepted by the computational people as good quality. The typical views of both control volume and hydro rotor are revealed in Figs. 5 and 6.

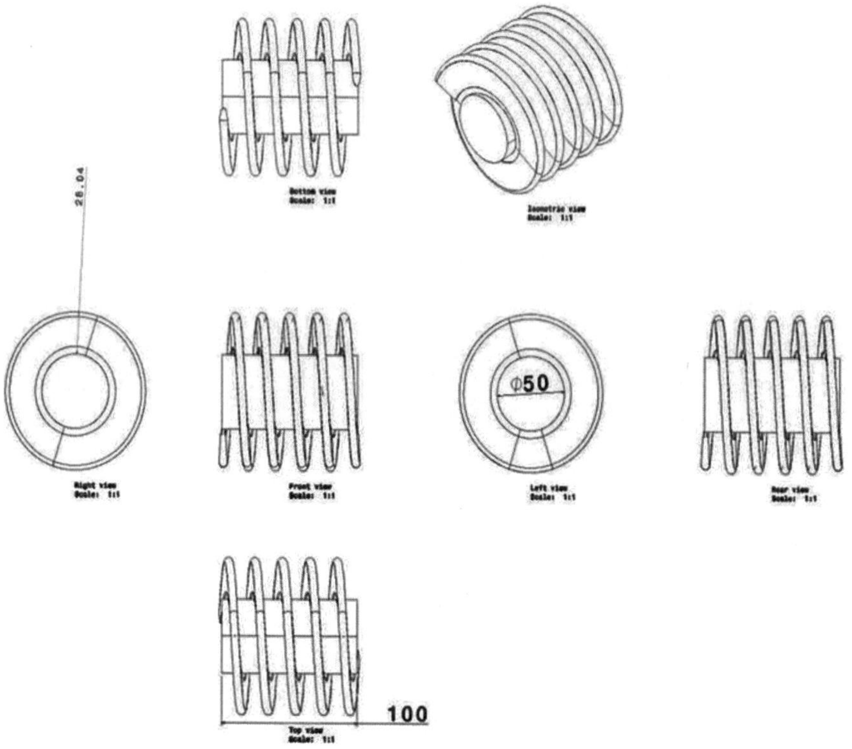

Fig. 3 Conceptual design of hydro rotor—Model-III

3.3 Boundary Conditions

The platform of this work is hydro environment so pressure based solver is imposed in this computation because of the hydro fluid's nature. The density of this working fluid is quite high so the possibility of turbulence formation at the flow separation regions is quite less so Spalart Allmaras is chosen as turbulence model with 5% precision. The targeted location of this hydro rotor in the sea or river is 1000 mm depth so the operating pressure is calculated as 111,380.3 Pa and the same value is given in this comparative computation. Through instrument, the average flow velocity is measured in the river, which is 0.1 m/s so the same input is give as inlet velocity. The coupled scheme is used in between the pressure and velocity interaction owing to the complicated dynamic nature of the hydro conditions. All other equations are implemented with second order accuracy, which can increase the reliability on the computation.

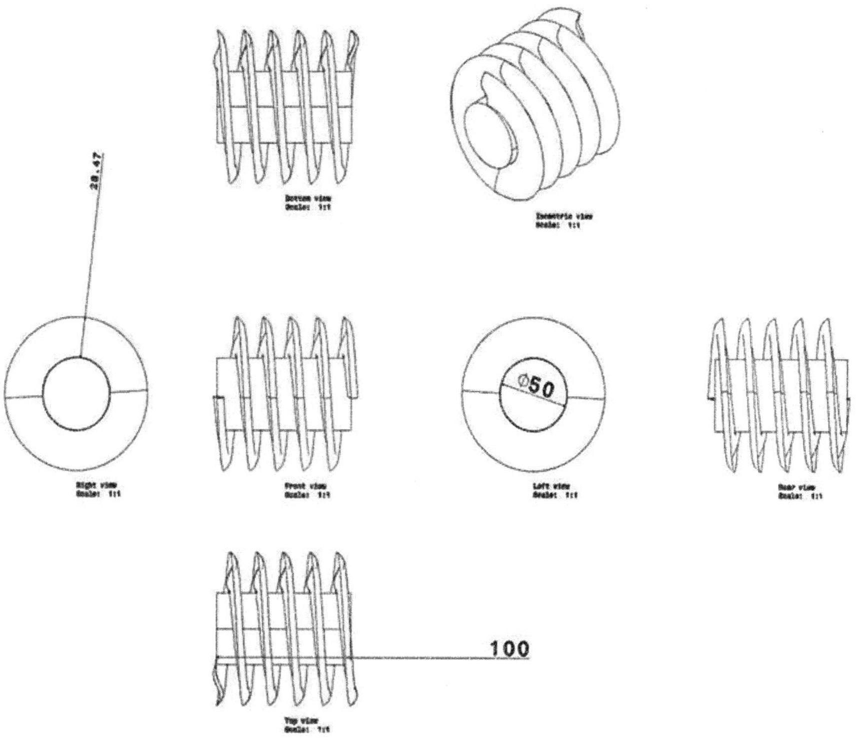

Fig. 4 Conceptual design of hydro rotor—Model-IV

4 Computational Hydrodynamic Simulation Results

4.1 Horizontal Axis Hydro Rotor—Results

Figures 7, 8, 9, 10, 11, 12, 13 and 14 are shown the clear representations of fluid properties such as velocity and pressure over various horizontal axis hydro rotors. The comparative performance outcomes of all the hydro rotors are revealed in Fig. 15.

4.2 Comparative Analysis of Vertical Axis Hydro Rotor

The pressure and velocity variations on the different Vertical Axis Hydro Rotor Models are revealed in Figs. 16, 17, 18, 19, 20, 21, 22 and 23. The comprehensive velocity increment due to the presence of vertical axis hydro rotors is shown in Fig. 24.

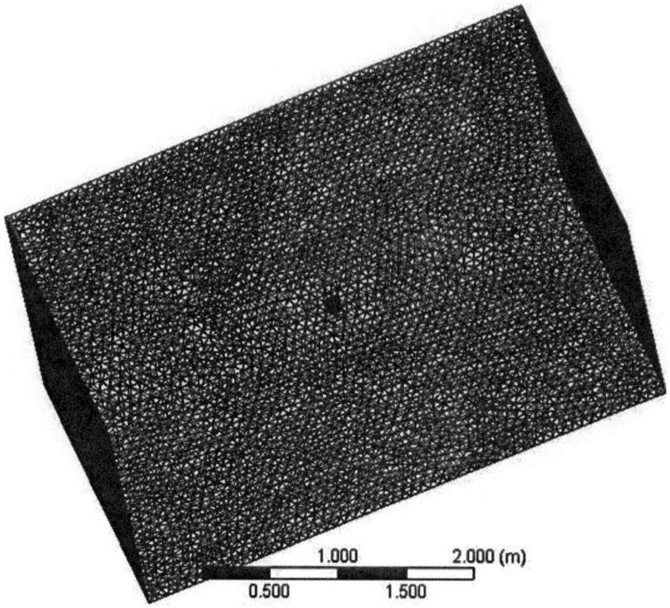

Fig. 5 The discretized structure of the entire control volume

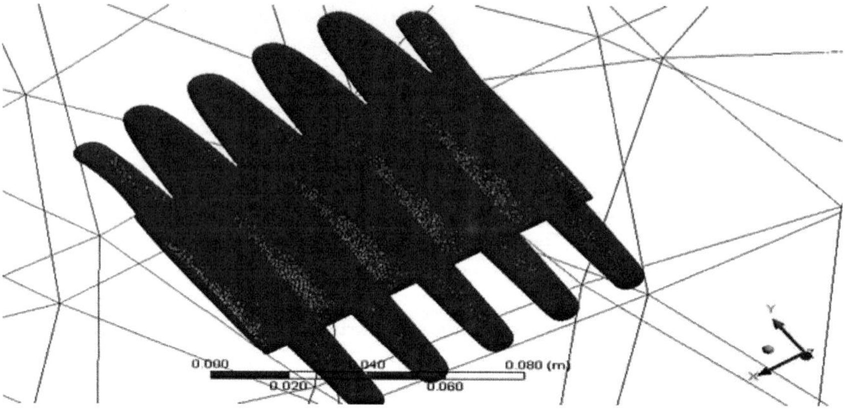

Fig. 6 The typical view of hydro rotor loaded on fine mesh

5 Conclusions

The computational hydrodynamic analyses on different Horizontal axis and Vertical axis based hydro rotors under same boundary conditions. From these comparative results, it is proposed that two hydro rotors are capable to extract high amount of hydro energy from hydrodynamic environments. The wing twisted over horizontal

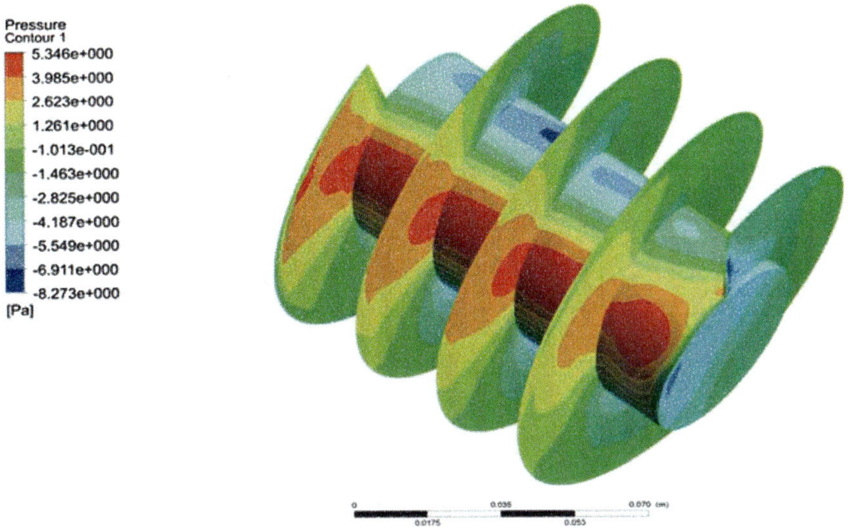

Fig. 7 Pressure variations on horizontal axis hydro rotor-I

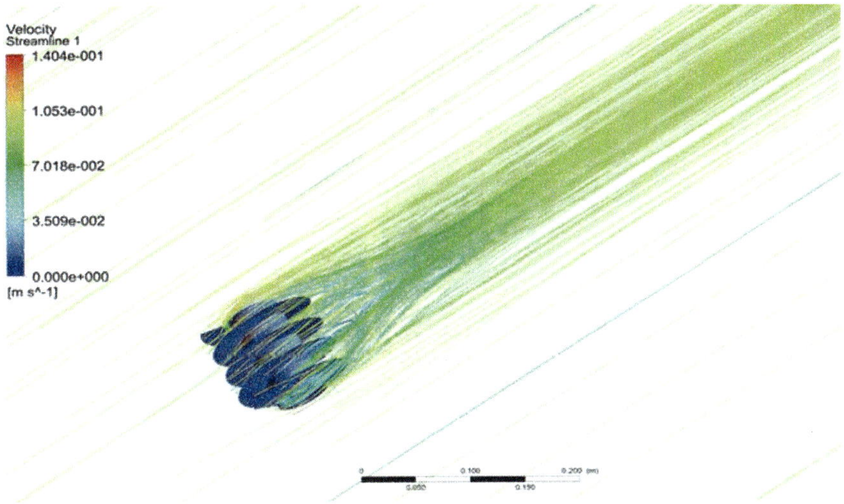

Fig. 8 Velocity generations over horizontal axis hydro rotor-I

axis hydro rotor (model-II) is increased the hydro fluid's velocity by 57.5% than the inflow fluid velocity. Also the hydro fluid's velocity is increased by 50.1% than inflow fluid velocity with the help of thick plate twisted over Vertical axis hydro rotor (model-III). The four predominant shapes are twisted over the Horizontal axis and Vertical axis hydro rotors with the support of CATIA. All of the hydro rotors are modelled with same dimensions for the purpose of effective comparison. The

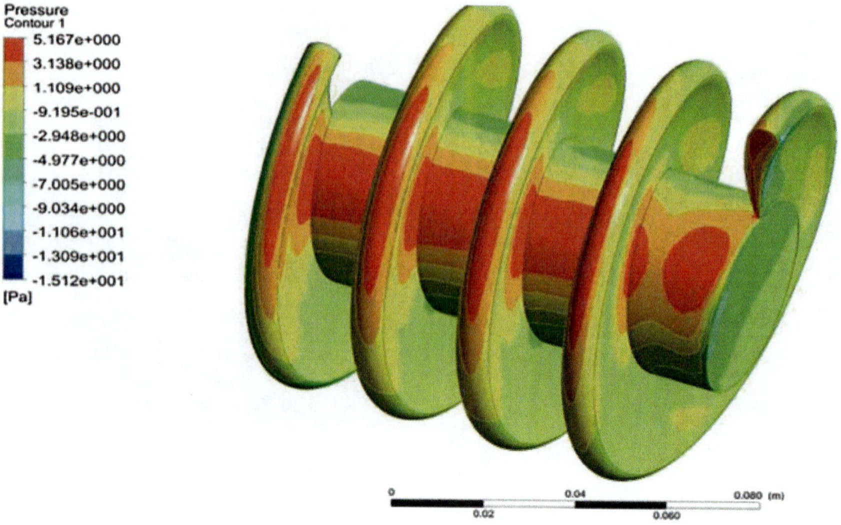

Fig. 9 Pressure variations on horizontal axis hydro rotor-II

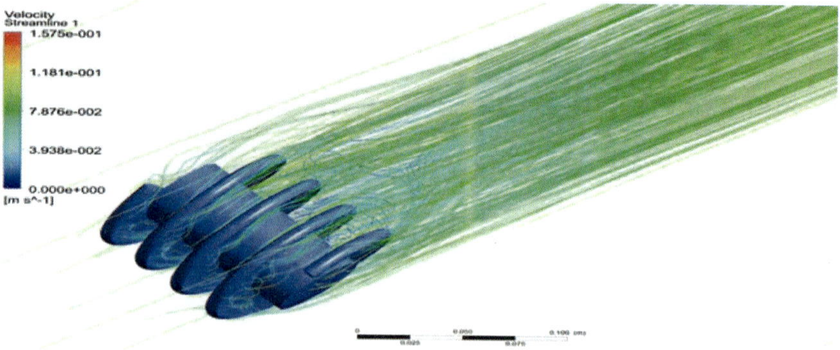

Fig. 10 Velocity generations over horizontal axis hydro rotor-II

advanced CFD tool is employed for this design optimization on hydro rotors and thus the optimized model is finalized, which is hydrofoil based wing twisted over the horizontal axis shaft. The fluid velocity has the capacity to increase the rotational force of the hydro rotor so the hydro energy extraction is quite high in the optimized models than other kind hydro rotors.

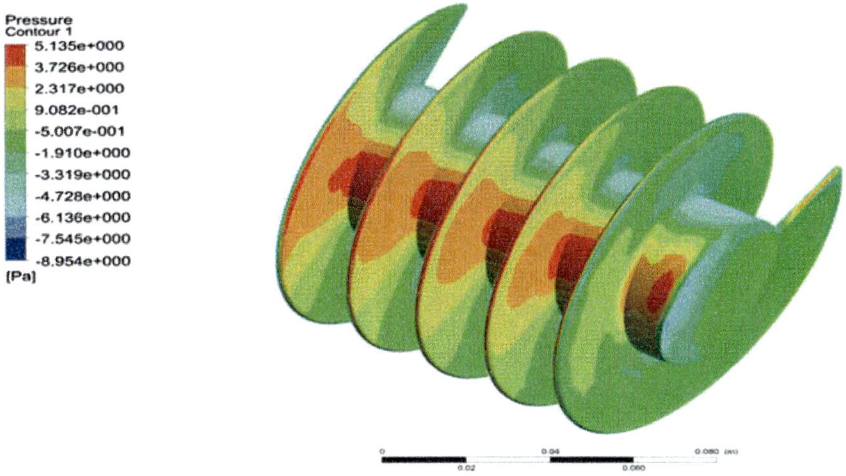

Fig. 11 Pressure variations on horizontal axis hydro rotor-III

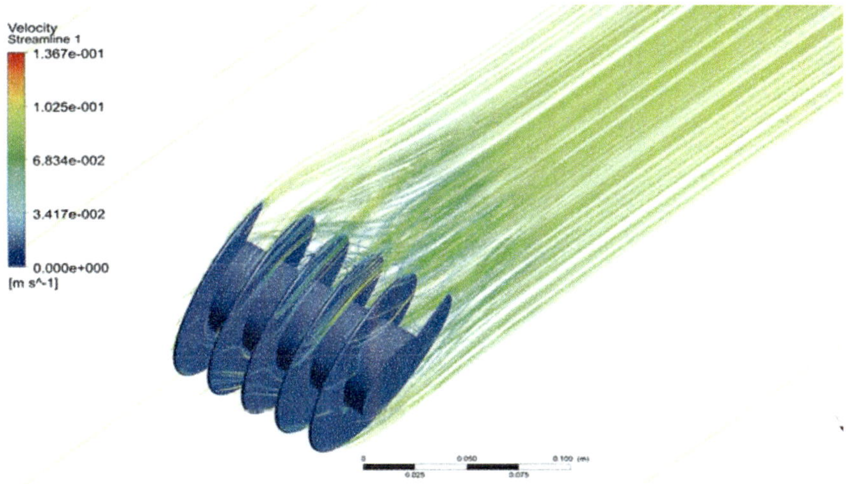

Fig. 12 Velocity generations over horizontal axis hydro rotor-III

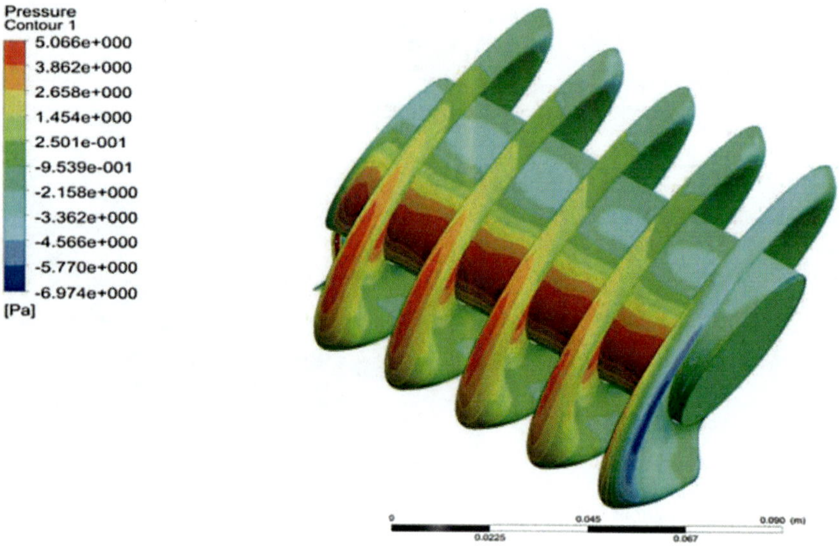

Fig. 13 Pressure variations on horizontal axis hydro rotor-IV

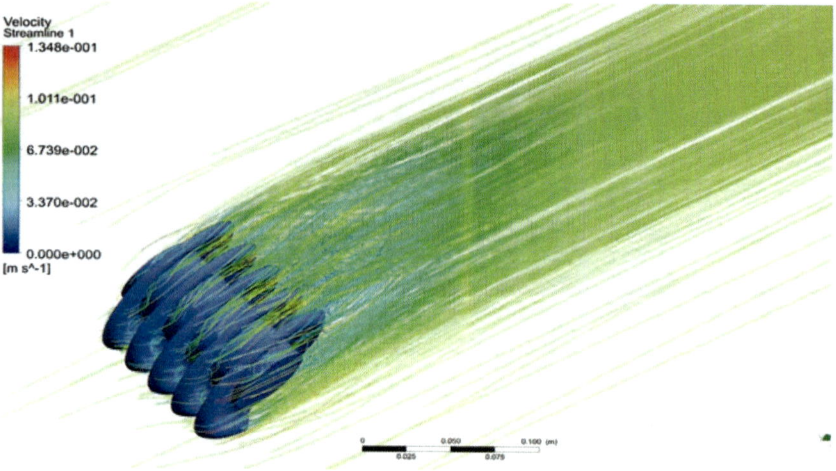

Fig. 14 Velocity generations over horizontal axis hydro rotor-IV

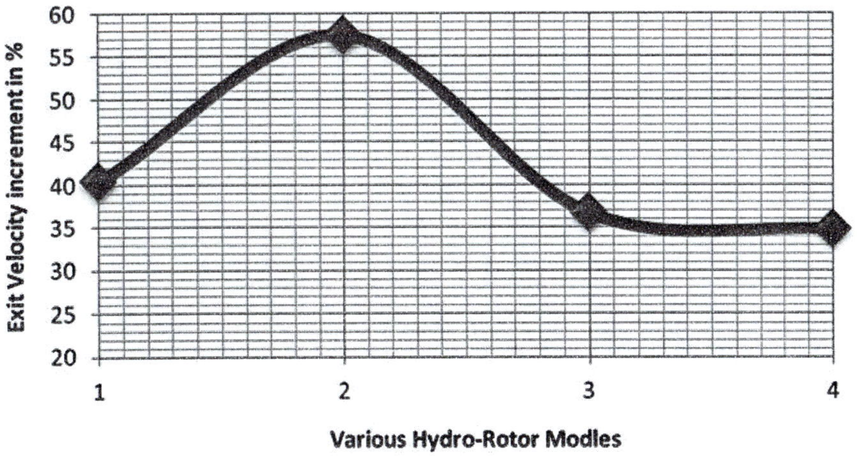

Fig. 15 Comparative velocity increment by the horizontal axis hydro-rotors

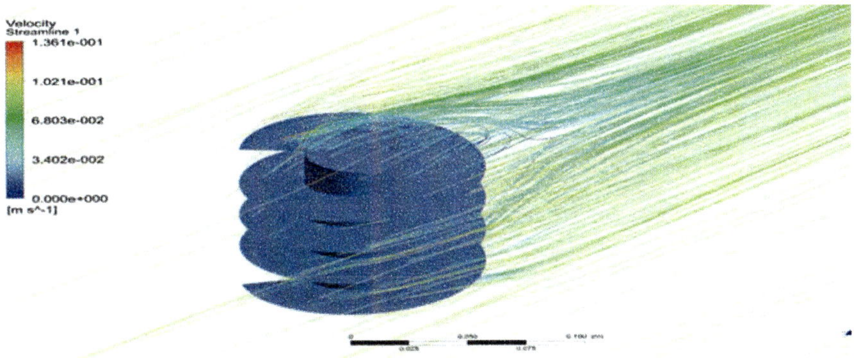

Fig. 16 Velocity generations over vertical axis hydro rotor-I

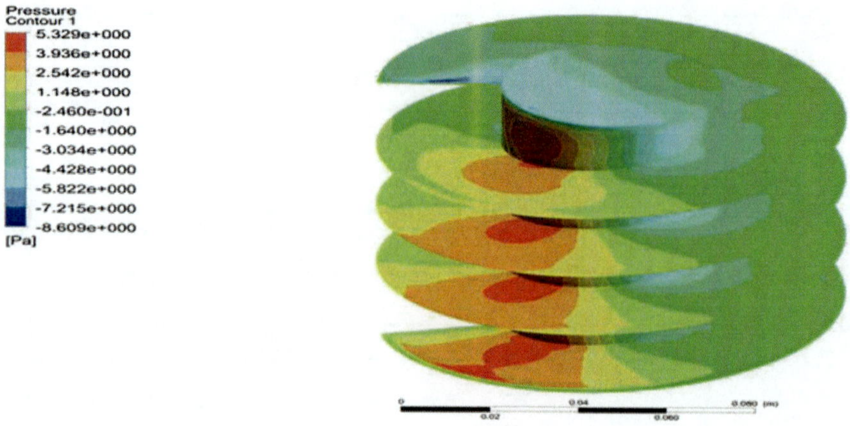

Fig. 17 Pressure variations on vertical axis hydro rotor-I

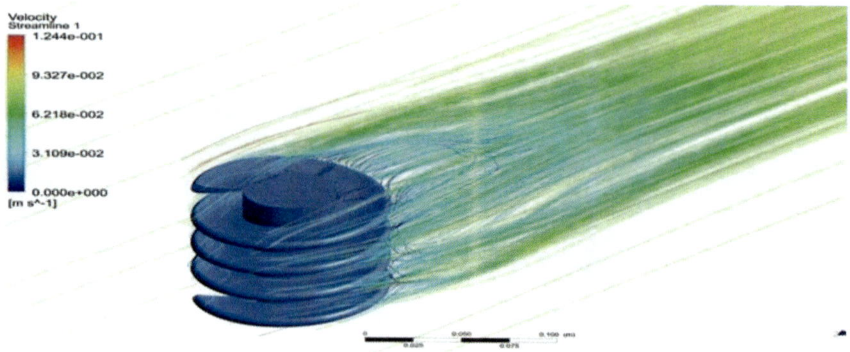

Fig. 18 Velocity generations over vertical axis hydro rotor-II

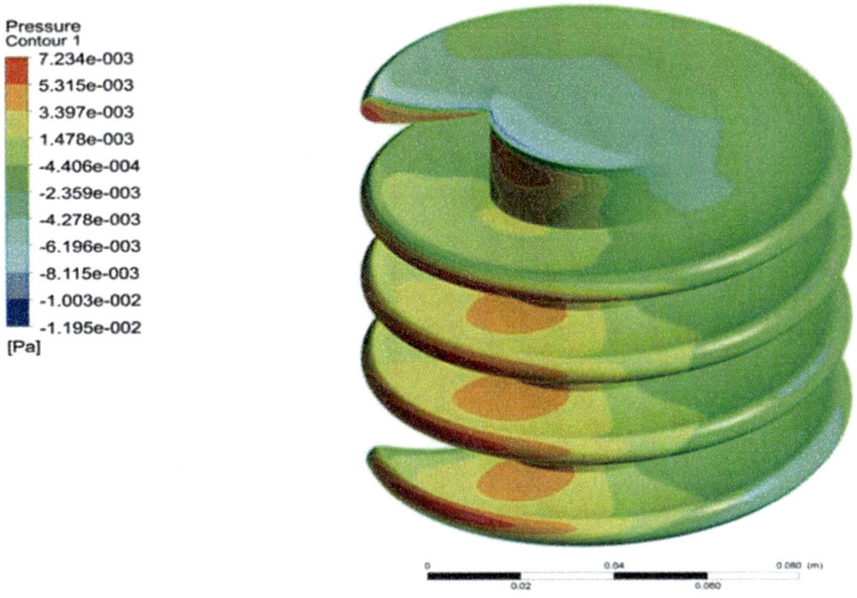

Fig. 19 Pressure variations on vertical axis hydro rotor-II

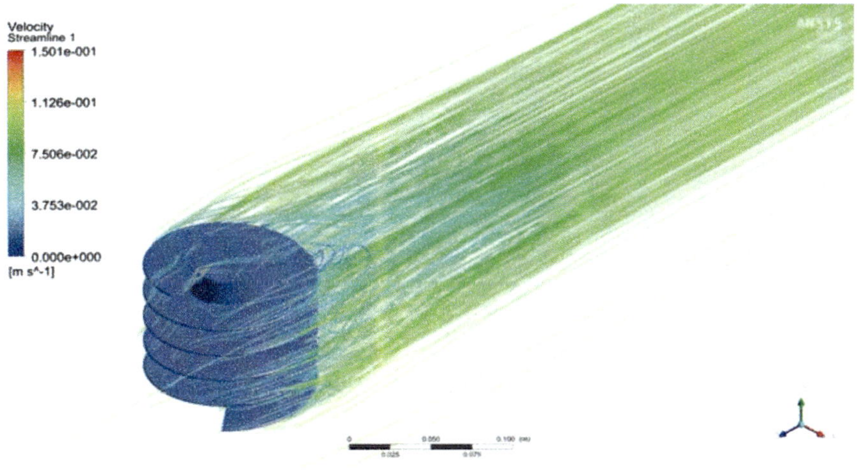

Fig. 20 Velocity generations over vertical axis hydro rotor-III

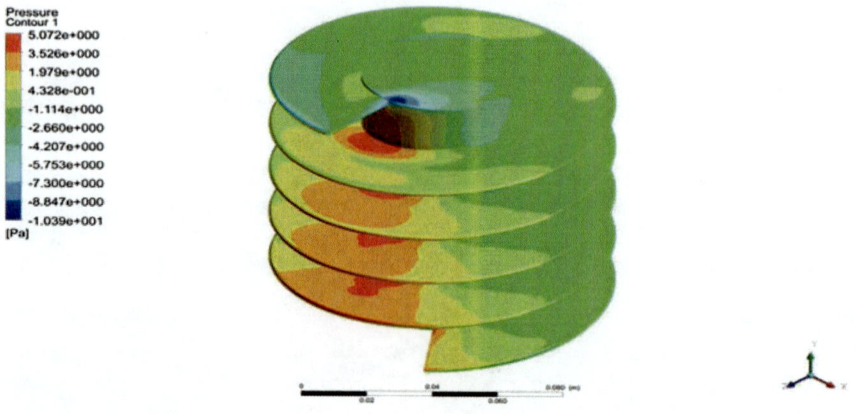

Fig. 21 Pressure variations on vertical axis hydro rotor-III

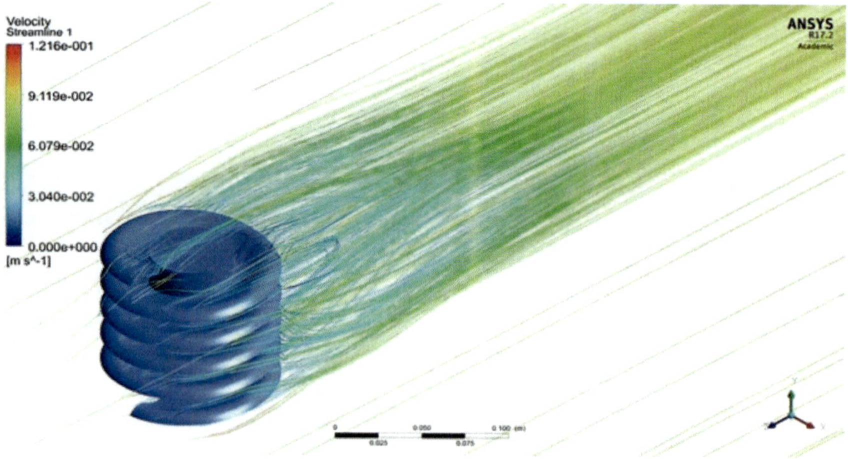

Fig. 22 Velocity generations over vertical axis hydro rotor-IV

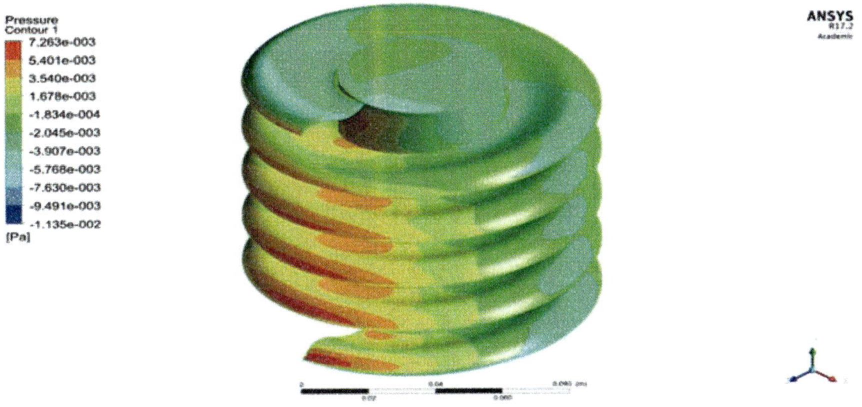

Fig. 23 Pressure variations on vertical axis hydro rotor-IV

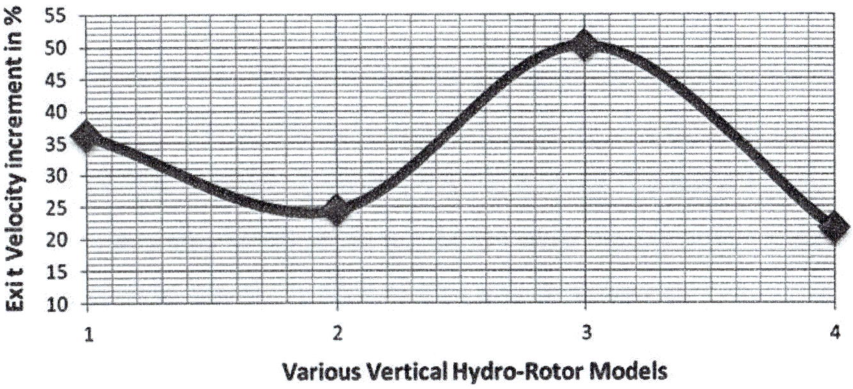

Fig. 24 Comparative velocity increment by the vertical axis hydro-rotors

References

1. Jingchun Wu, Katsumasa Shimmei, Kiyohito Tani, Kazuo Niikura, Joushirou Sato, CFD-Based Design Optimization for Hydro Turbines, Journal of Fluids Engineering, Vol. 129/159, 2007. https://doi.org/10.1115/1.2409363.
2. Nitin Kolekar and Arindam Banerjee, A coupled hydro-structural design optimization for hydrokinetic turbines, Journal of Renewable and Sustainable Energy, 5, 053146, 2013, pp. 053146-1–053146-22, https://doi.org/10.1063/1.4826882
3. W. Nuantong and S. Taechajedcadarungsri, Optimal Design of VLH Axial Hydro-Turbine using Regression Analysis and Multi-Objective Function (GA) Optimization Methods, Journal of Applied Fluid Mechanics, Vol. 9, No. 5, pp. 2291–2298, 2016.
4. Abdul Muisa, Priyono Sutikno, Aryadi Soewono, Firman Hartono, Design optimization of axial hydraulic turbine for very low head application, Energy Procedia, 68, 2015, pp. 263–273.
5. Abdullah Muratoglu, and M. Ishak Yuce, Design of a River Hydrokinetic Turbine Using Optimization and CFD Simulations, Journal of Energy Engineering, 04017009-1–04017009-11, https://doi.org/10.1061/(ASCE)EY.1943-7897.0000438.

6. Cheng Liu, Changle Xiang, Qingdong Yan, Wei Wei, Cori Watson & Houston G. Wood, Development and validation of a CFD based optimization procedure for the design of torque converter cascade, Engineering Applications of Computational Fluid Mechanics, 13:1, 128–141, 2019, https://doi.org/10.1080/19942060.2018.1562383
7. Jonathan Aguilar, Ainhoa Rubio-Clemente, Laura Velasquez and Edwin Chica, Design and Optimization of a Multi-Element Hydrofoil for a Horizontal-Axis Hydrokinetic Turbine, Energies, 2019, 12, 4679; pp. 1–18, https://doi.org/10.3390/en12244679
8. Tilahun Nigussie Abraham Engeda and Edessa Dribssa, Design, Modeling, and CFD Analysis of a Micro Hydro Pelton Turbine Runner: For the Case of Selected Site in Ethiopia, International Journal of Rotating Machinery, Volume 2017, Article ID 3030217, 17 pages, https://doi.org/10.1155/2017/3030217
9. Balaji S, Prabhagaran P, Vijayanandh R, Senthil Kumar M, Raj Kumar R, Comparative Computational Analysis on High Stable Hexacopter for Long Range Applications, Lecture Notes in Civil Engineering, ISSN: 2366-2557, book title "Proceedings of UASG 2019", Book Subtitle: Unmanned Aerial System in Geomatics, eBook ISBN: 978-3-030-37393-1, https://doi.org/10.1007/978-3-030-37393-31.
10. Vijayanandh R, Senthil Kumar M, Rahul S, Thamizhanbu E and Durai Isaac Jafferson M, Conceptual Design and Comparative CFD Analyses on Unmanned Amphibious Vehicle for Crack Detection, Lecture Notes in Civil Engineering, ISSN: 2366-2557, book title "Proceedings of UASG 2019", Book Subtitle: Unmanned Aerial System in Geomatics, eBook ISBN: 978-3-030-37393-1, https://doi.org/10.1007/978-3-030-37393-14.
11. Vijayanandh R, Kiran P, Indira Prasanth S, Raj Kumar G, Balaji S, Conceptual Design and Optimization of Flexible Landing Gear for Tilt-Hexacopter Using CFD, Lecture Notes in Civil Engineering, ISSN: 2366-2557, book title "Proceedings of UASG 2019", Book Subtitle : Unmanned Aerial System in Geomatics, eBook ISBN: 978-3-030-37393-1, https://doi.org/10.1007/978-3-030-37393-15.
12. V Praveen Kumar, S Kishor Kumar, K R Sankaresh Pandian, E Ashraf, K Thanga Tamil Selvan, Vijayanandh R, Conceptual Design And Hydrodynamic Research On Unmanned Aquatic Vehicle, International Journal of Innovative Technology and Exploring Engineering, ISSN: 2278-3075, 8, 11S, September 2019, https://doi.org/10.35940/ijitee.K1027.09811S19, pp. 121–127.
13. Vijayanandh R, Ramesh M, Raj Kumar G, Thianesh U K, Venkatesan K, Senthil Kumar M, Research of Noise in the Unmanned Aerial Vehicle's Propeller using CFD, International Journal of Engineering and Advanced Technology, ISSN: 2249-8958, Volume-8 Issue-6S, August 2019, https://doi.org/10.35940/ijeat.F1031.0886S19, pp. 145–150.
14. Arul Prakash. R, Sarath Kumar. R, Vijayanandh. R, Darsi Venkata Praveen Raja Sekar. K & Ananda Krishnan. C, Design Optimization Of Convergent—Divergent Nozzle Using Computational Fluid Dynamics Approach, International Journal of Mechanical and Production Engineering Research and Development, ISSN(E): 2249-8001 Vol. 9, Special Issue 1, Jan 2019, 220–232
15. K Naveen et al 2019 J. Phys.: Conf. Ser. 1355 012012, Design Optimization of Nozzle and Second Throat Diffuser System for High Altitude Test using CFD
16. Balaji Sonaimuthu, Prabhagaran Panchalingam, Vijayanandh Raja, Comparative Analysis of Propulsive System In Multi-Rotor Unmanned Aerial Vehicle, ASME 2019 Gas Turbine India Conference, GTINDIA 2019, pp. 1–8, https://doi.org/10.1115/GTINDIA2019-2429.
17. Ramesh Murugesan, Vijayanandh Raja, Acoustic Investigation On Unmanned Aerial Vehicle's Rotor Using CFD MRF Approach, ASME 2019 Gas Turbine India Conference, GTINDIA 2019, pp. 1–7, https://doi.org/10.1115/GTINDIA2019-2430.
18. P. Jagadeeshwaran, Dr. V. Natarajan, Vijayanandh R, Senthil Kumar M, Raj Kumar G, Raajadurai R, Abdulkader M, Harimeyyan M M, Mathan Kumar S, Naveen A, Numerical Estimation Of Ultimate Specification Of Advanced Multi-Rotor Unmanned Aerial Vehicle, International Journal Of Scientific & Technology Research, ISSN 2277-8616, Volume 9, Issue 01, January 2020, pp. 3681–3687.

Sand-Gravel Mining as a Threat to Macro-benthic Assemblage and Habitat Parameters: A Case Study of River Ganga, India

Nitin Kamboj and Vishal Kamboj

Abstract The present study was conducted for assessing the impact of sand-gravel mining on the habitat parameters, diversity and abundance of macro-benthic species in the mining impacted area of the Ganga river from April 2017 to March 2018. Samples were collected from four different zones i.e. zone A as reference zone while zones B, C and D are mining impacted area of Ganga river correspondingly. During the study, habitat parameters such as substratum type, pH, dissolved oxygen, biochemical oxygen demand, turbidity and TDS showed a significant difference at selected zones throughout the study. Besides, during the survey, a total of twenty-nine macro-benthic species belonging to ten groups mainly *Oligochaeta* (2 species), *Turbellaria* (2 species), *Hirudinea* (3 Species), *Odonata* (2 species), *Ephemeroptera* (4 species), *Trichoptera* (3 species), *Diptera* (6 species), *Gastropoda* (3 species), *Decapoda* (1 species) and *Coleoptera* (3 species) were found respectively. The *Diptera* group was found a maximum of 30.77% throughout the study period at selected zones. Besides this, the other major groups were such as *Coleoptera* 12.73%, *Gastropoda* 10.80% and *Ephemeroptera* 9.65%. The result showed that the zone A contains the higher number 1205 ind./m^2 while zone B, zone C, and zone D contain 899 ind./m^2, 714 ind./m^2 and 497 ind./m^2 respectively throughout the year. The canonical correspondence analysis (CCA) analysis showed a strong relationship with habitat parameters mainly substratum structure, pH, water temperature, dissolved oxygen with the macro-benthic species. Besides this, the reason for the decline in species number in respected zones (B, C & D) is the removal of the sand and gravel material. The removal of these materials causes habitat destruction in the form of increasing depth, slope, channelization of river and water quality.

Keywords Macro-benthic species · Mining-impacted area · CCA · Ganga river

N. Kamboj · V. Kamboj (✉)
Department of Zoology and Environmental Science, Gurukula Kangri Vishwavidyalaya, Haridwar 249404, India

N. A. Siddiqui et al. (eds.), *Advances in Environment Engineering and Management*, Springer Proceedings in Earth and Environmental Sciences,
https://doi.org/10.1007/978-3-030-79065-3_35

Abbreviations

WT	Water temperature
DO	Dissolved oxygen
TDS	Total dissolved solids
TU	Turbidity
BOD	Biochemical oxygen demand
CCA	Canonical corresp.ondence analysis
MB1	*Tubifexsp.*
MB2	*Limnodrilussp.*
MB3	*Planariasp.*
MB4	*Dugesiasp.*
MB5	*Glossiphoniasp.*
MB6	*Barbroniasp.*
MB7	*Hemiclepsissp.*
MB8	*Libellulasp.*
MB9	*Enallagmasp.*
MB10	*Ephemera sp.*
MB11	*Baetissp.*
MB12	*Heptageniasp.*
MB13	*Caenissp.*
MB14	*Hydroptilasp.*
MB15	*Triaenodessp.*
MB16	*Rhyacophilasp.*
MB17	*Chironomussp.*
MB18	*Chaoborussp.*
MB19	*Procladiussp.*
MB20	*Polypedilumsp.*
MB21	*Glyptotendipessp.*
MB22	*Forcipomyiasp.*
MB23	*Gyraulussp.*
MB24	*Promenetussp.*
MB25	*Lymnaeasp.*
MB26	*Potamansp.*
MB27	*Hyphoporussp.*
MB28	*Bidessussp.*
MB29	*Helocharessp.*

1 Introduction

The macro-benthic species are a vital part of aquatic organisms which are used as an indicator to assess the functional status of the river. The benthic organisms live in the bottom of the aquatic system part such as sediments, and rocks [1]. They are the part of the aquatic food chain which link the lower trophic level by feed plankton i.e. phytoplankton and zooplankton with higher tropic level by act as a food source of the fishes. In freshwater, the macrobenthos act as a keystone organism which circulate the ecosystems process such as food chain and nutrient cycling [2]. They interact with the various environmental condition and are distributed along with the water characteristics, substrate, food utility and quantity [2, 3]. They are sensitive to the characteristics and pattern of the habitat and substratum, water quality and types of riparian vegetation along with the aquatic system [4, 5]. Nowadays, all the aquatic systems are under pressure due to the dependency of humans. The involvement of human activities such as sand-gravel mining, disposal of wastewater, tourism activities degrades the quality of the hydrological parameters which directly affect the quality and quantity of the aquatic organisms [6, 7]. Out of these activities, sand-gravel mining activity i.e. the extraction of raw materials such as gravel and sand from the river can directly affect aquatic organisms [8, 9]. The removal of the substratum changes the quality of water, habitat characteristics and channelization of the river [7, 10].

In India, the river Ganga is the well-known river system which supports a large variety of aquatic organisms. The river Ganga is the combination of two streams mainly Bhagirathi and Alaknanda originate from the glacier of the Uttarakhand state. Uttarakhand is the developing state, and most of the cities such as Tehri Garhwal, Chamoli, Devprayag, Rishikesh, Haridwar are depend on the river Ganga for their daily needs. In these cities, the river Ganga is the main source for the construction materials such as stone, sand and gravel. The enhanced extraction of these materials directly or indirectly disturbed the ecology of river Ganga. However, for keeping this view the present study was conducted to assess the impact of sand-gravel mining on the habitat and macro-benthic species of river Ganga.

2 Materials and Methods

2.1 Sampling Stations

Haridwar is situated at the bottom of the Shivalik Himalaya endowed with natural resources. The Haridwar is the famous and holy place of India and also a developing city of the Uttarakhand state. In Haridwar, the river Ganga is the main source which provides the construction materials such as stone, sand and gravel. During the study period, four sampling zones were selected in a 30 km stretch of the river Ganga based on mining activities such as zone A which is taken as reference zone i.e. no mining activity done here while zone B, C & D are the mining intruded zones (Table 1 and Fig. 1). The zone B is situated at approximate 8 km from the zone A in the left bank while zone C and zone D is situated at approximately 12 km and 10 km respectively far from the zone A in the right bank of the river Ganga.

Table 1 Geo-coordinates of sampling zones of Ganga river

Sampling zones	Landmark of sampling zone	Types of zone	Geo-coordinates of sampling sites	Average elevation (m)	Bank condition
Zone A	Near Bairaagi camp	Reference zone	29°55′08.50″ N, 78°09′45.30″ E	274	Good riparian zone
Zone B	Near Kangri Village	Mining + sewage effluent	29°53′51.05″ N, 78°10′06.18″ E	270	Degraded riparian zone
Zone C	Near Ajitpur Village	Mining + sewage effluent	29°52′34.30″ N, 78°08′42.56″ E	266	Degraded riparian zone
Zone D	Near Tanda Village	Mining + sewage effluent	29°46′48.98″ N, 78°10′58.73″ E	243	Degraded riparian zone

effluent zone

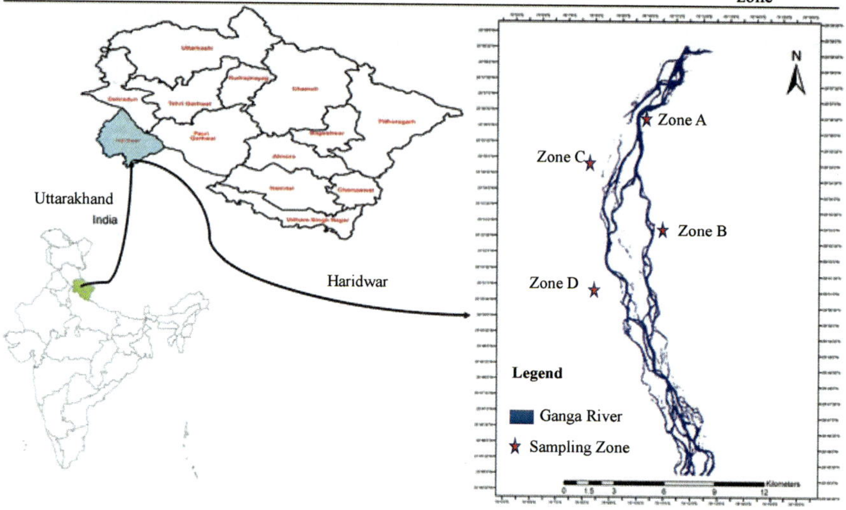

Fig. 1 Study map of the sampling zones of Ganga river at Haridwar, Uttarakhand (India)

2.2 Collection of Habitat Parameters and Benthic Species

The samples of habitat parameters and macro-benthic species were collected in monthly intervals (n = 24, 2 × 12) from April 2017 to March 2018 at the selected zones. The analysis of habitat parameters mainly substratum types using visualization method developed by [11]. However, water temperature, pH, total dissolved solids, turbidity, and salinity was estimated by digital instruments while dissolved oxygen and biochemical oxygen demand were performed using the analytical techniques provided by standard method [12]. Besides, macro-benthic species were collected using the kick method [13] and preserved in 5% formalin. The identification was done using standard methods developed by [14–16].

2.3 Statistical Analysis

The results of habitat parameters were statistically analyzed with the mean, standard deviation, and one-way ANOVA to reduce the statistical error. After that, canonical correspondence analysis (CCA) was performed to assess the relationship between the habitat parameters and macro-benthic species [17].

3 Results

3.1 Habitat Parameters

The annual variation of the habitat parameters at selected sampling zones is depicted in Table 2. The composition of the substratum of river Ganga is varied from sampling zone A to zone D. The higher percentage of the substrate was found at zone A in comparison to zone B, C & D respectively. At zone B, C, & D, the percentage of substrata such as pebbles, gravel and sand decrease due to the extraction process of these materials. Besides this, water temperature increased from zone A i.e.19.67 ± 5.34 °C to zone D i.e. 23.37 ± 4.51 °C, pH varied from 7.40 ± 0.57 at zone A to 7.67 ± 0.51 at zone C; TDS ranged from 103.61 ± 22.83 mg L^{-1} at zone A to 145.41 ± 31.61 mg L^{-1} at zone C, TU ranged from 141.27 ± 169.73 NTU at zone A to 206.86 ± 212.61 NTU at zone D, DO change from 8.01 ± 0.54 mg L^{-1} at zone A to 7.06 ± 0.38 mg L^{-1} at zone D; BOD varied from 1.58 ± 0.54 mg L^{-1} at zone A to 2.89 ± 0.73 mg L^{-1} at zone D and salinity varied from 7.83 ± 1.74 ppm at zone A to 15.24 ± 4.95 ppm at zone B at significant level $p < 0.05$,0.01 and 0.001 respectively.

Table 2 Annual variation of habitat parameters at selected zones from April 2017 to March 2018 (mean ± S.D.; one-way ANOVA)

Parameters		Sampling zones			
		Zone A	Zone B	Zone C	Zone D
Composition of substratum	Boulder	54.33 ± 2.67	41.67 ± 3.98	42.92 ± 2.11	53.08 ± 2.31
	Cobbles	12.83 ± 3.93	28.33 ± 1.07	31.67 ± 1.44	23.5 ± 1.24
	Pebbles	20.17 ± 3.43	15.00 ± 1.65	13.18 ± 1.64	9.33 ± 2.71
	Gravel	5.92 ± 1.38	7.25 ± 2.30	6.58 ± 1.62	9.33 ± 1.56
	Sand	6.75 ± 1.66	7.75 ± 1.48	5.67 ± 1.37	4.75 ± 1.76
Water temperature (WT; °C)		19.67 ± 5.34	21.46 ± 4.60 ns	22.42 ± 4.41 ns	23.37 ± 4.51 ns
pH		7.40 ± 0.57	7.51 ± 0.50 ns	7.67 ± 0.51 ns	7.40 ± 0.55 ns
Total dissolved solids (TDS; mg L^{-1})		103.61 ± 22.83	136.15 ± 25.29**	145.41 ± 31.61**	140.74 ± 28.33**
Turbidity (TU; NTU)		141.27 ± 169.73	199.41 ± 227.46 ns	201.38 ± 211.27 ns	206.86 ± 212.61 ns
Dissolved oxygen (DO; mg L^{-1})		8.01 ± 0.54	7.19 ± 0.38***	7.07 ± 0.38***	7.06 ± 0.38***
Biochemical oxygen demand (BOD; mg L^{-1})		1.58 ± 0.54	2.59 ± 0.74**	2.86 ± 0.76***	2.89 ± 0.73***
Salinity (S; ppm)		7.83 ± 1.74	15.24 ± 4.95***	11.02 ± 2.38**	10.67 ± 2.12**

ns Non significant

*, **, ***Significant at $F > p$ (0.05, 0.01, 0.001)

3.2 Taxonomic Collection of Macro-benthic Species

The taxonomic list of macro-benthic species collected at selected sampling zones is presented in Table 3. During the study, a total of twenty-nine macro-benthic species belonging to ten groups mainly *Oligochaeta* (2 species), *Turbellaria* (2 species), *Hirudinea* (3 Species), *Odonata* (2 species), *Ephemeroptera* (4 species), *Trichoptera* (3 species), *Diptera* (6 species), *Gastropoda* (3 species), *Decapoda* (1 species) and *Coleoptera* (3 species) respectively were found. In selected sampling zones, the *Diptera* group was dominant with a higher number of 1020 ind./m^2. However, the other groups such as *Coleoptera* contain 422 ind./m^2, *Gastropoda* contains 358 ind./m^2 and *Ephemeroptera* contains 320 ind./m^2 are depicted in Table 3. Besides this, all the ten groups of macro-benthic species were found a maximum number of species at zone A followed by zones B, C & D (Fig. 2a). However, *Forcipomyia sp.* contains the higher number 214 ind./m^2 followed by *Glyptotendipes sp.*194 ind./m^2, *Chironomus sp.* 175 ind./m^2, *Polypedilum sp.* 158 ind./m^2, *Hyphoporus sp.*154 ind./m^2 and *Chaoborus sp.* 147 ind./m^2 at selected sampling zones depicted in Fig. 2b and c.

3.3 Relation Between Habitat Parameters and Macro-benthic Species

Canonical correspondence analysis (CCA) was performed to assess the relationship of macro-benthic species and habitat parameters such as cobbles, pebbles, gravel, sand, water temperature, pH and dissolved oxygen were selected. The result obtained from the CCA axis 1 and axis 2 are depicted in Table 4 and Fig. 3. The axis 1 shows the 69.270% with eigenvalue 0.006 and axis 2 showed 21.03% with eigenvalue 0.002 relation of habitat parameters with macro-benthic species. In CCA analysis, the vector length of the parameters indicates the importance of the particular parameters. The vector length showed significant relation such as dissolved oxygen with zone A, sand and pebbles with zone A and zone B, water temperature and gravel with zone D, while pH and cobbles showed a significant relation with zone C depicted in Fig. 3. Macro-benthic species such as *Tubifex sp.*, *Dugesia sp.*, *Enallagma sp.*, *Ephemera sp.*, showed a positive relationship with gravel while *Chaoborus sp.*, *Polypedilum sp.*, *Promenetus sp.*, showed a positive relation with sand and pebbles. Besides, *Limnodrilus sp.*, *Forcipomyia sp.*,*Glyptotendipes sp.*, *Chironomus sp.*, *Hyphoporus sp.*, showed a positive relation with pH while *Rhyacophila sp.* showed a significant relationship with water temperature. However, *Bidessus sp.*, and *Glossiphonia sp.*, showed substantial relation with cobbles.

Table 3 Spatial variation in macro-benthic species from April 2017 to March 2018

Groups	Species	Number of species	%age	Zone A	Zone B	Zone C	Zone D
Oligochaeta	*Tubifex sp.*	111	3.35	40	29	24	18
	Limnodrilus sp.	125	3.77	45	34	29	17
	Total	236	7.12	85	63	53	35
Turbellaria	*Planaria sp.*	71	2.14	27	18	14	12
	Dugesia sp.	86	2.59	30	20	18	18
	Total	157	4.74	57	38	32	30
Hirudinea	*Glossiphonia sp.*	89	2.68	30	25	20	14
	Barbronia sp.	55	1.66	20	13	11	11
	Hemiclepsis sp.	121	3.65	44	32	24	21
	Total	265	7.99	94	70	55	46
Odonata	*Libellula sp.*	125	3.77	43	33	29	20
	Enallagma sp.	81	2.44	28	19	17	17
	Total	206	6.21	71	52	46	37
Ephemeroptera	*Ephemera sp.*	65	1.96	22	15	14	14
	Baetis sp.	95	2.87	34	26	19	16
	Heptagenia sp.	120	3.62	41	32	28	19
	Caenis sp.	40	1.21	16	9	9	6
	Total	320	9.65	113	82	70	55
Trichoptera	*Hydroptila sp.*	42	1.27	17	10	9	6
	Triaenodes sp.	90	2.71	33	27	20	10
	Rhyacophila sp.	123	3.71	43	31	28	21
	Total	255	7.69	93	68	57	37
Diptera	*Chironomus sp.*	175	5.28	62	48	41	24
	Chaoborus sp.	147	4.43	57	41	32	17
	Procladius sp.	132	3.98	55	34	26	17
	Polypedilum sp.	158	4.77	62	47	32	17
	Glyptotendipes sp.	194	5.85	69	55	44	26
	Forcipomyia sp.	214	6.46	76	62	49	27
	Total	1020	30.77	381	287	224	128
Gastropoda	*Gyraulus sp.*	141	4.25	51	43	26	21
	Promenetus sp.	132	3.98	50	37	28	17
	Lymnaea sp.	85	2.56	32	20	17	16

(continued)

Table 3 (continued)

Groups	Species	Number of species	%age	Zone A	Zone B	Zone C	Zone D
	Total	358	10.80	133	100	71	54
Decapoda	*Potaman sp.*	76	2.29	31	23	12	10
	Total	76	2.29	31	23	12	10
Coleoptera	*Hyphoporus sp.*	154	4.65	53	44	36	21
	Bidessus sp.	142	4.28	50	39	32	21
	Helochares sp.	126	3.80	44	33	26	23
	Total	422	12.73	147	116	94	65
Grand total		3315	100.00	1205	899	714	497

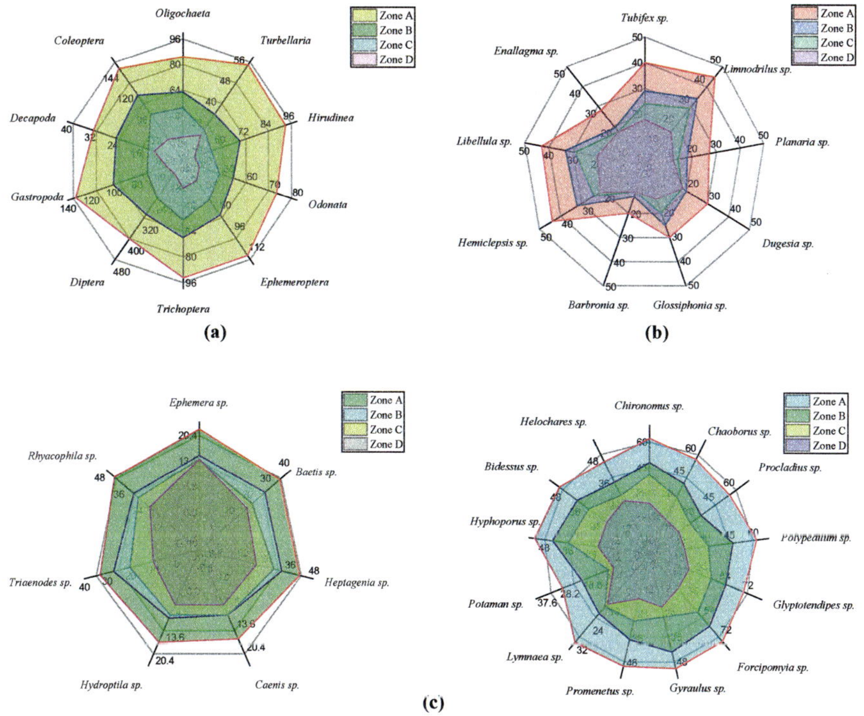

Fig. 2 **a** I.e. composition of macro-benthic groups; **b, c** I.e. composition of macro-benthic species

4 Discussion

During the study period, the outcome result of habitat parameters showed a significant change from zone A to zone D. It has been noticed that change in habitat parameters such as water quality and substratum structure of the river Ganga has deteriorated.

Table 4 CCA biplot shows the relationship between habitat parameters and macro-benthic species at selected zones

Codes	Parameters/species	Axis 1	Axis 2
	Zone A	−0.025	−0.044
	Zone B	−0.062	0.005
	Zone C	−0.007	0.073
	Zone D	0.182	−0.006
Cobbles	Cobbles	−0.054	0.895
Pebbles	Pebbles	−0.743	−0.473
Gravel	Gravel	0.857	−0.014
Sand	Sand	−0.875	−0.247
WT	Water temperature	0.699	0.564
pH	pH	−0.409	0.942
DO	Dissolved oxygen	−0.373	−0.723
MB1	*Tubifex sp.*	0.468	0.032
MB2	*Limnodrilus sp.*	−0.423	0.788
MB3	*Planaria sp.*	0.696	−1.230
MB4	*Dugesia sp.*	2.188	−0.229
MB5	*Glossiphonia sp.*	0.205	0.987
MB6	*Barbronia sp.*	1.834	−0.895
MB7	*Hemiclepsis sp.*	0.780	−0.779
MB8	*Libellula sp.*	0.419	1.050
MB9	*Enallagma sp.*	2.195	−0.129
MB10	*Ephemera sp.*	2.415	0.225
MB11	*Baetis sp.*	0.556	−0.533
MB12	*Heptagenia sp.*	0.350	1.171
MB13	*Caenis sp.*	0.318	−0.615
MB14	*Hydroptila sp.*	−0.027	−1.081
MB15	*Triaenodes sp.*	−1.445	0.423
MB16	*Rhyacophila sp.*	0.838	0.673
MB17	*Chironomus sp.*	−0.394	1.014
MB18	*Chaoborus sp.*	−1.174	−0.328
MB19	*Procladius sp.*	−0.657	−1.930
MB20	*Polypedilum sp.*	−1.594	−0.947
MB21	*Glyptotendipes sp.*	−0.574	0.727
MB22	*Forcipomyia sp.*	−0.867	0.872
MB23	*Gyraulus sp.*	−0.323	−1.070
MB24	*Promenetus sp.*	−0.763	−0.381
MB25	*Lymnaea sp.*	1.453	−1.155
MB26	*Potaman sp.*	−0.946	−3.135

(continued)

Table 4 (continued)

Codes	Parameters/species	Axis 1	Axis 2
MB27	*Hyphoporus sp.*	−0.494	1.268
MB28	*Bidessus sp.*	−0.064	0.676
MB29	*Helochares sp.*	1.112	−0.168
Eigenvalue		0.006	0.002
%		69.270	21.030
p		0.591	1.000

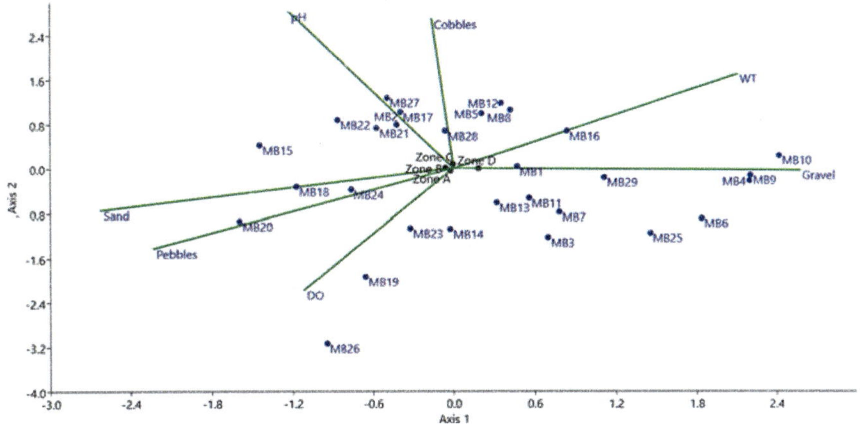

Fig. 3 CCA biplot shows the relationship between habitat parameters and macro-benthic species at selected zones

The reason behind for decline the water quality and substratum structure of river Ganga from sampling zone B to zone D is the extraction of the sand and gravel from the instream part of the river. The extraction of these materials directly or indirectly affects the substratum structure and increase the turbidity level. Although, it was observed that the changing in water quality parameters such as TDS, DO, BOD and salinity at zones B, C and D may be due to the adding of sewage water in that zones. Besides, The DO is the most vital parameters for surviving the aquatic organisms. If the oxygen level in water was decline than it directly showed some of the anthropogenic activities have happened. However, some previous studies also showed that due to anthropogenic activities such as river bed mining [7, 8], discharge of sewage water [18] and other activities such as tourism [19] was deteriorated the DO level and enhancing the turbidity, TDS and BOD level.

During the study period at selected zones, the macro-benthic species group was recorded in the order such as *Diptera* > *Coleoptera* > *Gastropoda* > *Ephemeroptera* > *Hirudinea* > *Trichoptera* > *Oligochaeta* > *Odonata* > *Turbellaria* > *Decapoda*, respectively. Some previous studies conducted by [3–5, 19] also found that *Diptera* group show a higher number of species in different aquatic systems. The other groups mainly *Coleoptera, Gastropoda, Ephemeroptera* showed a slight change in abundance of species in selected zones. However, species like *Tubifex sp., Limnodrilus sp., Hydroptila sp.* and *Triaenodes sp.* were a good indicator of clean and polluted ecosystem respectively. The variation in diversity and abundance of macro-benthic species showed the change in environmental and habitat condition at the selected zones. The diversity of macro-benthic species is increase with an increase in the substrate structure and favourable environmental condition [5]. Besides, they found less diversity of macrobenthic species in downstream as a comparison to upstream of Ken and Paisuni rivers because of the devoid condition of substrate structure in the downstream part. However, CCA analysis showed the relation of habitat parameters with macro-benthic species. The substrate structure and water quality parameters play an important role in the distribution of the benthic organisms. The substratum structure provides the habitat or niche, while parameters like water temperature, velocity, and dissolved oxygen provide the favourable environmental condition for surviving the aquatic biota. Some scientists also figured these parameters such as types and size of substratum, pH, dissolved oxygen, conductivity, and total dissolved solids, water temperature, velocity, channelization of the river in their study for distribution of the benthic organisms [2, 5, 19].

5 Conclusion

The present study indicated the diversity of macro-benthic species decreased along the mining impacted i.e. zones B, C, and D as a comparison to non-mining zone i.e. zone A. The variation in the composition of the benthic fauna varied along the selected zones indicated the impact of river bed mining and other anthropogenic activities. The removal of sand and gravel directly affects the habitat parameters such as substratum, channelization of river, and decline the water quality. Thus, the results of the present study indicated the variation in the diversity and abundance of macro-benthic diversity in mining intruded zones.

Acknowledgements Mr Vishal Kamboj highly grateful to the Department of Science and Technology, New Delhi (IN) for financial support through the INSPIRE Fellowship program (grant number IF160805) and also thanks to the Department of Zoology and Environmental Science, Gurukula Kangri Vishwavidyalaya, Haridwar for providing the lab facilities.

Conflict of Interest There is no conflict of interest during the study.

References

1. Wurdig, N.L., Cenzano, C.S.S., Motta, D.M.: Macroinvertebrate communities' structure in different environments of the Taim hydrological system in the state of Rio Grande do Sul, Brazil. Acta Limnologica Brasiliensia 19, 4 (2007) 57–68.
2. Malik, D.S., Sharma, A.K., Sharma, A.K.: Current status of macrobenthic diversity and their habitat ecology in river Ganga and its tributaries, Uttarakhand. Journal of Experimental Zoology India, 23,2 (2020) 1413–1420.
3. Silveira, M.P., Buss, D.F., Nessimian, J.L., Baptista, D.F.: Spatial and temporal distribution of benthic macrospecies in southeastern Brazilian river. Brazilian Journal of Biology 66, 2B (2006) 623–632.
4. Smith, R.F., Lamp, W.O.: Comparison of insect communities between adjacent headwater and main-stem streams in urban and rural watersheds. Journal of North American Benthological Society 27,1 (2008) 161–175.
5. Nautiyal, P., Mishra, A.S., Verma, J., Agrawal, A.: River ecosystems of the central highland ecoregion: spatial distribution of benthic flora and fauna in the plateau rivers (tributaries of the Yamuna and Ganga) in central India. Aquatic Ecosystem Health & Management 20,1–2 (2017) 43–58. https://doi.org/10.1080/14634988.2017.1296324.
6. Kamboj, V., Kamboj, N., Sharma, S.: Environmental impact of riverbed mining—a review. International Journal of Scientific Research and Reviews 7, 1 (2017) 504–520.
7. Kamboj, N., Kamboj, V.: Water quality assessment using overall index of pollution in riverbed-mining area of Ganga-River Haridwar, India. Water Science 33, 1 (2019) 65–74. https://doi.org/10.1080/11104929.2019.1626631.
8. Kamboj, V., Kamboj, N.: Spatial and temporal variation of zooplankton assemblage in the mining-impacted stretch of Ganga River, Uttarakhand, India. Environmental Science and Pollution Research, 27, 21(2020) 27135–27146. https://doi.org/10.1007/s11356-020-09089-1.
9. Padmalal, D., Maya, K.: Sand mining. Environmental impacts and selected case studies. Environmental Science and Engineering (2014) 1–161. https://doi.org/10.1007/978-94-017-9144-1.
10. Kamboj, N., Kamboj, V.: Riverbed mining as a threat to in-stream agricultural floodplain and biodiversity of Ganges River, India. In: Kumar, V., Kumar, R., Singh, J. and Kumar, P. (eds) Contaminants in Agriculture and Environment: Health Risks and Remediation, Volume 1, Agro Environ Media, Haridwar, India, (2019) 250–263, https://doi.org/10.26832/AESA-2019-CAE-0151-019.
11. Wentworth, C.K.: A Scale of Grade and Class Terms for Clastic Sediments. The Journal of Geology (1922).
12. American Public Health Association (APHA) The standard method for the examination of water and wastewater (22 editions) Washington D.C. (2012). ISBN 978-087553-013-0.
13. Victor, R., Ogbeibu, A.E.: Macrobenthic species of a stream flowing through farmland in Southern Nigeria. Environmental Pollution series A. 39 (1985) 333–347.
14. Pennak, R.W.: Fresh water Species of the United States, 2nd Edition. John Wiley & Sons, New York, (1953) 512–733.
15. Needham, J.G., Needham, P.R.: A guide to freshwater biology. Holden Day Ins., San Francisco (USA) 108 (1962).
16. Edmunds, G.F., Jensen, S.L., Berner, L.: The may flies of north and central America. University Minnesota Press, Mineapalis, (1976) 330.
17. Ter-Braak, C.J.F., Verdonscoht, P.F.M.: Canopy correspondence analysis and related multivariate methods in aquatic ecology. Aquatic Science 57, 3 (1995) 255–289. https://doi.org/10.1007/BF00877430.

18. Aswal RS, Singh P, Kamboj N, Singh R (2016) Chemometric techniques: a comparative study of drinking water sources of Dehradun and Haridwar, Uttarakhand (India). Advances in Health and Environment, Safety Select Proceedings of HSFEA, 345–352. https://doi.org/10.1007/978-981-10-7122-5_33.
19. Semwal, V.P., Mishra, A.S.: The distributional pattern of benthic macrospecies in a spring-fed foothill tributary of the Ganga River, western Himalaya, India. Journal of Threatened Taxa 11, 12 (2019) 14511–14517.

Organizational Safety Perception Survey—A Tool to Identify and Correct Organizational Contributors for Industrial Accidents

N. N. Pisharody, Kanchan Deoli Bahukhandi, Prashant S. Rawat, and R. K. Elangovan

Abstract Safety perception survey is a tool to understand organizational culture on accident prevention and control. It facilitates to extract weak safety culture attributes within the organization, which contribute to accidents. Identifying and addressing the weak attributes of safety culture will help organization to improve accident prevention and control culture. Effective implementation of corrective measures is expected to prevent accidents and support organization in its tread towards Zero Harm. This paper gives details of an organizational safety perception survey conducted in a Public Sector Power Company in India and corrective measures to address weak attributes.

Keywords Safety perception survey · Industrial accidents · Organizational safety culture

1 Introduction

Organizational safety culture is that assembly of characteristics and attitudes in organizations and individuals; which establishes that, as an overriding priority, plant safety issues receive the attention warranted by their significance [1]. Safety perception survey is a tool to understand safety culture prevailing in an organization. Corrective action on the weak attributes of safety culture will facilitate organizational safety culture to excel further. In an organization with strong safety culture and positive employees' perception on work place safety; accident preventive culture will also be strong. Such organizations are expected to have diminishing trend of accidents.

N. N. Pisharody (✉)
Nuclear Power Corporation of India Limited (NPCIL), Mumbai, India

K. D. Bahukhandi · P. S. Rawat
University of Petroleum and Energy Studies (UPES), Dehradun, India

R. K. Elangovan
Directorate General of Factory Advice and Labor Institute (DGFASLI), Mumbai, India

N. A. Siddiqui et al. (eds.), *Advances in Environment Engineering and Management*, Springer Proceedings in Earth and Environmental Sciences,
https://doi.org/10.1007/978-3-030-79065-3_36

This paper gives details of an organizational safety perception survey conducted in a Public Sector Power Company in India.

2 Taxonomy

The paper comprises of 9 sections. Section 1 is Introduction. Section 2 gives Taxonomy. Section 3 gives a brief account of Literature Survey carried out on Organizational Safety Perception Survey. Section 4 is on Safety Perception Survey Methodology. Section 5 is Discussion and Recommendations. Section 6 is Conclusion. Section 7 is on Future Research Directions. The conflicts of interest statement and acknowledgements are provided at the end of the section.

3 Literature Survey

3.1 Human Factors

Human mind is effective in detecting and eliminating potential problems and hence human factors have important role in establishing safety culture at grass root level in the organization. Human factors are environmental, organizational and job factors, which influence human behavior at the work place and affect health and safety. All jobs shall be performed correctly, with alertness, due thought, full knowledge, sound judgment and proper sense of accountability [2].

Overall objective of human factors is to design systems, jobs and organizations to match with human capabilities and limitations. Nature of issues related to human factors varies according to the type of systems. For example, issues of human factors in manual systems and automated systems will be different. Human factors should be integrated in the design and development process by considering following aspects [3]:

1. The most effective way to control hazards and risks associated with human factors is to address those issues at the design stage and continuation of it throughout the life cycle.
2. System design should support operators.
3. Addressing issues of human factors as afterthought will not be effective enough.
4. Human factors activities must proceed in parallel with technical development.

Five essential activities to be considered in the overall design for human factors are:

1. Plan human centered process: Ensure that specific human factors activities are built in and sufficiently resourced.

2. Understand and specify context of use: Identify the users and what they will be doing. Ensure that user characteristics and tasks are considered as the basis for design.
3. Specify user and organizational requirements: Specify characteristics of the system which affect users and organization.
4. Produce design solutions: Apply human factors expertise to generate design options to meet user requirements. Design iteratively. Use prototypes to clarify requirements.
5. Evaluate designs against user requirements: Test out requirements by involving target users and human factors specialists.

Majority of the accidents occur due to actions initiated by people and could have been prevented by correct actions at right time. About 80% of the accidents attributed to human factors are due to actions or omissions of people [2]. Even though modern controls, automated safety systems and structured documentation systems can facilitate safer operational environments; human intervention is crucial in accident prevention and control; hence responsibility of operators for ensuring safety shall not be ignored [4, 5]. Higher the management commitment to safety, lower the rates of accident. Failures in decision making are one of the major causes of accidents. Errors in decision making may remain dormant for many years and can cause accidents later on. Therefore, benchmarking of safety management strategy and safety climate among the organizations at regular intervals will help in ensuring safer work environment and improving safety climate [6–10].

3.2 *Organizational Safety Culture*

Organizational safety culture comprises of basic values, norms, perceptions and practices within the organization and among the individuals that, safety related issues are received the importance as warranted [1, 11, 12]. It is a culture that places a high level of importance on safety beliefs, values and attitudes; and those are shared by the majority of people within the organization. It can also be characterized as 'the way we do things around here'. This implies that safety culture in an organization is related to characteristics and attitudes of individuals as well as organization. It is not necessary that an organization with well documented procedures and regulations will have good safety culture; rather it depends on how well it is practiced even without any supervision.

Safety culture development is a management driven and worker supported activity. It consists of two components i.e., a frame work by means of organizational policy by the management and response of individuals to work within the frame work. Success of the safety culture frame work depends on the commitment by the management and compliance by the individuals. A positive safety culture will lead to improved workplace safety and organizational safety performance. It is permanent in nature and management commitment and workers' support can bring continual improvement in it. Causal factors for accidents and injuries are largely systemic and are reflections of organizational safety culture. In organizations where management commitment is strong, high priority is given to safety and hence number accidents would be less.

Key element of strong positive safety culture is an all-pervading safety thinking, which allows questioning attitude, prevention of complacency, commitment to excellence and fostering of both personal accountability and organizational self-regulation in safety related issues. Key characteristics of organizations with positive safety culture are [1, 6-10]:

1. Active managerial involvement and visible commitment, by giving overwhelming priority to safety and aligning all individuals and organization towards the common goal of safety.
2. Leadership and motivation to improve safety through setting of objectives and systems of rewards and reprimands; and through self-generated attitudes on the individuals.
3. Delegation of authority and responsibility, through effective personal communication, formal assignment and description of duties; and ensuring that those are understood by the individuals.
4. Supervision of jobs and development of supervisors to serve as role models, including audit and review practices, with readiness to respond to individuals' questioning attitudes.
5. Knowledge and competence to enhance awareness on the importance of safety; through trainings, instructions and by self-education to bring consistency in safety practices.

Five attitudes which may adversely affect organizational safety culture are:

1. Feeling among the employees that top management does not care about the safety issues.
2. Emphasizing on just getting the jobs done.
3. Safety issues are being considered second to financial issues and production.
4. Job descriptions do not include safety statements as part of important job activities.
5. Lack of on-going programs, which are indicators of commitment to health and safety.

Many serious injuries occur due to less than adequate conservative decision making and shortfalls in organizational safety culture. Latent weaknesses such as unsafe design, inadequate supervision and failure to identify and correct unsafe conditions, inadequate training and improper tools, in combination with weak organizational safety culture may cause accidents and injuries. Culture developed by top management has influence on organizational safety culture. As unsafe conditions in workplaces are easier to manage than human minds, top management shall become a driver for fostering strong positive safety culture within organization for addressing human factors. Since organizational safety culture is an assembly of both individual and organizational characteristics, the frame work for positive safety culture development should consist of human factors and organizational factors. The highest level of safety culture is achieved if individuals and organization are aligned and dedicated towards a common goal of establishing strong positive safety culture [1]. Safety culture in an organization can be assessed by conducting Safety Perception Survey.

3.3 Safety Perception

If safety performance is not measured, it is not being managed and hence the organization will miss the opportunity for future improvement. Safety perception survey is a tool to measure safety performance [13]. Accident statistics and safety audits are two commonly used methods for assessment of safety performance in an organization. However, accident statistics give information to take corrective measures in reactive manner. Though safety audit is a proactive approach, it is being conducted with a few performance indicators to verify adequacy of existing safety management systems and hence its scope is limited. In reality, safety performance of an organization depends on many more parameters such as management support for safety, employees' perception on safety, effectiveness of corrective and preventive action program, consideration given to human factors etc. Organizations often used to get puzzled; when accidents occur and recur, despite all efforts. Therefore, in addition to conventional methods for assessment of safety performance through accident statistics and safety audits, it is essential to go extra mile by obtaining direct feedback from employees, who are victims of the accidents. Safety perception surveys are conducted to get this information. Safety perception surveys support organization to tap employees' collective knowledge on accident prevention and control [14]. Safety perception surveys facilitate identification of factors that influence safety culture and safe behavior. Once these factors are known, actions to achieve excellence in safety can be decided. Following are the objectives of the safety perception surveys:

1. To assess human factors and safety culture in a quantitative manner.
2. To assess employees' perception on existing safety performance assessment methods.
3. To identify gap between safety perception of employees and management.
4. To identify employees' safety perception on effectiveness of organizational policy, documents and management comments.

4 Safety Perception Survey Methodology

Safety perception survey process involve following steps as depicted in Fig. 1:

1. Development of Survey Questionnaire
2. Selection of Sample Size
3. Conduct of Test Survey
4. Communication of Intentions of Survey

Fig. 1 Safety perception survey process

5. Administration of Survey Questionnaire
6. Analysis of Safety Perception Survey Data
7. Validation of Safety Perception Survey Results

4.1 Development of Survey Questionnaire

Survey questionnaire is developed in such a way that those are understood by all respondents and are clear to them. Each question consisted of multiple choices based on Likert scale with expected response from the most positive to the least positive as given below:

1. Strongly Agree
2. Agree
3. Do Not Know
4. Disagree
5. Strongly Disagree

The questions were framed to select any one of the multiple choices. An answer of 'Strongly Agree' and 'Agree' conveys perception of safety culture with higher and lower levels of agreement. Similarly, an answer of 'Strongly Disagree' and 'Disagree' conveys a higher and lower degree of disagreement. Any attribute with agreement less than 90% of the total samples needs corrective measures for improvement.

4.2 Selection of Sample Size

It is desirable to include all employees in the survey, so that everyone will get opportunity to express his/her feelings. This will give each employee a sense of participation. In addition, more the number of participants, higher the reliability. However, in a big organization, it is impractical to involve all employees in the survey and it will be time consuming. Analysis of sample will also be difficult due to large quantity of data. In such cases it is advantageous to go for sampling survey. Number of samples is decided statistically based on the required confidence level.

There are about 11,000 employees in the subject Public Sector Power Company. Sample size is determined by following two steps [15]:

1. Calculation of Sample Size for Infinite Population.
2. Adjustment of sample size to the required population.

Calculation of Sample Size for the Infinite Population

Calculation of sample size for infinite population is calculated by the following formula:

$$S = (\text{Z-Score})^2 * p * (1 - p)/(m)^2$$

where
S—Sample size for infinite population,
Z—Z-Score,
p—Population proportion assumed to be 50% or 0.50,
m—Margin of Error.

Z-Score is determined based on confidence level and confidence level is the probability of a parameter falls within the specified range of values. Table 1 gives Confidence Levels and Z scores. If we consider a Confidence Level of 95%, the Z score is 1.960.

Margin Error is a small amount allowed for miscalculation or change of circumstances. Generally, the Margin of Error is considered as 5% or 0.05.

Table 1 Confidence levels and Z scores

Confidence levels (%)	Z scores
90	1.645
95	1.960
99	2.576

With the above values, sample size for infinite population can be calculated as given below:

$$S = (\text{Z-Score})^2 * p * (1 - p)/(m)^2$$

Z—1.960,
p—0.50,
m—0.05.

$$S = (1.960)^2 * 0.5 * (1 - 0.5)/(0.05)^2 = 384.16$$

Sample size for infinite population is 384.16.

Adjustment of Sample Size to the Required Population

Adjustment of sample size for required population is determined by using the formula given below:

$$\text{Adjusted Sample Size} = S/[1 + ((S - 1)/\text{Population})]$$

$$S = 384.16$$

Population = Number of employees in the Public Sector Power Company = 11,000.

$$\begin{aligned}\text{Adjusted Sample Size} &= 384.16/[1 + ((384.16 - 1)/11000)] \\ &384.16/(1 + 0.03483) \\ &371.23 \\ &371\end{aligned}$$

Required sample size for 11,000 population is 371.

4.3 Conduct of Test Survey

Test survey was carried out in a group of *10 persons* to ensure that the questions are clearly understood and interpreted correctly by the respondents. Feedback on the test survey was utilized and necessary modifications were incorporated in the questionnaire before distribution.

4.4 Communication of Intentions of Survey

Before sending the questionnaire to the employees, its purpose was communicated to the prospective respondents. In order to obtain an unbiased opinion without any fear of reprisal, the information pertaining to accident prevention and control were sought in anonymous manner.

4.5 Administration of Survey Questionnaire

Administration of survey was carried out by sending questionnaire to the employees by emails. Though statistically estimated required number of samples were 371, total 410 filled questionnaires were received from the employees.

4.6 Analysis of Survey Data

All 410 filled questionnaires were analyzed with respect to number of positive agreements, neutral agreements and negative agreements. Employee's perceptions with 'Strongly Agree' and 'Agree' opinions are treated as positive agreements and those with 'Strongly Disagree' and 'Disagree' opinions are treated as negative agreements. Employee's perceptions with 'Neither Agree Nor Disagree' opinions are considered as neutral agreements.

Table 2 gives positive, negative and neutral agreements by the employees in each attribute as per the Likert Scale. Corrective actions need to be taken on questions in which positive response is less than 90%.

Table 2 indicates that following are the attributes with less than 90% positive agreements in employee perception:

1. Hazard Identification & Corrective Action Program—Safety Related Deficiencies are Corrected Promptly
2. Honesty—Employees work safely even when they know that they are not being watched
3. Importance to Safety—Employees consider that sometimes it is necessary to ignore safety issues to perform jobs
4. Importance to Safety—Supervisors insist and enforce safe work practices while performing jobs
5. Strive for Excellence—Management looks for engineering solutions to address safety issues.

Table 2 Positive, negative and neutral agreements by the employees

S. No.	Description of question	Employee perception					Total
		Strongly agree	Agree	Neither agree nor disagree	Disagree	Strongly disagree	
Employee safety involvement							
1	Employees can raise safety issues and hazards without fear	326	78	1	5	0	410
		98.54%		0.24%	1.22%		100%
Safe work environment							
2	Equipment and tools available at the site are safe for use	230	174	4	1	1	410
		98.54%		0.98%	0.49%		100%
Organizational safety leadership							
3	Managers are committed to take prompt corrective actions on safety issues	218	157	29	6	0	410
		91.46%		7.07%	1.46%		100%
Safety training and coaching							
4	JHAs and Safe Work Procedures are available for all hazardous jobs	204	165	32	7	2	410
		90.00%		7.80%	2.20%		100%
5	Employees are trained and coached on safe work procedures for the jobs	233	151	15	10	1	410
		93.66%		3.66%	2.68%		100%
6	Compliance to Safe Work Procedures and JHAs is ensured while performing jobs	195	181	20	12	2	410
		91.71%		4.88%	3.41%		100%

(continued)

Table 2 (continued)

S. No.	Description of question	Employee perception					Total
		Strongly agree	Agree	Neither agree nor disagree	Disagree	Strongly disagree	
Hazard identification and corrective action program							
7	**Safety related deficiencies are corrected promptly**	**208**	**157**	**19**	**24**	**2**	**410**
		89.02%		**4.63%**	**6.34%**		**100%**
8	Whenever any job is planned, hazards are identified and corrective measures are taken	185	188	20	15	2	410
		90.98%		4.88%	4.15%		100%
Accident investigation and corrective action							
9	All accidents are reported and investigated	241	154	10	5	0	410
		96.34%		2.44%	1.22%		100%
10	Corrective actions are taken on all accidents to prevent recurrence	241	150	16	3	0	410
		95.37%		3.90%	0.73%		100%
Honesty							
11	**Employees work safely even when they know that they are not being watched**	**163**	**186**	**38**	**22**	**1**	**410**
		85.12%		**9.27%**	**5.61%**		**100%**
Importance to Safety							

(continued)

4.6.1 Hazard Identification & Corrective Action Program—Safety Related Deficiencies Are Corrected Promptly

Safety Related Deficiencies (SRDs) are identified by the Industrial Safety professionals during plant safety inspection, by the job performers while performing jobs or by the independent agencies other than the job performers during job observation

Table 2 (continued)

S. No.	Description of question	Employee perception					Total
		Strongly agree	Agree	Neither agree nor disagree	Disagree	Strongly disagree	
12	**Employees consider that sometimes it is necessary to ignore safety issues to perform jobs**	**40**	**78**	**39**	**133**	**120**	**410**
		28.78%		**9.51%**	**61.71%**		**100%**
13	**Supervisors insist and enforce safe work practices while performing jobs**	**140**	**199**	**41**	**22**	**8**	**410**
		82.68%		**10.00%**	**7.32%**		**100%**
Strive for excellence							
14	**Management looks for engineering solutions to address safety issues**	**170**	**194**	**32**	**13**	**1**	**410**
		88.78%		**7.80%**	**3.41%**		**100%**
15	Required type and quantity of personal protective equipment are available to perform jobs	195	184	17	12	2	410
		92.44%		4.15%	3.41%		100%

programme. These deficiencies when interact with human being, could lead to accidents. SRDs are reported and corrected through Safety Related Deficiency Management System (SRDMS). However, if SRDs are not promptly corrected, deficiencies will remain at the work place, which increase the potential for accidents. Figure 2 indicates that 89.02% of the employees believe that SRDs are addressed promptly, which is marginally below the acceptance level of 90%. Therefore, SRDMS should be further strengthened for prompt correction of SRDs by tracking and follow-up.

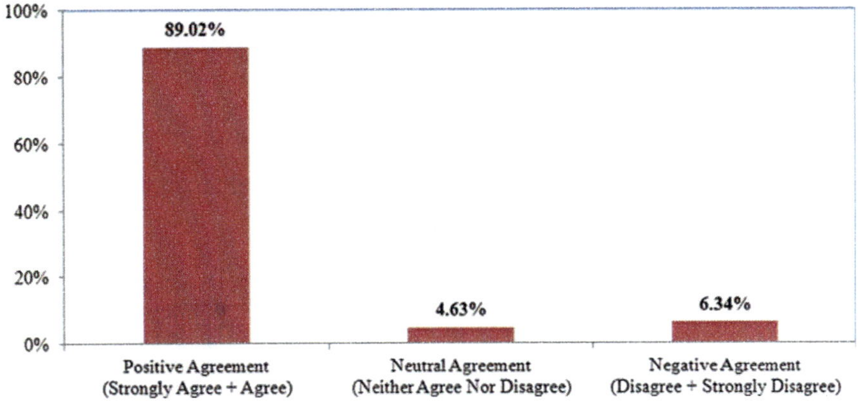

Fig. 2 Hazard Identification & Corrective Action Program—safety related deficiencies are corrected promptly

4.6.2 Honesty—Employees Work Safely Even When They Know That They Are Not Being Watched

Safety Culture is a characteristic in an organization, where the employees perform jobs in safe manner by following safe work procedure, in compliance with the recommendations of Job Hazard Analysis; even without any supervision. Figure 3 indicates that 85.12% of the employees believe that employees work safely even when they know that they are not being watched, which is below the acceptance level of 90%. Therefore, management should conduct more programs on safety culture to bring awareness among the employees to ensure compliance to safe work procedures and recommendations of Job Hazard Analysis. Since, this is a continual process, both line managers, employees and safety professionals shall work in cohesion towards this.

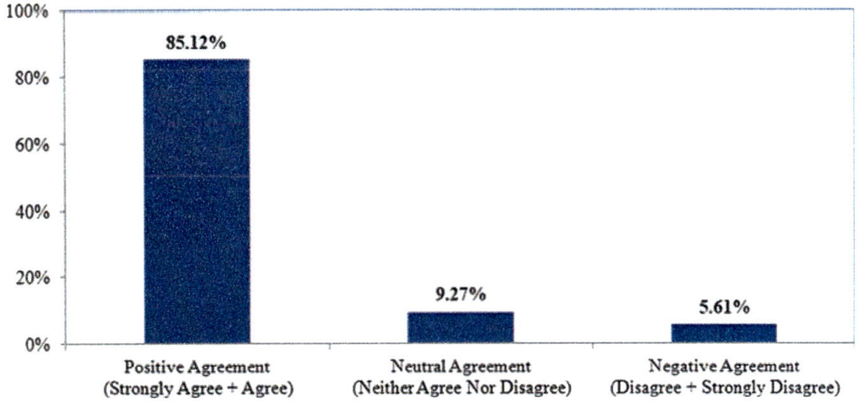

Fig. 3 Honesty—employees work safely even when they know that they are not being watched

Line managers and supervisors should reiterate necessity of compliance to safety requirements during pre-job briefing, pre-employment personal talks and toolbox talks; and during execution of jobs.

4.6.3 Importance to Safety—Employees Consider That Sometimes It is Necessary to Ignore Safety Issues to Perform Jobs

Safety culture necessitates an overriding priority to safety related issues. In principle, organizations follow 'Safety First and Production Next' policy. However, this policy is not adhered in spirit in some organizations. In such organizations, production related issues receive an overriding priority over safety related issues. This may lead to non-adherence to safe work procedures, non-compliance to recommendations of Job Hazard Analysis, delay in addressing safety related deficiencies, non-usage of requisite personal protective equipment, inadequate training and awareness program on safety requirements to employees, non-compliance to recommendations of safety committee meetings and other statutory requirements. In such organizations, investigation of accidents may not be rigorous enough to identify root causes, which may lead to recurrence of accidents. Figure 4 indicates that 28.78% of the employees consider that sometimes it is necessary to ignore safety issues to perform jobs. Being a negative attribute, its acceptance value is less than 10%. This indicates that the company should exert more efforts in enhancing safety culture and emphasize strict adherence to safety requirements during the training sessions, safety committee meetings and management meetings. The company may institute reward scheme on compliance to safety requirements during job execution to motivate employees to perform jobs in safe manner, without ignoring safety requirements.

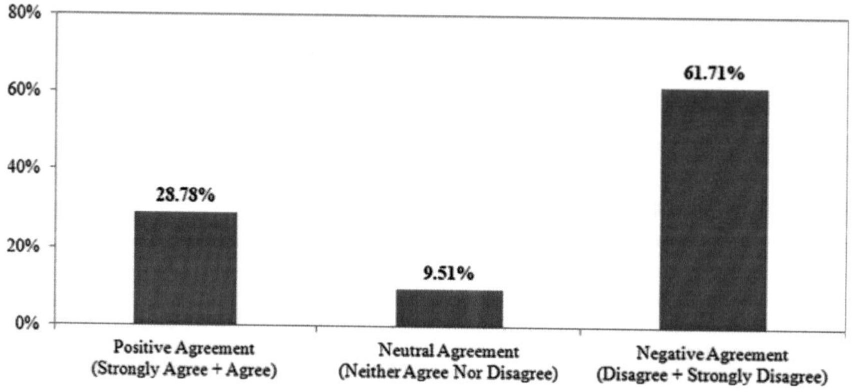

Fig. 4 Importance to safety—employees consider that sometimes it is necessary to ignore safety issues to perform jobs

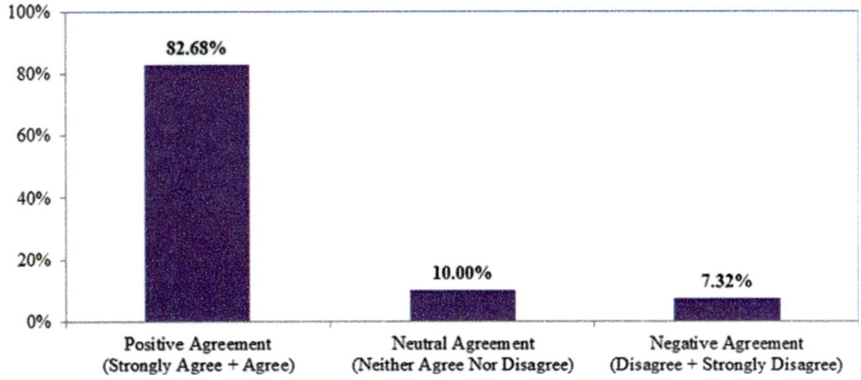

Fig. 5 Importance to safety—supervisors insist and enforce safe work practices while performing jobs

4.6.4 Importance to Safety—Supervisors Insist and Enforce Safe Work Practices While Performing Jobs

Work supervisors play very important role in enforcement of safety requirements during job execution. Supervisors shall insist compliance to safety requirements during pre-job briefing, pre-employment personal talks, toolbox talks and training programs. In addition, in case any employee violates safety requirements, supervisors shall stop him/her from such violations and enforce safety requirements to perform the job in safe manner. Figure 5 indicates that 82.68% of the employees believe that supervisors insist and enforce safe work practices while performing jobs, which is below the acceptance level of 90%. This indicates that supervisors shall give more importance to safety during job execution and enforce safety requirements during job execution. Management may organize training and awareness program on role of supervisors in enforcement of safety requirements for supervisors and line managers.

4.6.5 Strive for Excellence—Management Looks for Engineering Solutions to Address Safety Issues

Safety is a management driven and employee compiled function. Therefore, management commitment is very important in establishing a strong safety culture in the organization. Solutions to safety issues should be achieved by following hierarchy of hazard control. Efforts of management should be for implementing safety measures through engineering solutions. This can be achieved by 'Safety through Design', which includes elimination, substitution and engineering control. Figure 6 indicates that 88.78% of the employees believe that management looks for engineering solutions to address safety issues, which is below the acceptance level of 90%. Therefore, there is a scope in improvement in management approach to address safety issues by 'Safety through Design'.

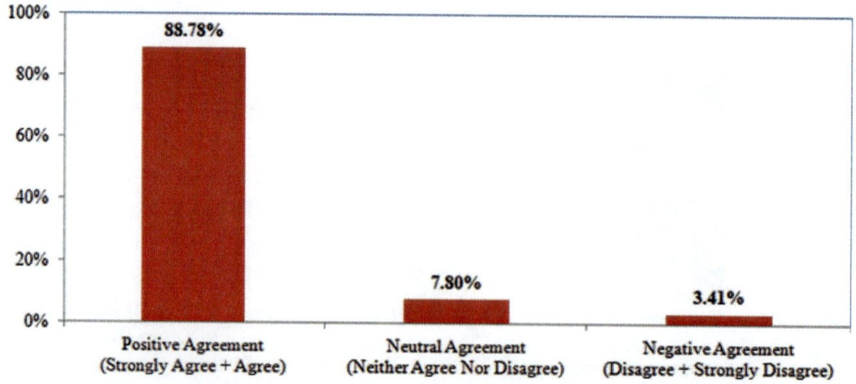

Fig. 6 Strive for excellence—management looks for engineering solutions to address safety issues

4.7 Validation of Safety Perception Survey Results

Results safety perception survey was discussed among a focused group of employees. Their observations on practices being observed at work locations were also corroborating with the survey results.

5 Discussion and Recommendations

Safety perception survey indicates the weak attributes in which corrective actions are to be taken to further improve safety culture and hence to develop a strong system for accident prevention and control in the Public Sector Power Company. These weaknesses should be addressed by means of 'Safety through Design', 'Administrative Controls' and 'Personal Protective Equipment" as detailed below:

Safety through Design

- Safety Related Deficiencies should be addressed promptly by considering hierarchy of hazard control at the design stage itself and throughout the life cycle through elimination, substitution and engineering controls.

Administrative Controls

- Safety Related Deficiency Management System (SRDMS) should be further strengthened by prompt correction of SRDs based on their safety significance. Critical and significant SRDs which have potential for loss of life or disabling injury should be addressed on priority, when compared with SRDs for safety improvement.

- Line managers and supervisors should reiterate necessity of compliance to recommendation of job hazard analysis and safe work procedures during pre-job briefing, pre-employment personal talks and toolbox talks; and during the execution of jobs.
- Strict compliance to safety requirements should be emphasized during the training sessions, safety committee meetings, management meetings.
- Company may institute reward scheme on compliance to safety requirements during job execution to motivate employees to perform jobs in safe manner, without ignoring safety requirements.
- Training and awareness program on safety culture and role of supervisors in enforcement of safety requirements should be conducted for the supervisors and line managers.

Personal Protective Equipment

- Training and awareness program on use of personal protective equipment should be conducted to develop a safety culture among the employees for adherence to safety requirements and use of personal protective equipment even in the absence of supervision.

6 Conclusion

Safety perception survey is a tool to understand organizational culture on accident prevention and control. It facilitates to extract weak safety culture attributes within the organization, which contribute to the accidents. Identifying the weak attributes will help the company to address the causes to improve accident prevention and control culture within the organization. Effective implementation of these measures is expected to prevent accidents and support organization in its tread towards **Zero Harm**.

7 Future Research Directions

Safety perception depends on the organizational safety culture. Hence, most of the solutions emerging out of safety perception survey will be based on management intervention. In order to develop an effective accident prevention culture, safety related issues should be addressed by 'Safety through Design'. This gap needs to be addressed through more efforts in this regard.

Acknowledgements The authors acknowledge the support of officials of the Public Sector Power Company for extending technical and administrative supports for carrying out Safety Perception Survey. The article has been written as a voluntary effort and has not of received any funding. Opinions expressed are those of authors and not of their organizations.

Conflicts of Interest None of the contributing authors have conflicts of interest to declare.

References

1. International Atomic Energy Agency (IAEA), Safety Culture, Safety Series No. 75, INSAG-4, 1991.
2. HSE, UK, Development of a working model of how human factors, safety management systems and wider organizational issues fit together, HSE, UK, 2007.
3. HSE, UK, Human Factors Integration: Implementation in the Onshore and Offshore Industries, Research report 001, Prepared by BAE Systems Defence Consultancy for the HSE, UK, 2002.
4. Rathnayaka Samith, Khan Faisal, Amyotte Paul., SHIPP Methodology: Predictive Accident Modeling Approach Part I: Methodology and Model Description, Process Safety and Environmental Protection, 89, 2011a, p 151–164.
5. Rathnayaka Samith, Khan Faisal, Amyotte Paul., SHIPP Methodology: Predictive Accident Modeling Approach Part II: Validation with Case Study, Process Safety and Environmental Protection, 89, 2011b, p 75–88.
6. HSE, UK, Factoring the Human into Safety: Translating Research into Practice (Vol. 1)—Benchmarking Human and Organizational Factors in Offshore Safety, Research report 059, Prepared by the University of Aberdeen for the HSE, UK, 2003a.
7. HSE, UK, Factoring the Human into Safety: Translating Research into Practice (Vol. 2)—The Development and Evaluation of A Human Factors Accident and Near Miss Reporting Form for the Offshore Oil Industry', Research report 060, Prepared by the University of Aberdeen for the HSE, UK, 2003b.
8. HSE, UK, Factoring the Human into Safety: Translating Research into Practice (Vol. 3)—Crew Resource Management, Research report 061, Prepared by the University of Aberdeen for the HSE, UK, 2003c.
9. HSE, UK, Good Practice and Pitfalls in Risk Assessment', Prepared by Drs. Sandra Gadd, Deborah Keeley, and Helen Balmforth of The Health and Safety Laboratory for the HSE, UK, 2003d.
10. HSE, UK, Out Of Control—Why Control Systems Go Wrong and How to Prevent Failure, HSE, UK, 2003e.
11. Biggs Herbert C, Dingsdag Donald P, Roos Colette R, A Practical Guide to safety Leadership, Department of Education and Employment and Work Place Relations, Australian Government, 2008.
12. Manuele F.A., Advanced Safety Management—Focusing on Z10 and Serious Injury Prevention, John Wiley & Sons, Inc, Hoboken, New Jersey, USA, 2014.
13. Ryan D, Safety Perception Survey, Professional Safety, December 2009, p 22–27.
14. Ryan D, 10 Reasons Why Safety Perception Survey Should be Your First Measurement Option, Safety Leadership, December 2019, https://ehstoday.com.
15. Cochran William G. Sampling Techniques, 3rd ed. Wiley, New York, USA, 2008.

Comparative Assessment of Performance and Emission Characteristics of DI Diesel Engine Fueled with Combined Blending of Biodiesel, Butanol and Conventional Diesel

Shashank Pal, S. M. Tauseef, J. K. Pandey, Amit Kumar Sharma, Sunil Kumar Tiwari, Tapas Goswami, and Tijendra Kumar

Abstract Seeing today's trend of petroleum product consumption and correspondingly its harmful effects on environment have forced researchers to investigate alternative and eco-friendly fuels. Biodiesel is emerging as most promising fuel to replace diesel in CI engines. However, the viscosity of biodiesel is generally higher than diesel fuels which lead to poor combustion. Therefore, to address above problems, biodiesel is blended with diesel fuel in different ratio. In addition, the presence of alcohols to diesel–biodiesel blend upgraded the fuel properties such as cloud point (CP) and cold filter plugging point (CFPP), while slightly decreased density, lower heating value, kinematic viscosity, cetane number and flash point. Hence, this investigation deals with emission reductions in CI engines by using different blends of biodiesel, butanol and diesel fuel blends. To perform the experiments, Jatropha and pongamia biodiesel fuels were mixed with conventional diesel fuel and butanaol in different combinations—Jatropha (J), Pongamia (P), Butanol (But), Diesel (D): (J10But5D85), (P100But5D85), (J15But5D80), (P15But5D85), and (J15P15But5D65). The experiments were conducted with single cylinder 3.7 kW diesel engine at different loads. For all combinations of blends, NO_x, CO, CO_2, HC, Smoke and BSFC were recorded and compared with diesel fuels at different loads. All the fuel variants worked well in engine and no difficulty was found. Best results were found in combine blending of Jatropha biodiesel and butanol in ratio of J15But5D80. With this blend the NO_x, CO_2 content was found minimum and fell down drastically, smoke content was lowest with J15P15But5D65. However, BTE was observed to be slightly decreased with

S. Pal · S. M. Tauseef · J. K. Pandey · S. K. Tiwari
Department of Mechanichal, College of Engineering Studies, University of Petroleum and Energy Studies, Dehradun, India

A. K. Sharma (✉)
Department of Chemistry and Biofuel Research Laboratory, Centre for Alternative and Renewable Energy Research, College of Engineering Studies, University of Petroleum and Energy Studies, Dehradun, India

T. Goswami · T. Kumar
Department of Chemistry, University of Petroleum and Energy Studies, Dehradun, India

N. A. Siddiqui et al. (eds.), *Advances in Environment Engineering and Management*, Springer Proceedings in Earth and Environmental Sciences,
https://doi.org/10.1007/978-3-030-79065-3_37

all combination of biodiesel, butanol and diesel. Fuel consumption was observed maximum with J15P15But5D65 fuel blends due to its lower calorific value.

Keywords Biodiesel · Jatropha · Pongamia · Butanol · DI diesel engine · Performance and emissions

1 Introduction

Increasing population and energy demands are continuing to deplete fossil fuel resources. In addition, the combustion of fossil fuel results in to greenhouse gas emissions and leads to global warming [1, 2]. Hence, there is a need to search alternative and green resources of fuels. Various alternative fuels such as biodiesel, bioethanol, biogas and bio-hydrogen are emerging as most promising renewable energy sources [3]. India is the country where the consumption of diesel is higher than gasoline [1, 4]. Biodiesel is an ecofriendly and alternative energy sources which can be used in diesel engine without any modification [2, 3]. Biodiesel is a mixture of mono alkyl esters and can be produced from the different edible and non-edible vegetable oils such as sunflower, soybean, Jatropha, Pongamia, polanga, microalgae and animal fat through transesterification process [1, 2, 4–6]. Biodiesel is an oxygenated fuel and leads to complete combustion due to presence of oxygen resulting into emission reductions [2, 7, 8].

India's transportation fuel requirements are unique in the world. India consumes almost five times more diesel fuel than gasoline, whereas, almost all other countries in the world use more gasoline than diesel fuel is the best alternative. As per the report of ministry of petroleum and natural gas of India, India consumed 83.52 million tones diesel in 2018–19. Thus, in India, search for alternatives to diesel is of special importance and the use of biodiesel is comparatively much more important for us than for rest of the countries [3]. Biodiesel is one of the renewable, nontoxic and environmental friendly alternative biofuels and can be used in diesel engines with little or without modifications in engines (5). It can be blended with diesel in any proportion. Combustion of biodiesel in engines leads to lower smoke, particulate matter (PM), carbon monoxide (CO) and hydrocarbon (HC) emissions, but higher nitrogen oxides (NO_x) emission keeping engine efficiency unaffected or improved [2, 3].

Sharma et al. examined the effect of algae, jatropha and polanga biodiesel blends on the diesel engine performance and observed a significant deterioration in DI engine missions when the it run with algae biodiesel blends [2]. In addition, Srivastava and Verma (2008) performed the experiments using methyl ester of karanja oil. The authors reported that, the maximum thermal efficiency with methyl ester of karanja oil was about 24.9%, whereas that of the diesel was 30.6% at maximum power output. The authors concluded that, the methyl ester of karanja oil is a suitable substitute of diesel. Gopinath et al. (2014) run diesel engine on four types of fuel blends (100% Diesel, 90% Diesel + 10% Corn oil Methyl Ester, 80% Diesel + 20% Corn oil Methyl

Ester, 70% Diesel + 30% Corn oil Methyl Ester and 60% Diesel + 40% Corn oil Methyl Ester) to analysis emissions from engines. They found that 80% Diesel + 20% Corn oil Methyl Ester is the optimum blend for engine emission as well as performance of the engine. Kaimal et al. (2015) run engine studied the combustion and emission characteristics of Cotton Seed Methyl Ester (CSME) blends with diesel. They used different blends of biodiesel i.e. B10, B20, B30 (10, 20, 30% blend of CSME with diesel) and compared with diesel fuel performance. They concluded that B20 blend gives comparable performance and combustion characteristics with considerable reduction in emissions when compared to diesel, therefore b20 can be used for engine run. The fuel properties are generally depends on FAMEcomposition of biodiesel [4, 5, 9]. It has been observed that mixing of two or more types of biodiesel results into improved fuel properties of fuel blends [3]. Furthermore, if alcohols are also mixed with these blends, cold flow properties of fuel blends were also observed to be improved. Therefore, this manuscript investigates the effect of different blending ratio of diesel, biodiesel and butanol on fuel properties of blended fuels and its impacts on engine performance.

2 Material and Methods

2.1 Biodiesel Production and Blend Preparation

Biodiesel was synthesised in a lab scale biodiesel reactor which consists of heating mantle, reaction flask and mechanical stirrer. The flask capacity was 1 L. It was made of three necks: one for stirrer, and the others for condenser and inlet of reactant as well as for placing the thermocouple to observe the reaction temperature. A two-step, 'acid–base' transesterification process was performed to produce biodiesel from jatropha and karanja oil. During acid pre-treatment H_2SO_4 was used as acid catalyst and methanol as reagent to reduce FFA less than 1%. While transesterification reaction was performed using KOH (1%, w/v) as a base catalyst and methanol (1:6, molar ratio) as co-solvent at 60 °C for 3 h. After completion of reaction, biodiesel was washed with warm water (55 °C) to remove impurities and glycerol. Moisture was removed from purified biodiesel using silica gel as moisture absorbent. Then purified biodiesel was filtered via Whatmann 40 filter paper and biodiesel was used to make blends with diesel.

2.2 Fuel Properties of Fuel Blends

The fuel properties of fuel blends are shown in Fig. 1 as shown in Table 1, the calorific values of Pure diesel (D100), jatropha biodiesel (J100) and pongamia (P100) was observed as 43.96, 39.67 and 36.78 MJ/kg. However, with addition of butanol,

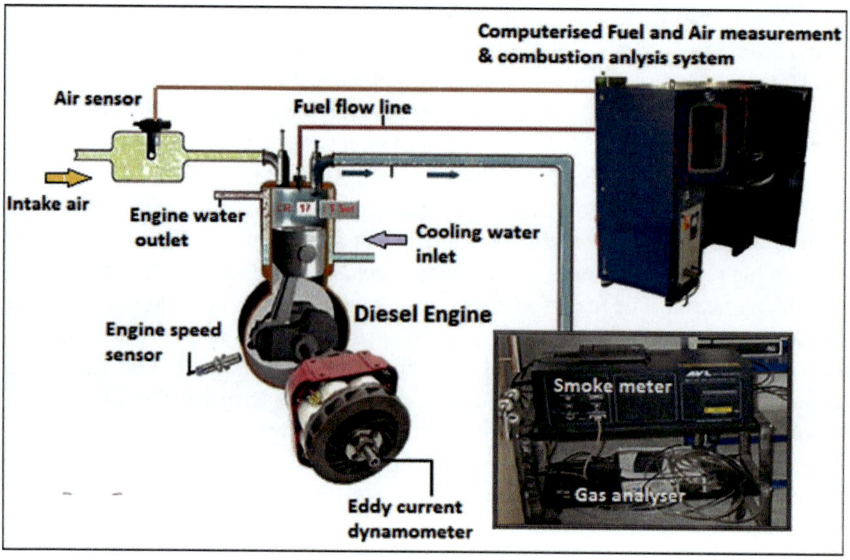

Fig. 1 Experimental setup with emission gas analyzer and smoke opacity meter

Table 1 Fuel properties of fuel blends

Fuel blends/fuel properties	CV (MJ/kg)	Density (g/cc)	Viscosity (CP)	Flash point (°C)	Pour point (°C)
D100	43.96	0.825	2.85	61	1
J100	39.67	0.878	5.12	124	-2
P100	36.78	0.895	6.32	154	3
But 100	32.8	0.81	2.64	29	< − 120.7
J10 But5 D85	42.57	0.83	3.24	54	− 1
P10 But5D85	41.68	0.835	3.76	53	− 2
J15But5D80	41.95	0.838	3.85	62	1
P15But5D80	42.28	0.836	3.56	65	− 2
J15P15But5D65	40.67	0.841	3.7	68	− 3

biodiesel and butanol resulted into lower values than diesel fuels but higher than pure biodiesel fuels. In addition, density, viscosity and cold flow properties was also found to be decreased.

Table 2 Technical specifications of the test engine

Particulars	Specifications
Engine	1-cylinder, 4-stroke, compression ignition engine
Bore × stroke	80.0 × 110.0 mm
Cubic capacity	0.530 L
Compression ratio	16.5:01
Rated speed (rpm)	1500
Power	3.7 kW
Specific fuel consumption	243 g/kWh

2.3 Experimental Set Up

The experimental study was based on emission and performance analysis of single cylinder diesel engine. The test engine fueled with Pongamia, jatropha, butanol and conventional diesel blends to run the engine. Experiment was conducted on 3.7 kW power engine at the speed of 1500 rpm. The water-cooled eddy current dynamometer used to measure the power of the tested engine. The specification of the engine depicted in Fig. 1 and Table 2. However emission characteristics such as carbon monoxide (CO), carbon dioxide (CO_2), Unburned hydrocarbon (UHC) and Nitrous oxides (NO_x) is analyzed by the NDIR based emission analyzer (Model AVL Digas 444). The smoke opacity of the engine is measured by the smoke meter in percentage. The tested engine is integrated with the computerized system for the measurement of air and fuel ratio during the injection.

To examine the experiments, the tested CI engine is fueled with the various blends of jatropha, Pongamia, Butanol, and Diesel (J10But5D85, P100But5D85, J15But5D80, P15But5D85, and J15P15But5D65), and compared with the conventional diesel at different loads (25, 50, and 100%).

3 Results and Discussions

3.1 Engine Performance with Diesel, Biodiesel and Butanol Blends

The effect of different fuel blends on engine brake thermal efficiency (BTE) at different loads is shown in Fig. 1a. It was observed from figure that BTE of the engine increased with increasing load. In addition, BTE of engine was found to be decreased with increasing share of biodiesel. On other hand, the fuel blends with jatropha were shown better efficiency in comparison to pongamia fuel blends. The results are with agreements to other studies [3]. This may be due to combine effect of higher viscosity, higher density and lower calorific value of fuel blends [3, 10].

BSFC of fuel blends at different loads are illustrated in Fig. 1b. The results revealed that BSFC decreases with increasing loads. Furthermore, BSFC was also found to be increased with increasing biodiesel concentration. Maximum BSFC was observed for J15P15But5D65. The higher fuel consumption of the biodiesel and their blends could be primarily related to lower heating value of the biodiesel. Another reason for the increase in BSFC with biodiesel may be a change in them combustion timing caused by the biodiesel's higher cetane number as well as injection timing changes [3, 11].

3.2 *Engine Emissions with Diesel, Biodiesel and Butanol Blends*

Figure 2a and b illustrates the comparison of hydrocarbon (HC) and carbon monoxide (CO) emission characteristics of the tested engine powered with various biodiesel fuel blends. Both types of emissions decreased significantly for all biodiesel fuel blends with increasing biodiesel blending share at lower loads. However, at higher loads, the emissions reduction was marginal. In addition, both hydrocarbon and co emissions were found to be increased with increasing load. This can be explained by the fact that biodiesel is an oxygenated fuel and the presence of oxygen in fuel blends leads to more clean combustion which results into lower emissions [2]. The CO emission formation during combustion in CI engines is illustrated with the mechanism given in Eqs. (1) and (2)

$$\mathrm{RH} \rightarrow \mathrm{R} \rightarrow \mathrm{RO2} \rightarrow \mathrm{RCHO} \rightarrow \mathrm{RCO} \rightarrow \mathrm{CO} \tag{1}$$

$$\mathrm{CO} + \mathrm{OH} \rightarrow \mathrm{CO_2} + \mathrm{H} \tag{2}$$

where, rate constant for this reaction is

$$\mathrm{kf, CO} = 6.76 \times 1010\, \mathrm{exp(T/1102)}$$

Figure 2c represents nitrogen oxide (NO_x) emissions for different diesel biodiesel and butanol fuel blends at various loads from the tested engine. The results showed that NO_x emission was higher with all the biodiesel fuels (i.e., J10But5D85, P100But5D85, J15But5D80, P15But5D85, and J15P15But5D65) as compared to base diesel fuel. Moreover, NO_x emissions were found to be increased with increasing loads. It is well known that the NO_x emission formation in a CI engine mainly depends on three parameters; (i) in-cylinder temperature, (ii) amount of oxygen present inside the combustion chamber, and (iii) combustion reaction time [2, 3]. Higher NO_x emission values in the present study could be due to the dominant effects of higher oxygen content and higher reaction time (higher combustion duration). The

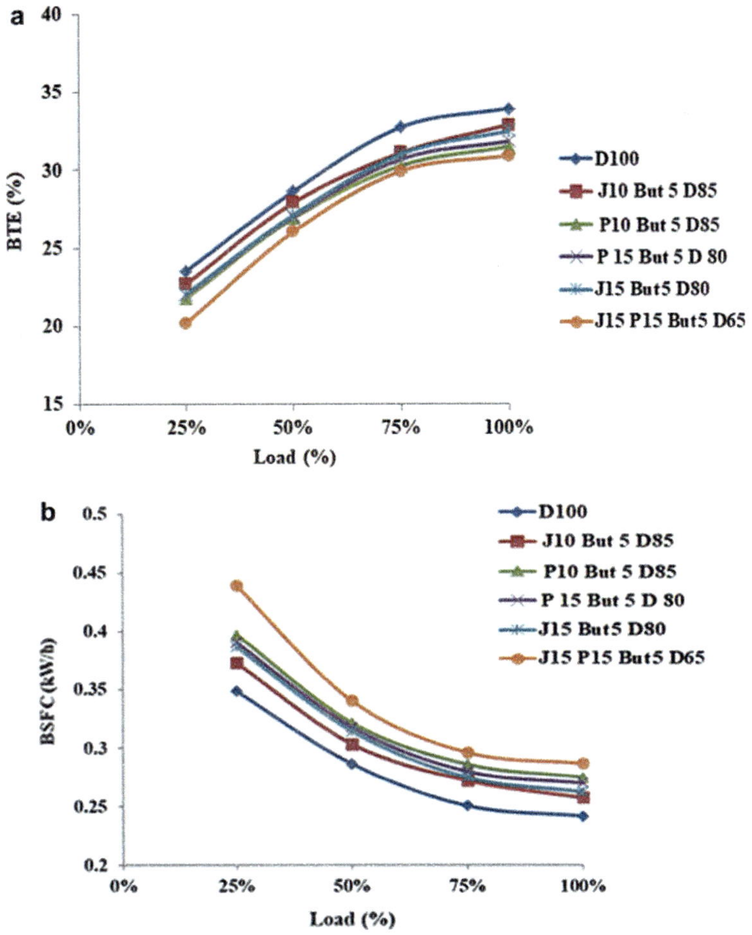

Fig. 2 Effect of different fuel blends on engine BTE and BSFC

in-cylinder temperature reduction with biodiesel blending might have insignificant effect on the emission formation (Fig. 3).

Figure 4 depicts the Variations of smoke opacity with load. Smoke opacity for biodiesels is lower than that for conventional Diesel at all loads. On increasing, the load smoke opacity increased due to increasing in the air-fuel consumption. However, Smoke comprises the visible by-products of poor combustion. Conventional diesel produced the higher amount of smoke because it has higher carbon content and lower oxygen as compare to the biodiesel, which led the improper combustion of the fuel. Therefore, J15P15B5D65 given the best result and reduced the amount of smoke opacity. Higher oxygen content and lower carbon led the better combustion inside the cylinder and decreased the smoke opacity.

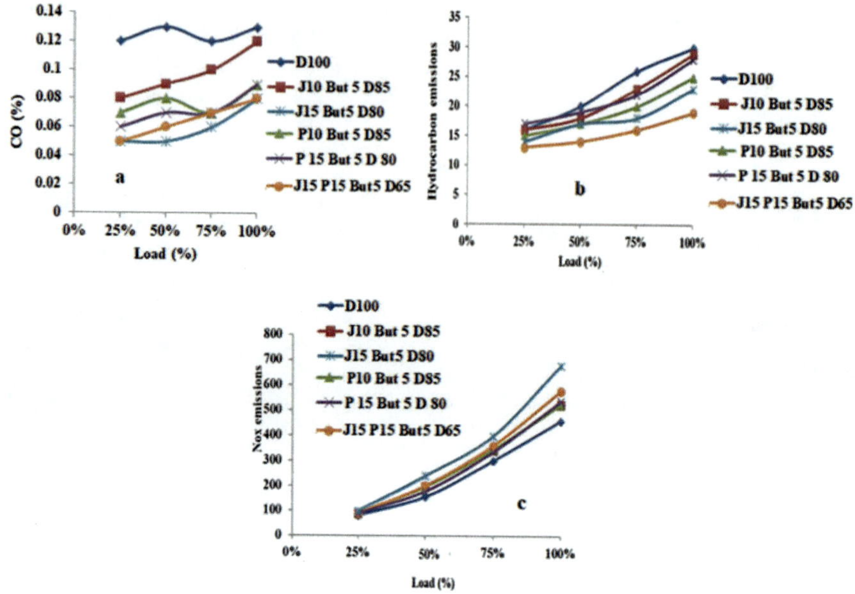

Fig. 3 Effect of different fuel blends on engine CO, unburnt hydrocarbon and NO_x emissions

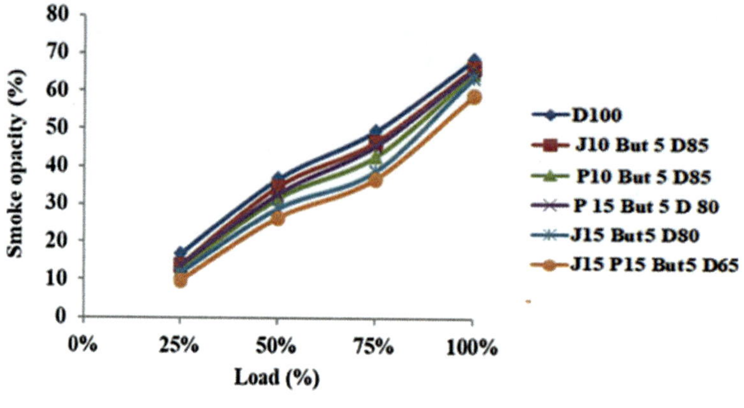

Fig. 4 Effect of different fuel blends on engine smoke emissions

4 Conclusions

The experiments tests were conducted with a single cylinder 37 kW DI diesel engine fuel with jatropha, pongamia butanol and Diesel fuel blends (J10But5D85, P100But5D85, J15But5D80, P15But5D85, and J15P15But5D65) at various loads (25, 50, 75, and 100% loads). The following conclusions are emerged from the comparative assessment.

1. Fuel properties of fuel blends (e.g. viscosity, density and cold flow properties) were improved significantly with addition of biodiesel and butanaol in diesel fuels when compared with pure biodiesel.
2. BTE of jatropha biodiesel based fuel blends were found better than pongamia biodiesel based fuel blends. In addition, BSFC was found to be increased with increasing share of biodiesel.
3. Emissions were found to be decreased with increasing share of biodiesel. However, NO_x emission showed vice versa and found to be increased with increasing biodiesel share.

References

1. Joshi G, Rawat DS, Sharma AK, Pandey JK. Microwave enhanced alcoholysis of non-edible (algal, jatropha and pongamia) oils using chemically activated egg shell derived CaO as heterogeneous catalyst. Bioresour Technol 2016;219. https://doi.org/10.1016/j.biortech.2016.08.011.
2. Sharma AK, Sharma PK, Chintala V, Khatri N, Patel A. Environment-friendly biodiesel/diesel blends for improving the exhaust emission and engine performance to reduce the pollutants emitted from transportation fleets. Int J Environ Res Public Health 2020;17. https://doi.org/10.3390/ijerph17113896.
3. Khan K, Kumar G, Sharma AK, Kumar PS, Mandal C, Chintala V. Performance and emission characteristics of a diesel engine using complementary blending of castor and karanja biodiesel. Biofuels 2018;9. https://doi.org/10.1080/17597269.2016.1256552.
4. Sharma AK, Sahoo PK, Singhal S. Comparative evolution of biomass production and lipid accumulation potential of Chlorella species grown in a bubble column photobioreactor. Biofuels 2016;7. https://doi.org/10.1080/17597269.2015.1138040.
5. Sharma AK, Sahoo PK, Singhal S, Patel A. Impact of various media and organic carbon sources on biofuel production potential from Chlorella spp. 3 Biotech 2016;6. https://doi.org/10.1007/s13205-016-0434-6.
6. Pali HS, Sharma A, Singh Y, Kumar N. Sal biodiesel production using Indian abundant forest feedstock. Fuel 2020;273:117781.
7. Agarwal S, Chhibber VK, Bhatnagar AK, Srivastav B, Kumar A, Singhal S, et al. Physico-chemical and tribological studies of Argemone biodiesel synthesized using microwave technique. Curr Sci 2017;113. https://doi.org/10.18520/cs/v113/i05/938-941.
8. Anand K, Sharma RP, Mehta PS. Experimental investigations on combustion, performance and emissions characteristics of neat karanji biodiesel and its methanol blend in a diesel engine. Biomass and Bioenergy 2011;35:533–41. https://doi.org/10.1016/j.biombioe.2010.10.005.
9. Sia CB, Kansedo J, Tan YH, Lee KT. Evaluation on biodiesel cold flow properties, oxidative stability and enhancement strategies: A review. Biocatal Agric Biotechnol 2020;24:101514.
10. Pal S, Chintala V, Sharma AK, Ghodke P, Kumar S, Kumar P. Effect of injection timing on performance and emission characteristics of single cylinder diesel engine running on blends of diesel and waste plastic fuels. Mater. Today Proc., vol. 17, 2019. https://doi.org/10.1016/j.matpr.2019.06.420.
11. Suresh Kumar P, Joshi S, Prasanthi Kumari N, Sharma A, Nair S, Chatterjee S. Reduction of emissions in a biodiesel-fueled compression ignition engine using exhaust gas recirculation and selective catalytic reduction techniques. Heat Transf 2020;49:3119–33. https://doi.org/10.1002/htj.21765.